The Routledge Companion to Actor-Network Theory

This companion explores ANT as an intellectual practice, tracking its movements and engagements with a wide range of other academic and activist projects. Showcasing the work of a diverse set of 'second generation' ANT scholars from around the world, it highlights the exciting depth and breadth of contemporary ANT and its future possibilities.

The companion has 38 chapters, each answering a key question about ANT and its capacities. Early chapters explore ANT as an intellectual practice and highlight ANT's dialogues with other fields and key theorists. Others open critical, provocative discussions of its limitations. Later sections explore how ANT has been developed in a range of social scientific fields and how it has been used to explore a wide range of scales and sites. Chapters in the final section discuss ANT's involvement in 'real world' endeavours such as disability and environmental activism, and even running a Chilean hospital. Each chapter contains an overview of relevant work and introduces original examples and ideas from the authors' recent research. The chapters orient readers in rich, complex fields and can be read in any order or combination. Throughout the volume, authors mobilise ANT to explore and account for a range of exciting case studies: from wheelchair activism to parliamentary decision-making; from racial profiling to energy consumption monitoring; from queer sex to Korean cities. A comprehensive introduction by the editors explores the significance of ANT more broadly and provides an overview of the volume.

The Routledge Companion to Actor-Network Theory will be an inspiring and lively companion to academics and advanced undergraduates and postgraduates from across many disciplines across the social sciences, including Sociology, Geography, Politics and Urban Studies, Environmental Studies and STS, and anyone wishing to engage with ANT, to understand what it has already been used to do and to imagine what it might do in the future.

Anders Blok is an Associate Professor of Sociology at the University of Copenhagen. He is the co-author (with Torben E. Jensen) of *Bruno Latour: Hybrid Thoughts in a Hybrid World* (Routledge 2011) and the co-editor (with Ignacio Farías) of *Urban Cosmopolitics: Agencements, Assemblies, Atmospheres* (Routledge 2016).

Ignacio Farías is a Professor of Urban Anthropology at the Humboldt University Berlin. He is the co-editor of *Urban Assemblages: How Actor-Network Theory Changes Urban Studies* (Routledge 2009, with Thomas Bender), *Technical Democracy as a Challenge for Urban Studies* (2016, with Anders Blok) and *Studio Studies: Operations, Topologies & Displacements* (Routledge 2015, with Alex Wilkie).

Celia Roberts is a Professor in the School of Sociology, Australian National University. She is the co-author, with Adrian Mackenzie and Maggie Mort, of *Living Data: Making Sense of Health Biosensors* (2019) and the author of *Puberty in Crisis: The Sociology of Early Sexual Development* (2016).

The Routledge Companion to Actor-Network Theory

Edited by Anders Blok, Ignacio Farías
and Celia Roberts

LONDON AND NEW YORK

First published 2020
by Routledge
4 Park Square, Milton Park, Abingdon, Oxon OX14 4RN

and by Routledge
605 Third Avenue, New York, NY 10017

First issued in paperback 2022

Routledge is an imprint of the Taylor & Francis Group, an informa business

Publisher's Note
The publisher has gone to great lengths to ensure the quality of this reprint but points out that some imperfections in the original copies may be apparent.

British Library Cataloguing-in-Publication Data
A catalogue record for this book is available from the British Library

Library of Congress Cataloging-in-Publication Data
A catalog record has been requested for this book

ISBN 13: 978-1-03-247548-6 (pbk)
ISBN 13: 978-1-138-08472-8 (hbk)
ISBN 13: 978-1-315-11166-7 (ebk)

DOI: 10.4324/9781315111667

Typeset in Bembo
by codeMantra

Contents

Contents

Contents

Figures

Contributors

Kristin Asdal is a Professor at TIK Center for Technology, Innovation and Culture at the University of Oslo. She has published widely across science and technology studies (STS), environmental studies and political and social theory. A key concern in her work is method issues in STS and beyond and she has co-edited several books and special issues on the subject. Her recent publications include *Humans, Animals and Biopolitics: The More than Human Condition* (co-edited with Tone Druglitrø and Steve Hinchliffe; Routledge 2017).

Yuri Carvajal Bañados was born in Valparaíso, Chile, in 1961. He studied at the University of Chile, Valparaíso. He was relegated in 1981, as a student opposed to the dictatorship, but then gained his MD in 1986. He was the Medical Chief in La Feria – public primary attention – in La Victoria (a lower-class township), Santiago during 1987–1993. He did a Master's in Public Health 1998 and worked as a Physician of Occupational and Environmental Health during 1997–2003. He became the Chief of the Llanchipal Health Service in 2003–2006 and then the Doctor of Public Health in 2011. He was the Editor of the *Chilean Journal of Public Health* 2011–2015 and the Director of Puerto Montt Hospital in 2016–2018. He is currently an Assistant Professor at the University of Chile and an Editor at Cuadernos Médico-Sociales.

Uli Beisel is an Assistant Professor of Culture and Technology at Bayreuth University (Germany) and holds a PhD in Human Geography from the Open University (UK). She has worked on human-mosquito-parasite entanglements in malaria control in Ghana and Sierra Leone and the translation of new health technologies in Uganda and Rwanda. Her new research is looking at trust in biomedicine and healthcare infrastructures after the Ebola epidemic in a comparative project in Sierra Leone, Ghana and Uganda. Uli's work has been published by *Science as Culture*, *Society and Space*, *Biosocieties*, *Medical Anthropology* and *Geoforum*.

Anders Blok is an Associate Professor of Sociology at the University of Copenhagen, Denmark. He has published widely within science and technology studies (STS), urban studies, environmental sociology and social theory. He is the co-author (with Torben E. Jensen) of *Bruno Latour: Hybrid Thoughts in a Hybrid World* (Routledge 2011) and the co-editor (with Ignacio Farías) of *Urban Cosmopolitics: Agencements, Assemblies, Atmospheres* (Routledge 2016).

Emma Cardwell is a Lecturer in Human Geography in the School of Geographical and Earth Sciences at the University of Glasgow. Her research explores the interrelationships between environment, materiality and knowledge practices, with a particular focus on the production of food. Using theories from science and technology studies, political economy and feminist science studies, she examines processes of financialisation, market-creation, equality and participation in fisheries management and marine conservation; and the inter-relationships of agriculture, land, economics and biochemistry, via the modes and practices of crop nutrition and fertiliser use.

Nigel Clark is a Professor of Social Sustainability and Human Geography at Lancaster University, UK. He is the author of *Inhuman Nature: Sociable Life on a Dynamic Planet* (2011) and the co-editor with Kathryn Yusoff of a recent *Theory, Culture & Society* special issue on 'Geo-social Formations and the Anthropocene' (2017). Current research revolves around questions of how to think *through* the Earth while also decolonising global thought. He is working on a book with Bronislaw Szerszynski entitled *The Anthropocene and Society: Toward Planetary Social Thought*.

Tomás Sánchez Criado is a Senior Researcher at the Chair of Urban Anthropology of the Department of European Ethnology, Humboldt-University of Berlin. His work, at the cross-roads of Anthropology and STS, experiments with forms of public and ethnographic engage-ment around urban and care infrastructures, and their technical democratisation through activist and pedagogic projects. He recently co-edited *Experimental Collaborations: Ethnogra-phy through Fieldwork Devices* (Berghahn, 2018, with Adolfo Estalella) and *Re-learning Design. Pedagogic Experiments with STS in Design Studio Courses* (DISEÑA, 2018, with Ignacio Farías).

Endre Dányi is a Guest Professor for the Sociology of Globalisation at the Faculty of Social Sciences at the Bundeswehr University in Munich. Growing up in Budapest in the 1990s, he became fascinated with the dual transformation commonly referred to as the Velvet Revo-lution and the Information Revolution. In the early 2000s, in London and Lancaster, he was strongly influenced by ANT, exploring new ways of thinking about *the technical* and *the polit-ical*. In the 2010s, in Berlin and Frankfurt, he has combined these with other strands of crit-ical thought, especially those associated with Walter Benjamin. Endre is currently working on his *habilitation*, provisionally titled 'Melancholy Democracy.' He is also busy co-running Mattering Press – an Open Access book publisher. E-mail: e.danyi@unibw.de.

David J. Denis is a Professor in Sociology at the Centre de Sociologie de l'Innovation (Mines ParisTech). He studies data labour in various organisations and explores maintenance practices, notably in urban settings. He is the co-founder of Scriptopolis, a collective scien-tific blog about writing practices (www.scriptopolis.fr/en).

Liliana Doganova is an Associate Professor at the Center for the Sociology of Innovation, MINES ParisTech. At the intersection of economic sociology and Science and Technology Studies, her work has focused on business models, the valorisation of public research and markets for bio- and clean-technologies. She has published in journals such as *Research Policy*, *Science and Public Policy* and the *Journal of Cultural Economy*, and she is currently preparing a monograph on the historical sociology of discounting and the economic valuation of the future.

Alexa Färber is a Professor of European Ethnology at the University of Vienna. She com-bines assemblage thinking and ethnographic urban research to explore the social implications of urban imagineering on everyday lives. Her recent project investigates the notion of the city as promissory assemblage. Before, she was the primary investigator for the 'Low-Budget Urbanity' research initiative at HafenCity University Hamburg (2012–2015). In her blog talkingphotobooks.net she explores the graspability of the city in photobooks and extroverted, shared reading. Her research interests include visual methodologies and ANT, historical dimensions of audit cultures, everyday representational work, dwelling and mobility.

Ignacio Farías is a Professor of Urban Anthropology at the Humboldt University Berlin. His research interests lie at the crossroads of urban studies, science and technology studies and cultural anthropology. His work explores the politics of urban disruptions, from disasters to noise, open forms of cultural and anthropological production, as well as participatory and pedagogical experiments with technical democratisation. He has co-edited *Urban Assemblages. How Actor-Network Theory Changes Urban Studies* (Routledge 2009, with Thomas Bender), *Technical Democracy as a Challenge for Urban Studies* (CITY 2016, with Anders Blok), *Urban Cosmopolitics: Agencements, Assemblies, Atmospheres* (Routledge 2016, with Anders Blok), *Studio Studies. Operations, Topologies & Displacements* (Routledge 2015, with Alex Wilkie) and *Re-learning Design: Pedagogical Experiments with STS in Design Studio Courses* (DISEÑA 2018, with Tomás S. Criado).

Arthur Arruda Leal Ferreira is a Postdoctoral Researcher in the History of Psychology from the UNED (Spain) and Janveriana University (Colombia), and is a Professor of the History of Psychology at the Institute of Psychology at the Federal University of Rio de Janeiro (UFRJ). He has recently edited the following books: *Psicologia, Tecnologia e Sociedade, A pluralidade do campo psicológico, História da Psicologia: Rumos e Percursos, Teoria Ator-Rede e a Psicologia* and *Pragmatismo e questões contemporâneas*. He has also contributed to the books: *Foucault e a Psicologia, Da metafísica moderna ao pragmatismo, Biosegurança e Biopolítica no Século XXI, Explicaciones em Psicología* and *Neoliberalism and Technoscience*.

Carolin Gerlitz is a Professor in Digital Media and Methods at the University of Siegen and a Member of the Digital Methods Initiative Amsterdam. Her research interests involve among others data-intensive media, platform and infrastructure studies, digital methodologies, issue mapping, apps, sensormedia, quantification and evaluation. Before joining Siegen, she completed her PhD in Sociology at Goldsmiths and was an Assistant Professor in New Media and Digital Culture at the University of Amsterdam. She is the co-speaker of the DFG-funded graduate school 'Locating Media' and the deputy speaker of the DFG cooperative research centre 'Media of Cooperation.' She holds an NWO Veni grant for the project 'Numbering Life.'

Michael Guggenheim is a Reader in the Department of Sociology, Goldsmiths, University of London and a Co-director of the Centre of Invention and Social Process (CISP). He works on disasters, buildings and food. He teaches mostly visual and inventive methods.

Francis Halsall is a Co-director of the Master's Programs, *Art in the Contemporary World*, at National College of Art and Design, Dublin and a Research Fellow at the Department of Art History and Image Studies, University of the Free State, Bloemfontein, South Africa. His research involves three main areas: (1) Modern and contemporary art; (2) Philosophical aesthetics; (3) Systems Thinking. He has published and lectured widely in all areas. He is currently working on several projects under the broad heading of 'Systems Aesthetics.' More information at: www.alittletagend.blogspot.com.

Kregg Hetherington is a Political and Environmental Anthropologist based at Concordia University in Montreal where he also runs the Concordia Ethnography Lab. He is the author of *Guerrilla Auditors: The Politics of Transparency in Neoliberal Paraguay*, and the editor of *Infrastructure, Environment and Life in the Anthropocene*. His current ethnographic work on the relations between soybean monocultures and the regulatory state develops an analysis of regulation at the intersection of ANT, biopolitics and feminist STS.

Kohei Inose is a Professor of Volunteer Studies at Meiji Gakuin University, Japan. He has conducted ethnographic research on disability and public movement in Japan. His research interest includes disability, agriculture, development and regional history. Among his publication is 'Living with Uncertainty: Public Anthropology and Radioactive Contamination' (*Japanese Review of Cultural Anthropology* 15, 2015). Currently, he is exploring the relationship between disability movements and economic development in Asian societies.

Casper Bruun Jensen is a specially appointed Associate Professor at Osaka University and Honorary Lecturer at Leicester University. He is the author of *Ontologies for Developing Things* (Sense, 2010) and *Monitoring Movements in Development Aid* (with Brit Ross Winthereik) (2013, MIT) and the editor of *Deleuzian Intersections: Science, Technology, Anthropology* with Kjetil Rödje (Berghahn, 2009) and *Infrastructures and Social Complexity* with Penny Harvey and Atsuro Morita (Routledge, 2016). His present work focuses on knowledge, infrastructure and practical ontologies in the Mekong river basin.

Sonja Jerak-Zuiderent is an Assistant Professor of Science and Technology Studies at the Department of General Practice at the Amsterdam University Medical Centres, Netherlands. Her overarching research interest lies in the question on 'how to care' in accountability and evaluation practices in healthcare and the sciences broadly defined; particularly in relation to the more neglected, socio-material aspects of 'good(s)' and 'bad(s)', studying the everyday marginalities that get produced. She has published on accountability devices in healthcare, like guidelines and performance indicators, on social studies of patient safety, is currently developing publications on trans★ care in Argentina, ethnographically exploring 'good science' practices across and within disciplines and is finalising her manuscript 'Generative Accountability: Comparing with Care' (forthcoming with MatteringPress).

Ericka Johnson is a Professor of Gender and Society at the Department of Thematic Studies, Linköping University. Her research interests include the simulated body, medical technologies, pharmaceuticals and the production of anatomical knowledge about the (sometimes) healthy subject and body. She is currently leading the Bodies Hub, a collaborative research environment that explores tensions between the social and medical body. Her most recent books are *Refracting through Technologies. Seeing and Untangling Bodies, Medical Technologies and Discourses* (Routledge, 2019) and *Gendering Drugs. Feminist Studies of Pharmaceuticals* (Palgrave Macmillan, 2017).

Shuhei Kimura is an Associate Professor of Cultural Anthropology at the University of Tsukuba, Japan. He has conducted ethnographic research on disaster in Turkey and Japan.

His research interest includes disaster, infrastructure, temporality and public anthropology. Among his publication is 'When a Seawall Is Visible: Infrastructure and Obstruction in Post-tsunami Reconstruction in Japan' (*Science as Culture* 25(1), 2016). His current project addresses multi-temporal interactions between human and nonhuman actors in the post-tsunami coastal Japan.

Wen-Yuan Lin is a Professor at the National Tsing-hua University, Taiwan. He uses STS material semiotic approaches to explore emerging alternative knowing spaces and the politics of empirical ontology in medical practices. One of his current projects is exploring the possibility of mobilising alternative mode of knowing in Chinese medical practices to provincialise the frameworks of EuroAmerican social sciences. His webpage is at http://cge.gec. nthu.edu.tw/faculty/wylin/.

Daniel López-Gómez is an Associate Professor in the Department of Psychology and Education at the Universitat Oberta de Catalunya and a Researcher at the Internet Interdisciplinary Institute (IN3) of the same university. He is an STS Scholar and Ethnographer interested in the study of care, ageing and disability, and of the emergence of grass-roots innovations in contexts of crisis and austerity.

Amade M'charek is a Professor of Anthropology of Science at the Department of Anthropology of the University of Amsterdam. Her research interests are in forensics, forensic anthropology and race. M'charek is the Principle Investigator of the ERC-Consolidator Project *RaceFaceID*, http://race-face-id.eu, a project on forensic identification, face and race. Her work has appeared in various peer-reviewed journals, such as STHV, Science as Culture, Theory, Culture and Society and Cultural Anthropology.

Adrian Mackenzie (Professor in the School of Sociology, ANU) researches how people work and live with sciences, media, devices and infrastructures. He often focuses on software and platforms. He has done fieldwork with software developers in making sense of how platforms are made, managed and maintained (see *Cutting Code: Software and Sociality*, Peter Lang 2006). He has tracked infrastructural experience (*Wirelessness: Radical Empiricism in Network Cultures*, MIT Press 2010). His most recent book *Machine Learners: Archaeology of a Data Practice* (MIT Press, 2017) describes changes in how science and commerce use data to make knowledge. He has a keen interest in the methodological challenges of media and data platforms for sociology and philosophy.

Noortje Marres is an Associate Professor and Director of the Centre of Interdisciplinary Methodologies at the University of Warwick (UK). She studied Sociology and Philosophy of Science and Technology at the University of Amsterdam, and conducted her doctoral research at the Centre de Sociologie d'innovation at MinesTech in Paris, on issue-centred concepts of participation in technological societies, in Pragmatism and ANT. She published two monographs, Material Participation (2012) and Digital Sociology (2017), and one edited collection Inventing the Social (with M. Guggenheim and A. Wilkie, 2018). More info at www.noortjemarres.net.

Alvise Mattozzi has PhD in Semiotics and is a Research Fellow in Sociology at the Faculty of Design and Art of the Free University of Bozen-Bolzano. He works at the crossroads of Science and Technology Studies and Design Studies, using semiotics as descriptive-analytical methodology. His main research focus regards the development of a method for the description of the social mediation of artefacts, by recovering and expanding the *script* model as elaborated by Madeleine Akrich and Bruno Latour. He has used a revised version of the *script* model in research works addressing the use of artefacts by final users as well as in research works addressing designing practices.

Derek P. McCormack is a Professor of Cultural Geography at the University of Oxford. He has written about non-representational theory, affect, atmospheres and the elemental. He is the author of *Refrains for Moving Bodies: Experience and Experiment in Affective Spaces* (2014), and *Atmospheric Things: On the Allure of Elemental Envelopment* (2018), both published with Duke University Press.

Brett Mommersteeg is a PhD student at the University of Manchester in the Architecture Programme. His research focuses on Actor-Network Theory, ethnography and architectural projects. Currently, his dissertation follows the making of a building in Manchester, UK, called Factory, and traces the complex ecology of actors and their modes of relating through the design and construction stages. He has a recent publication about his dissertation in *Ardeth* entitled, 'The Garden of Bifurcating Paths: Towards a Multi-Sited Ecological Approach to Design.'

Atsuro Morita is an Associate Professor of Science, Technology and Culture at Osaka University. His work concerns the travels of scientific knowledge and technology between Thailand and Japan, postcolonialism and Asianism in relation to the recent rise of Asian STS, and water infrastructures and global hydrology in the context of the crisis of climate change. His publications include *Infrastructure and Social Complexity* (co-edited with Penny Harvey and Casper Bruun Jensen), *The World Multiple* (co-edited with Keiichi Omura, Grant Otsuki and Shiho Satsuka) and *Engineering in the Wild* (in Japanese). His article 'The Ethnographic Machine' (*Science, Technology and Human Values* 39(2)) experiments with a notion of internal comparisons embedded in machines.

Fabian Muniesa, a Researcher at the Centre de Sociologie de l'Innovation (Mines ParisTech, PSL University, in Paris, France) since the mid-2000s and an ANT Practitioner since the mid-1990s, is the author of *The Provoked Economy: Economic Reality and the Performative Turn* (Routledge, 2014) and the co-author of *Capitalization: A Cultural Guide* (Presses des Mines, 2017). His past and current research includes work on the politics of valuation, quantification, simulation, automation and organisation, a series of topics he has been approaching from a constructivist, pragmatist perspective.

Irene van Oorschot is a Postdoctoral Researcher in the RaceFaceID project (PI: Prof. Amade M'charek). Having defended her dissertation on judicial and social-scientific ways of truth- and fact-making in 2018 (cum laude), she is currently focused on the mobilisation and contestation of forensic knowledge within legal settings, paying attention especially to the way relationships between individuals and populations are made and unmade within these practices.

Robert Oppenheim is a Professor of Asian Studies and Anthropology at the University of Texas at Austin. His primary work on the 'ontological politics' of development, heritage and place in the city of Kyŏngju, South Korea appears in the 2008 book *Kyŏngju Things: Assembling Place*. Other articles on place and space through an ANT lens appear in *Anthropological Theory* and the volume *Objects and Materials: A Routledge Companion*.

José Ossandón is an Associate Professor in the Organization of Markets, Department of Organization, Copenhagen Business School. His current research focuses on two main areas. The first area is the organisation of markets. In particular, he studies expert knowledge and technologies developed with the intention to make markets to solve matters of collective concerns. The second is the management of households' financial economies. Here, he expects to help bridging between studies of mundane economic practices and analyses of the calculative devices produced by financial firms. He is an Associate Editor at the *Journal of Cultural Economy* and *Valuation Studies*.

Jérôme D. Pontille is a Researcher in the French National Center for Scientific Research and works at the Centre de Sociologie de l'Innovation (Mines ParisTech). His research focuses on writing practices, scientific authorship and evaluation technologies, the maintenance of urban infrastructures. He is the co-founder of Scriptopolis, a collective scientific blog about writing practices (www.scriptopolis.fr/en).

Kane Race is a Professor of Gender and Cultural Studies at the University of Sydney. Motivated by the capacity of bodies and pleasures to intervene in the disciplinary production of knowledge, subjects, technologies and forms of life, he has published widely on questions of HIV infection, sexuality, biomedicine, drug use, digital culture, risk and care practices. He is the author of *Pleasure Consuming Medicine: The Queer Politics of Drugs* (Duke University Press, 2009), *Plastic Water: The Social and Material Life of Bottled Water* (co-authored with Gay Hawkins and Emily Potter, MIT Press, 2015), and *The Gay Science: Intimate Experiments with the Problem of HIV* (Routledge, 2017).

Celia Roberts is a Professor in the School of Sociology at the Australian National University, Canberra. She previously worked at Lancaster University and was an active member of the Centre for Science Studies there for many years. She is currently working on health biosensing, stress and reproduction, and is the co-author, with Adrian Mackenzie and Maggie Mort of a forthcoming book entitled *Living Data: Making Sense of Health Biosensors*. Her previous books include *Puberty in Crisis: The Sociology of Early Sexual Development* (Cambridge, 2015) and *Messengers of Sex: Hormones, Biomedicine and Feminism* (Cambridge, 2007).

Israel Rodríguez-Giralt is an Associate Professor/Researcher at CareNet, Internet Interdisciplinary Institute (IN3), Open University of Catalonia (UOC). His research focuses on technoscientific activism and new forms of social experimentation, mobilisation and public engagement, particularly in the field of social care and disaster situations. He co-edited *Disasters and Politics: Materials, Experiments, Preparedness* (Wiley/Blackwell, 2014).

Marcelo C. Rosa is an Associate Professor of Sociology at the University of Brasilia, Brazil and a Researcher of National Council for the Scientific and Technological Development (CNPq). In the last two decades, he has led, participated and supervised research projects on landless social movements in the global south. The research has developed to a theoretical and methodological agenda on the possibilities of the Southern theories and existences to challenge the hegemonic social sciences. He is the Head of the Research Lab on Non-Exemplary Sociology (www.naoexemplar.com) and the Founder of the *Agrarian South: Journal of Political Economy*.

Martin Savransky is a Lecturer and Director of the Unit of Play in the Department of Sociology, Goldsmiths, University of London, where he teaches courses in philosophies of difference, cultural theory and pluralistic politics. He is the author of *The Adventure of Relevance* (2016), co-editor of *Speculative Research: The Lure of Possible Futures* (2017) and editor of *Isabellle Stengers and the Dramatization of Philosophy* (2018). He is currently working on a new monograph titled *Around the Day in Eighty Worlds: Politics of the Pluriverse*.

Michael Schillmeier is a Professor of Sociology at the Department of Sociology, Philosophy and Anthropology at the University of Exeter (UK). He received his PhD from Lancaster University (UK). He had Schumpeter-Fellowship between 2010 and 2015 (funded by VolkswagenStiftung), is an honorary Senior Member of EGENIS, the Center for the Studies of Life Sciences, Exeter University, and the Co-Editor of *Space & Culture*. He has widely written on the eventful dynamics of the heterogeneity of societal orderings, outlining the relevance of embodied and affective relations, material objects and technologies. His research includes Science and Technology Studies, Dis/ability Studies and the Sociology of Health and Illness. His work is cross-disciplinary and links Sociology with Philosophy, Anthropology and Art. Publications include *Eventful Bodies: The Cosmopolitics of Illness* (Ashgate), *Rethinking Disability: Bodies, Senses, and Things* (Routledge), *Agency without Actors? New Approaches to Collective Action* with Jan-Henrick Passoth and Birgit Peuker (Routledge), *Un/knowing Bodies* with Joanna Latimer (Wiley-Blackwell), *New Technologies and Emerging Spaces of Care* with Miquel Domenech (Ashgate), *Disability in German Literature, Film, and Theater* with Eleoma Joshua (Camden House). With Juliane Sarnes, he has translated Gabriel Tarde's *Monadology and Sociology* into German.

Claire Waterton is a Professor of Environment and Culture in the Department of Sociology at Lancaster University. She uses STS theory and methods to explore some of the troubling consequences of contemporary systems of production, consumption and living. For example, the disposal of nuclear wastes, the diffuse pollution of soils and water, escalating biodiversity loss. She wants to open up new ways of knowing/feeling/connecting/making sense around these environmental issues and hence to find new ways of enacting nature-culture relations. She is the co-author of *Care and Policy Practices* (Sage 2017), *Barcoding Nature* (Routledge 2013) and *Nature Performed* (Blackwell 2003).

Esther Weltevrede is an Assistant Professor of New Media and Digital Culture, Coordinator of the Digital Methods Initiative at the University of Amsterdam and a Member of the App Studies Initiative. Her research interests include digital methods, platform and infrastructure studies, automation, apps, fake news circulation and software studies. In her dissertation on 'Repurposing digital methods: The research affordances of platforms and engines,' she has developed the notion of 'research affordances' to understand the action possibilities within software, from the perspective of, and aligned with the interests of, the researcher.

Alex Wilkie is a Reader in Design and a Sociologist at Goldsmiths, University of London, where he also directs the Centre for Invention and Social Process hosted by the Department of Sociology. His work involves experimental design and empirical research to explore social and technological future – making practices and more-than-human sociality. Alex's work cuts across and ties together science and technology studies, empirical philosophy, process theory and speculative reasoning as well as human–computer interaction, practice-based design research and design theory. He has co-edited *Studio Studies* (Routledge), *Speculative Research* (Routledge), and *Inventing the Social* (Mattering Press) and co-authored *Energy Babble* (Mattering Press).

Brit Ross Winthereik is a Professor in the Technologies in Practice Research Group at the IT University of Copenhagen. Her research focuses on public sector digitalisation in Denmark, energy infrastructures, digital data and accountability. She has published widely in Science and Technology Studies and anthropology journals. Her book *Monitoring Movements in Development Aid: Recursive Infrastructures and Partnerships* (MIT Press, 2013) with Casper Bruun Jensen and her research on wave energy investigate lateral ethnography as a mode of analysis. She is a co-founder of the *Danish Association for Science and Technology Studies* (dasts. dk), and of *the ETHOS Lab* (ethos.itu.dk), and is heading the research project *Data as Relation: Governance in the age of big data* (2016–20).

Albena Yaneva is a Professor of Architectural Theory at the University of Manchester, UK. She gained her ANT training at *Ecole des Mines* (Centre of Sociology of Innovation) with Bruno Latour, Antoine Hennion and Madeleine Akrish. She is the author of a number of books: *The Making of a Building* (2009), *Made by the OMA: An Ethnography of Design* (2009), *Mapping Controversies in Architecture* (2012) and *Five Ways to Make Architecture Political* (2017). Her work has been translated into German, Italian, Spanish, French, Portuguese, Thai, Polish and Japanese. Yaneva is a Visiting Professor at Princeton, Lise Meitner Visiting Chair at Lund and is the recipient of the RIBA President's award for research (2010).

Acknowledgements

Producing a Companion of this size is challenging. As editors, selecting and inviting authors was an enjoyable privilege, and it was wonderful to meet many of them in person at our two workshops in Berlin and Copenhagen. Keeping track of versions of texts, size of images and US vs UK spelling was perhaps less fun. We would like to thank all the authors for their generous and friendly engagement with this project, for attending the workshops if they could, producing such great pieces of writing and forgiving our administrative slip-ups.

Some authors had to drop out along the way, and we want to thank them too for the work they did. Lots of people – including the unnamed proposal reviewers – were encouraging about this project, and sent us excellent ideas and names of potential contributors. While too numerous to list, we want to thank all of them warmly! Thanks are particularly due to Noortje Marres for early conversations about the book's framing.

We were supported in this project by Routledge – thanks are due particularly to Egle Zigaite – and by the Department of Sociology, University of Copenhagen, as well as the Department of European Ethnology, Humboldt University.

The cover image was made by Mirja Busch. We feel lucky to have collaborated with her on this.

As always, thanks are also due to family, friends and colleagues for support and love during the making of this book.

Actor-network theory as a companion

An inquiry into intellectual practices

Ignacio Farías, Anders Blok and Celia Roberts

Near ANT: 20 years after 'actor-network theory and after'

Welcome to our actor-network theory (ANT) companion. The companion is both the figure and the forum that this volume seeks to assemble, in order to continue and to interrupt current conversations about the past, present and future of ANT. This volume aims to make an intervention in the ways we speak, do inquiry and think with ANT, paying attention to what it means to invoke this acronym to describe a specific intellectual practice. As such, the curating of the following 38 contributions should be understood as an attempt to articulate a critical and constructive response to the enthusiastic 'reception' that ANT has enjoyed in the social sciences, humanities, design and artistic disciplines in the last 10–20 years in particular (see Jacobs & Frickel, 2009: 51), a reception that, in our eyes, amounts to an ambivalent sign of ANT's simultaneous success and demise, a mark of its continued livelihood and a signal of a potential death-by-exhaustion.

Since everything in this volume is about ANT, one way or the other, an initial characterisation is called for. Put (too) simply, ANT denotes a family of conceptual and methodological sensibilities that grew out of French-British studies in the sociology of science and technology from the late 1970s onwards (see among others Akrich & Latour, 1992; Callon, 2001; Latour, 2005; Law, 2009; Michael, 2016; Mol, 2010). While often ascribed to Bruno Latour, ANT is nothing if not a collective achievement, with Michel Callon, Madeleine Akrich, Antoine Hennion, Vololona Rabeharisoa, John Law, Annemarie Mol, Vicky Singleton and others central to its emergence and subsequent trajectory. Among a host of alternative labels, ANT is known sometimes as a 'material semiotics' (Law & Mol, 1995), in that it entails a basic ontological claim that all entities in the world – from nanoparticles to bodies, groups, ecologies and ghosts – are constituted and reconstituted in shifting and hybrid webs of discursive and material relations (Latour, 1988). Accordingly, ANT has been at the forefronts of conceptualising new predicaments beyond the 'modern divides' between nature and culture, the technical and the social (Latour, 1993). In ANT, processes and relations run deep and literally stretch the world, also undoing scalar modes of conceiving the social as a dialectical articulation of agency and structure, the local and the global, the big and the small (Callon & Latour, 1981).

While we come back in the second section of this introduction to unfold some more of ANT's complicated intellectual genealogy, it is not our aim here, or in the companion writ large, to add more layers to a purportedly 'canonical' ANT (Domènech & Tirado, 1998; Inglis & Thorpe, 2019; Schüttpelz, 2013). Similarly, we do not aim in this introduction to unpack all the translations and equivocations that have catapulted ANT into such an influential intellectual project in the current intellectual landscape. Still, as a mode of emphasising the timeliness of this companion, it might be important to begin by at least venturing some ideas about why the most heterogeneous fields of intellectual and creative practice have of late embraced ANT as a new semantics and analytical repertoire. Let us advance three hypotheses:

The first one concerns the widely misunderstood principle of generalised symmetry proposed by early ANT as a methodological strategy to remain agnostic concerning the question of what counts as an actor in a given setting and course of events (Callon, 1986; Callon & Latour, 1992; Latour, 1999a). The alleged symmetry between humans and non-humans that this principle is oft taken to universally authorise remains perhaps the strongest selling point of ANT. It allows not just to radically challenge the grand social theories that have shaped our intellectual landscape during the second half of the 20th century, but also to align the ethico-political commitments of sociocultural researchers in widely different fields with the varying ontologies of their interlocutors, be they environmentalists, technology developers, indigenous people, designers, accountants, economists or whichever. Generalised symmetry, in short, contributes to ANT success by seemingly making it endlessly versatile and extendable into new domains.

The second hypothesis pertains to a more specific methodological injunction, namely, the need to attend to materials, devices and artefacts, that is, to the various objects that shape the social (Latour, 1991, 1992, 1996b; Law, 1987; Law & Singleton, 2005). Besides the sociocultural sciences, ANT has become extremely important for the various disciplines concerned with 'thing-making,' especially design, architecture and contemporary arts (Björgvinsson, Ehn, & Hillgren, 2012; Halsall, 2016; Yaneva, 2009). Here, it provides a language, a conceptual repertoire to think with and about things, in terms of the 'sub-political' agendas inscribed in their designs but also the different ways in which things overtly politicise states of affairs (Fogué & Rubio, 2017). Consequently, ANT has played an important role in rethinking the democratic challenges associated with the making of our socio-material worlds and made major intervention in current conceptualisations of 'participation' (Callon, Lascoumes, & Barthe, 2009; Latour, 2004; Marres, 2012). It is no surprise that ANT is currently becoming influential among design disciplines, not to mention its influence in the arts, where ANT is widely referenced when speaking about the agential capacity of things.

The third hypothesis is historical. If Ulrich Beck (1992) was crucial to think the techno-scientific risk culture of late modernity, the latest work of Bruno Latour (Latour, 2017, 2018), in particular, has become critical to think our current predicament in the (so-called) Anthropocene. Latour, one of ANT's key founders and theorists, is nowadays a highly influential public intellectual of the contemporary, actively participating in current debates on climate change and planetary ecological crises. There is much irony and more than a little intellectual wriggling involved in this widening embrace of a thinker who, 30–40 years ago, found himself widely accused of the cardinal sin of 'relativism' amongst the scientific establishment (see e.g. Dawkins, 1998). There is also, more importantly, a real concern that Latour's public presence may come to overshadow ANT's collective achievements.

The increasing popularity of ANT in a wide array of fields is startling and ambivalent to say the least, especially considering that already two decades ago, key ANT scholars declared an end to what was first called ANT. Indeed, the volume *Actor-Network Theory and After* (Law & Hassard, 1999) marked an end of ANT as a theoretical position fully embedded in the field of science and technology studies (STS) and thus mostly concerned with problems of technoscience and biomedicine. Concomitantly, it marked the beginning of broader and more open-ended inquiries into other fields of practice, such as the economy (Callon, 1998), cultural taste (Hennion, 1993), social movements (Rabeharisoa & Callon, 1999), transnational government (Barry, 2001), design (Callon, 1996) and urbanism (Latour & Hermant, 1998). Similarly, emerging inquiries drew inspiration from conceptual figures different from that of the classic actor-network, such as fractality (Law, 2002), multiplicity (Mol, 2002) and performativity (MacKenzie, Muniesa, & Siu, 2007) – moves that were later dubbed 'post-ANT' (see Gad & Bruun Jensen, 2010).

Thus, while in the last 20 years ANT has tended to become itself an 'immutable mobile' (Latour & Woolgar, 1986), in the sense of a highly mobile label for a stabilised conceptual repertoire concerned with generalised symmetry, networks and non-humans, ANT scholars have tended to become after- or post-ANT scholars, experimenting with ANT as a much more 'mutable mobile' (de Laet & Mol, 2000). In the latter mode, scholars have coined a widening array of after- and post-ANT conceptual repertoires, to the extent that the very concept of actor-networks has long lost its centrality, as ANT-informed analysts these days are more likely to speak, say, of assemblages and enactments. All along, however, actor-networks remain good company to think with and beyond.

In this spirit and against the backdrop just drawn up, the challenge for this volume is to reclaim ANT not as a theory, but as an open-ended intellectual project that can only remain successful if it keeps reinventing itself. In other words, the question we need to ask is how to respond to the 'after ANT' (or post-ANT) predicament and the trajectories of intellectual development that embody it? Which kind of approach and book would not hasten the death but rather *extend* ANT as a vibrant intellectual practice for the future?

It is with these questions in mind that we have put together this Companion to ANT, aiming to make visible the many companions of ANT who are out there, across many different settings, doing research that extends ANT in new and interesting directions. Indeed, for us as editors as well as for our many contributors – albeit in importantly varying ways – this also involves affirming that ANT may well be one of the best companions around for contemporary intellectual practices committed to think, explore and engage current and oftentimes worrying sociocultural, politico-ecological and technoscientific predicaments writ large. In this conjuncture, ANT can help us to spark curiosity and open inquiries, to unpack modernist and colonial ontologies, to encounter a multitude of human and more-than-human interlocutors, to experiment with research devices, to invent new conceptual repertoires and to care for the accounts we give of others' practices and worlds.

While creating space and visibility for some of the more original extensions of ANT presently at work around the world is indeed an ambition of this companion volume, ours is then not simply a book *about* current ANT-inspired research. Rather, it is a collective exploration into what it takes and entails to think 'near ANT.' We take this figure from Abdoumaliq Simone (2014) and his reflections on the 'near South,' an interstitial space that challenges standard geopolitical constructs of 'North' and 'South,' yet still remains close to them. Indeed, Simone deploys the term advisedly to propose also to 'keep near' concepts of neo-liberalism, democracy and even development, that is, to not quite deploy such terms but also not discard them entirely when accounting for (near-) South liveabilities (see also Minh-Ha's notion of 'speaking nearby,' in Chen, 1992). In a similar mode, this volume explores modes of thinking and speaking near ANT, that is, not simply deploying the existing ANT canon of concepts, research strategies and writing experiments, but keeping them near as a source of questions, problems and inspiration.

Moving from an 'after ANT' to a 'near ANT' intellectual project, this volume aims to demonstrate and help to further articulate two conditions of an intellectual space in which ANT may continue to thrive. The first one is providing an anti-exceptionalist account of ANT. Indeed, thinking ANT as a companion, thinking near ANT, demonstrates that this is not an exceptionally powerful intellectual apparatus, but rather part of quite specific enactments of a more-than-human social inquiry. It is important to recognise the existence of a whole field of neighbouring intellectual practices (epitomised by authors such as Donna Haraway (1991), Isabelle Stengers (1997), Tim Ingold (2000) and Karen Barad (2007)), where ANT has never been alone. This is not to say that we do not recognise qualities that make ANT excel (why else put together a Companion?). Rather, it means that we are operating in an expanded field of post-, non- or more-than-humanist problematisations of the common world(s) we construct and inhabit. ANT, we believe, needs to feed on and explore its capacities and limitations in this wider ecology.

The second condition for this intellectual project is the continuing 'ex-titutionalisation' of ANT. If the figure of 'after ANT' proposes a stabilisation of what 'ANT' was and what 'post-ANT' would be – leading to dichotomous distinctions concerning key conceptual figures of complicatedness/complexity, networks/fluids or translation/multiplicity – then the notion of 'near ANT' refuses distinctions that might stabilise, institutionalise or canonise what ANT *is* in any finalised or teleological sense. This does not mean that ANT has no identity whatsoever; indeed, the willingness to accept and endorse its own continual translation and re-specification across time and space might be said to belong to just such an ANT identity. Yet, as Annemarie Mol (2010) remarks, this is less the identity of a 'theory' or a coherent framework, and more the identity of an adaptable and open repository: More a notebook than a central perspective.

To be more specific, we would claim that ANT's identity is not defined by any kind of conceptual or methodological core. Rather, it is defined through a certain mode of doing intellectual work, certain kinds of intellectual practices and sensitivities, which include ways of doing conceptual resemblances and differences differently. The ANT we are interested to continue pursuing is thus a moving target that is constantly allying with and making visible actors and articulations that challenge otherwise stabilised conceptualisations of 'the social.' The ANT we are interested in is an intellectual project that is always in beta: It does not construct edifices, but amounts to a continuous production of prototypes, partially broken yet productive intellectual devices that open up further questions and the need for further prototypes (Corsín Jiménez, 2014). Committing to 'ex-titutionalise' ANT is to cultivate ANT as an open-ended and experimental becoming, sensitive to variations in its conceptual and empirical repertoire (see Jensen, 2014).

In articulating this agenda, we are not claiming for ourselves any kind of radical or even moderate novelty. Indeed, as hinted, we are quite closely following Annemarie Mol (2010) in her account of the 'sensitive terms' and the 'enduring tensions' of ANT, right down to the actor, the network, the theory and the hyphen (see also Latour, 1999b). In proposing the analogy of ANT as a 'notebook,' Mol is gesturing towards the kinds of family resemblances – not identity, not sameness – that exist among the various 'pages' (or 'versions') of ANT's growing repertoire. There is, in ANT, no singular theoretical or methodological vision, she affirms, but rather 'a rich array of explorative and experimental ways of attuning to the world' (Mol, 2010: 265). Questions of translation, ordering and coordination in diverse socio-material practices and worlds sit centrally in this array; yet, analysts continually refract and respecify such questions under novel conditions of inquiry. The present volume projects this image well, we believe, in that it traces and indeed stages such intellectual dialogues across a wide range of ANT's settings and versions.

Mol's invocation of the many family resemblances in the ANT notebook is also fitting to describe what holds this volume together – a volume for encountering ANT's many companions. Here, however, the slight refraction in metaphor may matter: While 'family' connotes kinships and ordered lineages, 'companion' carries connotations of evolving acquaintances and friendships of a more mutualistic kind. This is the sense in which Donna Haraway (2008) – herself, clearly, a core ANT companion – speaks of dogs as companion species. While families quarrel, sometimes over petty details, companions feed, Haraway suggests, on recognising the 'significant otherness' of their partner in crime. If readers find in this volume reasons for making ANT their companion in any of these senses, drawing near to it if only sometimes to recognise its otherness to their own intellectual concerns, then we will consider this volume a success.

Alongside ANT: probing into an intellectual practice

The question this volume sets out to explore is what ANT entails as an intellectual practice or, perhaps, as a group of companionate intellectual practices. More specifically, we are interested in probing how ANT's intellectual practices are and should be changing to keep their relevance in the future. In this sense, this volume is interested in exploring the ANTs to come, in extending important trajectories and seeing how they might bend or twist further. In contrast to the notion of an 'after' ANT, the intellectual practices we would like to explore is thus better imagined as being 'alongside' ANT or, in some senses, even 'before' ANT, in the sense that they engage with ANT as something unfinished, always in the making.

We speak then of an 'intellectual practice' to stress two angles from which we explore what ANT is or could be. The first angle is relatively straightforward: We are interested in ANT as a 'practice' in the broadly pragmatist sense of the concept, that is, as a *doing* not defined by an underlying logic, but by a transactional engagement in a specific ecology of relations and by the production of certain sets or types of effects (Bateson, 1987; Dewey, 1925). In the first half of this section, we discuss some of the elements that shaped the ecology in which ANT as an intellectual practice unfolded. By doing this, we hope to make apparent the radically different intellectual, political and planetary ecologies in which ANT is practised today. The second angle from where we propose to look at ANT is its articulation as an 'intellectual' practice, by which we understand it not just as a mode of knowing, but also of intervening in the world. At the end of this section, we will briefly point to some of the public, ethical and political commitments of ANT, thus anticipating a concern that crosses the whole volume.

Situating ANT in the ecologies that made it possible requires us to begin with the back then embryonic field of STS, which, as we have noted already, is where ANT was developed in the 1980s and 1990s by a small group of scholars based in Western and North European universities. Doing this is important, in order to avoid unrealistic myths of origins. Just to be clear: ANT was not born in a Renault Five with Latour behind the wheels (see Harman, 2009), but rather in the context of a broader scholarly movement committed to turning age-old philosophical questions concerning the nature of knowledge, science and truth into questions amenable to historical and empirical inquiries (Bloor, 1991 [1976]). Indeed, prior to ANT coming into existence, the first major intervention of Bruno Latour and Steve Woolgar (1979/1986) was in co-creating the field of the so-called laboratory studies (Knorr-Cetina, 1981; Lynch, 1985; Traweek, 1988), proposing to look at and account for how scientific knowledge is produced in practice by ethnographic or anthropological means. Yet, characteristically, what came to constitute ANT was not the ethnographic method per se (although this remains important), nor any single-minded commitment to the sociology or philosophy of science, which it would soon enough overflow. Rather, the key contention was a debate about 'the social' in the field of STS, which famously led Latour and Woolgar to take out the adjective 'social' from the 1986 edition of their *Laboratory Life* (which was then subtitled 'the construction of scientific facts'). Here, ANT came into its own via something like an open declaration of war against otherwise widespread ideas of social constructivism in the study of science and technology, epitomised by the title of the oldest journal in the field *Science, Technology and Human Values* (1967), but also across the board of sociocultural sciences.

To be sure, the writings of Bruno Latour, in particular, raised some flags among analytical philosophers of science, in a 1990s climate of 'science wars' pitting (so-called) 'realists' against (so-called) 'post-modernists' (see Latour 1999c). However, the real 'war' of ANT was always against the idea that reality is purely and simply a social, symbolic or discursive construct. In this sense, all of early ANT was involved in bringing back in the 'missing masses' of sociocultural studies, whether in the shape of scallops and scallop-hatching equipment contributing to regional fisheries development (Callon, 1986); machines that travel as part of technology transfer schemes to developing countries (Akrich, 1992) or long-distance vessels, winds and currents co-determining the fate of European colonialism (Law, 1986).

The philosophical import of such claims about the hybrid, semiotic-material and technoscientific nature of 'the social' has not been lost on ANT's 'first wave.' A foundational ANT text, for instance, Latour's *Irreductions* (1988) – published as an annex to his detailed historical case study of Louis Pasteur and his microbes – reads as a metaphysical treatise on the ontological irreducibility of all 'existents' and the processual and constructivist (but not 'socially constructed') nature of 'the real.' Some years on, Annemarie Mol (2002) would add to and further such ontological interests when she deployed her ethnographic work into the diagnostics and treatment of atherosclerosis in a Dutch hospital to argue that 'reality' is not singular, but rather consists in a multiplicity of partly overlapping, partly disjunctive realities or worlds. In such steps, ANT gradually came to be identified as a style of 'empirical philosophy' (see Gad & Jensen 2010).

While the intellectual radicalness and originality of such propositions should not be in doubt, we also want to register a certain hesitation, at this point, as to whether such an identification of ANT is by itself adequate to its present and future flourishing. On the one hand, amidst still more frequent claims to 'new materialist,' 'speculative' and 'ontological' turns in philosophy and sociocultural studies (Bryant, Srnicek, & Harman, 2011; Dolphijn & van der Tuin, 2012; Holbraad & Pedersen, 2017), shaped in part by ANT, one might be forgiven for thinking that a mere continuation of ANT's original impetus would nowadays amount to pushing against open doors (although, we add, important philosophical differences do remain). On the other hand, as ANT has spread widely and to divergent effects across vast landscapes of sociocultural study and fields of practice around the world, it is perhaps less than self-evident why one would want to insist on its *philosophical* orientation and pedigree, in particular? Beyond science and technology, would we not want to keep the *stakes* of ANT's interventions a bit more open-ended?

A second important characteristic of the intellectual ecology within which ANT emerged is the fact that STS were a radically interdisciplinary or even anti-disciplinary endeavour, witnessing a constant experimentation with slightly different approaches and a playful proliferation of what are, today, rather unrelatable acronyms (SSK, EPOR, SCOT, SST, LTS, etc.). Only ANT and STS, it seems, stuck. This does not mean, of course, that early STS were a constraint-free space. Quite to the contrary, STS (and with it ANT) came of age in the midst of institutional and biographical attachments to the natural and engineering sciences.

The number of influential STS scholars with a background in science and technology is significant. Notably, the scholar who actually invented the notion of 'actor-network theory,' Michel Callon (1981), was himself a retrained engineer – like his colleagues Madeleine Akrich and Antoine Hennion (also a musicologist) – working at the *École des Mines*, an elite institute of technology situated in Paris. Attendant commitments were also practical, in that this French branch of ANT developed amidst a particular ecology of practice oriented to problems of technoscientific innovation (Akrich, Callon, & Latour, 2002a, 2002b) and situated amidst various collaborations with technoscientific disciplines, private corporations and governmental agencies.

Our point here is not to rehearse well-worn debates as to whether or not this particular trajectory of emergence of ANT led to a certain 'critical' blind spot towards questions of power and domination (de la Bellacasa, 2011; Star, 1991) – although we recognise the continued importance of these debates and will indeed revisit them intermittently throughout the volume. Rather, in the spirit of thinking ANT as intellectual practice, the question to be asked today, we believe, is about the many influences on and effects of various ANTs across the whole span of settings and domains to which it has subsequently travelled. Such travel clearly entails that other problems, other questions, other accounts and other interventions have been emerging under the collective banner of 'ANT,' even as the shifts and turns and proliferations involved have perhaps not always attained (enough) visibility and acknowledgement. Our volume indeed seeks to address and rectify this issue, even as we remain acutely aware of its own limitations (more on this in sections 4 and 5).

There is every reason to suggest that, in moving in many directions at once away from its early implication into engineering-heavy settings, ANTs of many hues retain and redo some of their interdisciplinary and indeed anti-disciplinary inclinations. Such is likely to be the case, for instance, when sociocultural scholars friendly to ANT engage with medical doctors and other health professionals over clinical and corporeal realities (Cussins 1996); when they discuss with designers and architects on the construction of buildings and entire cityscapes (Latour & Yaneva, 2008) or when they intervene amongst environmental scientists, activists and concerned citizens embroiled in controversy over 'natures' (Blok, 2010). Yet, few have attempted to take stock of attendant shifts and reconfigurations in intellectual practice, whether in terms of the minutia and politics of methods, the craft and skill of collaboration or the very question of how constraints and possibilities of world-making interventions differ across sites and settings.

Finally, in this initial line-up, it strikes us as particularly important to emphasise that early ANT also involved an inventive writing practice that became possible in a very unique ecology of scholarly publishing. The work of Bruno Latour has been a constant experimentation with literary devices: The male anthropologist of *Laboratory Life* (Latour & Woolgar, 1986) and the female anthropologist of *An Inquiry into Modes of Existence* (Latour, 2013); the imaginary figures of *Science in Action* (Latour, 1987); the polyphony of voices, documents and dialogues in *Aramis, or the Love of Technology* (Latour, 1996a); the parables and allegories collected in *Pandora's Hope* (Latour, 1999c), etc. The list is indeed infinite. Relatedly, John Law has not just proposed to follow the figure of mess as a method for the social sciences (Law, 2004), but also to experiment with fractal and baroque modes of description (Law, 2002; Law & Ruppert, 2016) to develop ways of relating to the complexities and multiplicities of practices that modernist cultures and languages do not permit. He has also cultivated a radically direct and apparently colloquial but carefully crafted mode of writing about profound philosophical and theoretical problems. So has Annemarie Mol, whose *Body Multiple* (2002) experiments with empirical philosophy as textual practice.

Yet, important as these creative writing practices of 'first wave' ANT scholars have been and are, they remain only part of the palette of inspirations and imaginative aspirations which scholars near ANT might nowadays want to mobilise. Indeed, the whole question of the modes and media of ANT-informed scholarship 'besides' or alongside writing has itself emerged as an important topic of exploration, as inquirers search for other and possibly more engaging ways of intervening in and 'inventing' the social (Marres, Guggenheim, & Wilkie, 2018). In this light, it is indeed striking that the most successful aspects of Latour's own art exhibitions (*Iconoclash*, *Making things public*, *RESET Modernity*) have arguably been the book-catalogues published on their occasion (see Blok & Thorsen, 2016). While texts and literary devices arguably retain a certain privileged position in the world of scholarship for good reasons – as evinced, of course, by the choice of medium for this very companion volume – a more radical commitment to multimodal making as a mode of knowing (see Dattatreyan & Marrero-Guillamón, 2019) is increasingly shaping contemporary intellectual practices near ANT as some contributions to this volume will show.

These and other elements that have shaped ANT as an intellectual 'practice' can be understood as what is alongside, and in some sense prior to, the set of propositions that we otherwise call ANT. As such, they are indeed 'before' ANT, in the specific sense of being prior to any *particular* way of doing ANT. Through the previous discussion, it also becomes apparent that while we seem to have a clear understanding of the ecologies that produce what we called the 'first wave' of ANT scholarship, we know little about what ANT as an intellectual practice might imply today or in the future. In a related spirit, as noted, this volume seeks to explore some key elements of contemporary intellectual practices that are shaping new ANTs. The volume thus does not make one programmatic proposition for the future of ANT. Rather, it aims at probing and shaping intellectual spaces for future ANT engagements and thus to allow ANT to keep its currency of intellectual curiosity for the years to come. It is in this sense that we like to think of this volume as not simply grasping a current moment in social theorisation, but as an open project of future-making, where at stake is not what ANT is or was but what it could become.

This brings us to the second angle on the notion of 'intellectual' practice that we suggested earlier: Taking intellectual work to mean primarily and inherently an engagement with matters of common concern and thus with publics and politics, featuring ANT as an intellectual practice allows us to stress that ANT involves more than a family of epistemic practices, narrowly construed. It is rather that ANT's epistemic practices are also shaped by more-or-less specific ethical commitments, senses of political urgency and public concerns of varying kinds. This is important for our volume in a double sense. First, in line with a core ANT claim, the volume aims to display and discuss the subtle 'ontological politics' (Mol, 1999) that co-shapes every epistemic practice, ANT included. On top of this, we are interested in exploring the currency of ANT as an intellectual practice of giving shape to and articulating matters of public concern and thus of intervening in the various domestic to world affairs that its protagonists are concerned with. In short, we are interested in exploring ANT as a *public* intellectual practice.

Again, there are many precursors of such 'intellectual' practices in the genealogy of 'first wave' ANT scholarship, too many to recount. Let us, however, point to three ethico-political problematisations that have gone hand in hand with ANT scholarship. The first one turns around the problem of 'technical democracy' in technoscientific innovation, economic markets and regulatory bodies. In this context, ANT's commitment to symmetry acquires a different meaning: How to create conditions for the shared engagement and co-inquiry by technical experts and by publics of materially affected and concerned people with a view to their joint articulation of matter of common concern? By reworking the divide between facts and values (Latour, 2004) and engaging the machineries of public participation (Callon et al., 2009), ANT has been practised as a project of 'civilizing' markets, of bringing matters of concern into democracy and thus ultimately as a project for democratising existing forms of representative democracy.

Along a related yet different trajectory, ANT is sometimes mobilised to pursue a fundamental critique of detached, modernist understandings of 'the political' in order to foreground alternative ontologies of attachment, material participation and care. Annemarie Mol's (2008) engagement with the problems of neo-liberal 'choice' and questions of care in contemporary health is a case in point. By exploring the tensions involved when patients are configured in a logic of choice as either paying customers or rights-bearing citizens, Mol critically exposes the exclusion of the body from conventional market and democratic figurations as well as delineates an alternative logic of care based on a collaborative attempt to attune medical knowledge and technologies to diseased bodies and complex lives. On a related line of exploration, Noortje Marres (2012) has proposed to bring back the everyday ecologies of artefacts and spaces into the centre of current imaginations of political participation, which does not just imagine citizens without bodies, but also without attachments to things. While exploring people's reflexive engagements with the material settings shaping their everyday life, Marres aligns with and contributes to feminist theories that vindicate the intimate, the private and the domestic as key sites of political participation.

Finally, Bruno Latour's (2018) more recent engagement with climate change and ecological crises has gone hand in hand with a critical reflection on the current rise of right-wing and nationalist forms of populism not just in Europe and the US, but also in Russia, China and the global South. By unpacking the connections between climate denialism, populism and socioecological inequalities, as noted, Latour is becoming an increasingly influential voice in current public debates in the Euro-American world.

These are only three examples of how ANT works and translates jointly as epistemic and public practice. Such public interventions, of course, come with no guarantees and with plenty of risks. As the ANT saying goes: Translation is also betrayal, and the fate of ideas is always in the hands of future users. Such realisation, however, is no excuse for not making the attempt, and indeed only serves in this volume to motivate a stronger and more self-reflexive interest in the variable modes, conditions and effects of ANT-inspired interventions into public–political affairs.

Around ANT: the volume as a hybrid forum

If we agree to think of ANT as an intellectual practice, in its twin epistemic and public senses, then a companion to ANT needs to be designed as a device for probing into the contemporary ecologies, versions and experiments with this intellectual practice in all its concrete situatedness. Rather than a blueprint of any kind, the challenge is to map out, showcase and seek to further the whole array of solidly ANT-informed practices at work across the globe and which contain within them the seeds of interesting new developments. Such, at least, is the challenge we have posed to ourselves as editors of this volume. This has already proven a productive one, we believe, in that it has triggered many conversations among and beyond our group of contributors about the affirmative and possibly detrimental implications of pursuing the endeavour. Let us briefly articulate some of the principles we derived from the aforementioned.

The first and in some sense core principle, hinted to in the aforementioned, is that the volume should not include the authors that first invented ANT as we know it, the 'first wave' as we have termed it. To put this differently, the volume should not follow only or even primarily the well-known actors, the many 'fathers' and many 'mothers' of ANT, but rather attempt to map the equally plentiful new voices that are adding different bits and pieces to the conversation. We want to be very clear, lest we be misunderstood, that we mean by this move no disrespect whatsoever towards this 'first wave,' in relation to whose rich intellectual legacies everyone in this volume (editors included) are grateful inheritors. Were we in the business of 'handbook-ing' ANT, so to speak, we would surely summon the texts and voices of precisely the scholars cited in this introduction but not included in the volume. Yet, as noted, for a companion volume to do its job, we believe, it needs to keep track of where the action is going, to tack alongside and near to the many movements of ANTs.

One way of thinking of this is to imagine a present-day intellectual 'issue public' for whom the future of ANT is a key matter of concern, and whose members are engaged in exploration and inquiry 'in the wild' as to what kind(s) of effect(s) ANT has and might have for intellectual practice. In this sense, as editors, we are in some sense following ANT's own 'messy' methods of trying to articulate a multiplicity (of ANTs) and to map issue publics (of engaged, ANT-inflected intellectual practices). Yet, from this follows an immediate afterthought: How could one even begin to map what is going on with ANT these days? How could we produce an even remotely reliable map of people, sites and issues that have been opening up new and interesting ANT-related conversations? What principles of mapping could we follow?

Assuming and accepting the impossibility of looking at the field from the outside, we decided to initially go by but then also to carefully challenge, stretch and extend our own connections into various interrelated fields and regions. We were happy to recognise that we already enjoyed strong connections to brilliant colleagues (and oftentime friends, companions indeed) active in quite divergent intellectual and practical fields, but also situated in different geographical regions within and indeed beyond Europe. Starting from such connections, we attempted to make a few 'jumps' in the process, using also bibliometric and other tools, to follow the 'weaker ties' allowing us to engage and invite sympathetic co-travellers who will be able to also challenge orthodoxies in interesting ways. In the process, mapping turned into more of a collective endeavour, as prospective authors and interested colleagues would offer up new suggestions, and as we ourselves learned about adventures of ANT research 'elsewhere.'

As a companion or, perhaps, a companion of companions, we fully recognise that this volume is inherently unfinished, and not only due to constraints of space and our incapacities as co-mappers. Rather, the volume is unfinished in the more principled sense that it is meant to bring new travellers on board and to inflect new conversations. In this sense, in the end, the companion is *not* in fact a map, but rather a set of possible routes through the sometimes inscrutable and in any case variegated landscape of contemporary and potentially future engagements with ANT. Or, conversely, it *is* a map, but only once we agree to take a map to entail a set of signposts in a vastly more extended landscape, and indeed as itself folded up in a 'notebook' style where new pages should be added as more territory is traversed.

The second overarching principle to which we have adhered as editors is that the volume should avoid the otherwise very real risk of reifying what is in fact a never-ending and quite overwhelming conceptual productivity of ANT intellectual practices. Well beyond 'actors' and 'networks,' any list of ANT-inflected concepts would quickly grow long. For starters, we would have agencements, agnosticism, assemblies, attachments, Baroque epistemology, circulating references, centres of calculation, complicatedness, controversies, cosmopolitics, distributed action, Dingpolitik, earthbound, ecology of practice, factishes, generalised symmetry, heterogeneous engineering, immutable mobiles, interobjectivity, inscriptions,

irreductions, issue publics, matters of concern, mediation, messy methods, modes of existence, multiplicity, non-humans, object-topologies, obduracy, oligopticons, ontological
politics, orderings, overflows, parliament of things, performance, plasma, qualculation,
spokespersons, sociologist-engineers, translation, treason. And the list could go on, at the risk
of transforming ANT into a self-referential theoretical system.

The trouble here, we believe, is that such a list *by itself* poses no real intellectual challenge, even as each of the concepts represents a powerful conceptual production and interventions developed, in most cases, in close relationship to specific empirical engagements
with different fields and sites of practice. Indeed, apart from science and technology, many
of these conceptual productions have arisen when studying music, medicine, global health,
economics, ecology, politics, law, religion, visual and performing arts, cities and so on. And
while it might thus seem tempting to deploy such thematic or *field-based* markers in the service of generating order and overview in ANT's many encounters and developments, such a
move would run the serious risk of reinstating that very 'domain ontology' of the social (as
consisting in economy, politics, science, religion and so on) that ANT scholars have always
sought to resist. Much as with the conceptual list, traversing through each one of these 'fields'
represent little by way of productive challenge.

So how to think of a more challenging and, eventually, more productive forum for the extension of ANT into the future? Perhaps the most important decision in this regard has been
to think of the volume as an opportunity to join and further extend a range of ongoing, but
not yet quite charted or stabilised conversations about ANT's capacities. Accordingly, instead
of starting with what we already know 'about ANT,' as editors, we decided to focus on all
the issues that lie 'around ANT' and which we do not quite know yet, even if this collective
'we' is taken (concretely or figuratively) as having been near and alongside (versions of) ANT
for a while or indeed all along – to focus then on the kinds of things, specifically, that we, as
editors, would like to ask to some of our companions out there.

This editorial decision implies that each of the 38 contributions to this volume provides
a uniquely expert response to a very specific question. In each case, while we as editors developed the initial questioning based on our reading of the author's existing work, as set in
the wider ANT realm, the final question and hence chapter titles developed in close dialogue
between editors and authors, with plenty of negotiation, revision and tinkering, as well as
the occasional tensions and disagreements. Traces of this process still hover in the back- or
foreground of many contributions, in the hope of doing justice to the dialogical imagination
with which we have striven to assemble this volume as a hybrid forum.

Looking back at the long back-and-forth exchanges with our authors, in some cases including more than a handful of versions, we reckon a kind of 'abductive' method of inquiry,
in the sense that the final chapters, and thus the ideas contained in each response, may not
always be quite identical to what editors *or* authors alike had in mind upon embarking on the
writing process. This is true, even as the roles played imply that while we as editors asked
the initial questions, by way of responding to them, authors also become responsible – and
literally 'response-able,' as Donna Haraway (2016) would say – for the direction traced by
their ideas. Indeed, as anthropologist Annelise Riles (2006) notes in a related context, the
very form of the intellectual response, which we simulate and play with in this volume, is an
interesting and demanding epistemic *and* ethical endeavour, full of questions of author- and
colleagueship. And while we make no claim to having fully mastered questions and responses
as a form of intellectual collaboration, we hope to have achieved with this volume some of
the 'more subtle, creative, and careful genres of empathy and intellectual appreciation among
colleagues' to which Riles (2006: 26) rightly gestures.

All of this implies that one might think of this companion as an attempt to chart out the collective journey of ANT so far, and to elicit pointers for new routes ahead. In developing the questions needed for such a tour around ANT, certain cross-cutting organising principles started to suggest themselves, gradually coalescing into what is now the section structure of the volume. While we do not wish to overemphasise the 'trick' of central perspective over what has, of course, also been a rather messy process of ordering chapters, the final structure and the route it suggests is also not entirely random or haphazard. Indeed, it pertains to what one might think of as *itself* an idealised version of an intellectual practice, this time concerned with the different aspects or facets of ANT's intellectual practices which one would conceivably want to visit, sooner or later, in the process of drawing near to ANT.

Each of the six sections of the book will have an introduction written by some combination of the three editors, motivating further the framing of the section and previewing each of the contributions within it. For present purposes, then, we will be very brief. Section 1 gathers a series of questions and explicit reflections concerning some key elements of the ANT paradigm(s), focusing on the modes of weaving ANT's conceptual–empirical inquiries. Authors' contributions variously reflect upon modes of problematising, inquiring, comparing, conceptualising, writing and criticising with or near ANT.

Sections 2 and 3 focus on the specific position of ANT in larger intellectual fields. For section 2, we asked authors to critically explore and discuss some of the many sources of ANT, genealogically speaking, paying also special attention to silenced, forgotten or potential sources that can still enrich ANT as an intellectual practice. In section 3, in turn, we are concerned with friendly problematisations of former and current developments of ANT. Thus, in short, while section 2 is about practices of sourcing ANT, section 3 is rather about dialogues, controversies and sympathetic differences with and in ANT.

Sections 4 and 5 explore ANT as a moving intellectual practice. In section 4, we explore how moving beyond the study of science and technology to inquire into various other socio-material phenomena and fields has led to the production of new conceptual, methodological and ethical challenges. Relatedly, in section 5, we explore how moving through different sites, in this term's emplaced and virtual senses, may both challenge and change ANT to take such sites' spatio-temporal specificities and scalar economies seriously. Finally, section 6 addresses some of the public and professional uses of ANT directly, by exploring what happens to ANT as an intellectual practice when it travels outside academia and works as a resource for public intellectuals, for architects, for designers, for communication consultants and so on.

To reiterate, as editors we are quite ready to accept and embrace the suggestion that ours is an incomplete tour near, alongside and around ANT, also potentially when it comes to our joint ability to imagine and encode the many aspects of this intellectual practice into sections and questions. Prosaically, some things and issues (besides 'ANT first wavers'!) simply did not end up in the book due to the unavailability of or the unforthcoming circumstances of prospective authors (which explains, for instance, why section 2 contains no entry on either of the two Michel's, Serres and Foucault, as otherwise important ANT sources). In other instances, imaginable questions searched in vain for their author-to-come. In still other instances, no doubt, our own situated limitations will be the more obvious. It is hardly the case, for instance, that no interesting ANT work is undertaken in 'Africa' by 'African' or 'Africa-based' scholars, and all the more likely that we have simply failed as editors in this case to sufficiently counteract postcolonial forces at work in the academy writ large. We hope to have succeeded on other accounts.

On that note, time now to hand over the word to the real protagonists of this volume, that is, to the authors of the following 38 chapters who kindly and courageously took up the challenge of responding ('response-ably') to our many ('questioning') questions. We invite the reader to explore this volume's many contributions in whichever order deemed appropriate for one's own intellectual and practical purposes. Hopefully, as reading progresses, there will be ample opportunity to digress and to follow new leads through other chapters or, equally, through other ANT, near-ANT or not-so-near-ANT sources (to which reference is made throughout). As editors of this Companion to ANT, we can only hope that readers will want to put its resources to work in ways we have not and could not imagine – and in doing so, to become themselves the new companions of ANT.

References

Akrich, M. (1992). The De-scription of Technological Objects. In W. E. Bijker & J. Law (Eds.), *Shaping Technology / Building Society: Studies in Sociotechnical Change* (pp. 205–224). Cambridge, MA: MIT Press.

Akrich, M., Callon, M., & Latour, B. (2002a). The Key to Success in Innovation. Part I: The Art of Interessement. *International Journal of Innovation Management, 6*(2), 187–206.

Akrich, M., Callon, M., & Latour, B. (2002b). The Key to Success in Innovation. Part II: The Art of Choosing Good Spokepersons. *International Journal of Innovation Management, 6*(2), 207–225.

Akrich, M., & Latour, B. (1992). A Summary of a Convenient Vocabulary for the Semiotics of Human and Nonhuman Assemblies. In W. E. Bijker & J. Law (Eds.), *Shaping Technology / Building Society: Studies in Sociotechnical Change* (pp. 259–264). Cambridge, MA: MIT Press.

Barad, K. (2007). *Meeting the Universe Halfway: Quantum Physics and the Entanglement of Matter and Meaning.* Durham, NC: Duke University Press.

Barry, A. (2001). *Political Machines: Governing a Technological Society.* London, New York: The Athlone Press.

Bateson, G. (1987). *Steps to an Ecology of Mind : Collected Essays in Anthropology, Psychiatry, Evolution, and Epistemology.* Northvale, N.J.: Aronson.

Beck, U. (1992). *Risk Society: Towards a New Modernity.* Los Angeles, CA, London, New Dehli: Sage.

Björgvinsson, E., Ehn, P., & Hillgren, P.-A. (2012). Design Things and Design Thinking: Contemporary Participatory Design Challenges. *Design Issues, 28*(3), 101–116.

Blok, A. (2010). War of the Whales: Post-Sovereign Science and Agonistic Cosmopolitics in Japanese-Global Whaling Assemblages. *Science, Technology & Human Values*, 36(1), 55–81.

Blok, A., & Thorsen, L. (2016). Reset Latour! *EASST Review, 35*(2), 23–26.

Bloor, D. (1991). *Knowledge and Social Imagery.* Chicago, IL: University of Chicago Press.

Bryant, L., Srnicek, N., & Harman, G. (2011). *The Speculative Turn: Continental Materialism and Realism.* Melbourne: re.press.

Callon, M. (1981). Pour une sociologie des controverses technologiques. *Fundamentia Scientia, 2*(3/4), 380–399.

Callon, M. (1986). Some Elements of a Sociology of Translation: Domestication of the Scallops and the Fishermen of St-Brieuc Bay. In J. Law (Ed.), *Power, Action, and Belief: A New Sociology of Knowledge?* (pp. 196–233). London: Routledge and Kegan Paul.

Callon, M. (1996). Le travail de la conception en architecture. *Situations. Les Cahiers de la recherche architecturale, 37*(1), 25–35.

Callon, M. (Ed.) (1998). *The Laws of the Market.* Oxford, Malden, MA: Blackwell Publishers/Sociological Review.

Callon, M. (2001). Actor-Network Theory. In N. J. Smelser & P. B. Baltes (Eds.), *International Encyclopedia of the Social and Behavioral Sciences* (pp. 62–66). Oxford: Pergamon.

Callon, M., Lascoumes, P., & Barthe, Y. (2009). *Acting in an Uncertain World: An Essay on Technical Democracy.* Cambridge, MA: MIT Press.

Callon, M., & Latour, B. (1981). Unscrewing the Big Leviathan: How Actors Macro-Structure Reality and How Sociologists Help Them To Do So. In K. Knorr-Cetina & A. V. Cicourel (Eds.), *Advances in Social Theory and Methodology. Towards an Integration of Micro and Macro-Sociologies* (pp. 277–303). London: Routledge.

Callon, M., & Latour, B. (1992). Don't Throw the Baby Out with the Bath School! A Reply to Collins and Yearley. In A. Pickering (Ed.), *Science as Practice and Culture* (pp. 343–368). Chicago, IL, London: University of Chicago Press.

Chen, N. N. (1992). "Speaking Nearby": A Conversation with Trinh T. Minh–ha. *Visual Anthropology Review, 8*(1), 82–91.

Corsín Jiménez, A. (2014). Introduction: The Prototype: More Than Many and Less Than One. *Journal of Cultural Economy, 7*(4), 381–398.

Cussins, C. (1996). Ontological Choreography: Agency Through Objectification in Infertility Clinics. *Social Studies of Science, 26*(3), 575–610.

Dattatreyan, E. G., & Marrero-Guillamón, I. (2019). Introduction: Multimodal Anthropology and the Politics of Invention. *American Anthropologist, Online First.* doi:10.1111/aman.13183

Dawkins, R. (1998). Postmodernism disrobed. *Nature, 394*, 141–143.

de la Bellacasa, M. P. (2011). Matters of Care in Technoscience: Assembling Neglected Things. *Social Studies of Science, 41*(1), 85–106.

de Laet, M., & Mol. A. (2000). The Zimbabwe Bush Pump: Mechanics of a Fluid Technology. *Social Studies of Science, 30*(2), 225–263.

Dewey, J. (1925). *Experience and Nature*. Chicago, IL: Open Court.

Dolphijn, R., & van der Tuin, I. (Eds.). (2012). *New Materialism: Interviews & Cartographies*. Ann Arbor, MI: Open Humanities Press.

Domènech, M., & Tirado, F. J. (1998). *Sociología simétrica. Ensayos sobre ciencia, tecnologia y sociedad.* Barcelona: Gedisa.

Fogué, U., & Rubio, F. D. (2017). Unfolding the Political Capacities of Design. *What Is Cosmopolitical Design? Design, Nature and the Built Environment* (pp. 163–180). London: Routledge.

Gad, C., & Bruun Jensen, C. (2010). On the Consequences of Post-ANT. *Science, Technology, & Human Values, 35*(1), 55–80.

Halsall, F. (2016). Actor–Network Aesthetics: The Conceptual Rhymes of Bruno Latour and Contemporary Art. *New Literary History, 47*(2), 439–461.

Haraway, D. (1991). *Simians, Cyborgs, and Women: The Reinvention of Nature*. New York, London: Taylor & Francis.

Haraway, D. J. (2008). *When Species Meet*. Minneapolis: University of Minnesota Press.

Haraway, D. J. (2016). *Staying with the Trouble: Making Kin in the Chthulucene*. Durham, NC: Duke University Press.

Harman, G. (2009). *Prince of Networks. Bruno Latour and Metaphysics*. Melbourne: re.press.

Hennion, A. (1993). *La Passion Musicale*. Paris: Métailié.

Holbraad, M., & Pedersen, M. A. (2017). *The Ontological Turn: An Anthropological Exposition*. Cambridge: Cambridge University Press.

Inglis, D., & Thorpe. C. (2019). The Actor-Network Theory Paradigm. In D. Inglis & C. Thorpe (Eds.), *An Invitation to Social Theory*. Cambridge: Polity.

Ingold, T. (2000). *The Perception of the Environment: Essays on Livelihood, Dwelling and Skill*. London: Routledge.

Jacobs, J. A., & Frickel, S. (2009). Interdisciplinarity: A Critical Assessment. *Annual Review of Sociology, 35*, 43–65.

Jensen, C. B. (2014). Continuous Variations: The Conceptual and the Empirical in STS. *Science, Technology, & Human Values, 39*(2), 192–213.

Knorr-Cetina, K. (1981). *The Manufacture of Knowledge: An Essay on the Constructivist and Contextual Nature of Science*. Oxford: Pergamon Press.

Latour, B. (1987). *Science in Action*. Milton Keynes: Open University Press.

Latour, B. (1988). *The Pasteurization of France*. Cambridge, MA: Harvard University Press.

Latour, B. (1991). Technology is Society Made Durable. In J. Law (Ed.), *A Sociology of Monsters: Essays on Power, Technology and Domination* (pp. 103–131). London: Routledge.

Latour, B. (1992). Where are the Missing Masses? The Sociology of a Few Mundane Artifacts. In W. E. Bijker & J. Law (Eds.), *Shaping Technology / Building Society: Studies in Sociotechnical Change* (pp. 225–258). Cambridge, MA: MIT Press.

Latour, B. (1993). *We Have Never Been Modern*. Cambridge, MA: Harvard University Press.

Latour, B. (1996a). *Aramis, or the Love of Technology*. Cambridge, MA: Harvard University Press.

Latour, B. (1996b). On Interobjectivity. *Mind, Culture, and Activity, 3*(4), 228–245.

Latour, B. (1999a). For David Bloor… and Beyond: A Reply to David Bloor's' Anti-Latour'. *Studies in History and Philosophy of Science, 30*(1), 113–130.

Latour, B. (1999b). On Recalling ANT. In J. Law & J. Hassard (Eds.), *Actor Network Theory and After* (pp. 15–25). Oxford: Blackwell.

Latour, B. (1999c). *Pandora's Hope: Essays on the Reality of Science Studies.* Cambridge, MA: Harvard University Press.

Latour, B. (2004). *Politics of Nature: How to Bring the Sciences into Democracy.* Cambridge, MA: Harvard University Press.

Latour, B. (2005). *Reassembling the Social: An Introduction to Actor-Network-Theory.* Oxford, New York: Oxford University Press.

Latour, B. (2013). *An Inquiry into Modes of Existence.* Cambridge, MA: Harvard University Press.

Latour, B. (2017). *Facing Gaia: Eight Lectures on the New Climatic Regime.* Hoboken, NJ: John Wiley & Sons.

Latour, B. (2018). *Down to Earth: Politics in the New Climatic Regime.* Cambridge: Polity Press.

Latour, B., & Hermant, E. (1998). *Paris Ville Invisible.* Paris: La Découverte.

Latour, B., & Woolgar, S. (1986). *Laboratory Life. The Construction of Scientific Facts.* Princeton, NJ: Princeton University Press.

Latour, B., & Yaneva, A. (2008). "Give Me a Gun and I Will Make All Buildings Move": An ANT's View of Architecture. In R. Geiser (Ed.), *Explorations in Architecture: Teaching, Design, Research* (pp. 80–89). Basel: Birkhäuser.

Law, J. (1986). On the Methods of Long-Distance Control: Vessels, Navigation and the Portuguese Route to India. In J. Law (Ed.), *Power, Action and Belief: A New Sociology of Knowledge* (pp. 234–263). London: Routledge and Kegan Paul Books.

Law, J. (1987). Technology and Heterogeneous Engineering: The Case of Portuguese Expansion. In W. E. Bijker, T. P. Hughes, & T. J. Pinch (Eds.), *The Social Construction of Technological Systems: New Directions in the Sociology and History of Technology* (pp. 111–134). Cambridge, MA: MIT Press.

Law, J. (2002). *Aircraft Stories: Decentering the Object in Technoscience.* Durham, NC: Duke University Press.

Law, J. (2004). *After Method: Mess in Social Science Research.* London: Routledge.

Law, J. (2009). Actor Network Theory and Material Semiotics. In B. S. Turner (Ed.), *The New Blackwell Companion to Social Theor y* (pp. 141–158). Oxford: Blackwell.

Law, J., & Hassard, J. (1999). *Actor Network Theory and After.* Oxford: Blackwell, The Sociological Review.

Law, J., & Mol, A. (1995). Notes on Materiality and Sociality. *The Sociological Review, 43*(2), 274–294.

Law, J., & Ruppert, E. (2016). *Modes of Knowing: Resources from the Baroque.* Manchester: Mattering Press.

Law, J., & Singleton, V. (2005). Object Lessons. *Organization, 12*(3), 331–355.

Lynch, M. (1985). *Art and Artifact in Laboratory Science: A Study of Shop Work and Shop Talk in a Research Laboratory.* London, Boston: Routledge.

MacKenzie, D., Muniesa, F., & Siu, L. (Eds.). (2007). *Do Economists Make Markets?: On the Performativity of Economics.* Princeton, NJ: Princeton University Press.

Marres, N. (2012). *Material Participation: Technology, the Environment and Everyday Publics.* Houndmills, Basingstoke, Hampshire, New York: Palgrave Macmillan.

Marres, N., Guggenheim, M., & Wilkie, A. (2018). *Inventing the Social.* Manchester: Mattering Press.

Michael, M. (2016). *Actor-Network Theory: Trials, Trails and Translations.* Los Angeles, CA, London, New Dehli: Sage.

Mol, A. (1999). Ontological Politics. A Word and Some Questions. In J. Law & J. Hassard (Eds.), *Actor Network Theory and After* (pp. 74–89). Malden, Oxford: Blackwell Publishing.

Mol, A. (2002). *The Body Multiple: Ontology in Medical Practice.* Durham, NC: Duke University Press.

Mol, A. (2008). *The Logic of Care: Health and the Problem of Patient Choice.* London, New York: Routledge.

Mol, A. (2010). Actor-Network-Theory: Sensitive Terms and Enduring Tensions. *Kölner Zeitschrift für Soziologie und Sozialpsychologie,* Sonderheft, 50(1), 253–269.

Rabeharisoa, V., & Callon, M. (1999). *Le Pouvoir des malades: l'association française contre les myopathies et la recherche.* Paris: Presses des MINES.

Riles, A. (2006). *Documents: Artifacts of Modern Knowledge.* Ann Arbor: University of Michigan Press.

Schüttpelz, E. (2013). Elemente einer Akteur-Medien-Theorie. In T. Thielmann & P. Gendolla (Eds.), *Akteur-Medien-Theorie* (pp. 9–70). Bielefeld: Transcript Verlag.

Simone, A. (2014). *Jakarta, Drawing the City Near.* Minneapolis: University of Minnesota Press.

Star, S. L. (1991). Power, Technologies and the Phenomenology of Conventions: On Being Allergic to Onions. In J. Law (Ed.), *A Sociology of Monsters. Essays on Power, Technology and Domination* (pp. 26–56). London, New York: Routledge.

Stengers, I. (1997). *Power and Invention: Situating Science* (Vol. 10). Minneapolis: University of Minnesota Press.

Traweek, S. (1988). *Beamtimes and Lifetimes: The World of High Energy Physicists*. Cambridge, MA: Harvard University Press.

Yaneva, A. (2009). *The Making of a Building. A Pragmatist Approach to Architecture*. Bern: Peter Lang.

Section 1

Some elements of the ANT paradigm(s)

Ignacio Farías, Anders Blok and Celia Roberts

It cannot be repeated enough: ANT is neither a theory nor a method. So, what is it? A productive answer has been that ANT entails a sensitivity for engaging with the world – not a central perspective, but an open repository of terms and texts, concepts and accounts, enacting and testing modes of attuning to the social life of things, to what an actor might be and to how things and actors coexist, clash, differ and associate. The figure of the repository evokes Thomas Kuhn's first understanding of scientific paradigms as sufficiently unprecedented and sufficiently open 'exemplars' to which scholars come back looking for inspiring problematisations. ANT could be then understood as a repository of paradigms that have shaped the intellectual practices of a whole community of scholars.

The contributions to this section deepen this understanding of ANT as a repository of paradigms by paying attention to what lies 'besides' (*para*) what is said and 'shown' (*deiknynai*) in such exemplars – the most literal meaning of the concept of paradigm. This also corresponds to the second and most influential way in which Kuhn deployed the concept of a scientific paradigm: A 'matrix' of concepts, methods, sensibilities, procedures, values, commitments and assumptions permitting the collective and collaborative advancement and articulation of scientific inquiries. So, while what is said and shown in paradigmatic ANT exemplars resembles a plural repository of concepts and accounts, what lies besides these exemplars is a certain mode (and skilled practice) of relating theory and methods, concepts and descriptions, problems and valuations, and epistemic devices and critical commitments. Exploring ANT as a scientific paradigm in this second sense requires us to focus less on ANT terms and texts than on ANT as a skilled practice or a mode of doing.

With this aim in mind, we invited authors to reflect upon modes of problematising, inquiring, comparing, conceptualising, writing and criticising with or near ANT. We certainly do not think that these elements cover all key aspects of ANT nor that the responses offered by our authors reflect an uncontested *doxa*: There are more questions to be asked and lot to discuss. Yet, our authors offer conceptually powerful and truly original engagements with at least some elements of ANT paradigm(s).

The *tour de force* begins with Daniel López-Gómez's chapter addressing this section's concerns straight on. 'What if ANT wouldn't pursue agnosticism but care?', he asks, answering with a story of initiation into the arts and crafts of ANT. Drawing on his work during the last 15 years on care arrangements for older people, López explores the epistemic and ethico-political effects of two modes of problematising with ANT. If the agnostic mode was essential to problematising the sociotechnical setting-up of a telecare system, the care mode forced itself upon the researcher through experiences of disconcertment resulting from encounters with the fragile daily life arrangements of telecare users. López explores the different types of ethico-political commitments for ANT researchers resulting from agnostic and caring modes of doing ANT.

We engage secondly with practices of making concepts. Adrian Mackenzie sets off his discussion of such practices with two observations: Firstly, concepts are not fundamentally different from any other objects; and, secondly, by often using the suffixes -ant and -ent, ANT has a tendency to create concepts denoting transitive relations. ANT concepts, he then suggests, might be better understood as 'conceptants' – a term that figures concepts as actants. Accordingly, the key question is how to make concepts more real. Mackenzie argues that there are some important lessons to be learned from how marmalade, clothes, code and salto-flips are made. Drawing on personal experience, ethnographic research and media accounts, Mackenzie discusses how ANT concepts can be made sticky, strong, trackable and lively.

Is this the radical empiricism that ANT aims to cultivate? And is this in any sense ethnographic? Britt Ross Winthereik's contribution is the first of many in this volume that reflects upon current encounters of ANT and anthropology. In her case, the focus is on how the ethnographic is and has been practised and conceptualised in both fields. Drawing on fieldwork research in a technology fair, Winthereik describes how she encountered not just interlocutors and technical artefacts, but also concepts that reflected and diffracted the concepts she would bring into the field. She suggests that ANT's radical empiricism – epitomised in the dictum 'follow the actors' – could be productively reinvented through an ethnography of conceptual relations and recursions.

Atsuro Morita engages with a similarly productive and equivocal question: Can ANT compare with anthropology? By recounting the history of comparison in anthropology and situating ANT in relationship to this history, Morita makes an important case to think ANT's self-description as an infralanguage – where the concepts and problems of a practice need to be stronger than those of the analyst – as related (and thus comparable) to the deployment of recursive comparisons in anthropology – where anthropological modes of thinking and inquiring are compared with the modes of thinking and inquiring encountered in the field. Following a technology transfer project from Japan to Thailand, Morita complicates anthropology's intellectual history by showing how travelling machines trigger practices of comparison 'in the wild,' which again invite the analyst to juxtapose comparisons and seek for lateral connections among them. Morita proposes the notion of lateral comparison as a mode of rethinking ANT's infralanguage as something that is never neutral and always entangled in the trajectories of peoples and things.

The question is, indeed, who writes these stories. Or, as José Ossandón argues in his contribution, which are the conceptual personas that articulate ANT theoretical, methodological and political sensibilities. Ossandón explores this issue in relationship to recent ANT work on markets and by means of the question of how to write after Michel Callon's performativity. The project starts with an exploration of the implicit instructions and sometimes rather explicit slogans aimed at formatting research personas (and, as it has turned out, whole armies of Callonian researchers) for the study of markets – instructions that invite to make making-offs, inversions and patchworks. In a second step, Ossandón introduces the readers to the characters or personas that some of the most renowned Callon followers enact in their texts not the least to comply with Callon's instructions. Importantly, the 'performative detectives,' the 'philosophical ethnographers' and the 'market reformists' the reader of such texts will encounter are not idiosyncratic articulations of singular authors, but resonate with some of the key conceptual personas of ANT writ large.

This is all fine, fellow social scientists might argue, but can ANT be a critique of capital? This is precisely the question Fabián Muniesa addresses. In his contribution, he revises the conceptual repertoires used by Latour and Callon to analyse the making of scientific knowledge and technology development, the study of accounting, economics and marketisation, as well as the exploration of the economic theologies of the Anthropocene. For all three moments, Muniesa demonstrates the extent to which capital is unpacked by ANT scholars as an operation of translation, that is, as a mode of conversion of different things into assets that can be mobilised for certain purposes. Continuing with this problematisation of capital, Muniesa proposes a more systematic research programme not of capitalism, but of capitalisation and the epistemic practices, political configurations and valuation regimes that make conversions into capital possible. Critique is here to be understood as a method for examining tensions in current practices of capitalisation.

Michael Guggenheim has written the last contribution to our first section. In it, he features a colourful palette of modalities that critique can take in and with ANT. The question he poses – namely, 'how to use ANT in inventive ways so that its critique will not run out of steam?' – involves a double critique of Latour's critique of critique. It does not just imply that ANT has always been in the business of critique, but also that the way ANT has done critique might require reinvention. To make this argument, Guggenheim speculatively expands Latour's suggestion that critical theorists resemble blasé tourists capable of going anywhere mounted on their budget flights. Exploring critique as a mode of displacement, Guggenheim reconstructs how the critical expeditions of ANT into different fields and sites have enacted at least three further modes of travel: Ecotourism among natural scientists, spa-holidays in the studios of architects and designers, and executive business travel above the clouds of modernity. As an alternative, Guggenheim proposes vindicating the speculative, playful and multisensorial modes of exploration and learning of the exchange student, who jumping into unforeseen situations articulates critique not as the examination, but as the experimental modification of practices.

1

What if ANT wouldn't pursue agnosticism but care?

Daniel López-Gómez

The question that opens this chapter does not seek to introduce new theoretical arguments to the passionate and long-standing debate concerning the epistemic, methodological and political attributes of ANT's agnosticism (see, for instance, Callon & Latour, 1992; Gad & Bruun Jensen, 2010; Pels, 1996).[1] As is now well known, the radicalisation of agnosticism led to a post-social shift with well-known consequences for Science and Technology Studies (STS) and beyond.[2] Neither does the opening question aim to present care as an alternative paradigm, ethos or frame that would reveal and overcome the problems to which the adoption of agnosticism might lead. The question is certainly triggered by some sort of discomfort with ANT's agnosticism, but points towards a more speculative yet practical concern. The paper attempts to appraise the pragmatic consequences that ANT's repertoires have in our inquiries. As the notion of repertoire suggests, there is no appreciation here of the structural consistency of agnostic and care repertoires, neither is there a discussion about their intrinsic attributes. What I am interested in testing here are their pragmatic affordances, what these repertoires can do to our empirical engagements and what these engagements can do to them, that is, the differences these repertoires can contribute to making in our empirical inquiries.

This type of testing exercise is not very common in STS and ANT literature,[3] which is quite surprising if we bear in mind that the most recognisable figures in ANT literature have repeatedly insisted since its inception that ANT should not be treated as a ready-made theory or method (Callon, 1999; Latour, 1999a), but as a repertoire (Mol, 2010, p. 261) that is diasporic (Law, 1998), increases sensibility and is shaped by the convoluted multiplicities of our case studies.[4] But this lack of attention to the pragmatic affordances of these repertoires is even more striking if we acknowledge that doing ANT is not about producing 'innocent' descriptions but rather about intervening (through these descriptions too) in the ontological becoming of the actors we 'follow' and with whom we also become entangled during our inquiries. ANT repertoires do intervene in the framing of empirical engagements since they affect the establishment of settings and relationships with the actors and the way their accounts and concerns shape our own and are articulated by the repertoires we try to articulate. In fact, ANT scholars usually commit to nurturing their repertoires with challenging and complex empirical cases in the aim of producing conceptual, political, methodological and even aesthetical displacements.

Due to all these considerations about how ANT should be understood and handled, Annemarie Mol (2010, p. 261) suggested that ANT scholars should be considered 'amateurs of reality.' This would mean, following Hennion, that ANT scholars are 'experts in the *consequential testing* of objects they are passionate about,' and that 'they confront them' and 'thus accumulate an experience that is always challenged by the way in which these objects deploy their effects' (Hennion, 2007, p. 75). I do think that ANT scholars should be considered as such, but I also think that, much more often, we should confront our repertoires in the same way that amateurs confront their passionate objects. Without incorporating amateurs' appreciating gestures towards our own repertoires, ANT will continue to be too easily treated as a ready-made theory or methodology regardless of the threatening messages of burying ANT if its amateurish character is not respected (see, for instance, *Actor Network and After*, Law & Hassard, 1999).[5]

This contribution is a very modest remedy to this situation. As amateurs do, it seeks to pursue a reflexive and pragmatic confrontation with the ANT repertoires we articulate in order to test how they shape our case studies and are shaped by them. The aim of this paper is precisely to contribute to doing so by performing a *consequential testing* of the agnostic and care repertoire in a particular case study. To do so, I will draw on an ethnography case study I conducted on a telecare service some time ago. It is a case study I have written much about and to which I did not wish to return, I must confess.[6] However, after several attempts, I have finally come to terms with the idea that retracing my steps might be instructive of the pragmatic reflection I would like to pursue in this volume. In this regard, I will examine the 'disconcertments' (see also Jerak-Zuiderent, this volume) that the encounters with the telecare users produced in my fieldwork, which shifted the focus of my analysis from the sociotechnical setting-up of the telecare system to the fragility and affordances of the older peoples' daily life arrangements. As a result of this, the pragmatic differences between the agnostic and care repertoires will be sketched out and discussed in light of the classic ANT literature and current debates on care and STS (Martin, Myers, & Viseu, 2015; Latimer & Puig de la Bellacasa, 2013; Puig de la Bellacasa, 2017).

Doing agnosticism with telecare

> Last year, 76 people aged 65 or over died alone at home in Madrid. So far this year, a further 53 have died and the total figure for 2002 may well exceed one hundred. The numbers can be extrapolated to other major Spanish cities, where there are increasingly more elderly people living alone, either as a result of a decision they have made or due to inevitable circumstances. The most painful aspect of these figures is the certainty that many of these people could have been helped in time to save their lives or that they suffered long hours of agony without medical support or the comfort of a friendly companion. It is genuinely annoying to know that the vast majority of such cases can be avoided through simple telecare systems, in use in many cities around the world. These programmes provide the elderly person living alone with a wristband or pendant and a special telephone device which allows them to immediately call for assistance from anywhere in the home. A fall or fainting loses its deadly potential and instead becomes a treatable medical condition.
>
> *(Editorial, 2002, p. 8)*

This *El País* editorial, published after the 2002 heat wave, offers a perfect snapshot of telecare when I began to study it. Telecare was seen as a technology that could offer 'security' to older people living on their own and their family members. It was a system that could make the provision of elderly care in ageing societies more sustainable, as well as more accessible and ethical because institutionalisation could be delayed, and older people given greater autonomy. In contrast to this, some critics considered that telecare was indeed reinforcing the problem by outsourcing care to their recipients and their families and impeding a fair redistribution of care across social class, ethnicity and gender. While in the former approach, technology was emphasised as an agent of transformation, in the latter, transformation takes place elsewhere and technology merely facilitates or hampers it.

At that time, coming across the classic ANT texts was an invitation to construct an agnostic repertoire with respect to both approaches. According to the programmatic text about the scallops (Callon, 1986), the ANT analyst must not 'change registers when moving from the technical to the social aspects of the problem' (Callon, 1986, p. 200). In other words, the analyst must set aside the 'religious respect for purity' (Latour, 1981, p. 210) and stop thinking in terms of two ontological regions, each with its own characteristics that demand differing explanations. Either as a technological cause or as an epiphenomenon of social transformations, telecare should be explained using the same repertoire. There is no single repertoire for doing so, but the analyst must take on two methodological principles when choosing one.

The first of these obliges the analyst not to consider his/her repertoire as more valid than others *per se* (Callon, 1986, p. 200). Initially, an analyst should not take any ontological position with regard to the explanations given by those involved in the case, including those given by the analyst. At that time, this methodological requirement was highly significant for me, as it meant putting the way in which the case had been theorised in quarantine.

In the beginning, I considered telecare as relevant because it had the potential to empirically feed a theoretical discussion we were having around the notion of 'extitution,' a concept from Serres (1995) we draw on to understand the material configuration of power in 'virtualized' institutions (Tirado Serrano & Domènech i Argemí, 2013). Telecare became the perfect example case to address the virtualisation of care and I took ANT as a theory that revealed the topological materiality of extitutions. For this reason, this first ANT principle was indeed not fully taken up. ANT was supposed to favour an agnostic approach with respect to any ready-made explanations, including those based on ANT studies. As Latour warned, 'ANT is first of all a negative argument. It does not say anything positive on any state of affairs' (Latour, 2005, p. 141). Thus, it turned out to be crucial for me to learn how to situate myself in the centre of the empirical generativity of the case study. I shifted away from trying to develop explanatory concepts and towards adopting concepts that would enable me to capture the various and complex ways in which the telecare service was set up. In other words, a more thorough reading of ANT's programmatic texts – such as Callon (1986) – made me approach telecare in a more modest and exploratory manner, as a multiplicity and collective endeavour that required me to follow and learn from the actors, including those that had been silenced or were apparently less concerned.

Besides not imposing a grid of categories on the actors, the other principle that any ANT repertoire must fulfil is not to repeat the analysis the actors suggest (Callon, 1986, p. 200). That is the reason why controversies for ANT are the precise situations that make agnosticism the most heuristically wise attitude. As there is no single definition of the issue at stake but rather multiple and unstable ones, being agnostic is in fact being realistic towards the actors involved in the controversy and their different accounts. However, the telecare case did not resemble the classic sociotechnical controversy and it was quite hard to refrain myself from either repeating the actors' repertoires or imposing my own. Telecare was already a stabilised technology and the service was established quite successfully ten years ago. Also, most of the actors I encountered (managers, technicians, users, relatives and alarm centre agents) tended to draw on very stabilised and 'powerful' explanations regarding the autonomy and capacities of older people and the peace of mind the service gave to their relatives. Their actions did not seem controversial and there was therefore no need to produce alternative accounts.

Due to this, my analytical strategy was based on paying greater attention to those frictional situations in which the service was put to the test – that is, 'trials of strength' (Akrich & Latour, 1992) through which the invisible, embedded and silenced work that endows telecare with its functional attributes was brought to the fore, similar to what Star calls 'infrastructural inversions' (Star, 1999, p. 380). These frictional situations appeared, for instance, when users would not open the door to installers, would not wear the pendant or used it for other purposes; when installers would place the telecare equipment in noisy rooms because it was the only location that the user would accept; or when call centre operators would establish more personal 'relationships' with service users as a way of providing better care, even though this was explicitly discouraged by the managers to reduce the number of burnouts. These frictions became the heuristics of my agnostic repertoire. Firstly, they enabled questions of sociological interest to be placed on the same level as the practical problems the actors had to face without favouring any particular perspective. Secondly, the attentiveness to these frictions also brought to the fore the work undertaken by less visible actors such as installers, maintenance workers, volunteers and relatives and the role of protocols and all sorts of standardised procedures for organising their work that are key parts of the infrastructure for the provision of care (López, Callén, Tirado, & Domènech, 2010). Thirdly, the focus on these frictions paved the way for the deployment of a monistic material-semiotic repertoire that could render the delivery of care (López et al., 2010) and the 'autonomous' life of their older users (López & Domènech, 2008) accountable as temporary outputs of the same heterogeneous engineering processes (Law, 1992).

To understand the policies, the technological designs and the provision of services, the agnostic repertoire turned out to be extremely important. Despite the fact that care practices, organisational cultures, technical devices, bodies, spaces and subjective dispositions needed to be transformed for the telecare to be designed and implemented (Hyysalo, 2004; Pols, 2012), this was concealed by the usual representations of telecare as a technical fix for the ageing society (Mort et al., 2015). Thus, by putting forward a nuanced and empirically grounded comprehension of what these innovations did in practice and how they were unleashed and tamed (Pols & Willems, 2011), the agnostic ANT repertoire contributed to opening up a critical discussion about the social and political implications of framing telecare as a plug-and-play, cost-effective and user-friendly solution. It was crucial to bring to the fore its constitutive multiplicities and also the power struggles involved in their design and implementation. However, the downside of this was that the agnostic repertoire also seemed to make me quite insensitive towards the 'violence' of the aforementioned frictions, to its uneven distribution and consequences for the actors at stake; it was as if I could point out the struggles without being touched or moved by them.

Entangled with reluctant telecare users, shifting from agnosticism to care

My disconcertment started to grow as I spent more time with the telecare users at their homes. Unlike in the conversations I had with the telecare workers, the queries I put to them – mostly women around 80 living on their own – were just a pretext to talk about other issues. These tended to revolve around their day-to-day routines and experiences; life-story and biographical memories; their relationship with relatives, neighbours, friends; stories about the neighbourhood and other places where they used to live; personal beliefs, convictions, fears, etc. There was almost no trace of the telecare system in these conversations. I have to confess that initially it was slightly irritating to be unable to channel interviews towards the topics I wanted to talk about. Nevertheless, little by little I realised that these conversations, despite looking tangential to the topic of my interest, were in fact far more relevant to my research on telecare. It turned out to be abundantly clear not only that telecare had to be continuously arranged in particular ways to become a 'solution' but that this entailed rear-ranging, even straining, other arrangements that were already in place. What mattered was not the adoption or rejection of the telecare service, neither was it important whether or not the way in which the service was used could be problematic or lead to future developments. What became relevant were the multiple arrangements that the people I encountered had to devise to continue living as they wanted, as well as the fragility of those arrangements and the 'violence' that gerontechnological innovations such as the telecare system exerted on them.[7]

Either for good or bad, the lives of the users with whom I shared afternoons of long con-versation already relied on ecologies of other, more or less, dense and operative arrangements, which were usually more informal but also more embedded than the telecare system. These arrangements endow their daily lives with continuity and richness, as they enable them to have and perform various social identities. For instance, some of the users I interviewed during the installation process used to phone or meet up with their relatives or neighbours at least once a day. These meetings were important for keeping in contact with the people they care about, to help them or to get help from them without an explicit request. Some had even devised 'hacks' to let the neighbours know about their situation without having to contact them, for example, by rolling up the window blinds of the bedroom every morning to communicate that they were fine.

These 'little' arrangements were so important that some users had the feeling of 'abandoning' them, and what goes with them, with the incorporation of telecare into their daily life. The aforementioned hacks would become useless and the meetings and phone calls would be less frequent, shrinking the possibilities for reciprocal care and perhaps contribut-ing to defining them merely as recipients of care. That is why many users ended up leaving the pendant in the drawer or, in a situation of emergency, were more inclined to phone some-body with whom they had a day-to-day relationship (usually a relative) than asking for help from an anonymous system activated by an alarm button. However, these arrangements were not seen, in most of the cases, as innocent either. In most of the cases, social workers, service providers and many relatives and users welcomed telecare because they realised that these ar-rangements were far too demanding and could lead to situations of exploitation or negligence due to an uneven gender and class-based distribution of care work (Mort, Roberts, & Callén, 2012). Most of the users decided to have a telecare device at home because they wanted to stop being a burden for their close acquaintances.

The most striking thing was that, despite the importance of these mundane arrangements and the harmful or beneficial consequences associated with maintaining them, they were usually disregarded by managers, policymakers and technology developers. Nobody but the users seemed to care about them. It was as if acknowledging their existence would ruin the image of telecare as an innovation that came to fulfil a void in the user's life and as a tech-nology designed to cover necessities that could not be covered otherwise. The consequences a simple telecare system could have on these arrangements were usually left unattended and rather than considering the users' reluctance and 'misuse' of the service as ways of accommo-dating and maintaining those arrangements, managers, policymakers and technology devel-opers viewed them as mere barriers to overcome in future developments.

A serious consideration of the users' reluctance and aforementioned arrangements was at odds with the agnostic repertoire. I needed a repertoire that not only made the heterogeneous engineering process of the telecare service and its power struggles visible, but that also made us sensitive to the violence that this entails. As the work of Star (1991, p. 29) shows, to 'translate' is to exercise power: some elements are endowed with the capacity to delegate, coordinate and translate other elements into specific programmes of action, while others are translated, coordinated and become delegates of others instead. The way of making us aware of the violence of these translation processes seemed to be to consider the attachments we build in our inquiries, not only as potential entrapments that we need to put on hold to safeguard our agnostic analysis, but also as something that we need to learn to cultivate.

The precautionary measures which the agnostic repertoire of ANT encourages adopting enable the analyst to capture the agonistic multiplicity of world-making processes in their case studies (Latour, 1991). However, the violence that goes with the translation processes tends to be 'cushioned' by its deliberately formalistic and 'poor' semiotic repertoire – Star's critique is pretty incisive in this regard (Star, 1991) – which according to Latour is the most valuable attribute of ANT (see, Latour, 1999a). The 'poorer' the repertoire, the lower the risk of being entrapped by dominant network-explanations, and therefore the higher the chances that a plurality of actors can make their concerns and categories relevant in the description of the analyst, which is the main thrust of ANT.[8] However, what can be seen as a powerful diplomatic and scientific device to be able to freely move across domains without imposing reductionist accounts onto the ontological plurality of the world can also lead the analyst to become desensitised to the violence of the translation manoeuvres he/she is aiming to describe. Thus, instead of opting for a 'poor,' almost aseptic, repertoire, I believe that a more sensitive repertoire should be developed. Rather than putting any form of attachment, either to theories, actors or data, in quarantine, what we need is a repertoire to re-attune these attachments so as to generate new situated sensitivities that do care about the becoming of the beings with whom we have become entangled, which in my case were the small neglected arrangements of the telecare users.

What would a care repertoire do to ANT?

If we go back to the programmatic question about the conditions of ANT repertoires, a care repertoire for ANT would not be valued on the basis of its capacity to translate others. As was suggested by Callon (1986), becoming an obligatory passage point is indeed the felicity condition of the ANT repertoire that was proposed to renew the sociology of innovation, the so-called sociology of translation. Neither would a care repertoire for ANT be concerned with adding 'dimensions' to a problem or to increasing the recalcitrance of the object of inquiry so that more relations and actors are included in our accounts. This would certainly make the repertoire more objective and diplomatic, both in political and scientific terms, as Latour and Stengers would put it. From my experience in the telecare case, I would say instead that a care repertoire for ANT would be devoted to the task of attuning the conditions under which one becomes sensitive and able to respond to a situation, whether these be concepts, experiences, methods, etc. As Maria Puig de la Bellacasa (2017) has said, a matter of care 'is not only a matter of sensitivity, of ways of being affected and affecting,' it also entails a 'strong sense of attachment to something' and a 'material doing' (Puig de la Bellacasa, 2011, p. 90) – a crucial nuance that the feminist tradition brings into the discussion about the 'matter of concern' (Latour, 2004). Thus, a care repertoire for ANT would account for a situation based on the possible consequences that it might have for those affected by the situation, which necessarily includes the 'analyst.' It would not only describe but 're-scribe' the onto-normativities of a given situation (Mol, 2012; Pols, 2015).[9]

Therefore, while an agnostic repertoire is deployed to afford the analyst the ability to move freely across multiple constituents, a care repertoire, in contrast, implies both a never-ending work of tinkering (Mol, 2010) and a political/ethical commitment to the becoming of those with whom we become entangled during our inquiries and with whom we share a situation. It concerns attachments that cannot be disregarded, but also their making and unmaking, that is, 'being alongside' (Latimer, 2013) in order to explore and 'test' the good and bad they bring about. Therefore, a caring repertoire for ANT does not allow free movement. It does not seek to multiply the attachments; to express itself in multiple ways; to explore the myriad ways of being something. The problem a care repertoire attempts to address is not the lack of 'mobility' but the inability to appreciate, test and respond to the situated becoming that our accounts also contribute to making possible (and impossible) with those whom we become entangled within the course of our inquiries.

Given this, hesitation is therefore crucial for the pragmatics of ANT repertoires, where we can learn to hesitate in different ways. In the agnostic case, it is an epistemic and diplomatic mechanism of dis-identification, to become an absent presence, that seeks to bring other actors and thereby the possibility of other 'becomings' into the situation. In the case of care, to hesitate is to 'stay with the trouble' (Haraway, 2016), responding without any assurance and certainty to the situated becoming of the beings with which one becomes entangled. It entails inventing new ways of responding, which are not innocent but entail ontological interventions that may also convey forms of violence, exclusion and domination to which it is crucial to be attentive to, and, again, responsive to. This care repertoire could be considered 'generatively unsettling' (Murphy, 2015) and should contribute to 'take better care of how we care' (Martin et al., 2015, p. 631). It thus implies taking sides, participating, acting, making a choice, taking a position, but without taking for granted a general or fundamental principle on which these actions would safely and coherently be grounded. That would prevent us from troubling the arrangements and responses we produce in our accounts, that is, exploring and testing the consequential objections of our ANT repertoires.

Acknowledgements

I would like to thank the editors and participants of the ANT Companion seminar in Copenhagen for their very insightful comments, especially Ignacio Farías, Michael Schillmeier, Amade M'Chareck, Celia Roberts, Michael Guggenheim and José Ossandon. I am also very grateful to Alex Wilkie, Nairbis Sibrian, Tomás Sánchez Criado, Roser Beneito, Israel Rodríguez and the rest of my CareNet colleagues at the UOC for their very insightful feedback, which was crucial to shaping the main argument of this chapter.

Notes

1 In contrast to other agnostic approaches such as the Strong Programme, ANT enacted agnosticism not 'as the doubting of values, powers, ideas, truths, distinctions, or constructions, but as the doubt exerted against this doubt itself, against the notion that belief could in any way be what holds any of these forms of life together' (Latour, 1999b, p. 276).
2 With ANT's agnosticism, 'the social' shifted from being a distinctive reality, the realm of the intersubjective beliefs and social imaginaries, from which both truth and falsity, success and failure could be explained, to be considered something that should be explained symmetrically as the result of materially heterogeneous entanglements.

3 For a very interesting case of consequential testing of ANT repertoires, see Munk & Abrahamsson (2012). Aligned with this, this text also proposes a discussion that has resonances with the special issue *Unpacking Intervention in Science and Technology Studies*, edited by Zuiderent-Jerak and Jensen (2007), particularly Vikkelsø's contribution (2007) on ANT descriptions, as well as Zuiderent-Jerak's paper (Zuiderent-Jerak, 2016) about interventions as a method of producing both STS knowledge and new normativities. These contributions have been very inspiring.

4 As Law has insisted, 'we can only understand the approach if we have a sense of those case studies and how these work in practice' (Law, 2009, p. 141).

5 For example, even though agnosticism has been defined as 'a sturdy theoretical commitment to be taken so as to forbid the analyst to dictate to actors what they should do' (Latour, 1997, p. 375), the theoretical arguments waived to make this 'strong polemical stance' prevail in the academic arena seem to have gained more attention than the pragmatic consequences its undertaking might have had in shaping the empirical engagements of our case studies.

6 A large proportion of the work on telecare was undertaken collectively with Francisco J. Tirado, Nizaiá Cassián, Blanca Callén, Miquel Domenech and especially with Tomás Sánchez Criado, with whom I have co-authored several papers. Telecare has been a research topic in my PhD and later as part of the FP7 EU project EFORTT (Ethical Frameworks for Telecare Technologies for Older People at Home –http://www.lancaster.ac.uk/efortt/), which ended in 2011.

7 For a more detailed discussion about the possibilities the term "arrangement" brought to the case study, see López (2015, p. 93).

8 For Latour (1981), agnosticism was considered important so as not to be captured by the agonistic logic of technoscience, where the production of facts and technological innovations entails purification processes in which certain actors are completely silenced: lay people, refuted theories, failed artefacts, rejected innovations, etc. For this reason,

> from the very beginning, ANT has been sliding in a sort of race to overcome its limits and to drop from the list of its methodological terms any which would make it impossible for new actors (actants in fact) to define the world in their own terms, using their own dimensions and touchstones.
>
> *(Latour, 1999a, p. 20)*

9 In this regard, the reciprocal capture of STS and design practices that Alex Wilkie (see in this volume) identifies in the *pro-compositional* and *retroscriptive* practices of certain design practices might have resonances with a care repertoire if the exploration of situated "becomings" was oriented towards testing the consequential effects of our modes of engagement and intervening rather than pursuing the proliferation of the new.

References

Akrich, M., & Latour, B. (1992). A summary of a convenient vocabulary for the semiotics of human and nonhuman assemblies. In W. Bijker & J. Law (Eds.), *Shaping Technology/Building Society: Studies in Sociotechnical Change* (pp. 259–264). Cambridge, MA: MIT Press.

Callon, M. (1986). Some elements of a sociology of translation: Domestication of the scallops and the fishermen of St Brieuc Bay. In J. Law (Ed.), *Power, Action and Belief: A New Sociology of Knowledge?* (pp. 196–223). London: Routledge.

Callon, M. (1999). Actor-network theory—the market test. *The Sociological Review, 47*(1_suppl), 181–195.

Callon, M., & Latour, B. (1992). Don't throw the baby out with the bath school! A reply to Collins and Yearley. In A. Pickering (Ed.), *Science as Practice and Culture* (pp. 343–368). Chicago, IL: The University of Chicago Press.

Editorial (2002). Ancianos en soledad. *El País*.

Gad, C., & Bruun Jensen, C. (2010). On the consequences of post-ANT. *Science, Technology, & Human Values, 35*(1), 55–80.

Haraway, D. J. (2016). *Staying with the Trouble: Making Kin in the Chthulucene (Experimental Futures)*. Durham, NC: Duke University Press Books.

Hennion, A. (2007). Those things that hold us together: Taste and sociology. *Cultural Sociology, 1*, 97–114.

Hyysalo, S. (2004). Technology nurtured: Collectives in maintaining and implementing technology for elderly care. *Science Studies, 17*(2), 23–43.

Latimer, J. (2013). Being alongside: Rethinking relations amongst different kinds. *Theory, Culture & Society, 30*(7–8), 77–104.

Latimer, J. E., & Puig de la Bellacasa, M. (2013). Re-thinking the ethical: Everyday shifts of care in biogerontology. In N. M. Priaulx & A. Wrigley (Eds.), *Ethics, Law and Society* (pp. 153–174). Farnham: Ashgate.

Latour, B. (1981). Insiders & outsiders in the sociology of science; or, how can we foster agnosticism. In R. A. Jones & H. Kuklick (Eds.), *Knowledge and Society: Studies in the Sociology of Culture Past and Present* (Vol. 3, pp. 199–216). Connecticut: JAI Press.

Latour, B. (1991). Technology is society made durable. In J. Law (Ed.), *A Sociology of Monsters: Essays on Power, Technology and Domination* (Vol. 38, pp. 103–131). London; New York: Routledge.

Latour, B. (1997). The trouble with actor-network theory. *Soziale Welt, 47*, 369–381.

Latour, B. (1999a). On recalling ANT. *The Sociological Review, 47*(1_suppl), 15–25.

Latour, B. (1999b). *Pandora's Hope.* Cambridge, MA: Harvard University Press.

Latour, B. (2004). Why has critique run out of steam? From matters of fact to matters of concern. *Critical Inquiry, 30*(2), 225–248.

Latour, B. (2005). *Reassembling the Social: An Introduction to Actor-Network-Theory.* Oxford; New York: Oxford University Press.

Law, J. (1992). Notes on the theory of the actor-network: Ordering, strategy, and heterogeneity. *Systems Practice, 5*(4), 379–393.

Law, J. (1998). After ANT: Complexity, naming and topology. *The Sociological Review, 46*(S), 1–14.

Law, J. (2009). Actor network theory and material semiotics. In B. S. Turner (Ed.), *The New Blackwell Companion to Social Theory* (pp. 141–158). Chichester: Wiley-Blackwell.

Law, J., & Hassard, J. (1999). *Actor Network Theory and After.* Oxford; Malden: Wiley-Blackwell.

López, D. (2015). Little arrangements that matter. Rethinking autonomy-enabling innovations for later life. *Technological Forecasting and Social Change, 93*, 91–101.

López, D., Callén, B., Tirado, F. J., & Domènech, M. (2010). How to become a Guardian angel? Providing safety in a home telecare service. In A. Mol, I. Moser, & J. Pols (Eds.), *Care in Practice. On Tinkering in Clinics, Homes and Farms* (pp. 73–91). Bielefeld: Transcript-verlag.

López, D., & Domènech, M. (2008). Embodying autonomy in a home telecare service. *The Sociological Review, 56*(2_suppl), 181–195.

Martin, A., Myers, N., & Viseu, A. (2015). The politics of care in technoscience. *Soc Stud Sci, 45*(5), 625–641.

Mol, A. (2010). Actor-network theory: Sensitive terms and enduring tensions. *Kölner Zeitschrift für Soziologie und Sozialpsychologie. Sonderheft, 50*, 253–269.

Mol, A. (2012). Mind your plate! The ontonorms of Dutch dieting. *Social Studies of Science, 43*(3), 379–396.

Mort, M., Roberts, C., & Callén, B. (2012). Ageing with telecare: Care or coercion in austerity? *Sociol Health Illn, 35*(6), 799–812.

Mort, M., Roberts, C., Pols, J., Domenech, M., Moser, I., & The EFORTT team (2015). Ethical implications of home telecare for older people: A framework derived from a multisited participative study. *Health Expectations, 18*(3), 438–449.

Munk, A. K., & Abrahamsson, S. (2012). Empiricist interventions: Strategy and tactics on the ontopolitical battlefield. *Science Studies, 25*(1), 52–70.

Murphy, M. (2015). Unsettling care: Troubling transnational itineraries of care in feminist health practices. *Social Studies of Science, 45*(5), 717–737.

Pels, D. (1996). The politics of symmetry. *Social Studies of Science, 26*(2), 277–304.

Pols, J. (2012). *Care at a Distance: On the Closeness of Technology.* Amsterdam: Amsterdam University Press.

Pols, J. (2015). Towards an empirical ethics in care: Relations with technologies in health care. *Med Health Care Philos, 18*(1), 81–90.

Pols, J., & Willems, D. (2011). Innovation and evaluation: Taming and unleashing telecare technology. *Sociol Health Illn, 33*(3), 484–498.

Puig de la Bellacasa, M. (2011). Matters of care in technoscience: Assembling neglected things. *Social Studies of Science, 41*(1), 85–106.

Puig de la Bellacasa, M. (2017). *Matters of Care Speculative Ethics in More than Human Worlds*. Minneapolis; London: University of Minnesota Press.

Serres, M. (1995). *Atlas*. Madrid: Cátedra.

Star, S. L. (1991). Power, technologies and the phenomenology of conventions: On being allergic to onions. In J. Law (Ed.), *A Sociology of Monsters: Essays on Power, Technology and Domination* (pp. 26–56). London; New York: Routledge.

Star, S. L. (1999). The ethnography of infrastructure. *American Behavioral Scientist, 43*(3), 377–391.

Tirado Serrano, F., & Domènech i Argemí, M. (2013). *Lo social y lo virtual. Nuevas formas de control y transformación social*. Barcelona: Editorial UOC.

Vikkelsø, S. (2007). Description as intervention: Engagement and resistance in actor-network analyses. *Science as Culture, 16*(3), 297–309.

Wilkie, A. (2020). How well does ANT equip designers for socio-material speculations? In A. Blok, I. Farías, & C. Roberts (Eds.), *The Routledge Companion to Actor-Network Theory* (pp. 389–399). New York: Routledge.

Zuiderent-Jerak, T. (2016). If intervention is method, what are we learning. *Engaging Science, Technology, and Society, 2*, 73–82.

Zuiderent-Jerak, T., & Bruun Jensen, C. (2007). Editorial introduction: Unpacking 'Intervention' in science and technology Studies. *Science as Culture, 16*(3), 227–235.

2

How to make ANT concepts more real?

Adrian Mackenzie

Let us, according to characteristic ANT pedagogical practice, begin by following some cases of making broadly understood, asking of each – marmalade-making, sewing a seam, adding some code to software and training to do a backflip – how they do-ably become real. Marmalade-making, sewing flat-felled seams in denim, rebasing a branch in a git code repository and doing a backflip are all eminently doable, albeit extraordinarily ordinary transformations of the relations between people and things. These practical transformations – of Seville oranges and sugar, of thread and twill, of a history of work done by programmers or of a body jumping off the ground – condense, accumulate, collect, equip, institute and configure things in different ways. They are not simple or even mundane transformations, but actually change something in the world. They are not transcendent ideas, propositions or discursive formations. But like ANT concepts, they come into being as a way of inhabiting problems. They heterogeneously render some things inseparable (for instance, human <-> non-human; modern <-> non-modern; culture ↔ nature). To make or create an ANT concept should, at least for the purposes of experiencing something of the slowing down and buoyant immersion I associate with ANT, result in something as sticky, strong, reachable and mobile as marmalade, seams, code or flips.

How can a concept take on reality like marmalade, a seam, a code repository or a backflip? While I won't provide a recipe for cranking out concepts, I think it is possible to locate in the ferment of ANT a set of reality-inducing moves, some of which crystallise as concepts.[1] In order to address the reality or *physis* of such concepts, the writing of this piece seeks to construct or cobble together a concept of ANT concepts. The concept I'd like to make do with in writing this piece is the *conceptant*, a concept of the reality-doing of ANT concepts. I'm hoping that the word play – concept + *ant* – does not slide or bounce away from the problem of concocting the reality of ANT concepts.

Stickiness: the accumulated consistency of marmalade

In Britain, the term 'marmalade' is reserved for jam made with citrus fruits. In Danish, Finnish, French, German, Italian and Spanish at least, the term includes jams and preserves made from almost any fruit. In Portuguese, the term refers to quince paste. In this discussion,

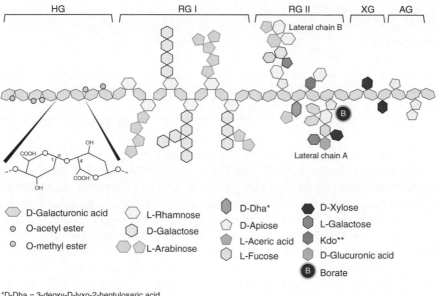

| HG | RG I | RG II | XG | AG |

Lateral chain B

B

Lateral chain A

	D-Galacturonic acid		L-Rhamnose		D-Dha*		D-Xylose
o	O-acetyl ester		D-Galactose		D-Apiose		L-Galactose
o	O-methyl ester		L-Arabinose		L-Aceric acid		Kdo**
					L-Fucose		D-Glucuronic acid
				B	Borate		

*D-Dha = 3-deoxy-D-lyxo-2-heptulosaric acid
**Kdo = 3-deoxy-D-manno-2-octulosonic acid

Figure 2.1   Pectin molecule

I refer to orange marmalade exclusively.[2] Despite reports over the last five years of a great decline in marmalade-eating at breakfast tables in England, the 3,000 entries at the possibly uncool annual Dalemain Mansion International Marmalade Competition in northern Cumbria suggest interest in marmalade-making simmers on (Figure 2.1).

Making marmalade is a relatively isolated and perhaps extraordinary annual event. It is one of the many cooking activities in which the process has a different reality to its products, even a product that endures in the glowing jars with their slices of orange suspended in translucent amber jelly. The recipe is short: cut oranges into thin or wide strips and soak them in water overnight, cook the oranges, add the sugar, boil and test if it has reached a setting consistency, ladle into sterilised jars standing them upside down so they vacuum seal themselves on cooling, and then check them the next day to see the bright jewels of fruit. Marmalade-making lightens the gales and darkness of January or February (in the Northern Hemisphere), and then carries through a year of eating or giving away.

Pectin, a polysaccharide whose dynamic molecular reconfigurations are still not fully understood, sets marmalade. In marmalade-making, pectins switch from their role as structural elements in fruit skin to an associative agent. During the boiling phase, branching and changing bonds of pectin soaked from the oranges form a network mesh that holds water, sugar and fruit in a colloidal solution. During cooking, sugar keeps water molecules away from pectin branches; citric acid from the oranges alters the negative charge on the pectin branches, so that they can bond with each other to form the associations that give marmalade colloidal consistency and translucency.

The consistency of ANT concepts depends, rather like the setting of marmalade in its combination of solids and liquids, on some different components rendered inseparable within them. ANT concepts tend to become molecular association machines and this machinic-becoming is characteristic of their antics. The tendencies to build ANT machines remain operational over decades: the Hume Machine, co-word analysis, or Tarde models, to name a few. In this setting, a grammatical suffix can operate as the setting agent. Rather than –ity, –ism, –ment, –ness or –isation, I start the train of association from the –ant/–ent suffix:

> –ant/–ent: (a) a personal agent, as agent, claimant, president, regent; (b) a material agent, as coefficient, current, ingredient, secant, tangent, torrent; esp. in Med., as aperient, astringent, emollient, expectorant.
>
> *(Oxford English Dictionary 2017)*

Core ANT concepts such as *actant* hack the –ant/–ent suffix. –ant/–ent channels agency through the deltas of persons, materials and movements. Given that ANT largely concerns agents in their differences (people, things, events, situations, configurations, etc.), the –ant suffix introduces a characteristic reversibility into ANT concepts. –ant and –ent work as a coefficient, mix as an ingredient and sometimes run a tangent through ANT work. Because –ant/–ent doesn't pick and choose between entities, or between person and things, concepts plugged into the suffix switch easily between people and things: president and aperient, and resident and repellant can be conceived together. Instability in the –ant/–ent fixation animates many ANT concepts, or conceptants. It imbues them with heterogeneity, exposes and sensitises them to mixtures and differences, and increases their stickiness in many situations. Symbolic and physical, legal and impersonal, ANT concepts, like the massive marmalade spill in the kitchen scene of *Paddington Bear*, get into a surprising number of places by means of such suffixes.[3]

Sewing seams in reality

It would be strange, given that ANT cut its teeth on the associations of people, devices, things and others, if this molecularity didn't shape up in some form or figures. A divergent series of para-discursive elements – diagrams, figures, graphics, software, techniques, experiments, exhibitions, events – swerve away from words and their propositional configurations. These extra-discursive events are not commonly found, if at all, in other concept-making practices. (For critical philosophies, the only things to be found in figures or diagrams, in events or exhibits, are the concepts preceding their construction. Kant even requires a mathematician that 'he must not ascribe to the figure anything save what necessarily follows from what he has himself set into it in accordance with his concept' (Kant 1963, 19; Figure 2.2).)

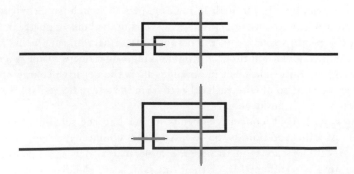

Figure 2.2 Flat-felled seam

Let us diagram the conceptant in order to attempt the swerve. Idling several hours at an inner Manchester skate park, I strike up conversation in the cafe with a former fashion retail manager about jeans. He trawls the city's charity shops buying all the Levi jeans he can find to send for resale in Japan. Jeans from the 1970s are especially valuable. While jeans wear out in the knees, crotch, cuffs or backside, and are sometimes deliberately ripped, distressed, bleached and washed with stones and chemical to create the appearance of rugged encounters with cattle and dirt, it is rare to find jeans coming apart at the seams. (Recent fashions in Japan of jeans consisting solely of seams attest not only to the durability of seams jeans.).

Many realities have seams. Jeans, along with other fabric-based equipment such as tents and backpacks, usually have flat-felled seams.[4] Flat-felled seams are made by laying the materials to be joined on top of each other, running a line of stitches near the edge, trimming one side of the seam, folding the wider seam allowance over and under the smaller seam and pressing, and then sewing a second line of stitches as close to the end of the fold as possible. Putting in a flat-felled seam requires at least two runs through a sewing machine, and two folds in the material. The resulting seam has no raw edges visible, and ripping the seam apart is very difficult because the folded fabrics pull in the same direction.

The lovely combination of stitching-folding-stitching that so durably joins materials in flat-felled seams differs from marmalade-making. Here too, there is a relative inseparability. The slow wear and tear of jean seams, and the difficulty in altering them (to narrow the width of legs on jeans means undoing the flat-felled seams running up the sides of each leg, an arduous task), is a figure for joints in concepts, and also for how we might think of giving them a more equipped reality.

The diagrammatic components of conceptants equip them by stitching language and figure together, folding the sayable and the visible in such a way that they pull together not apart. Like jeans, backpacks or tents, their equipmental workability depends on sewing such seams. The seam provides a way of thinking of a concept's empiricism. Para-discursive elements of conceptants are not always recognisable or admissible to Theory, since they resemble and are a form of empiricism, albeit a radical or relational one.[5] A component of the concept threads through moments in particular places (marmalade-making, a conversation of the skate park, at the sewing machine, etc.). An ethnographic empiricism of ANT functions as a material whose weave forms an observational fabric, full of threads to be picking up and folding over. Rather than a drone view, the diagrammatic component of concepts weaves or sews, criss-crossing and pinching together planes of the empirical. The many case studies – proteins, particle accelerators, arteriosclerosis, financial trading rooms, electric cars, muscular dystrophy patients, climate models, call centres and satellites – weave a twill, to which a concept adds seams, weaving in and out textures, textiles and texts in pursuit of a seam of inseparability.

Rebasing the collective in the distributed mode: 'everything is local'

Contemporary software making relies heavily on code version control systems to address the problem of distribution in its several senses. Software making is distributed across people, places, infrastructures (cloud compute, etc.) and time (indeed rendered continuous, as in 'continuous deployment' (Mackenzie 2016)). The git version control system allows distributed making in all three senses. Tracking and publishing each modification allow parallel, non-linear histories to happily coexist, and provisionally appear as a 'release' or a 'version' coming out of a repository (Figure 2.3).

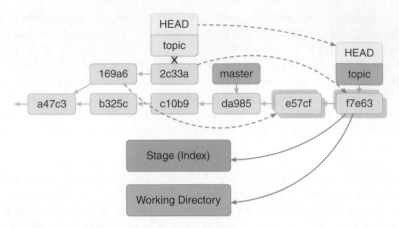

Figure 2.3 Rebasing a branch

Git, whose motto 'everything is local' has a distinctly ANTish resonance, is an extraordinarily clever arrangement. Take the example of the git rebase command. Given two branches of a code repository that have diverged (usually as a result of different programmers working on the code repository independently), a software project effectively exists in two different realities. The rebase command allows a form of time travel in which the changes in one branch are 'replayed' on another branch. Rebase replays the 'commits' – the operation of making more or less atomic snapshots of the state of a project permanent and visible to others (by cryptographically hashing them in the same way that Bitcoin does) – to construct a linear history from a process that is essentially non-linear, branching, collective and divergent. Rebasing is described as dangerous – 'using a chainsaw on LSD' in one colourful description (Tomakyo 2008) – because it 'alters history' and if the history is shared, then people will care about it. The Golden Rule of git rebasing is, therefore: 'never rebase on a public branch' (Atlassian 2017).

The case of git rebasing is instructive. At first glance, git is a technical implementation of ANT's worst nightmare: the flattening of richly non-linear histories of associations and references into a neat, linear history of progress. Or is it? The tracking of references (refs in git terminology), and the merging or rebasing of histories of changes entails a reversible accumulation and ordering of references. Other histories remain reachable within the repository. This is the whole point of git. The making of concepts also entails gathering and tracking of references, their movement and reshaping in ways that permits their tracing and retrieval. From the standpoint of the graph of hashed references, forming a richly brachiated history, logging collective entanglements of commitment and rebasing alters but retains a reachable non-linear history. Both -ant/-ent particle and the diagram with its ethnographic seams seek to construct a radical relationality, to underscore inseparability rather than separability, to hold the multiple chains of references that matter to actors rather than minimise them.

Conceptants differ from other concepts in large part in their relation to objects typically found in sciences and technologies. Sciences, technologies and their hardly mundane objects (positrons, microbes, arteriosclerosis, climate models, aircraft, wave turbines, etc.) differ from the mundane objects favoured by philosophy in its meditations on candles, stones, jugs and bridges. ANT concepts, it should be said, respond to the problems of agency associated with sciences, medicine and engineering in specific situations, but they are often 'rebasing public branches,' to speak gittishly, or 'reconstructing specific situations' to speak like Dewey (Dewey 1957, 91). Conceptants extend into other domains (markets, finance, institutions, education, health, etc.) by unravelling threads of calculation, experiment, device or inscriptions. But conceptants differ because they carry with them complicated, at times extraordinary or somewhat indeterminate things such as 'body multiples' (Mol 2003) or 'material participations' (Marres 2012).[6]

Rather than freebasing, conceptants rebase technical or scientific situations along with the various public knowings, working and living with these objects on the part of scientists, patients, users, engineers and others, all of which belong in the referential repository. The mode of existence of knowledge, as (Latour 2007) terms it, consists in all its correspondences and transfers between knowledge, people and things. There is no denying the flattening that occurs in conceptants, and this flattening can arouse indignation as well as fascination. Conceptants rebase, in the way git rebases, richly brachiated histories, always maintaining everything as local. In doing so, they revise a shared history but maintain the reachability of the chain of references that make it sharable.

Given the Cambrian explosion of knowledge-things we live with, can we expect lots of conceptants rebounding and rebasing sciences and technologies, incorporating their diverse commitments in a flatter history? Even though we can list many conceptants, they are quite rarely made. Like the Bitcoin blockchain, a lot of mining goes into them, and their coinage is finite. This again is a consequence of their recursive, switching character, their resolve 'not to jump' and their tendency to move along connected paths. The rarity of conceptants does not mean that they are comets, travelling in from outer realms, awaited and observed as stellar events (even if clever rebasing can give that impression). It would be better to thinking of conceptants as points of condensation or accumulation, consistent with other conceptants. Their creation, like Bitcoins, slows down as their power of condensing chains of references increases, and as they maintain consistency with other conceptants by making continuous paths, equipped with para-discursive elements, distributed across many people and places, and somewhat instituted (in conference, journals and other documents such as this handbook).

Conceptants have a powerful correlational or resonant aspect, despite the bad press correlation has recently received (in the context of representations thinking about things for speculative realism, and in context of causality for data science and predictive analytics). Correlation or resonance between concepts is not such a bad way of thinking about how conceptants cohere as it allows to comprehend the equipped, accumulated, collective and instituted features of conceptants in time, the results of which can be seen as not only engaging so many academics in debate but in existing as things in the world. To use the git analogy, conceptants are in continuous deployment.

Learning to salto

The *physis* of concepts in ANT abhors a void. ANT concepts do not affirm separations, divides, gaps, breaks, discontinuities, chasms or voids. Almost every pedagogy of ANT concepts warns against dividing or separating. ANT concepts deprecate attempts to jump, especially the forms of jumping – the salto mortale – from subject to object or culture to nature, or from humans to non-humans. They don't gravitate towards accounts of science and technology such as Hannah Arendt's that refer to 'freedom from the shackles of earth-bound experience' (Arendt 1998, 265; Figure 2.4).

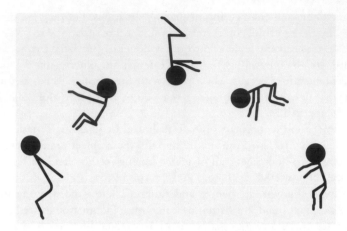

Figure 2.4   Salto

Jumping, however, can be a sign of great excitement, ebullient energy or an attempt to leap to save life in times of danger. According to YouTube exponents, where many how-to-backflip videos can be found, it is possible to learn to do a standing backflip in five minutes. Learning to backflip in such a short time – why not do it now, before you finish reading this? – might seem miraculous.[7] I've tried to do it, but didn't manage in five minutes. The elements of learning the backflip or somersault include a whole body jump, a power-backwards roll from sitting, a macaco (Caporeira single-handed back handspring), culminating in the full backflip. In each of these steps, practised for around 45 seconds, the backflipper is spending more and more time upside down and above the ground. For instance, the power-backwards roll remains fully on the ground, but enjoyably upside down for a good part of a second. While a full flip only lasts 0.5–0.75 seconds from go to woe, the transformation of body in relation to ground is striking (so to speak). The flip is an event, a gap-becoming, that temporarily constructs a bridge in the air for a muscle-skeletal body using rotational mass and elasticity as a central component.

The risk in talking about ANT concepts at all is doing an accidental flip. Concepts, by virtue of the history of their practice, their concretisation and their mode of existence, tend to jump, bounce and flip backwards into a mind position. They are prone to mentation. A subject–object gap opens up in the immediate vicinity of most concepts of the concept. Conceptants, by contrast, are somewhat mindless, in the sense that they enter at a lower level, avoiding the questions of mind, looking for arrangements that give rise to occasional mind-effects, or to mindedness.

It is necessary to risk something in order to benefit from the thrill of the backflip and arrival back on two feet. Despite their somewhat bad name in ANT literature, somersaults and flips do happen, can be learned and might on occasion be worth doing. They have a bad name because they imply gaps, chasms or discontinuities (nature-culture, premodern-modern, subject-object, etc.). As anyone who has learned to somersault or flip knows, they are fun, even exhilarating, and accomplished in a coordinated movement, shaping and positioning of bodies in relation to the always imminent return to ground, in fact to almost the same place. Conceptants might be seen as harnessing or stabilising jumps, drops, flips and the like in the same way that the preparatory jumps, power-rolls and macacos scaffold a nascent backflip. Conceptants sometimes flip, and perhaps need to for physico-aesthetic effect.

Conceptants are not immune to, and in fact rely on, moments of idealisation, moments of departure from the ground, associated with flip. They lack the regulatory form ascribed to them by much philosophy (e.g. Immanual Kant in his *Critique of Pure Reason* presents 'pure concepts' in the form of categories as aspects of the appearance of objects in general, prior to any experience). Although they are not the regulatory forms that govern and order experience, conceptants externally resonate with ideals ('the concept of freedom') even as they rebase specific situations.

The reality of conceptants

How are ANT concepts made? Every word of that question skids into problems and possibilities. You may have noticed that I did not answer the original question 'how to make ANT concepts?', altering it instead to 'how to make ANT concepts more real?' or better, 'how to make do with ANT concepts?' The elements I've assembled in the conceptant address the problem of the reality of ANT concepts: a grammatical particle – -ant/-ent – that switches recursively between humans and non-humans; a diagrammatic component that not only brings different fabrics together but finds ways of folding the sayable and the visible together; a workflow for wrangling versions, for rebasing repositories of shared references; and a jump, a *salto non-mortale*, a way of marshalling energy to spring from the ground, a tumble of idealisation. In slightly more conventional terminology, this account of conceptants concerns the forms of accumulating, equipping, collecting and instituting that make ANT concepts more real.

I don't know whether these elements exhaust the reality of ANT concepts. They may, due to their exoteric derivation from marmalade-making, sewing, coding and acrobatics, wash out of the mainstream of ANT work. Regardless of their derivation, and even their relation to other forms of thinking, what I have sought to connect and hold together in the conceptant is the combination of excitement and relief that it is possible to experience in reading ANT accounts (in my case, reading *Laboratory Life*, *Science in Action* and 'The De-scription of Technical Objects'). It might not be possible today for an existing or new conceptant to impress in the same way after several decades of accumulation. It should, however, be possible for the setting, joining, rebasing and flipping operations to set in motion vibrations of the concept, with the peculiar energies flowing through it, sometime akin to music, yet within a different horizon, resonating with states of affairs.

Notes

1 A point of *accumulation* of publications, of events, of stories, of a discipline, an institution, ANT concepts arise from a ferment, something undergoing transformation around it. Concepts surface from a broth, an archive of documents and publications associated with the term 'ANT,' as infrastructurally and documentarily constructed and practised in the publication of ANT books and articles on web-accessible platforms, and indexed through publication databases such as Thomson ISI or the ANT resource. In this discussion, I draw on an archive of documents and publications associated with the term 'ANT,' as infrastructurally and documentarily constructed and practised in the publication of ANT books and articles on web-accessible platforms. The complete data set of this object can be found at [https://github.com/rian39/ant/data]. Bibliographic objects have a special function in conceptants. It is not accidental that much ANT work, especially in the more document-oriented versions, has worked with publications (Latour et al. 2012). But they could also be websites, tweets or other inscriptions. The documentary component, whether in co-word analysis (Callon, Courtial, and Penan 1993) or in ANT-oriented digital sociology (Marres 2017), anchors analysis of accumulation, of density, or large numbers, of universals. Hence, it's an important component, like a probability distribution, because it comprises many variations.

2 EU Directive 79/693/CEE defines it according to the British usage. While the first large commercial batch of marmalade made with Seville oranges is usually attributed to a Dundee woman, whose merchant husband purchased a cargo of Seville oranges from a storm-damaged Spanish ship sheltering in Dundee, Scotland in 1700 (Scot-breakfasts first began to include marmalade in the late 18th century; England followed in the 19th century), I refer to non-commercial practice.

3 To attribute the consistency of ANT concepts to a grammatical particle such as -ant might seem incompatible with the ontological diversity of ANT. The special situation of the making of ANT concepts, however, has been documents and texts. The working through of the semiotic operations of shifting first explored by structural linguists such as Roman Jakobson was generalised by early ANT into heterogeneous networks of actants (Akrich 1992). What started out as a graphical analysis of sentence structures becomes networks of actants organised and operating in different ways around centres, through displacements and detours, and in enrolments. Inscription, the crossover entity, the trace-structure that criss-crosses the most formal, abstract or institutional formation with technical and commercial, physical and practical coinages, carried the shifting, recursive -ant/ent suffix across the different domains and settings into which ANT ebbed and flowed as it explored laboratories, clinics, engineering projects, machines, architecture, infrastructures, markets, laws and devices, or any of the other situations that might be listed on the ANT itinerary.

4 The French seam is closely related to the flat-felled seam, but has less visible stitching. Just as strong, it is more common in formal garments.

5 For all their anti-correlationalism, object-oriented philosophy and speculative realism seem reluctant to take steps towards any empirical conversion. A component of the conceptant passes through material practicalities towards an outside that it cannot be separated from. Maybe every conceptant has some extra-discursive elements, whether in the form of diagrams (network graphs, the many actor-network visual forms), in ways of writing outside the lines of statements (in stories, in images, in odd typography, paratextual forms, dialogues and artworks), in seams of technical practice (sewing seams with materials, cooking, dialogues, fabrication, exhibitions, etc.) and probably many other things I can't think of (activism?). A conceptant might even be characterised by its exteriority, by its own irreduction (Latour 1987) to discourse, to propositions enunciated in words.

6 New materialisms of recent years also embrace complicated or extraordinary entities, but they tend to mobilise them in more orderly conceptual configuration.

7 Backflips have also been done in wheelchairs, and as I write, the first humanoid robots are backflipping nicely. Meanwhile, some humans are doing standing double backflips (e.g. Aaron Cook).

References

Akrich, Madeleine. 1992. "The de-Scription of Technical Objects." In *Shaping Technology/Building Society*, edited by Wiebe E. Bijker and John Law, 205–24. Cambridge, MA: MIT Press.

Arendt, Hannah. 1998. *The Human Condition: With an Introduction by Margaret Canovan*. Translated by Margaret Canovan. Chicago, IL; London: University of Chicago Press.

Atlassian. 2017. "Merging Vs. Rebasing | Atlassian Git Tutorial." *Atlassian*. 2017. www.atlassian.com/git/tutorials/merging-vs-rebasing.

Callon, Michel, Jean-Pierre Courtial, and Hervé Penan. 1993. "La Scientométrie–Que Sais-Je." *PUF, Paris*.

Dewey, John. 1957. *Reconstruction in Philosophy*. Boston, MA: Beacon Press.

Kant, Immanuel. 1963. *Critique of Pure Reason*. Translated by Norman Kemp-Smith. 2nd ed. Houndmill: Macmillan Publishers.

Latour, Bruno. 1987. *Science in Action: How to Follow Scientists and Engineers Through Society*. Cambridge, MA: Harvard University Press.

Latour, Bruno. 2007. "A Textbook Case Revisited. Knowledge as Mode of Existence." In *Handbook of Science and Technology Studies*, edited by Edward J Hackett, Olga Amsterdamska, Michael Lynch, and Judy Wacjman, 3rd ed. Cambridge, MA: MIT Press.

Latour, Bruno, Pablo Jensen, Tomasso Venturini, Sébastian Grauwin, and Dominique Boullier. 2012. "The Whole Is Always Smaller Than Its Parts. How Digital Navigation May Modify Social Theory." *British Journal of Sociology* 63 (4): 590–615.

Mackenzie, Adrian. 2016. "28 Infrastructures in Name Only?" In *Infrastructures and Social Complexity: A Companion*, edited by Penny Harvey and Casper Bruins Jensen, 379. London & New York: Routledge and K. Paul.

Marres, Noortje. 2012. *Material Participation Technology, the Environment and Everyday Publics*. Basingstoke: Palgrave MacMillan.

Marres, Noortje. 2017. *Digital Sociology: The Reinvention of Social Research*. 1 ed. Malden, MA: Polity.

Mol, Annemarie. 2003. *The Body Multiple: Ontology in Medical Practice*. Durham, NC: Duke University Press.

Oxford English Dictionary. 2017. "–Ent, Suffix." 2017. www.oed.com/view/Entry/62756?

Tomakyo, Ryan. 2008. "The Thing About Git." Ryan Tomayko. April 8, 2008. https://tomayko.com/blog/2008/the-thing-about-git.

3

Is ANT's radical empiricism ethnographic?

Brit Ross Winthereik

In a more-and-other-than-human world, we need an ecological epistemology, argue Christina Hughes and Celia Lury, and propose a methodology of returns (Hughes & Lury, 2013). In this chapter, I argue for an epistemology that follows concepts as they enter into conversation and convince each other of a different being. Below, I return to the radical empiricism of actor-network theory (ANT), more specifically the 'follow the actor' dictum. This was a key element in ANT's methodology and the one that has been instrumentalised. As a simple instruction, 'follow the actor' does not outline any specific kind of action, nor does it explain anything. It is an invitation for in situ sense-making and sorting out relations and attachments. Extending the suggestion put forward by Hughes and Lury that we need an ecological rather than a social epistemology, i.e. one that is relational and lateral, my ambition in this chapter is to *upcycle* the ANT dictum as it may prove generative for new generations of fieldworkers.[1]

Initially, 'follow the actors, whose concerns you share' was a radical proposition (Blok & Jensen, 2011; Callon, 1987). It was radical, because it required the bracketing of all *a priori* assumptions that the researcher could possibly have about the chosen field of study. It meant that a fieldworker would need to acknowledge that she would never be able to *bound* her field in advance of the study. The proposition implied accepting that not only would the empirical field continuously emerge as a consequence of researcher engagement, but so would the conceptual framework that framed and nurtured this engagement.

In the years that have passed, 'follow the actor' has often been translated into a soft requirement in STS to use the ethnographic method in ways that allow researchers to attend to *practices*. While ethnography can surely be seen as a method that is suited for observing practices, we need to attend to the possibilities in fieldwork for analytical and conceptual regeneration. The anthropological concept does not function 'as a kind of surplus value extracted by the "observer" from the existential labour – the life – of the "observed"' (Viveiros de Castro, 2003). Instead, following actors around, attending to practices, is all about attending to emergent concepts and theories in ways that allow for determining 'the problems posed by each culture' (ibid.). Fieldwork requires attending to experience and participation, as well as to apparatuses for categorising, formalising and knowing the world, informants' as well as the ethnographer's own.

In the relations between anthropological ethnography and ANT ethnography ('follow the actor'), there are differences and shared histories. In this chapter, I will attend to some of these when telling an ethnographic story, in which *concepts* (technological and analytical) play an important role. I do so to be able to ask about how fieldworkers might at one and the same time be radically empirical and radically analytical in their descriptions as we use ethnography in ways that allow for engaging with a more-and-other-than-human world. I conclude by arguing that ANT has something to offer to those, who would like to play with cross-overs between the empirical and the conceptual as they are brought together in fieldwork and analysis. I add to ANT's dictum the possibility of analytical regeneration through thinking about concepts as fieldworkers' companions. To benefit from this offer, we must subject ourselves to the more-and-other-than-human agency of the concepts that surround us when we engage with the world.

Blurred lineages and separations

Ethnography is the name of a practice of immersion into ways of life that hold an assumed foreignness. It is not a 'timeless' practice (Dalsgaard & Nielsen, 2016), but a mode of immersion that nevertheless anticipates a point in time, the time of writing, in which immersion in an empirical field is no longer a possibility (Strathern, 1999). According to Marilyn Strathern, a main characteristic of ethnography is the paradox that the ethnographic method tends to produce an overwhelming, but insufficient amount of data, as the ethnographer aims to 'observe everything,' while knowing that at the time of writing, she is likely to be missing data. Since ethnography, like other data-intensive practices, is predicated upon the production of both too much and too little data, ongoing analysis is needed. This is happening in many different locations, also when ethnographers don't have their minds set on doing analysis. Analysis is a formalisation practice that is systematic, yet open enough to let the ethnographic material form unexpected relations and draw in contexts that were unavailable during fieldwork. Thus, ethnographic analysis involves navigating a constant threat of collapse between a situation of data collection and a situation of conceptualisation and analysis (Strathern, 2018).

I refer to Strathern's analytical ethnography here because she presents an extensive critique of the concept of the network that forms the theoretical rationale for the method of following actors around. Her article *Cutting the Network* formed a contact zone between anthropology's ethnography and ANT's version of it (Strathern, 1996). She makes the point that there are different kinds of collapse to be found between empirical and conceptual worlds, but that these should never be assumed to be shared with informants, who will have other conceptualisations, of what 'a relation' should be taken to mean, for example.

Historically, anthropological ways of thinking and doing have inspired Science and Technology Studies – and vice versa (Hess, 2001), and over the years their connections and detachments have multiplied (Andrea Ballestero, 2017). Now, let us have a look at how fieldwork has been accounted for in the past.

In 1997, David Hess wrote a piece in *Cyborgs & Citadels* for 'perplexed anthropologists' advising them on how to 'live in Science and Technology Studies.' The chapter reads as a field guide for anthropologists of science and technology in an emerging interdisciplinary field of STS. His aim was to identify familiarity and difference between anthropology and STS and make sense of what he saw as certain differences in their epistemologies and methodologies. In so doing, however, fieldwork is flattened and ethnography accounted for as the practice of going to a foreign place with an empty mind. 'Science and technology' is not like a remote village, Hess tells anthropologists, and is already being studied by STS researchers (Hess, 1997). By framing his argument as a matter of 'living' in either STS or anthropology, Hess turns ethnography into a battlefield where disciplines demonstrate their expertise. Hess also points out that while ethnography may be anthropology's preferred method, ownership does not belong solely to anthropologists. Hess outlines three research strands where he finds certain affinity between anthropologists' and STS scholars' interests, and where he surmises that cross-fertilisation has already happened. Here, fieldwork is taken hostage in an identity political game.

This problematisation takes immersion to be the main instrument through which an ethnographer can interact with and get *closer* to the discourse and practices of her interlocutors. Early ANT scholars puzzled over this problem of distance, or rather lack of distance. Latour and Woolgar were concerned that laboratory culture was too familiar. According to Hess, this was a problem that anthropologists who work in cultures unlike their own were less likely to face (Hess, 2001). Latour and Woolgar wanted a certain distance from the sciences and scientists under study, and they appealed to the idea of 'anthropological strangeness' for that sense of distance (Hess, 1997). One way in which distance was achieved was by a playful rendering of 'the anthropologist' in the text (Latour, 1993; Latour & Woolgar, 1979). This way of placing the anthropologist front and centre in the text was one way of coming to terms with the problem of representation. This was a construction of epistemic positions and practices of positioning. Latour and Woolgar did not aim for full representation of the scientists, whose work they studied. This was seen as a provocation by some and as a refreshing new way of accounting for the difference between researcher and researched.

Ethnography was taken up by ANT as a suitable method for knowledge production and fieldwork became a key element in the development of ANT as an analytical–methodological approach (Latour, 1983, 1987, 1993; Latour & Woolgar, 1986). The notion that analysts of science and technology should aim to follow actors around has later been taken up as akin to ethnographic immersion, with the difference that ethnography in the ANT tradition was never about making thick descriptions (Baiocchi, Graizbord, & Rodríguez-Muñiz, 2013). As Latour jokingly told in an interview, ANT is about *seeking* to do ethnomethodology, semiotics and metaphysics all at once (Ihde, Crease, Jensen, & Selinger, 2003).

ANT's relational redescriptions of laboratory work and other technologically mediated practices happened around the same time as sociologists elsewhere were making inquiry into the complicatedness of analysing modern technoscience (Knorr-Cetina, 1981; Knorr-Cetina, 1983). Only very few anthropologists would study science and engineering at the time, and ANT studies could have been taken up as part of a growing body of literature of anthropology 'at home,' i.e. conducted in Euro-America. It seemed to be an enduring familiarity between the ethnographer and the object of study that got in the way for ANT's transformation into ANThropology. Despite the use of ethnography and despite questions pertaining to scientific *cultures*, ANT studies never really became part of the anthropological canon. The 'scrupulous detour through the empirical' that ANT argued for (Ang, 2011; Latour, 2005) was not recognised by anthropologists as sufficiently anthropological, or as scrupulous in just the right way (Jensen, 2017). For the critics, ANT was not sufficiently articulate about its own grounds for claiming to know about the knowledge practices of others as well as the politics of its (non)methods (Holbraad, Pedersen, & Viveiros de Castro, 2014).

Lateral ethnography

One of the texts in which ANT's radical empiricism is coined is Latour's rendering of scientists' work on the edge of the Boa Vista rain forest, a text that became world-famous as a chapter in the greatest hits collection *Pandora's Hope*. Here, Latour argues that one should not look for a correspondence between words and things as the ultimate standard of truth. There is truth and there is reality, but there is neither correspondence nor *adequatio*. To attest and to guarantee what we say, there is a much more reliable movement – indirect, crosswise, and crablike – through successive layers of transformations.

(Latour, 1999: 64)

Knowing, as expressed here, is a result of attending to transformations between 'words and things.' Knowing reliably, then, is to follow the actors (words and things) through successive layers of transformations. Moving across in a crablike manner means doing the work of detailing the empirical as well as paying attention to and detailing the ontological transformations happening underway. The difficult part is staying on the same empirical–conceptual plane as those whose existence, lives and deeds are studied, including the Boa Vista forest and its material representations. Where this turns social theory into ontology akin to the cultures under study became a much-debated issue (Blaser, 2010; Gad, Jensen, & Winthereik, 2015; Holbraad & Pedersen, 2017; Maurer, 2005; Mol, 2002). The trouble arises because moving sideways implies laterality in the data-collection that puts concepts and theories on the same plane as those whose practices the concepts are about. This means that the research attends to and participates in both empirical worlds and in the shaping of concepts and theories that are already part of these worlds (Gad & Jensen, 2016). ANT assumes that if power, inequality, etc., are already part of the social worlds under study, and not something to be unveiled, what the fieldworker brings to the table is re-articulation though translation rather than revelation (Latour, 1984, 1994).

What might such a translation look like? Coming to STS from anthropology, Atsuro Morita offers a version of lateral ethnography in his analysis of technology as a comparative device, in his case, a harvester (Morita, 2014; this volume). Morita studies a case of technology transfer, where a Japanese harvester is put to use in Thailand. Involved in the maintenance and repair of the harvester, Morita argues, is the attendance to how the machine was originally imagined and designed. Thai mechanics need to both be able to imagine what fields and crops the harvester was designed for and compare these imaginings to the fields and crops where the harvester is now put to use. A lateral, comparative mode of comparison is made possible by the conceptualisation of the harvester as a conceptual machine that is flexible enough to work in different contexts, yet embeds elements from both.

'The ethnographic machine' avoids comparisons that would need a third referent outside the practices that the comparison aimed to elucidate, in order to work (see Helmreich, 2011). The comparison is embedded in machinic parts, user imaginations, harvesting practices, fields in Thailand and Japan and the conceptualisation of their relation. We here see how 'following actors' is never simply a feat but includes attending to comparisons made in the field as well as concepts embedded in technological devices.

We can now build on these insights about relations between the empirical and the conceptual and move on to another machinic experience, one that I had with a technological prototype, which allowed me to revisit the 'follow the actor' methodology. The following description of my encounter with an analytical–empirical prototype is done to respond to the question if ANT's radical empiricism is indeed ethnographical. My description of a technological prototype forms an entry point into the discussion of the empirical and the conceptual and demonstrates the openings that occur when the analysts can no longer uphold the 'I say, they say' distinction that grew out of an understanding that our analytical concepts were fundamentally different from the concepts 'found' during fieldwork. This distinction is fundamental in ethnographic writing and in practising ethnography, and while it has already been problematised by works on laterality, I offer another example of their entwinement. What this chapter adds to existing work on lateral ethnography is a focus on *concepts as companions*. I take companions to mean fellow travellers that help open our worlds, but which can also be overly talkative in joint exploration of the already crowded spaces of humans and things in which we move as ethnographers.

A relational predisposition

I am at Bella Center, at the Energy Europe fair, to meet with two of my interlocutors, who are visiting the fair as part of a publicity event for the marine renewable energy sector in Denmark. The purpose of the meeting is to discuss a collaborative exhibition focusing on marine renewable energy and the social and political practices around prototyping.[2] I have a few hours on my hands before the meeting, so I stroll around the impressive halls where technologies for harnessing energy from renewable sources are on display. This fair is huge and many different artefacts, including electric cars and small wind turbines, fill the enormous halls. The audience is clearly international. Visitors stop by the various booths and ask for information or browse the pamphlets for details about what is on show. As I enter into one of the smaller halls, a familiar face greets me. It belongs to the inventor Snorre Rastrup, an invented name, who is manning the booth occupied by the Danish Wave Energy Association. He shakes my hand eagerly and talks a little about some recent travelling he has done. He seems surprisingly uninterested in my questions about the Wave Energy Association's booth and its visitors. After the chat, I move on and enter into a much larger exhibition hall. Compared to the previous hall, which had no clear theme, this hall is dedicated entirely to solar energy.

I am surprised to see a white metal construction moving smoothly up and down amidst the black, shiny surfaces of hundreds of solar panels. It is only when I get closer that I recognise this machine as Snorre Rastrup's wave-to-energy conversion prototype. Next to me, a man is in conversation with a retailer. They discuss the possibility of installing Rastrup's converter near his house by the Black Sea. The retailer assures him that this is indeed possible and that electricity production for his mansion could then come from a mix of renewable sources, including wind and solar power. The wave converter appears alien amidst the solar panels, but what puzzles me the most is finding this prototype located among solar panels and far apart from its inventor. During technical demonstrations, inventors and prototypes are usually near to each other in case something should go wrong.

What made me focus on the detachment of Snorre Rastrup and his prototype in this moment? They were in different exhibition halls, the inventor and his prototype, but still close enough for the inventor to come by if anything happened. Rastrup did not want to talk about prototypes with me, so I had not expected that he had brought one. Being taken by surprise that there was indeed a prototype was thus partly enacted by the space and partly by him not wanting to 'talk shop' with me as an ethnographer. I found the separation puzzling, as if Snorre should have been standing next to his prototype. Was this interpretation based on a failure on my part to, in the moment, suspend an existing theoretical and conceptual apparatus that assumed an inventor and a prototype to be visibly connected. A relational bias, so to speak.

It was also puzzling to see the wave energy convertor in the solar panel exhibition hall; I was unable to make sense of it; yet, the display captured my full attention. I recall the feeling of being taken by surprise by something that seemed quite normal by everyone else. Marilyn Strathern writes about being dazzled by the experience an ethnographer can have, when she realises that she cannot let go of a particular observation upon returning from fieldwork (Strathern, 1999). I could not let go of the image in my mind of the prototype surrounded by solar panels. But what was equally captivating was my sense that the prototype had pressed upon me a particular analytical disposition.

Alberto Córsin Jiménez argues that what some prototypes do during demonstrations is that they entrap the imagination of their audience (Jiménez, 2017). This conceptualisation is akin to the conceptualisation made by another inventor in our study. He described the purpose of a prototype as that of demonstrating its natural place in the world. To him, this entailed building trust. The inventor described how a good prototype was able to make him trust that he could 'make a future' for it. Similarly, if at a demonstration the audience become trusting champions for the prototype, it is more likely to become a commercial product.

Of course, the wave energy converter was not entirely on its own. As I said, there was a retailer, who was also demonstrating solar technologies. But would he be able to care for and nurture the prototype to perform at its best? I worried about this, but then recalled Snorre's relaxed smile and his small talk. After all, he had been just a few hundred meters away, and he could check up on the device whenever he wanted to. I tried to make sense of the incident by thinking of it as a practical doing of ontological politics.

During the demonstration, the wave converter namely also demonstrated its capacity to work politically. Snorre Rastrup did not stand next to it to explain the technical function of the prototype; he was not present to explain how solar panels and wave energy conversion might work together in the future. Among dissimilar technologies, the prototype both demonstrated the capacity of its hydraulic system to smoothly follow the movements of imaginary waves, *and* a green transition that would involve a number of different renewable technologies. With this placement, the wave converter was not just another potential product on display. It was an artefact that made the proposal that in the future, energy consumers will get their energy from a mix of sources, wave energy being one of them, wind another, solar power a third and so on. The prototype subtly transformed the organisation of the fair that separated an emerging industry (wave energy) from established ones (like solar).

Thinking of a prototype as a political intervention is informed by Andrew Barry's notion of technological demonstration sites as political machines (Barry, 2001). What is important here is not whether political principles or ideologies formed the reason for the placement of the wave energy prototype, but the effect of the placement. It was the possibility and specificity of a dazzling 'sight' that could turn a technical concept into a political question because it addressed the question of how society should be fuelled in the future (see Political Machines ch. 8).

At the Energy Fair bodies and things were acting jointly in achieving an effect: The prototype, the inventor, the retailer, the Russian visitor, the imagination of a dacha at the Black Sea, the exhibition halls, the solar panels, the ethnographer and so much more. In an ANT ethnography, a disposition would be to see networks and infrastructures (black-boxed networks). This was exactly what happened at first. It puzzled me to think that physical detachment – even if this was just momentarily – could be a means to the end of creating a future for an object. At last, I realised that the puzzle was less about the separation of inventor and prototype than it was about my difficulty as an ethnographer to suspend my previous knowledge and theories that assumed co-location.

Concepts are devices that help us deal with the puzzlements generated by fieldwork. In that sense, they help us make the strange familiar. But concepts also work in curious ways. They help us establish relations with unfamiliar phenomena, and as such help extend our understanding of the world and of the concepts and their limitations (Jimenez & Willerslev, 2007). At the same time, and this is where the radical empiricism of ANT is helpful, by accepting that concepts are *of* the world, they become part of what is *in* the world, as well as what might become. The encounter at the Energy Fair demonstrated that this is the case for the concepts that are 'analytical' as well as for the concepts that are 'technological.' If they are good at capturing ethnographers' imagination, both the abstract and the material concepts are prototypes for things we would like to make in the future.

Companion concepts

In specifying companionship, Haraway shows that companionship is a form of sym-biology. It is about being implicated in one another in ways that are caring and often very complicated (Haraway, 2016).[3] Natural science has traditionally worked with foundational definitions of concepts that new scientific findings can then expand or overturn and replace. Concepts are extended or rejected. Based on studies in natural science laboratories, early ANT made it clear that concepts may be considered foundational by their users, but are also always inscribed in objects (Latour, 2000). 'Follow the actor!' was thus a method to detect the multiple ways in which 'words' and 'things' were assembled in practice. The dictum has been taken by many as an invitation to describe and analyse technological agency, but there have never been 'just' artefacts that act. There will be inventors and there will be ethnographers, and other care-takers as well. Still, the uptake of ANT's empiricist credo that social scientists should follow actors in their world-making activities has sometimes been interpreted as if an ANT approach would imply putting less emphasis on the conceptual aspects of such activities than on the materialities that are also present. As anthropologist Mateo Candea has pointed out, the lab ethnographies might have been empirically more complicated, if, for example, Latour and his colleagues had attended to how scientific–biological concepts are both modifiable and are also to some extent set in time and space; they come with histories and are part of both scientists and ethnographers' embodied history (Candea, 2013).

To be taken from this is that analysts as well as interlocutors both bring concepts into the ethnographic encounter. While both active, these concepts will not be the same. The inventor brought a concept-prototype, a wave energy converter; the ethnographer brought the concept-prototypes of the network and the infrastructure (black-boxed network). My concept-prototypes did not materialise in the social field of the technical demonstration. Nevertheless, encountering an inventor both with and detached from his concept at the Energy Fair created a puzzle: Were the concepts of network and infrastructure that the ethnographer brought to the social field to be trusted? Was there a future for those concepts? Could anything be made out of them? Perhaps only through continuous detachment and relation-making. Like Snorre Rastrup needed to *detach* from his concept to make an intervention in local energy politics, I had to detach from my preferred concepts to write about ANT's radical empiricism as ethnographic.

The point is not so much that ethnographers need to detach from the familiar to see something new, but rather that we must follow actors in many different ways. Ethnographers follow by following somebody's reasoning, by making an analogy, by adding a context that was not there initially or by cutting off something else as irrelevant. This way 'follow the actor whose concerns you share' is about both recognising technological agency and about the possibility of novel exchanges with concepts that help suspending what ethnographers already know. Following in multiple ways could then allow for the concepts that are already present (e.g. the technological concept) to engage in a conversation with the concepts we bring (network, infrastructure, detachment). Together, they might convince each other of a different becoming (de la Cadena, 2015).

What this means is that the concept-prototypes that ethnographers bring are in a state of testing in a way that can be compared to technological concepts or prototypes that are in a similar provisional state. We bring theories and concepts to the field and they interact with all that which surrounds them, and with that which they are expected to describe or explain. Concepts have in them the capacity to play with us, but also create persistent troubling. They are our companions in fieldwork and analysis, and we have to trust that there is a future for them as they transform into something quite other.

I set out to *upcycle* ANT's radical empiricism asking if it is ethnographic. The answer to this question is that as we follow how things, people and projects become what they are, we encounter concepts. These concepts are part and parcel of the empirical-material world, and if we are willing to let them work as prototypes, they can become our companions as we parse relations and detachments. Taking 'follow the actor!' seriously as an actor on that scene requires us to attend to the many possible meanings of 'following.'

Notes

1 Wikipedia: Upcycling, also known as creative reuse, is the process of transforming by-products, waste materials, useless or unwanted products into new materials or products of better quality or for better environmental value.
2 One year after this first meeting, the exhibition was realised through the design of a 'digital walking stick' (Winthereik, Watts, Maguire 2019). The exhibition was carried out as part of the research project titled 'Marine Renewable Energy as Alien: Social Studies of an Emerging Industry,' conducted in collaboration with Laura Watts and James Maguire between 2013 and 2016, and funded by DFF-FSE grant no. 0602-02551B.
3 Donna Haraway in her Companion Species Manifesto writes about the intense companionship between her and her dogs (Haraway, 2003). Inventor and ethnographer have become what they are through engaging with concepts, but are they companion species? Do inventors and ethnographers exchange properties with concepts. The greyish metal construction that was later attached to a car does certainly not part-take in Snorre's biological system in just the same way as Donna Haraway's dogs leave biological traces in her body. Yet, they do take up space in each other's worlds, like, analogously companion concepts take up space in ethnographers' worlds.

References

Ang, I. (2011). Navigating complexity: From cultural critique to cultural intelligence. *Continuum: Journal of Media & Cultural Studies, 25*(6), 779–794.

Baiocchi, G., Graizbord, D., & Rodríguez-Muñiz, M. (2013). Actor-network theory and the ethnographic imagination: An exercise in translation. *Qualitative Sociology, 36*, 323–341.

Ballestero, A. (2017). Lineages and discomforting relatives. *Social Anthropology, 25*(4), 540–541.

Barry, A. (2001). *Political machines: Governing a technological society*. London and New York: Athlone Press.

Blaser, M. (2010). Political ontology: Cultural studies without 'cultures'? *Cultural Studies, 23*(5), 873–896.

Blok, A., & Jensen, T. E. (2011). *Bruno Latour: Hybrid thoughts in a hybrid world*. London and New York: Routledge.

Callon, M. (1987). Society in the making: The study of technology as a tool for sociological analysis. In T. Huges & T. Pinch (Eds.), *The social construction of technological systems: New directions in the sociology and history of technology* (pp. 83–103). London: MIT Press.

Candea, M. (2013). The fieldsite as device. *Journal of Cultural Economy, 6*(3), 241–258.

Dalsgaard, S., & Nielsen, M. (2016). *Time and the field*. New York: Berghahn Books.

de la Cadena, M. (2015). *Earth beings: Ecologies of practice across Andean worlds*. Durham, NC: Duke University Press.

Gad, C., & Jensen, C. B. (2016). Lateral concepts. *Engaging Science, Technology and Society, 2*, 3–12.

Gad, C., Jensen, C. B., & Winthereik, B. R. (2015). Practical ontology: Worlds in STS and anthropology. *NatureCulture, 3*, 67–86.

Haraway, D. J. (2003). *The companion species manifesto: Dogs, people, and significant otherness*. Chicago, IL: Prickly Paradigm.

Haraway, D. J. (2016). *Manifestly haraway*. Minneapolis: University of Minnesota Press.

Helmreich, S. (2011). Nature/culture/seawater. *American Anthropologist, 113*(1), 132–144. doi:10.1111/j.1548-1433.2010.01311.x

Hess, D. J. (1997). If you're thinking of living in STS: A guide for the perplexed anthropologist. In G. L. Downey & J. Dumit (Eds.), *Cyborgs and citadels: Anthropologisl interventions in emerging sciences and technologies* (pp. 143–164). Santa Fe, NM: School of American Research.

Hess, D. J. (2001). Ethnography and the development of science and technology studies In A. C. Paul Atkinson, S. Delamont, J. Lofland, and L. Lofland (Eds.), *Handbook of ethnography* (pp. 234–245). Thousand Oaks, CA: Sage.

Holbraad, M., & Pedersen, M. A. (2017). *The Ontological Turn: An Anthropological Exposition*. In *New Departures in Anthropology* (pp. 1 online resource (352 pages)). Retrieved from Cambridge. Restricted to UC campuses https://dx.doi.org/10.1017/9781316218907

Holbraad, M., Pedersen, M. A., & Viveiros de Castro, E. (2014, January). The politics of ontology: Anthropological positions. *Theorizing the Contemporary, Fieldsights*. Retrieved from https://culanth.org/fieldsights/the-politics-of-ontology-anthropological-positions

Hughes, C., & Lury, C. (2013). Re-turning feminist methodologies: From a social to an ecological epistemology. *Gender and Education, 25*(2), 786–799.

Ihde, D., Crease, R., Jensen, C. B., & Selinger, E. M. (2003). An interview with Bruno Latour In D. Ihde & E. M. Selinger (Eds.), *Chasing technoscience: Matrix for materiality* (pp. 15–27). Indianapolis: Indiana University Press.

Jensen, C. B. (2017). New ontologies? Reflections on some recent 'turns' in STS, anthropology and philosophy. *Social Anthropology, 25*(4), 525–545.

Jiménez, A. C. (2017). *Prototyping cultures: Art, science and politics in beta*. London and New York: Routledge.

Jimenez, A. C., & Willerslev, R. (2007). 'An anthropological concept of the concept': Reversibility among the Siberian Yukaghirs. *Journal of the Royal Anthropological Institute, 13*(3), 527–544.

Knorr-Cetina, K. (1981). *The manufacture of knowledge: An essay on the constructivist and contextual nature of science*. Oxford: Pergamon Press.

Knorr-Cetina, K. D. (1983). The ethnographic study of scientific work: Towards a constructivist interpretation of science. In K. D. Knorr-Cetina & M. J. Mulkay (Eds.), *Science observed: Perspectives on the social study of science* (pp. 115–140). London: Sage.

Latour, B. (1983). Give me a laboratory and I will raise the world. In K. M. Knorr-Cetina, Michael (Ed.), *Science observed: Perspectives on the social study of science* (pp. 141–170). London and Beverly Hills: Sage.

Latour, B. (1984). The powers of association. *The Sociological Review, 32*(S1), 264–280.

Latour, B. (1987). *Science in action*. Cambridge, MA: Harvard University Press.

Latour, B. (1993). *We have never been modern*. Cambridge, MA: Harvard University Press.

Latour, B. (1994). On technical mediation: Philosophy, sociology, genealogy. *Common Knowledge, 3*(2), 29–64.

Latour, B. (1999). *Pandoras hope: Essays on the reality of science studies*. Cambridge, MA: Harvard University Press.

Latour, B. (2000). The Berlin key or how to do words with things. In P. M. Graves-Brown (Ed.), *Matter, materiality and modern culture* (pp. 10–21). London and New York: Routledge.

Latour, B. (2005). *Reassembling the social: An introduction to actor-network-theory* (Vol. Clarendon lectures in management studies). Oxford and New York: Oxford University Press.

Latour, B., & Woolgar, S. (1979). *Laboratory life: The social construction of scientific facts* (Vol. Sage library of social research ; v. 80). Beverly Hills: Sage Publications.

Latour, B., & Woolgar, S. (1986). *Laboratory life: The construction of scientific facts*. Princeton, NJ: Princeton University Press.

Maurer, B. (2005). *Mutual life, limited: Islamic banking, alternative currencies, lateral reason*. Princeton, NJ: Princeton University Press.

Mol, A. (2002). *The body multiple: Ontology in medical practice* (Vol. Science and cultural theory). Durham, NC: Duke University Press.

Morita, A. (2014). The ethnographic machine: Experimenting with context and comparison in Strathernian ethnography. *Science Technology & Human Values, 39*(2), 214–235.

Strathern, M. (1996). Cutting the network. *Journal of Royal Anthropological Institute, 2*, 517–535.

Strathern, M. (1999). *Property, substance & effect*. London: the Athlone Press.

Strathern, M. (2018). Infrastructures in and of ethnography. Rivista dell'Associazione Nazionale Universitaria Antropologi Culturali, 7(2), 49–69.

Viveiros de Castro, E. (2003). Anthropology and science. *Manchester Papers in Social Anthropology, 7*.

Winthereik, B. R., Watts, L., and Maguire, J. (2019). The energy walk: Infrastructuring the imagination. In J. Vertesi and D. Ribes (Eds.), *digitalSTS: A field guide for science & technology studies* (pp. 348–363). Princeton and Oxford: Princeton University Press.

4

Can ANT compare with anthropology?

Atsuro Morita

Comparison occupies an interesting place in social science. While being the archetype of 'scientific method' to provide causal explanations (Krause 2016), comparison also highlights difference and diversity, thereby playing a critical role in provincialising seemingly universal or hegemonic viewpoints, particularly those of 'the West' (Gingrich and Fox 2002). Comparison thus often denotes different or even contradictory practices. This is particularly tricky when one reflects on recent debates on comparison inspired by actor-network theory (ANT). In these debates, the authors from different disciplinary traditions, particularly those from anthropology and sociology, tend to discuss different aspects of comparison inspired by ANT.

On the one hand, despite the constructionist critique that questioned the comparison of societies and nations as independent entities (Deville et al. 2016; Gingrich and Fox 2002; Sørensen 2008), this form of comparison is still entrenched in a commonsensical attitude in and around sociology: 'If someone studies a social movement in Finland and in Norway, it is assumed that something about these countries can explain the observed difference' (Krause 2016: 52). ANT involves a radical critique of that ontology of comparison. Madeline Akrich and Vololona Rabeharisoa (2016) have, for instance, recently discussed in detail their troublesome experience with comparison in their EU-funded research project on patient groups across Europe, where the idea of the nation state as the predominant 'context' for comparison repeatedly arose in discussions and hindered their effort to pursuit an alternative mode. In this context, ANT-inspired sociologists have turned to explore infrastructures of comparison – assemblages of conceptual, technical, institutional and bureaucratic elements that make social scientific comparison possible. This enables them to experiment with technicalities of comparison to envision their potentialities (Deville et al. 2016; Sørensen 2008).

On the other hand, anthropologists have historically been interested in comparison in relation to the Benjaminian understanding of translation as expansion and transformation of the language to which the text is translated (Asad 1986). In this tradition, comparison concerns making connections between indigenous (whatever this means) ideas and social scientific concepts originated in Europe and North America. In this translational connection, the former forces the latter to expand or transform themselves. For anthropology, comparison serves this kind of translation between 'us' and 'them' that enables the provincialisation of universal claims (Candea 2017; Strathern 1987; Viveiros de Castro 2004).

Because of these different histories, ANT's impact has been felt differently in these disciplines. While ANT has provided sociologists insights into experimenting with comparison as technoscientific practice, it has stimulated anthropologists, as I will discuss in detail in this article, to reconsider comparison's role for translation, the latter also a key idea in ANT. By making accessible ANT-inspired debates on anthropological comparison to the ANT community, I aim to contribute to concerns for provincialising the Western notions now increasingly relevant among ANT scholars (Law and Lin 2017; Nakazora 2018; Verran 2002) due to changing circumstances of its major academic base, namely, science and technology studies (STS) (Fu 2007; Medina 2018).

The growing interest in the multiplicity of knowledge practices in the non-West raises some important challenges to STS. Cultures and societies imagined as autonomous entities played a central role in the imagination of the geography of diversity in the 20th century (Strathern 1992; Yanai 2018). But STS scholars have elucidated much more complex arrangements of difference that require a new vocabulary. New currents of 'non-Western' STS, which define themselves as intersections of Western-origin STS and other thoughts and activism, explore not only the diversity of knowledge practices, but also the heterogeneity of the practice of STS itself (Fu 2007; Law and Lin 2017; Morita 2017). This also concerns unequal relationships between languages and knowledge traditions, an issue that has long been discussed in postcolonialism and anthropology (Asad 1986; Clifford 1997; Verran 2002). Therefore, just like for feminism (Strathern 1988; Haraway 1991), for non-Western STS, multiplicity concerns not only the subject of study, but also the researcher's subjectivity and further complicates the topology of difference in non-Western STS (Morita 2017).

These debates on non-Western STS concern some of the central claims of ANT. In a classic paper, Latour (1988a) criticises explanation in social science for reducing phenomena to a predetermined list of causes. He notes: '(a)ctors have [...] very strong ideas about what framework is, who is responsible, what counts as an explanation and who is innocent' (1988a: 174). He eventually rearticulated this point by arguing that ANT should use an *infralanguage*, a 'banal' and 'meaningless' language in order to allow actors to have stronger voice than the researcher (Latour 2005: 30). ANT's call for infralanguage seems quite relevant to non-Western STS's concern about how to articulate local problematisations and languages that are often marginalised in international academia. Moreover, ANT's commitment to infralanguage brings together a new vocabulary to imagine multiplicity. By denying the hierarchical relationship between the analyst and the analysed, ANT allows to see 'explanation' in terms of movements and connections made by researchers. Latour argues that ANT should delineate the movements of scientists and science studies scholars symmetrically (Latour 1988a).

Following this recommendation, the rest of this chapter traces, first, comparison's complex journey from experimental biology, the field where comparison as scientific method had spectacular success, to anthropology, where comparison of societies became a tool to provincialise the West. In a second step, taking into account the vanishing self-evident character of societal units, the chapter turns to another form of comparative practice and imagination. In this context, ANT's imagination of networks of intersecting trajectories of travelling actors seems to emerge as *the* new reality that allows us to see a new form of multiplicity. I will then discuss the new mode of comparison called lateral comparison (Gad and Jensen 2016b) that puts the analyst's and informants' comparison side by side. In the end of this chapter, I will argue how lateral comparison allows us to imagine new topologies of difference that concern not only the world that ANT scholars and anthropologists explore but also the heterogeneous and partially connected nature of their practices.

Contexts of comparison

It is hard to imagine a better starting point for an ANT reflection on comparison than Latour's (1983, 1988b) well-known analysis of Louis Pasteur's vaccination experiment. Indeed, Latour's story climaxes at a point where Pasteur performed a public comparison at a farm in Pouilly le Fort. For the experiment, Pasteur had divided the farm's animals into two groups and vaccinated one with a weakened strain of anthrax bacillus. Then, he inoculated both groups with fatal anthrax. The outcome was spectacular. A contemporary observer noted: 'The same number of animals, which had been covered with the palladium of the new vaccine, remained invulnerable to fatal inoculation, and were shown, very much alive, surrounded by corpses' (Bouley 1881: 548 cited in Latour 1988b: 87).

It is not very surprising that this kind of powerful comparison captured the hearts and minds of many social scientists at the time (Krause 2016). While such stories of comparison as 'the scientific method' are well-known, Latour draws attention to some important and hitherto overlooked aspects of Pasteur's experimental practice.

Latour insists that Pouilly le Fort was only one specific moment in Pasteur's movement to connect farms and his laboratory. Earlier, he set up a temporary laboratory in a farm to collect anything that deemed relevant to the outbreak of anthrax. Then, he brought some of these elements back to his laboratory in Paris. With his well-rehearsed technique of microbe cultivation, Pasteur successfully isolated anthrax bacillus and identified their different strains. Then, he compared the toxicity of these strains in miniature epizootic events in the laboratory farm. Through these numerous comparisons, Pasteur gradually understood the relationship between the mutation of anthrax bacillus, its varying toxicity and animal immunity in his experimental settings. In other words, Pasteur created experimental settings in the laboratory 'so as to reproduce inside its walls an event that seems to be happening only outside' (Latour 1983:154).

At Pouilly le Fort, Pasteur brought these laboratory phenomena back to the farm again for public demonstration. The successful experiment created a situation in which laboratory elements became able to explain and allow for practical intervention in epizootic events in farms. Latour argues that Pasteur's experimental success rested on 'translation,' creating an equivalence between the laboratory and the farm. This translation is a rather selective process:

> Only a few elements of the macroscopic epizootics [in farms] are captured in the lab, only controlled epizootics on experimental animals are done in the lab, only specific inoculation gestures and vaccine inoculant are extracted out of the laboratory to be spread to farms.
>
> *(Latour 1983: 154)*

Importantly, the equivalence of phenomena in the laboratory and in the farm rested on the negotiation among involved actors including anthrax bacillus themselves.

Recently, Isabelle Stengers (2011) revisited this topic and pointed out the central feature of comparison in the experimental sciences: Scientists, she argued, create 'situations that authorize them to claim that the subject matters that they address lend themselves to quantitative comparison' (49). In the above-mentioned case, Pasteur created a situation in the laboratory in which anthrax bacillus dictated the conditions of the comparison at the farm through their action on the petri dish and in animals. However, not all the activities called science follow this form of comparison. Some of the sciences indeed build centres of calculation by collecting samples, reports and statistics, but engage in a different kind of comparison – what Stengers (2011) calls 'unilateral comparison,' where the subject scientists address is not involved in setting the terms of comparison. The early history of anthropology testifies how the experimental form of comparison turned to unilateral comparison through displacement from biology to social science.

Nineteenth-century anthropology: from experimental to unilateral comparison

In the 19th century, European anthropologists and sociologists such as Herbert Spencer, E. B. Tylor and James Frazer enthusiastically adopted the 'comparative method' from experimental science. However, these scholars are now seen as estranged precursors of social science. The 'comparative method' is precisely one of the reasons they became so obsolete.

Although it was called 'the comparative method,' there is nothing in common between the 19th-century approach and comparative social research as we know it today. Frazer, for example, one of the most prominent figures in this tradition, collected data not only from ethnographic reports of contemporary non-Westerners but also from the Bible, Roman and Greek classic texts and so on. As the anthropologist Marilyn Strathern (1987) notes, this style of comparison looks so wild that it is almost incomprehensible for those trained in modern social science. In mid-20th century, Evans-Pritchard noted that 19th-century anthropologists 'thought of exogamy, totemism, matrilinity, ancestor worship, slavery, and so forth as customs – *things*,' or stand-alone entities (1951: 40, emphasis added). Based on this peculiar ontology, 19th-century anthropologists plotted customs and institution taken from various places and times for their comparative endeavours. In stark contrast with the imagination of societies as natural contexts for comparison in the 20th and 21st centuries, there was practically no consideration for social context understood as interrelated customs and institutions in a given society, nation or geographic area. Rather than societies and nations as autonomous units, 19th-century anthropologists saw a single evolutionary scale of the human as the natural context of comparison.

While looking strange to contemporary eyes, 19th-century anthropology was a sustained effort to build a centre of calculation. It tried to account for the vast variety of customs of peoples by constructing evolutionary stages based on comparison of ethnographic reports sent by missionaries, traders and colonial administrators. Like Pasteur's, this effort connected two locations: The desks from which the anthropologists composed tabulations and texts, and the fields where 'primitive people' outside Europe lived. The two locations were mediated by a movement of knowledge-enabling artefacts: In this case the colonial traffic of ethnographic reports and other forms of documents.

However, there was one fundamental difference. While Pasteur's experimental assemblage was designed to confer anthrax bacillus the power to set the conditions of comparison, 19th-century anthropology's comparative assemblage did not allow non-Western people to have a say on them. This material assemblage was rather an arrangement for 'unilateral comparison' (Stengers 2011: 48), in which anthropologists impose the terms of comparison on peoples all over the world. Such anthropological project replicated and drew on the political geography of colonialism. As with the colonial administration, the 'natives,' who are the central matters of concern for the practice of comparison, were totally excluded from that very practice.

Modern anthropology: from context-making to comparison for/in translation

The kind of modern anthropology that Evans-Pritchard advocated defined itself against this 19th-century form of comparison. While not themselves committed to anticolonial critique, modern anthropologists tried to mimic scientific comparison where 'objective' methods seemed to prevent the researcher from imposing unilateral comparison. This effort was crucial for the rise of the bounded units of societies and cultures as the new *natural context* for comparison.

This new form of comparison was initiated by Bronislaw Malinowski, the iconic figure of modern anthropology. In his first ethnographic monograph *Argonauts of the Western Pacific* (1922), he redefined ethnography as description and analysis of a whole society based on long-term direct observation by a well-trained anthropologist. While Malinowski himself presented his method as scientific and empirical, he faced the same challenge as the 19th-century anthropologists: How to convey to a Western audience the meanings of unfamiliar customs and ideas from non-Western worlds. While both needed some textual artifice to make such customs understandable to their readers, their solutions were almost opposites. Nineteenth-century anthropologists appealed for knowledge their readers already had. Thus, stories from the Bible and popular accounts of social evolution served as a context for making the customs of 'primitive peoples' comprehensible. In this sense, evolutionary comparison was also a textual device to persuade readers by creating a *shared context*. In contrast, Strathern (1987: 254) argues, Malinowski

> insisted that practices were to be related to other practices [...] [He] turned not to practices found in other cultures but to other aspects of this one culture. [...] [T]his led to a view of individual societies as entities to be interpreted in their own terms, so that both practices and beliefs were to be analysed as intrinsic to a specific social context.

Malinowski's insistence on social context introduced a new form of comparison between 'us' and 'them' as the central analytic of modern anthropology. In contrast with Frazer who frequently connected English readers and non-Westerners, Malinowski clearly separated the two contexts. Accordingly, the figure of the fieldworker played the central role in mediating between these contexts. By witnessing the local life, the fieldworker induces readers to 'accept the naturalness of Trobriand ideas in their context' (Strathern 1987: 261). At the same time, he constantly compares ideas and customs from the Trobriand Islands with those from the life of British readers, so that they can understand the former in analogy with the latter. Malinowski's literary device is thus essentially comparative.

However, the point of comparison is not simply to connect ideas and customs in Trobriand Islands to those in England. Rather, it also brings about transformation of English ideas. This transformative effect, which has been commonly called 'translation of culture,' became the trademark of anthropology's conceptual innovation. Talal Asad notes its working by citing Walter Benjamin:

> The language of a translation can—in fact must—let itself go, so that it gives voice to the *intentio* of the original not as reproduction but as harmony, as a supplement to the language in which it expresses itself, as its own kind of *intentio*.
>
> *(Benjamin 1969: 79 cited in Asad 1986: 156)*

Eduardo Viveiros de Castro echoes this vision: '[a] good translation is one that allows the alien concepts to deform and subvert the translator's conceptual toolbox' (2004: 3).

Comparison for translation does not neatly fit Stenger's distinction between experimental and unilateral comparison. In contrast to experimental comparison, the 'natives' are not in a position to directly dictate the terms of comparison. They are always mediated by the text written by the fieldworker. But comparison for translation is not just imposing the analyst's terms either. Rather, it works reflexively on the researcher's own conceptual apparatus.

ANT: infralanguages of comparison?

Anthropology's comparison for translation is similar to Latour's 'infralanguage, which remains meaningless except for allowing displacement from one frame of reference to the next.' Like in a 'good ANT account,' comparison for translation aims to make 'the concepts of the actors to be stronger than that of the analyst' (Latour 2005: 29–30).

However, there is one notable difference. Defining infralanguage as 'meaningless,' Latour depicts ANT as somewhat neutral. While Malinowski indeed assumed the neutrality of his own language (Strathern 1987), anthropologists committed to comparison for translation usually questioned this kind of neutrality. Accordingly, what is at stake is not so much getting a grip on the other's language as pushing one's own language close to, or beyond, the limit. More than 50 years ago, Godfrey Lienhardt noted that in this kind of translation, 'it is not finally some mysterious "primitive philosophy" that we are exploring, but *the further potentialities of our own thought and language*' (1954: 97, emphasis added). Such concern ironically led anthropology to move away from a line of research in which nation, region or culture offered the *natural context* for comparison, bringing about a crisis of identity in anthropology, which now seeks to reclaim its capacity for conceptual innovation without drawing on social context (Yanai 2018). In this context, ANT provides a vocabulary and imagination of 'actor-networks' that eventually has proven to be useful to tackle the challenge anthropology is faced with. What has become visible is a new form of relationality that connects different practices – precisely the kind of connection that Latour elucidated about Pasteur (Strathern 1995). Here, comparison takes on yet another significance. While modern anthropologists turned to distinct and bounded societies, or what Mol and Law call 'regions,' the rising interest in the network topologies of travelling practices has led some anthropologists to imagine the space of fieldwork not as bounded local context, but as complex entanglement of trajectories of travelling actors (Maurer 2005; Mohácsi and Morita 2013; Morita 2013). Following ANT's call for symmetry between the analyst and the analysed (Latour 1988a), this imagination also highlighted the fact that those whom the fieldworker studies also engage with travelling knowledge making (Clifford 1997). Here, comparison became a focus for exploring the complex interrelationship between those trajectories.

Comparing comparison

In my fieldwork on technology transfer from Japan to Thailand in the early 2000s, I found that comparisons of technologies, organisations, cultures and environments between the two countries were constitutive elements of the phenomenon I studied (Morita 2013, 2014). Japanese engineers in Bangkok repeatedly compared the two countries in formal and informal occasions. Some of them even drew on anthropological and area studies literature written by Japanese scholars. Since my project was about social dimensions of technology transfer, this put my project in complex complicity with the engineers' comparisons.

The entanglement of the engineers' comparative interests and my own thus became my subject of research when I adopted the figure of the network to see my project as a certain kind of travel whose trajectory crosses those of others. Through this shift in imagination, trajectories of Japanese engineers emerged as an important part of the web of cross-cutting paths along which not only them but also Thai mechanics, myself and various artefacts travelled. This shift of perspective also allowed me to see local context as an effect of the incessant comparisons that the Japanese engineers engaged.

Confronted with unexpected reactions from Thai counterparts, the Japanese engineers often imagined invisible social and cultural factors influencing their counterparts' practices (cf. Rottenburg 2009). Comparing education systems, traditional values and Buddhism in Thailand and Japan, the Japanese engineers mobilised whatever was available to explain those reactions. Like Malinowski's comparison, here comparison created two distinct contexts of 'us' and 'them.' In a sense, my own fieldwork became an attempt to transverse the space created by the engineer's comparison. I travelled to Nakhon Ratchasima, the largest city in the Northeast Thailand, to look into small- and medium-scale factories manufacturing agricultural machinery. These factories constituted one of the target groups of the project. But also, the engineers and I saw these factories as appropriate sites to explore the entanglements of Thai social relations and engineering practice because of their direct ties with village life and agriculture.

However, what I found in Nakhon Ratchasima was not so much a distinctive context, but nodes of travelling trajectories of factory owners, machines and second-hand parts (Morita 2012). Some of the factory owners had business partners in Japan and visited them occasionally. The factory floors were full of second-hand tractors, engines and parts imported from Japan that mechanics routinely fixed, tinkered, remodelled and recombined. They also frequently talked about Japanese technology. Notably, there was a notable difference in their practice of comparison. While the Japanese engineers' discourse resembled comparative social research in which cultural values and social relations are seen as causes of observed practices (Krause 2016), Thai mechanics' knowledge about Japan looked somewhat strangely fragmented. Comparisons on the factory floors focused on perplexingly small details of Japanese environment and society, which were closely tied with certain designs of machines (Morita 2014).

On a fine day at the end of the 2003 dry season, I encountered some farmers complaining about a Japanese second-hand rotary cultivator on the shop floor of a factory located in the outskirts of the city. The factory had recently imported the rotary cultivators from Japan, and the owner of the factory was trying to introduce them to farmers in the region. The farmers I encountered were among the first customers to buy the rotary cultivators from the factory. They came to the factory to complain that the machine had stopped working because weeds had entwined around the rotary blades. In response to this complaint, the owner and the chief mechanic of the factory started checking the machine. They asked the farmers about the condition of their field, and they all discussed the possible causes of the problem for a while. After a while, the chief mechanic pointed out that the angles of the blades were too flat at the moment when the blades contact the ground. He then proposed to bend all the blades to about 90 degrees. The farmers agreed to his proposal, and the mechanic and one of the farmers began bending all the blades by using a gas burner and pincers. In the following test run on the farmer's own field, the rotary cultivator proved to work efficiently.

After the event, the chief mechanic mentioned that it was the difference between Japanese and Thai environments that had caused the trouble. According to him, Thai weeds are taller and stronger than Japanese ones because of Thailand's tropical climate. Although the narrow blades of the Japanese rotary cultivator were able to cultivate Japanese paddy fields, they were not able to cut the stronger weeds one finds in the tropical environment of Thailand.

In this episode, it was the machine and its breakdown that evoked comparison of the two environments. As Michel Callon notes in his classic article on French electric vehicle development, engineers designing a new machine try to ally aspects of the design with heterogeneous entities such as its users, infrastructure and the market (Callon 1987). The smooth functioning of the rotary cultivator required a specific, stable relationship between the machine and the entities in the field, such as the weeds, the soils and the farmer operating it. This was also an assumption of the mechanics. Therefore, they saw that the rotary cultivator's relations with other entities in the Japanese field had already been inscribed in the design of the machine.

Accordingly, the travel of the rotary cultivator from Japan to Thailand surfaced as the major context of the problem in the mechanics attempts. Because a machine works by becoming a part of the heterogeneous connections surrounding it, its travel requires new connections between its place of origin and the destiny. The mechanics and farmers saw the weeds entangled around the blades as what hindered this connection. At the same time, it also embodied the difference between the Japanese environment inscribed in the machine and the actual environment in the farmers' field. The blades entangled with the weeds produced a double vision in which the Thai and the Japanese environments were seen *at once* through their difference.

The episode of machinic comparison exemplifies an important moment for innovation for mechanics in local factories (Morita 2012). At the same time, it also provides an opportunity for innovation for the anthropologist, as it alludes to another mode of comparison that implies neither the human as its central agent nor separate contexts as its prerequisite, but draws on the imagination of trajectories and more than human practice (Morita 2014).

Ethnography: from machinic to lateral comparison

The reflection on my travel in Thailand illuminates the complex connections between diverse forms of comparisons. For the Japanese engineers and Thai mechanics, comparison works as practical means to engage with intersecting trajectories of travelling humans and non-humans. In this landscape, the ethnographer's comparison becomes part of this web. It was my travel from the technology transfer project to the local factories that connected the two forms of comparison as well as that of anthropology. Here, my travel also became a sort of comparison that enabled to see these comparisons.

The anthropologist Bill Maurer (2005) calls this kind of knowledge making 'lateral reasons.' In his study on Islamic banking and alternative currency movements in the United States, he also found that his informants use anthropological concepts, such as that of gift by Marcel Mauss, to explain their practice. In response, he has proposed to place the anthropologist's and informants' own modes of analysis side by side rather than aiming to elucidate the latter by the former.

My ethnographic take on travelling comparisons (Mohacsi and Morita 2013) closely allies with Maurer's lateral reasons. Putting my own comparison along with the others allows me to redescribe the complex workings of the comparisons from my situated trajectory. At the same time, this move also reveals how my comparison itself was composed. For example, the Japanese engineers' comparison that saw culture as an external factor influencing technology was not what I was after. Nevertheless, it shaped my own travel and, thus, comparison. Equally, my encounters with travelling machines and Thai mechanics also shape my own analytical take. The juxtaposition of these comparisons became the central tenet of my analytical move as well as a means to make explicit how that move was constituted.[1]

Because of this dual effect, lateral comparison seems suitable to articulate complex topologies of difference in non-Western STS. It allows to locate the researcher's analytic movement at the intersection of different and often incommensurable practices and interests. The ethnographic machine destabilises the notion of comparison of distinct contexts and offers its possible expansion. Such transformation is exactly what ANT and anthropology are after. In this sense, comparison for translation, which is itself going through transformation through the encounter with ANT, also holds significance for ANT. By foregrounding the new topology of difference, which the very vocabulary of ANT allowed to articulate, comparison for translation invites ANT to explore new forms of multiplicity and their wide-ranging political implications including possible transformations of the relationship between the West and the non-West.

Conclusion

About 30 years ago, Strathern (1987) asked what kind of world would emerge if the self-evident-ness of social context was coming to an end. In a sense, the encounter between ANT and anthropology evolved around this question. Thus, while introducing the anthropological concern of comparison for translation, this chapter has demonstrated the usefulness of ANT in exploring the changing landscape of comparison. I took ANT's imagination of intersecting trajectories of travelling actors as a ground upon which various transformations of comparison have been staged. Transformations of comparison from experimental to evolutionary to translational and machinic ones became visible because of the use of this imagination as the context for my argument.

One might argue that this testifies that ANT is, in the end, the right ontology of comparative practice, and that its presumption about what a thing, act or relation is allows one to see something that stays invisible in other frames. This seems to be what Latour's 'banal' and 'meaningless' infralanguage implies (Latour 2005). One of Latour's classic papers alludes to this. He notes:

> we shall always look for weak explanations rather than for general stronger ones. [...] Instead of explaining everything with the same cause and framework, and instead of

abstaining from explanation in fear of breaking the reflexive game, we shall provide a one-off explanation, using a tailor-made cause.

(Latour 1988a: 174)

Christopher Gad and Casper Bruun Jensen (2010: 64) recently questioned this claim as a 'positivist ideal.' They point out that it assumes 'the existence of two metalanguages ("researcher's" and "informant's"), which are opposed and struggling to get voice in socio-logical discourse.' In their view, ANT is now already shifting away from this early formulation by transforming itself through successive encounters with other practices including anthropology. They call for a new imagination about ANT, what they call post-ANT, 'to see ANT as allowing for the coexistence of *several* infralanguages, including [that of] the researchers, which may change and transform precisely because of their partial connection.' (Gad and Jensen 2010: 64).

The lateral comparison depicted in this chapter follows this path. Because there is no neutral position, the coexistence of infralanguages is only possible, as Gad and Jensen (2016b) note, by transforming them through the making of partial connections. Therefore, it requires a conscious effort to make explicit the artifice of how the analytical frame is weaved through the encounters with others. By gaining inspiration from ANT, as well as feminist anthropology and STS (Haraway 1991; Strathern 1988), lateral comparison attempts to make such an explicit mode of analysis by putting analyst's and interlocutors' comparisons side by side.

Lateral comparison also seems relevant to the calls from non-Western STS and anthropologies that seek ways to articulate its heterogeneous constitution. Being unsatisfied with and/or unable to occupy a neutral viewpoint, they are struggling to foreground the heterogeneity and situatedness of their practice as a legitimate matter of academic concern. This chapter argues that it might be possible to turn comparison – a notion that has always been pertinent to non-Western scholars and practices – into a springboard to move towards that direction. ANT's encounter with other forms of scholarship and practice would possibly open up new space for multiplicity that allows further conversation across difference.

Acknowledgement

I am grateful to Ignacio Farias and Anders Blok for their comments. My thanks go as well to Casper Bruun Jensen and Émile St-Pierre for commenting on the draft and correcting my English.

Note

1 Gad and Jensen (2016a, b) also discuss the relevance of lateral reasons to ANT and STS.

References

Akrich, Madeline and Vololona Rabeharisoa. 2016. Putting ourselves out of the traps of comparison. In *Practising comparison: logics, relations, collaborations*, edited by Joe Deville, Michael Guggenheim, and Zuzana Hrdličková, 130–165. Manchester: Mattering Press.

Asad, Taral. 1986. "The concept of cultural translation in British. Social anthropology." In *Writing culture: the poetics and politics of ethnography*, edited by James Clifford and George E. Marcus, 141–164. Berkeley: University of California Press.

Callon, Michel. 1987. "Society in the making: the study of technology as a tool for sociological analysis." In *The social construction of technological systems: new directions in the sociology and history of technology*, edited by Wiebe E. Bijker, Thomas Parke Hughes and T. J. Pinch, 83–103. Cambridge, MA: MIT Press.

Candea, Matei. 2017. "We have never been pruralist." In *Comparative metaphysics: ontology after anthropology*, edited by Pierre Charbonnier, Gildas Salmon and Peter Skafish, 85–106. London: Rowan and Littlefield.

Clifford, James. 1997. *Routes: travel and translation in the late twentieth century.* Cambridge, MA: Harvard University Press.

Deville, Joe, Michael Guggenheim, and Zuzana Hrdličková. 2016. *Practising comparison: logics, relations, collaborations.* Manchester: Mattering Press.

Evans-Pritchard, E. E. 1951. *Social anthropology.* Westport, CT: Greenwood Press, 1987.

Fu, Daiwie. 2007. "How far can East Asian STS go?" *East Asian Science, Technology and Society: An International Journal* 1 (1): 1–14.

Gad, Christopher, and Casper Bruun Jensen. 2010. "On the Consequences of Post-ANT." *Science Technology and Human Values* 35 (1): 55–80.

Gad, Christopher, and Casper Bruun Jensen. 2016a. "Lateral concepts." *Engaging Science, Technology and Society* 2: 3–12.

Gad, Christopher, and Casper Bruun Jensen. 2016b. "Lateral comparisons." In *Practising comparison: logics, relations, collaborations*, edited by Joe Deville, Michael Guggenheim and Zuzana Hrdličková, 189–219. Manchester: Mattering Press.

Gingrich, Andre, and Richard G. Fox. 2002. *Anthropology, by comparison.* London: Routledge.

Haraway, Donna Jeanne. 1991. *Simians, cyborgs, and women: the re-invention of nature.* London: Free Association.

Krause, Monika. 2016. "Comparative research: beyond linear-causal explanation." In *Practising comparison: logics, relations, collaborations*, edited by Joe Deville, Michael Guggenheim and Zuzana Hrdličková, 45–67. Manchester: Mattering Press.

Latour, Bruno. 1983. Give me a laboratory and I will raise the world. In *Science Observed*, edited by Karin Knorr-Cetina and Michael Mulkay, 141–170. London: Sage.

Latour, Bruno. 1988a. "The politics of explanation: an alternative." In *Knowledge and reflexivity: new frontiers in the sociology of knowledge*, edited by Steve Woolgar, 155–176. London; Newbury Park, CA: Sage.

Latour, Bruno. 1988b. *The pasteurization of France.* Cambridge, MA: Harvard University Press.

Latour, Bruno. 2005. *Reassembling the social: an introduction to actor-network-theory, Clarendon lectures in management studies.* Oxford; New York: Oxford University Press.

Law, John, and Wen-yuan Lin. 2017. "Provincializing STS: postcoloniality, symmetry, and method." *East Asian Science, Technology and Society* 11 (2): 211–227.

Malinowski, Bronislaw. 1922. *Argonauts of the western Pacific.* London: Routledge.

Maurer, Bill. 2005. *Mutual life, limited: Islamic banking, alternative currencies, lateral reason.* Princeton, NJ; Oxford: Princeton University Press.

Medina, Leandro Rodriguez. 2018. "Editorial welcome." *Tapuya: Latin American Science, Technology and Society* 1 (1): 1–6.

Mohácsi, Gergely, and Atsuro Morita. 2013. "Traveling comparisons: ethnographic reflections on science and technology." *East Asian Science, Technology and Society* 7 (2): 175–183.

Morita, Atsuro. 2012. *Yasei no Enjiniaringu.* Kyoto: Sekaishiso-sha.

Morita, Atsuro. 2013. "Traveling engineers, machines, and comparisons: intersecting imaginations and journeys in the Thai Local Engineering Industry." *East Asian Science, Technology and Society* 7 (2): 221–241.

Morita, Atsuro. 2014. "The ethnographic machine: experimenting with context and comparison in Strathernian ethnography." *Science, Technology & Human Values* 39 (2): 214–235.

Morita, Atsuro. 2017. "Encounters, trajectories, and the ethnographic moment: why "Asia as Method" still matters." *East Asian Science, Technology and Society* 11 (2): 239–250.

Nakazora, Moe. 2018. "Temporalities in translation: the making and unmaking of 'folk' Ayurveda and bio-cultural diversity. In *The world multiple: quotidian politics of knowing and generating the entangled worlds*, edited by Keiichi Omura, Grant Otshuki, Shiho Satsuka, and Atsuro Morita, 140–154. London: Routledge.

Rottenburg, Richard. 2009. *Far-fetched facts: a parable of development aid.* Cambridge, MA; London: MIT.

Sørensen, Estrid. 2008. "Multi-sited comparison of 'Doing Regulation'." *Comparative Sociology* 7 (3): 311–337.

Stengers, I. 2011. "Comparison as a matter of concern." *Common Knowledge* 17 (1): 48–63.

Strathern, Marilyn. 1987. "Out of context: the persuasive fictions of anthropology." *Current Anthropology* 28 (3): 251–281.

Strathern, Marilyn. 1988. *The gender of the gift: problems with women and problems with society in Melanesia.* Berkeley: University of California Press.

Strathern, Marilyn. 1992. *Reproducing the future: essays on anthropology, kinship, and the new reproductive technologies.* New York: Routledge.

Strathern, Marilyn. 1995. Afterward. In *Shifting contexts: transformations in anthropological knowledge*, edited by Marilyn Strathern. London: Routledge.

Verran, Helen. 2002. "A postcolonial moment in science studies: alternative firing regimes of environmental scientists and aboriginal landowners." *Social Studies of Science* 32 (5–6): 729–762.

Viveiros de Castro, Eduardo. 2004. "Perspectival anthropology and the method of controlled equivocation." *Tipití: Journal of the Society for the Anthropology of Lowland South America* 2 (1): Article 1.

Yanai, Tadashi. 2018. *Imeji no jinruigaku.* Tokyo: Serika Shobou.

5

How to write after performativity?

José Ossandón

The question

The question this chapter asks is 'how to write after performativity?' What is this chapter about?

Performativity

This chapter is not about performativity in general. It is not about the philosophy of language of Austin, it is not about Butler or Derrida. It is about the extension of Actor-Network Theory initiated by Michel Callon to the study of markets. The chapter focuses on what Callon has termed – in part in order to distinguish his own emphases from the many other branches of the 'performativity turn' – 'performation.'

After performativity

'After,' writes Sloterdijk in a different context, 'is the name for a break, an *epoché*, in the traditional sense of the word, which indicates both the caesura and also the time following it' (Sloterdijk 2017: 50). After Callon's performativity is not against, versus or even beyond performativity. It refers to the possibilities that have been opened after the breach introduced by Callon's research programme. It is, as it were, about the performativity of performativity.

Some famous lines by Callon read: '*homoeconomicus* does exist, but it is not an a-historical reality; he does not describe the hidden nature of the human being. He is the result of a process of configurations' (Callon 1998a: 22). In Callon's view, the relation between theoretical constructs and the reality they refer to is not about representation but, to use Hacking's (1983) terms, intervention. Just like the model of a new bicycle path prefigures that its users will cycle on the right side of the lane and not in zigzag, social scientific theories prefigure particular forms of acting. From this perspective (Callon 2007), a theory is successful when it is actualised in practical interventions that make those involved in their use to act differently. Homoeconomicus is the particular type of economic agency performed with economic theory; economic actors *become* homoeconomicus as they act the roles prefigured in the theories used to organise markets.

This chapter is about the success of Callon's theory. This chapter, though, does not inspect the ways in which economic agencies have been transformed after they have been intervened with tools that mobilise concepts proposed by Callon. New theories are successful not only as they transform their object of study. Actually, most of the times, theories do not even reach their objects. Social theories are successful also; they *perform*, as they transform the researchers that use them. To use the term Callon borrows from Deleuze and Guattari, theories are *agencements*. Theories distribute agencies, but not only in the sense that they mobilise scripts prefiguring the actions of those whose actions are explained, but as they prefigure the particular characters that will use them. Economic theories do not only construct the homoeconomicus, they also create the economist.

To write after performativity

New philosophical concepts, Deleuze and Guattari explain (1994), do things: They open new horizons for thought and create conceptual personae. Conceptual personae are not the philosophers but their alter egos. It is not that any rhetorical character is a conceptual persona and only a few philosophical theories create new personae. Conceptual persona is the character that inhabits the position a new philosophical concept invents. Social theory produces new characters too, but these are characters of a different kind. Social theory composes what could be called 'research personae.' Social theories do not only provide explanations, they teach their readers how to relate with their object of study in new ways. When social researchers use social scientific theories, they have to enact these theories' research personae.

The question 'how to write after performativity' asks about the research personae enacted after Callon's performativity thesis. In order to answer this question, what ought to be done first is to clarify the scripts of Callon's research persona. To write after Callon, however, is not simply to repeat Callon's guidelines. It is also about how uses of his theory develop new research personae. The second step is to identify the different personae enacted *after* (but not against, versus, or beyond) Callon's performativity thesis. This is precisely what the two remaining parts of this chapter try to accomplish.

The following section revisits Callon's landmark introductory piece to *The Laws of the Markets* in order to identify the particular instructions his approach set for future researchers interested in studying markets. This requires dealing with a particular feature of Callon's writing. As Deleuze (2004) pointed out, some philosophical works, for instance those of Kierkegaard and Nietzsche, are explicitly dramaturgical. These are texts that explicitly instruct their readers how to enact and inhabit their proposed conceptual personae. The same is the case with some social theories. Garfinkel's (1999) *Studies in Ethnomethodology*, for instance, is as much a theory about its objects as a set of instructions that researchers should follow in order to become ethnomethodologists. Becker's (2008) *Tricks of the Trade* and Latour's (2005) *Re-assembling the Social* are other examples of texts that teach their readers how to enact their proposed research personae. Most of the times, however, the characters featured in philosophical and social scientific narratives are those whose actions are to be explained or modelled. Thus, conceptual and research personae can and many times remain tacit. This is, in fact, the case in most of Callon's writings. Callon's writing is in this sense very different from Latour's (Ossandón 2015). While Latour craftily uses devices such as polemics, dialogues and diagrams to illustrate the particularities of the research persona his theory creates, Callon presents new theories but tends to leave the instructions of how to enact its particular research persona implicitly. In order to identify the research persona in Callon's performativity thesis, we therefore have to devise a method. What the next section attempts is to identify Callon's instructions by analysing the specific modifications he introduced in relation to the case he uses as the main empirical illustration of his performativity thesis, namely Garcia-Parpet's analysis of a strawberry market. The second section revises works that have used Callon's performativity thesis but that enact, I argue, different research personae. The section revises three different forms of writing *after* Callon. I term these personae a performative-detective, a philosophical-ethnographer and a market-reformist.

The instructions

In 1981, a new market place for the trading of table strawberries was set up in the commune of Fontaines-en-Sologne in France. This strawberry market became part of the sociological discussion in 1986, when a paper about the case by Marie-France Garcia-Parpet was published in *Actes de la Recherche en Sciences Sociales*. In 1998, Garcia-Parpet's strawberries began a second life when Callon used her study as the central empirical evidence in his famous introduction to the edited collection *The Laws of the Markets*. In fact, Garcia-Parpet's piece only appeared in English in 2007 when it was included in the book *Do Economists Make Markets?* which consolidated the international academic influence of Callon's performativity thesis. What Callon did not make explicit is that while using Garcia-Parpet's basic insight, his conclusions are radically different to hers. In what follows, Callon's modifications are used as entry points to identify the script – how his theory teaches researchers to relate to markets – for the research persona developed with Callon's performativity thesis.

Strawberry exchanges forever

The case at Fontaines-en-Sologne, Garcia-Parpet points out, presents characteristics that make it of special interest. She says: 'our data suggests, that this market is, in some sense, a concrete realization of the pure model of perfect competition' (2007: 20). This is an exchange that – like in economic textbook models – features atomised actors, homogenous products, fluid entry and transparent information. This fact, in turn, challenges traditional sociological approaches to the economy. Sociologists, Garcia-Parpet explains, would normally look for issues that deviate from the ideal market. They would search for evidence to prove the economists' idealised view of the economy wrong. What to do with a case without such *deviations*? Garcia-Parpet's article is an important contribution because it proposes a new type of task for future social researchers. Sociologists, now, can explain *how* a situation like the market in Fontaines-en-Sologne ended up organised in this particular way. Sociologists can study what Garcia-Parpet labels the 'social construction of the perfect market.'

When Callon revisited Garcia-Parpet's case in his 1998 piece, he seems to depart from a very similar starting point. Perfect competition, Callon explains, is not merely an abstract ideal invention of economists. It is also concrete, a particular mode of organising economic exchange. Accordingly, sociologists and other social scientists interested in studying the economy do not have to limit their work to assessing whether the model is or is not realistic; they should study how this is sometimes practically achieved. Garcia-Parpet and Callon, then, share a basic starting point. As Garcia-Parpet's put it: 'the practices which constitute the market are not market practices.' The real story at Fontaines-en-Sologne is not the one about the actors that participate in the strawberry exchange as Garcia-Parpet found them in her fieldwork. The situation – with merchants, traders, suppliers, buyers and standardised goods – is what has to be explained. The real story is not there; it lies elsewhere. It is in this 'elsewhere,' however, that the differences start to appear. How does Callon *callonise* the strawberry market case?

Modification 1: from prequel to making-off

These are Garcia-Parpet's instructions:

> The practices which constitute the market are not market practices' [...] 'The creation [of a perfect market] should be seen as a social innovation resulting from the work of a number of individuals interested, for different reasons, in changing the balance of power between the growers and the buyers' (37) [...] 'Thus the market is better conceived as a field of struggle (46).'
>
> *(Garcia-Parpet 2007)*

To explain the practical organisation of a situation such as the strawberry market, analysts should expand their focus to the story of power struggles in which the relevant participants of the exchange have been involved for years. The situation in 1981 is a particular and, as we learn in the postscript Garcia-Parpet added to the 2007 version of her article, momentary stage in this larger drama. Garcia-Parpet's study displays a convincing empirical analysis that explains how these power struggles ended up configuring the conditions for the 'perfect market.'

While Callon seems to agree with Garcia-Parpet's analysis, what he points out as crucial is very different. In his words:

> As the [strawberry market] example clearly shows, the crucial point is not that of the intrinsic competences of the agent but that of the equipment and devices (material: the warehouse, the batches displayed side by side; metrological: the meter, and procedural: regressive biding) which gives her or his action a shape' (20–21) [...] 'Beyond the material procedures, legal and monetary elements which facilitate the framing and construction of the space of calculability, there is a capital, yet rarely mentioned, element: economic theory itself (22).
>
> *(Callon 1998a)*

To use a cinematographic analogy, it could be said that what Garcia-Parpet's analysis does is to construct a prequel. In order to explain the extraordinary situation, you need to understand how the actors ended up playing the parts in which you met them. If Callon were making a movie, his movie would not be a fully dramatised prequel. It would rather be something like a documentary, a bit like those 'making-off' films where the attention goes to the technical details behind what we see. His instruction is: You should pay more attention to equipment, devices and economic theory itself.

Modification 2: de-contextualise and re-contextualise

The first modification Callon introduced concerns the focus of empirical attention. The second modification is theoretical. Remarkably, what Callon does, but not explicitly, is to completely reject Garcia-Parpet's theoretical explanation. His message for future researchers could be summarised as: Don't do as Garcia-Parpet suggests you to do! Interestingly, though, Callon does not construct an alternative empirical story. He does not present new evidence to challenge Garcia-Parpet's explanation. Callon's argumentation does not work at this level. What he does is to construct a new theoretical context to situate the case. The new context is the outcome of three operations.

The first theoretical operation is what Abbott (2004) calls an 'inversion': A type of heuristic that discovers new explanations by inverting the direction between *explanandum* and *explanans*. While previous research in economic sociology and economics would disagree about almost everything, they share the common assumption that economic agencies preexist market exchange. Callon, instead, argues that it is calculability (the fact that things are calculable and actors can calculate) that has to be explained. Of course, Callon's inversion is not entirely surprising. It is a logical consequence of the extension of Actor-Network Theory. If you will study the economy with ANT, you cannot start from preformed actors; you must start from the sociotechnical network and explain agencies as their outcome. It is as if at this moment, even though he has been one of its original developers, Callon shows that extending ANT to a new area has its price. You must follow its rules. As Callon vividly puts it: 'if we are to avoid the temptation of dualism, we need to banish any explanation separating the agents from the network, and, in particular, avoid the usual concepts of resources or social capital' (Callon 1998a: 12).

The second theoretical operation is not a consequence of ANT but of Callon's particular style of theorising. Callon constructs theory by knitting pieces of existing theory into a new patchwork. To use the film analogy yet again, what Callon does is like editing and montage. A bit like filmmaker Harun Farocki, who constructed some of his movies by montaging pieces of existing films, Callon cuts and pastes elements of existing theories in order to construct a new story. The pieces that Callon mobilises here include elements of social network analysis, sociological theory (Goffman, Simmel and Zelizer), economic anthropology (discussions on gift and time by Bourdieu and Thomas), social studies of accounting and economics (Guesnerie's definition of markets).

Callon, though, does not tell the reader 'do theory as I do.' He does not make a method explicit, he does not show how he chooses what he picks and what he doesn't pick. He does what he does and leaves it for future researchers to wonder how *callonising* works. A clue, however, can be found elsewhere. The quotation below summarises Callon's and Muniesa's explanation of the trajectory of economic goods:

> The process of individualization or singularization consists in a gradual definition of the properties of the product, shaped in such a way that it can enter into the consumer's world [...] The transfer can then take place. The good leaves the world of supply, breaks away from it (which is possible since it has been objectified) and slots into another world, that of the buyer, which has been configured to receive it. It becomes entangled in the networks of sociotechnical relations constituting the buyer's world.
>
> *(Callon & Muniesa 2005: 1233)*

What Callon does to Garcia-Parpet's case is not that different. A piece of evidence is 'objectified,' cut from its original context to be re-entered into a different world, where it becomes entangled in a completely different network.

What is the new network for the strawberry market? The third theoretical operation is to set a new theory context to relocate the assembled pieces. This operation relates to an element that is not always associated with ANT as an intellectual practice. As other chapters in this book point out, there is a tension between the usual self-description of ANT, which normally stresses a descriptive empiricist stance, and the practice of ANT, which, since the beginning, has relied on highly abstract formulations. It is as if the T in the acronym meant a particular mode of abstracting. Take, for instance, Callon's (1984) 'Some elements of a sociology of translation.' The Actor-Network theorist elaborates concepts – technical notions used to label things which do not necessarily respond to the way in which those things are named by those who are studied (i.e. 'problematization, interessement; enrolment, mobilization') – and the theorist assembles the series of labels into a larger model that works like a flow or cycle diagram ('the four moments of translation'). It is perhaps here where the engineer and the sociologist meet!

The model in the *Laws of the Markets* works at two levels. The first level is set to explain the relation between 'devices, equipment and economic theory' and 'calculative agencies.' Calculable goods and calculable agents are the outcome of framing and formatting, what Callon calls 'performation.' Performation is, in turn, located in a larger cycle that Callon unfolds in his second chapter in the book (Callon 1998b). Calculable agencies are challenged by issues (for instance, a particular contaminant material and the people affected by it) that were not accounted for in the existing frame. Callon uses the metaphor of 'overflows': Calculative frames are overflown and these overflows trigger new processes of framing and formatting. In other words, the field of struggle, suggested by Garcia-Parpet, is replaced with this other broad encompassing theoretical dynamic of framing and overflowing.

To sum up, in 1998, Callon introduced his performativity thesis. This theory produced a new research persona: It taught future researchers a new form of approaching markets. What this section has tried to do is to make the script of this persona explicit. In order to do that, the modifications Callon introduced in relation to the main empirical illustration of his theory have been reconstructed. The first modification was a matter of focus. While Garcia-Parpet constructed a detailed account of the past of the actors we find in the strawberry market, Callon asks us to look at the equipment, devices and economic theory. The second modification is theoretical. The strawberry market case is *cut* from its theoretical context and placed in a different model. This model is an assemblage of various pieces of existing theories and it tells future researchers to look at the intersections between the overall dynamic of framing and overflows. These are the main instructions Callon's 1998 pieces set for future researchers interested in the study of markets.

The characters

The previous section was about the research persona created with Callon's performativity thesis. It showed that Callon used Garcia-Parpet's case to create a new position to approach markets. This section explores work conducted *after* Callon. It revises work that is not set against or beyond but that follows Callon's performativity thesis, and that, a bit like Callon did with Garcia-Parpet, has enacted different research personae. The following lines distinguish three different characters, three different sets of instructions of how to write after Callon's performativity.

Before moving on, there are two disclaimers to make. Callon's performativity thesis has inspired thousands of papers in several sub-disciplines (Cochoy 2014, McFall & Ossandón 2014). The distinction between the three different ways of writing after Callon proposed here is informed by years of close reading of this literature, but it cannot claim to be exhaustive. The typology should be read as a tentative classificatory hypothesis. Second, it is worth mentioning that some of the questions posed here have been asked before. Inspired by Ian Hunter's (2006) critical historical analysis of recent humanities, Du Gay (2010) identified a tension in the work of Callon and colleagues. Sometimes, this work is descriptive and empirically oriented, while other times, it is populated by empirically untestable statements. Jenle (2015) picked the label Du Gay uses, the 'theoreticist,' to characterise the stance of work informed by Callon's performativity programme. He identifies two features: 'a primary commitment to or prioritization of the development of generally applicable conceptualizations of markets' and 'a lack of concern with the object of study as constituted by an empirical state of affairs' (Jenle 2015: 216). The exercise here is certainly inspired by these discussions. It will be argued, for instance, that Callon's theory has enabled the development of different personae and that these have different stances in relation to empirical inquiry. The point here, however, is not to evaluate whether the orientation of the performativity thesis is empiricist enough. Neither is it to identify this theory's overall stance. The point is rather to identify the type of research personae, the implicit characters and the rules set to them, enacted with and after Callon's approach to markets.

A performative-detective

If the strawberry market was the main empirical example Callon used to illustrate the 1998 formulation of his performativity thesis, undoubtedly the most famous empirical study informed by this thesis is MacKenzie's work on derivatives' markets. MacKenzie's derivatives' studies are oftentimes seen as a straightforward empirical operationalisation of Callon's insight that 'economics performs markets.' In fact, this is how MacKenzie himself often describes his own work. But, unsurprisingly, the translation is not that straightforward. The research persona in MacKenzie's work is both clearly influenced by Callon's performativity thesis and proposes a distinctively different empirical stance. It is, it could be said, *after* Callon, but not Callonian.

The most influential paper in the series is MacKenzie and Millo's (2003) piece in the *American Journal of Sociology*. There, MacKenzie and Millo revise different accepted theories in economic sociology and show how these help to partially explain the evolution of their case study, the derivatives market in Chicago. Their argument is that there is a missing link, a particular formula – the Black–Scholes equation widely used in the valuation of derivatives – whose empirical relevance in the case cannot be accounted for with the commonly accepted theories in economic sociology. The empirical situation of the market is *also* the outcome of the practical use of economic theory.

The research persona presented in MacKenzie and Millo's paper explicitly follows some of Callon's instructions – 'look at devices and economic theory' – but does not follow Callon's theorising style. In fact, when the economic situation that has to be explained changes, for instance, in his work on the credit crisis (MacKenzie 2011), Mackenzie works with different theoretical tools in order to sustain his particular explanation. What MacKenzie and Millo do is to transform Callon's performativity from a general model of markets into an empirical hypothesis, or conjecture, a thesis among others to explain a particular outcome. The argument is constructed a bit like a jury case, a carefully crafted defence to prove a missing link in front of a sceptical jury. The researcher in this context, unlike Callon's ANT theorist, is not oriented to construct a general model of the market. Like a detective producing evidence for a court case, the researcher appears as a character that has to identify the clues and construct proof in order to demonstrate the empirical connection between a specific economic theory and the explained economic situation.

A philosophical-ethnographer

Fabian Muniesa has co-authored some of Callon's work on markets. In his own work, though, Muniesa has also introduced important modifications. It is not that Muniesa has produced an 'anti-program,' e.g. set an agenda against Callon's performativity. What his work does is to enact a research persona for the performativity thesis that is different to Callon's and MacKenzie's.

Callon's performativity thesis was a model to explain a particular outcome, marketisation, understood in terms of calculable agents and calculative goods. In Muniesa's work (2014), instead, performativity becomes an overall perspective to study a wider variety of modes of economic performances – for instance, case-based teaching in business schools, perfume testing and financial valuation. This movement becomes more explicit in Muniesa and colleagues' (2017) more recent book, which makes capitalisation – a situation composed of investors, investee and assets – and not calculability, the central outcome to be explained. As mentioned, in MacKenzie's work, performativity became an empirical conjecture. The fact that the Black–Scholes formula performs the economy is demonstrated because the studied market starts to resemble the assumptions of the formula that informs the calculative devices used by market practitioners. In Muniesa's work, performativity signals a particular *stance*, a mode of relating to the world of business that is not interested in studying whether market A looks like the content of theory B (Ossandón & Pallesen 2016). Empirical research, instead, inquires how business reality is constructed through different sets of performances.

Performative, in this context, acquires a broader sense. It does not refer only to 'performativity' in the philosophy of language sense. It also means performance, in the dramaturgical sense, and *agencement* in the sense of situated economic instances in which specific ways of acting are distributed. The research persona proposed in Muniesa's work does not resemble a detective. Muniesa's research stance is ethnographic and descriptive, but not empiricist. He does not look for clues that can demonstrate a particular empirical link. It is, as in Callon's case, more speculative. However, unlike Callon's position, Muniesa's theorising is not oriented at constructing a model of its empirical object. Theorising is oriented at conceptualising, identifying and naming the particular ways in which different performances are *realised*. The research persona is a philosophical-ethnographer: A producer of concepts informed by ethnographic work (Ossandón 2015).

A market-reformist

A third persona produced *after* Callon's performativity has been developed by Callon himself. As mentioned already, Callon's 1998 pieces outlined a model on two levels, a descriptive layer in which researchers study how devices and economics work in the organisation of specific markets, and a second level that situates the process of formatting markets in the dynamic of framing and overflows. An important part of Callon's recent work has been oriented at developing a new persona whose work is at the intersection of both levels. This position already features, although not centrally, in the texts from 1998.

> This is the point at which it would make sense to draw up a new contract between sociology and economics [...] Thus AST [The anthropology of science and technology] can help with the work of framing interactions by improving the visibility of various efforts to keep track of overflows as well as the visibility of the disagreements or agreements to which they give rise. Like those satellite imaging systems that enable navigators to keep track of their relative positions at all times, the anthropology of science and technology can provide the actors with a cartographical outline of overflows in progress, thereby paving the way for preliminary negotiations.
>
> *(Callon 1998b: 263)*

The new persona marks an important modification in relation to Callon's own early ANT work. As Callon explains in his early pieces (for example, Callon 1984), the task of the ANT researcher was to reconstruct the traces of how engineers and other technicians problematise

53

and construct alliances, and to map and conceptualise these processes. In Callon's recent pieces, anthropologists of science and technology are not only invited to provide social scientific descriptions of the work of other actors. The new persona is somehow closer to what economists attempt to do than to early ANT. From this position, markets are not only the object the experts that ANT's scholars *study* construct, markets also become an object of ANT intervention. Like when economists do policymaking, this new position assumes that well-functioning markets are desirable collective outcomes and that the public role of social scientists is justified on the basis of the identification and repair of market failures. What Callon does, more clearly in his essay 'Civilizing Markets,' is to define a new type of market failure: Markets should be assessed in relation to their ability to deal with the overflows they might produce.

> To be considered as efficient, a market should pay very careful attention to the numerous matters of concern that it creates, and to the groups that express and promote them, thus becoming economic agents in their own right.
>
> *(Callon 2009: 546)*

The new task for social researchers, in this context, is to map controversies, identify overflows and help give voice to those affected or orphaned by the development of new markets. This new persona is not simply a researcher; it is a market-reformist.

Performativity and ANT as an intellectual practice

To paraphrase Sloterdijk (2013), again, new theories are like secessionist movements, anthropotechnic programmes that transform those that exercise them. In the terms used in this chapter, new social theories do not only provide new explanations. They produce new research personae: They teach and retrain future researchers in new ways of relating with their object of study. The object of interest here is not an exception in this regard. Callon's performativity thesis has no doubt been immensely successful. It introduced a new approach to the study of markets which has been enacted by an army of *callonised* researchers.

What this chapter has tried to do is to identify performativity's research personae. The process of answering this question has been like an exercise in the 'sociology of translation.' In order to identify the research persona of Callon's performativity thesis, his instructions of how to approach markets, I inspected his translation of García-Parpet's work. To delimit the research personae enacted *after* Callon's thesis, the instructions to approach markets informed by Callon's work, I revised how his work has been translated by MacKenzie and Millo, Muniesa and Callon himself.

What does this exercise say about ANT as an intellectual practice? Two things. First, performativity appears like a fractal of ANT. Like other chapters in this book show, ANT cannot be reduced to a single intellectual stance. Sometimes, ANT is a particular type of modelling heterogeneous networks, other times, it is a more descriptive type of ethnographic research, on other occasions, it refers to a form of philosophical conceptual speculation and yet at other times, it is a particular political position. Callon's performativity thesis does not only combine these different elements, it has been enacted in all these *keys*. Second, this exercise shows that ANT has become a particular type of critical intellectual practice. Doing ANT is not simply following the explicit instructions set by canonical texts. To use ANT is also a *performance,* a creative form of acting the instructions that previous theories set to us. Future will tell if the new personae created in this process are still *after* or are beyond or against performativity and ANT.

References

Abbott, A. (2004). *Methods of Discovery: Heuristics for the Social Sciences*. New York: W. W. Norton & Company.

Becker, H. S. (2008). *Tricks of the Trade*. Chicago, IL: University of Chicago Press.

Callon, M. (1984). Some Elements of a Sociology of Translation: Domestication of the Scallops and the Fishermen of St Brieuc Bay. *The Sociological Review*, 32(1_suppl), 196–233.

Callon, M. (1998a). Introduction: The Embeddedness of Economic Markets in Economics. In Callon, M. (Ed.), *The Laws of the Markets*. Oxford: Blackwell.

Callon, M. (1998b). An Essay on Framing and Overflowing: Economic Externalities Revisited by Sociology. In Callon, M. (Ed.), *The Laws of the Markets*. Oxford: Blackwell.

Callon, M. (2007). What Does it Mean to Say that Economics Is Performative? In MacKenzie, D., Muniesa, F., & Siu, L. (Eds.), *Do Economist make Markets? On the Performativity of Economics*. Princeton, NJ: Princeton University Press.

Callon, M. (2009). Civilizing Markets: Carbon Trading between in Vitro and in Vivo Experiments. *Accounting, Organizations and Society*, 34(3–4), 535–548.

Callon, M., & Muniesa, F. (2005). Peripheral Vision: Economic Markets as Calculative Collective Devices. *Organization studies*, 26(8), 1229–1250.

Cochoy, F. (2014). A Theory of 'Agencing': On Michel Callon's Contribution to Organizational Knowledge and Practice. In Adler, P. S., Du Gay, P., Morgan, G., & Reed, M. I. (Eds.), *The Oxford Handbook of Sociology, Social Theory, and Organization Studies: Contemporary Currents*. Oxford: Oxford University Press.

Deleuze, G. (2004). *Difference and Repetition*. London: Continuum.

Deleuze, G. & Guattari, F. (1994). *What Is Philosophy?* NYC: Columbia University Press.

Garfinkel, H. (1999). *Studies in Ethnomethodology*. Cambridge: Polity Press.

Garcia-Parpet, M-F. (2007). The Social Construction of a Perfect Market: The Strawberry Auction at Fontaines-en-Sologne. In MacKenzie, D., Muniesa, F., & Siu, L. (Eds.), *Do Economist make Markets? On the Performativity of Economics*. Princeton, NJ: Princeton University Press.

du Gay, P. (2010). Performativities: Butler, Callon and the Moment of Theory. *Journal of Cultural Economy*, 3(2), 171–179.

Hacking, I. (1983). *Representing and Intervening*. Cambridge: Cambridge University Press.

Hunter, I. (2006). The History of Theory. *Critical Inquiry*, 33(1), 78–112.

Jenle, R. P. (2015). *Engineering Markets for Control*, PhD Thesis. Frederiksberg: Copenhagen Business School.

Latour, B. (2005). *Reassembling the Social*. Oxford: Oxford University Press.

MacKenzie, D., & Millo, Y. (2003). Constructing a Market, Performing Theory: The Historical Sociology of a Financial Derivatives Exchange. *American Journal of Sociology*, 109(1), 107–45.

MacKenzie, D. (2011). The Credit Crisis as a Problem in the Sociology of Knowledge. *American Journal of Sociology*, 116(6), 1778–1841.

McFall, L. & Ossandón, J. (2014). What's New in the 'New, New Economic Sociology' and should Organisation Studies Care? In Adler, P., du Gay, P., Morgan, G. and Reed, M. (Eds.), *Oxford Handbook of Sociology, Social Theory and Organization Studies: Contemporary Currents*. Oxford: Oxford University Press.

Muniesa, F. (2014). *The Provoked Economy. Economic reality and the Performative Turn*. London: Routledge.

Muniesa, F., et al. (2017). *Capitalization: A Cultural Guide*. Paris: Presses des Mines.

Ossandón, J. (2015). ¿Cómo escribir lo social después de la performatividad y sus obstrucciones? *Cuadernos De Teoría Social*, 2(1), 8–32.

Ossandón, J., & Pallesen, T. (2016). Testing 'The Provoked Economy'. *Journal of Cultural Economy*, 9(3), 310–315.

Sloterdijk, P. (2013). *You Must Change Your Life*. Cambridge: Polity Press.

Sloterdijk, P. (2017). *Not Saved: Essays After Heidegger*. Cambridge: Polity Press.

6

Is ANT a critique of capital?

Fabian Muniesa

Can Actor–Network Theory be considered as, transformed into, or used for a critique of capital? And, subsequently, of the capitalist, the capitalistic or capitalism altogether? The extent to which a contribution from ANT can be recognised as a contribution to the critique of capital depends on multiple parameters. The explicit purpose of the contribution, first, will obviously be most centrally determined by what the critical author wants to do, irrespective of the claimed intellectual apparatus. But then, more decisively, the changing meanings of the terms in question – ANT, capital and critique – precipitate a situation in which no clear rule of interpretation can be established.

It should be noted from the outset that ANT posits itself, in its canonical sources (Latour 1993), as both a radical critique of the categories of the modern mind and as a denunciation of modern critique as perhaps the most salient of such categories (Lezaun 2017; see also Guggenheim, this volume). The scruple put forward by the latest Latour (2013) regarding that demystifying achievement, and the concomitant urge to conserve or eventually repair such modern categories, is not without interest in this respect. The fact that such preoccupation can be transferred to capital is certainly interesting. This is surely a category which is central to the operations of the modern mind, and which also operates as the prime vehicle for the production of the world as we know it today. How is it addressed, then, within the ANT mind?

Multiple strategies are available for an assessment of the intellectual and investigative positions of ANT – or ANT contributors – in relation to the subject matter. The one adopted here, rather modest and incomplete, consists in examining the way in which capital (the notion but also its incessantly malleable reality) features in several paradigmatic contributions from Michel Callon and Bruno Latour, two authors who, in their distinctive ways, are credited for having established the ferment of ANT (this piece unjustly leaving aside, it should be noted, many other relevant contributors). In doing so, this presentation relies on an approach that interprets ANT in the light of its situation in a particular historical configuration (Muniesa 2015). Three periods are identified in what follows, each characterised by a defining orientation (or lack thereof) towards capital and its problems: An early ANT period preoccupied with issues of scientific production; a middle period focused on the sociology of economics and the construction of markets; and a recent period oriented towards issues of ecological 'disinhibition' and its implication for questions of valuation, economic and otherwise.

The notion of capital, it should be noted straightaway, is taken here primarily in its mundane sense (the financier's), that is, as whatever can be considered as an asset and hence work as a vehicle for economic power. The notion of markets, also crucial in what follows, is also considered as an ordinary term (the merchant's), that is, as an exchange in which things are traded at a price. Supporting the task of accompanying ANT with a prospective hope, the piece concludes with a call for an ANT-inspired anthropology of capitalisation.

Competition, technology transfer and scientific capitalisation

The 1970s provided ANT with one of its most characteristic concerns: Competition. The category features prominently, for example, in the writings in which Callon first offered its singular articulation of the notion of translation, a crucial element of the ANT repertoire. The inception of this notion is often referred to a study on aquaculture published in the 1980s (Callon 1986), but the problem of translation and its basic definitional tenets were already present in Callon's work a decade before. Callon (1976) suggested that 'operations of translation' are the key for the understanding of the struggles that form the ambits of scientific research, technological innovation and industrial policy, and their multiple interconnections.

Market competition was at the heart of the early version of Callon's notion of translation, which was instrumental in the examination of the battle for the electric vehicle in France that he was carrying out in that period. The intellectual elaborations found there were already guided by the critique of mainstream idioms in both the sociology and the economics of science. Attention to the particulars of the semiotic knots that link 'problematic statements' to one another was already explicit in this early work. The empirical concern that nurtured the reflexion was on markets, though: The 'operations of translation' that were examined there are, most typically, those involved in technology transfer and the circulation and transformation of problematic statements from scientific research to industrial commerce (and vice versa).

There was nothing congratulatory, at that time, in Callon's view on competition within the web that links scientific work and economic value. That view, on the contrary, stood as the ground from which later sprang his political stand on science as a public good (Callon 1994). The purpose was rather to imprint in his reader's mind a sense of realism that would allow escaping a vision too heavily marked by the idea of separate spheres or domains – science and the market – that would only communicate through interference. On the contrary, the two sets of practices were shown to operate in close tandem (with the tensions this entailed). Capital, though, was a notion with little presence in Callon's analyses. Money appeared as a medium for competition and calculation rather than as full-fledged object of inquiry, and the categories of financial operation (investment, credit, discounting) were marginalised in favour of that of commercial operation (demand, supply, price), certainly more prominent – if not invasive – in Callon's analytical repertoire.

The idea of capital proper, it can be argued, was more emphatically approached (and theorised within what would become ANT) in the late 1970s and early 1980s by Bruno Latour. The particular blend of semiotic analysis that was introduced in the ethnographic analysis of an endocrinology laboratory in the 1970s (Latour and Woolgar 1979; see Mattozzi, this volume) included as a crucial element the analysis of the several forms of 'conversion' – a notion close to that of 'translation' – between scientific credibility and economic credit, that could propel several forms of accumulation and rent. Considering scientific credit, for example, as a form of capital (and scientific publication as a form of asset management) was an intuition that was already articulated by Latour in that period, with capital explicitly considered not as a thing but as an operation of conversion, that is, as power – with an explicit acknowledgement of the debt towards Marx's *Capital* in this respect (Latour 1983, 1984).

'Capitalisation' indeed, understood as the process of converting something into (or sub-suming under, one could say in Marxian jargon) a capital form, became a central notion in Latour's *Science in Action*: A central piece (the apex) in the theoretical engine known as 'centres of calculation' (Latour 1987: 215–257). The fact that this insight could have turned into an explicit examination of the operations of money, of the financial industry and of the several forms of conversion or translation that govern a capital appraisal of reality might seem obvious today. Capitalisation – that is, in short, the process of turning things into assets and considering them in their capacity to generate a return on investment – is indeed the rule that determines today, to a considerable extent, the existence and form (read: the financing or not) of almost everything, particularly of science (Muniesa et al. 2017). But in that period, arguably because of the particulars of scientific and industrial policies in France, these authors were more inclined to consider the capitalistic determi-nations of science from the angle of public policy and public budgeting (Arvanitis, Callon and Latour 1986).

Accounting, economics and the construction of markets

The 1990s and 2000s were in part marked, in these two authors' work, by a drastic engage-ment with economics' paradoxes, blind spots and consequences. While recognising the eco-nomistic drawbacks of economic science for a properly realistic account of economic reality and abundantly calling, in this respect, for a renewal of the anthropological approach, Callon and Latour did also demand the inclusion of the efficacy of economics among the objects that an economic anthropologist ought to scrutinise (Callon and Latour 1997). Accounting oper-ations were certainly featured as a critical target for such an approach (Latour 1996), therefore opening a window for an ANT's take on capital: The valuation methods that it prefers, its penchant for monopoly, its grip on the future.

Yet, markets (not the same as capital) were still unequivocally situated at the forefront of the ANT proposal instead: Markets and their taste for liberal trade, for competition and for the rule of the here and now. Callon's (1998) reinterpretation of a study of the economic design of an auction market (Garcia 1986) proved particularly effective in incentivising a research programme attentive to the calculative configuration of markets (see also Callon, Millo and Muniesa 2007; MacKenzie, Muniesa and Siu 2007; see Doganova, this volume). Latour's visual ethnography of the central wholesale food market in the Paris region was also meant to fuel the ANT imagination in that direction (Latour and Hermant 1998).

In part due to the polemical tone of Latour and Callon's formulations, but also of the wider challenges posed by ANT for the modern critical mind (Latour 1993), the idea of suspending a purely epistemic or purely sociological critique of economic truth in favour of a pragmatist examination of economic verification was received in the 2000s with mixed feelings in crit-ical circles (Muniesa 2016). The contentions, however, largely revolved around the problem of the disciplinary imperialism of economics, and of how partial or not to it were Callon and Latour. Not around the absence or not of a critique of capital in their writings, though. The idea that the conversions or translations capital relies on (things into assets, knowledge into credit, prospect into return, time into money) could be dealt with from the intellectual perspectives opened by these two authors was not central to the discussion. This is the more remarkable in a period in which both Callon and Latour were also acknowledging the in-fluence of Gilles Deleuze and Félix Guattari, two authors who had laid out in the 1970s an approach to capital as a semiotic, performative assemblage of clear potential ANT resonance (Alliez 1996; Deleuze and Guattari 1987; see also Muniesa et al. 2017; Jensen, this volume).

Latour's own venture into the terrain of economic theory in the 2000s was controlled by his interest in Gabriel Tarde, an author that, for essentially philosophical reasons, could now be presented as a founder of ANT (Latour 2002). Tarde was the author of a treatise of economics – or rather 'economic psychology' – whose rediscovery Latour contributed to (Latour and Lépinay 2009). Of clear interest for a venture into the styles of thought that the categories of political economy require (including capital, considered as a factor of production), Tarde's treatise proves however of little value for a way out of them – well, out of them in the direction pointed out here at least. Tarde indeed fuelled a renewed emphasis on an intersubjective theory of value, on social influence and the role of opinion, on emotional behaviour and economic affect, on marketing, on the economic nature of cognition and on the condition of novelty, invention, innovation and valuation within the economic process (see Lazzarato 2002). Potentially interesting connections with behavioural economics and the sociology of innovation could be laid out from there, but fell short from providing, except for a few notable exceptions (Lépinay 2007), the instruments that would have allowed for a refinement of the ANT intuition, expressed earlier, of capital as an operation rather than a productive factor.

Economic theology, ecological disinhibition and the enigma of value creation

The 2010s inescapably witnessed, within the circumscribed ANT landscape examined here, Latour's turn towards the critique of environmental eschatology (Latour 2017). The discovery of the debates in the tradition of political theology (especially Carl Schmitt and Eric Voegelin) served the purpose of reassessing the impasses of the modern mind already examined two decades before (Latour 1993, 2013). The notion of capital could quite naturally be situated more centrally as an operational motor for that mind, if considered from that perspective. Latour is not proceeding explicitly in that direction. But his role in the establishment of an anthropological assessment of the anthropogenic condition of geological and climatic transformations is indeed part of a movement that situates capital at the centre of what is meant by 'anthropogenic' (Bonneuil and Fressoz 2016) – discussions on how the Anthropocene should be better called 'Capitalocene' are symptomatic in this respect (Moore 2016). Latour's contribution within these directions (which contain important theoretical divergences) resides in situating the sense of 'disinhibition' that characterises the modern mind's approach to anthropogenic catastrophe in the wider cosmology – or theological unconscious – that defines modernity (see also Fressoz 2012). The extent to which this can be recognised as part of an ANT perspective is debatable. But it is possible to argue that the 'conversion to capital' that Latour identified in early analyses of the operations of capitalisation within the medium of modern science could provide a useful template for the articulation of an empirical response to the environmental preoccupation.

Why is not capital featuring explicitly as a fully fledged 'regime of enunciation' (or 'mode of existence') in Latour's signature *Inquiry* (Latour 2013), then? This is certainly a good question, which was repeatedly raised in early collective discussions on the project – perhaps most vocally (or vociferously) by Isabelle Stengers in the discussion on the first draft of the manuscript held at the Colloque de Cerisy in June 2007. After a few meanders, issues of economic existence were in the end left, in the version finally published in 2012 (2013 in English), to the interface between the rule of desire, need and attachment on the one hand and the rule of rationing, organising, allocating and calculating on the other, augmented with the rule of the moral elicitation of the optimum when the arrangement of economic things is to correspond to a doctrine of justice (ATT, ORG and MOR, in the *Inquiry*'s official dialect). This turned handy, but certainly left unfinished both the task of identifying in investment a distinctive mode of enunciation, and of signalling its moral, political and eventually theological (if need be) limits.

Callon's work in this period does certainly not go down that path. Following a rather coherent orientation, Callon endeavours to further strengthen the programmatic call he had been articulating since the 1970s on the sociology of markets. The main focus of intellectual preoccupation is on the definitional quandaries on what markets amount to, and on how their formation can be understood in terms of political compromise (Callon 2009, 2016, 2017). This includes, then, a political critique of, or rather engagement with, how markets are organised concretely, and a – tentative (Blok 2011) – call for a study of how markets are contested and transformed, or not, through contestation. In a sense, Callon takes seriously, and courageously endorses, a liberal viewpoint that sees in markets, precisely, a critical site, that is, a site in which the value of things is submitted to critical consideration. He also takes seriously a concomitant sociological viewpoint that sees in markets a battlefield for a battle whose rules are dictated by conflicting positions, valuations and situations (see Geiger, Harrison, Kjellberg and Mallard 2014). Overall, though, the analytical focus is primarily on how 'economic actors' achieve 'coordination' in this medium, and the political one on how 'competition' and 'innovation' are or shall be organised for the better.

But where are the nuts and bolts of capital proper in these latest developments? If 'marketization' is the name of the paradigmatic problem in Callon's latest writings, the problem of capital is, by default, considered as a component to it (e.g. markets for... money?), not as a distinctive, singular reality. Latour's latest considerations on the subject matter are different, but the question of the market remains central. It features centrally in the discussion of the regimes of enunciation that lie in both political economy's ideals of optimum and equilibrium, in marketing's arts of capture and desire and in the calculative organisation of allocation (Latour 2013). The fact that capital – i.e. value considered from the financial perspective – could constitute a regime of enunciation different from that of the market price, and far more determinant in the composition of the modern condition, is left open, but not addressed.

Conclusion

Attempts at situating capital at the centre of an ANT inquiry count now on some precedent (esp. Muniesa et al. 2017) – which this piece blatantly aims at advertising, for the progress of discussion. A possible guiding strategy, it is suggested, would consist in considering capital as a semiotic operation (one of conversion or translation), a strategy that is in line with the main tenets of the approach elaborated in part by Latour and Callon in the 1970s, and which certainly complies with the relational intuition once brilliantly offered by Marx (Muniesa et al. 2017: 159–161). 'Capitalization' is thus the rhetoric programme, the enunciation apparatus, the semiotic assemblage that is to be deciphered, first and foremost by characterising the political configuration of things that it both requires and prompts.

As the key to this configuration lies in the figuration of 'the investor' and its capacity to value things properly, the most crucial aspect for this strategy's investigative felicity is a proper engagement with the anthropology of finance and with the particulars of financial valuation. The fact that the valuation complex on which the financial industry relies resides not on the idea of market competition but rather on its cancellation is now abundantly documented (Ortiz 2014), as is the fact that the crux of financial analysis points to considering things from the viewpoint of a financier or an investor rather than of a merchant or a trader. Falling outside the perimeter of ANT is something that a properly cogent anthropology of finance is very well entitled to do (Hart and Ortiz 2014; Ortiz 2013). Keeping track of an ANT sense – or of a comparable constructivist, pragmatist sensibility – requires, though, emancipating in the most radical possible way from the vernaculars of 'value' and 'value creation' that still populate the scholarly critique of capital (Muniesa 2011, 2014, 2017).

The reader may have noticed by now that this assessment focuses on the critique of capital rather than of capitalism. There is no contention here on the usefulness of a concept – 'capitalism' – that attempts at characterising a particular form of economic organisation in which the means of economic activity largely rely on private ownership and profit. There is rather recognition of the fact that its most distinguishable feature is capital, nonetheless: That is, capital understood as the operation of capitalistic procedures (as opposed to, say, capital understood as an amorphous chimera). Narrowing down the investigative lens to the vagaries of financial analysis, discounted cash flows, capital budgeting, investment policy, business models and such rather amounts, it can be claimed, to an expansion of the wideness of the critical scope. These are indeed the techniques through which the 'proper value' of things is expected to be expressed in all kinds of matters today, and then of those things brought into existence (or not) altogether (Birch 2017; Doganova 2014, 2015, 2018; Doganova and Muniesa 2015).

The fact that the directions 'competition' or 'innovation' can take today are largely determined from the perspective of the cost of capital and that the dominant investment rationale explicitly requires the eradication of the risks involved in (competitive and innovative) markets – e.g. from the logic of portfolio insurance to the premium on monopolistic concentration – certainly offers promising avenues for research in that direction. The equally remarkable fact that protecting things (especially 'nature') from the whims of speculative markets and the perils of dilapidation insistently translates today into calls for revaluing those things properly as assets from which one is expecting a future, that is, transforming them into capital ('natural capital'), is certainly part of the syndrome that a semiotic take on the 'conversion to capital' can elucidate (Muniesa 2017).

So, is ANT a critique of capital? It certainly can be, potentially. With critique, that must be said, not in the sense of saying if something is right or wrong (which obviously depends on the point of view) but in the philosophical sense of examining the tensions in the meaning of a concept from all possible angles, including that from which it does not make sense. This ANT potentiality is still to be developed, though. The philosophical lineaments once put forward by Deleuze and Guattari (1987) on their take on capitalisation as a distinctive form of performative enunciation apparatus, and later by Alliez (1996), for example, on capitalisation as the engine that produces the very idea of the future (see also Guattari and Alliez 1984), are certainly only missing the empirical, punctilious, investigational nerve that ANT requires in order to fuel an inquiry in that direction. Capitalisation can then feature centrally among the 'operations of translation' (i.e. translation to capital) that Callon's programme brilliantly set out to study, once released from the intellectual spell of the market. It can be revisited anew, properly characterised as a distinctively financial 'regime of enunciation,' as a crucial constituent of the 'centres of calculation' that Latour saw populating our ordinary modernity and forming its distinctive 'forked tongue.' And it can finally be rightly observed as a most significant element of the syndrome Latour's environmental eschatology attempts at capturing, with the entire idea of the Earth *as capital* being at once, forcefully unnoticed, both the problem and the (outstandingly problematic) 'solution.'

References

Alliez, É. (1996), *Capital Times: Tales from the Conquest of Time*, Minneapolis: The University of Minnesota Press.

Arvanitis, R., M. Callon, and B. Latour (1986), *Evaluation des Politiques Publiques de la Recherche et de la Technologie: Analyse des Programmes Nationaux de la Recherche*, Paris: La Documentation Française.

Birch, K. (2017), 'Rethinking value in the bio-economy: finance, assetization, and the management of value', *Science, Technology, & Human Values*, 42(3), 460–490.

Blok, A. (2011), 'Clash of the eco-sciences: carbon marketization, environmental NGOs, and performativity as politics', *Economy and Society*, 40(3), 451–476.

Bonneuil, C. and J.-B. Fressoz (2016), *The Shock of the Anthropocene: The Earth, History and Us*, London: Verso.

Callon, M. (1976), 'L'opération de traduction comme relation symbolique', in C. Gruson (ed.), *Incidence des Rapports Sociaux sur le Développement Scientifique et Technique*, Paris: Maison des Sciences de l'Homme, 105–141.

Callon, M. (1986), 'Some elements of a sociology of translation: domestication of the scallops and the fishermen of St. Brieuc Bay', in J. Law (ed.), *Power, Action and Belief: A New Sociology of Knowledge?* London: Routledge, 196–233.

Callon, M. (1994), 'Is science a public good? Fifth Mullins Lecture, Virginia Polytechnic Institute, 23 March 1993', *Science, Technology & Human Values*, 19(4), 395–424.

Callon, M. (1998), 'Introduction: the embeddedness of economic markets in economics', in M. Callon (ed.), *The Laws of the Markets*, London: Blackwell, 1–57.

Callon, M. (2009), 'Civilizing markets: carbon trading between *in vitro* and *in vivo* experiments', *Accounting, Organizations and Society*, 34(3–4), 535–548.

Callon, M. (2016), 'Revisiting marketization: from interface-markets to market-agencements', *Consumption Markets & Culture*, 19(1), 17–37.

Callon, M. (2017), *L'Emprise des Marchés: Comprendre leur Fonctionnement pour Pouvoir les Changer*, Paris: La Découverte.

Callon, M. and B. Latour (1997), '"Tu ne calculeras pas!" Ou comment symétriser le don et le capital', *Revue du MAUSS*, 9, 45–70.

Callon, M., Y. Millo, and F. Muniesa (eds.) (2007), *Market Devices*, London: Blackwell.

Deleuze, G. and F. Guattari (1987), *A Thousand Plateaus: Capitalism and Schizophrenia*, Minneapolis: The University of Minnesota Press.

Doganova, L. (2014), 'Décompter le future: la formule des flux actualisés et le manager-investisseur', *Sociétés Contemporaines*, 93, 67–87.

Doganova, L. (2015), 'Que vaut une molécule? Formulation de la valeur dans les projets de développement de nouveaux médicaments', *Revue d'Anthropologie des Connaissances*, 9(1), 17–38.

Doganova, L. (2018), 'Discounting the future: a political technology', *Economic Sociology*, 19(2), 4–9.

Doganova, L. and F. Muniesa (2015), 'Capitalization devices: business models and the renewal of markets', in M. Kornberger, L. Justesen, A. K. Madsen and J. Mouritsen (eds.), *Making Things Valuable*, Oxford: Oxford University Press, 109–125.

Fressoz, J.-B. (2012), *L'Apocalypse Joyeuse: Une Histoire du Risque Technologique*, Paris: Le Seuil.

Guattari, F. and É. Alliez (1984) 'Capitalistic systems, structures and processes', in F. Guattari, *Molecular Revolution: Psychiatry and Politics*, London: Penguin, 273–287.

Garcia, M.-F. (1986), 'La construction sociale d'un marché parfait: le marché au cadran de Fontaines-en-Sologne', *Actes de la Recherche en Sciences Sociales*, 65, 2–13.

Geiger, S., D. Harrison, H. Kjellberg and A. Mallard (eds.) (2014), *Concerned Markets: Economic Ordering for Multiple Markets*, Cheltenham: Edward Elgar.

Hart, K. and H. Ortiz (2014), 'The anthropology of money and finance: between ethnography and world history', *Annual Review of Anthropology*, 43, 465–482.

Latour, B. (1983) 'Le dernier des capitalistes sauvages: interview d'un biochimiste', *Fundamenta Scientiae*, 314(4): 301–327.

Latour, B. (1984), *The Pasteurization of France*, Cambridge, MA: Harvard University Press.

Latour, B. (1987), *Science in Action: How to Follow Scientists and Engineers Through Society*, Cambridge, MA: Harvard University Press.

Latour, B. (1993), *We Have Never Been Modern*, Cambridge, MA: Harvard University Press.

Latour, B. (1996), 'Foreword: the flat-earthers of social theory', in M. Power (ed.), *Accounting and Science: Natural Inquiry and Commercial Reason*, Cambridge: Cambridge University Press, xi–xvii.

Latour, B. (2002), 'Gabriel Tarde and the end of the social', in P. Joyce (ed.), *The Social in Question: New Bearings in History and the Social Sciences*, London: Routledge, 117–132.

Latour, B. (2013), *An Inquiry into Modes of Existence: An Anthropology of the Moderns*, Cambridge, MA: Harvard University Press.

Latour, B. (2017), *Facing Gaia: Eight Lectures on the New Climatic Regime*, Cambridge: Polity.

Latour, B. and E. Hermant (1998), *Paris Ville Invisible*, Paris: La Découverte.

Latour, B. and V.-A. Lépinay (2009), *The Science of Passionate Interests: An Introduction to Gabriel Tarde's Economic Anthropology*, Chicago, IL: Prickly Paradigm Press.

Latour, B. and S. Woolgar (1979), *Laboratory Life: The Social Construction of Scientific Facts*, London: Sage.

Lazzarato, M. (2002), *Puissance de l'Invention: La Psychologie Économique de Gabriel Tarde contre l'Économie Politique*, Paris: Le Seuil.

Lépinay, V.-A. (2007), 'Economy of the germ: capital, accumulation and vibration', *Economy and Society*, 36(4): 526–548.

Lezaun, J. (2017), 'Actor-Network Theory', in C. E. Benzecry, M. Krause, and I. A. Reed (eds.), *Social Theory Now*, Chicago, IL: University of Chicago Press, 305–336.

MacKenzie, D., F. Muniesa and L. Siu (eds.) (2007), *Do Economists Make Markets? On the Performativity of Economics*, Princeton, NJ: Princeton University Press.

Moore, J. W. (ed.) (2016), *Anthropocene or Capitalocene? Nature, History, and the Crisis of Capitalism*, Oakland, CA: PM Press.

Muniesa, F. (2011), 'A flank movement in the understanding of valuation', *Sociological Review*, 59(s2): 24–38.

Muniesa, F. (2014), *The Provoked Economy: Economic Reality and the Performative Turn*, London: Routledge.

Muniesa, F. (2015), 'Actor-Network Theory', in J. D. Wright (ed.), *The International Encyclopedia of Social and Behavioral Sciences, 2nd Edition*, Oxford: Elsevier, 80–84.

Muniesa, F. (2016), 'The problem with economics: naturalism, critique and performativity', in I. Boldyrev and F. Svetlova (eds.), *Enacting Dismal Science: New Perspectives on the Performativity of Economics*, London: Palgrave Macmillan, 109–219.

Muniesa, F. (2017), 'On the political vernaculars of value creation', *Science as Culture*, 26(4), 445–454.

Muniesa, F., L. Doganova, H. Ortiz, A. Pina-Stranger, F. Paterson, A. Bourgoin, V. Ehrenstein, P.-A. Juven, D. Pontille, B. Saraç-Lesavre and G. Yon (2017), *Capitalization: A Cultural Guide*, Paris: Presses des Mines.

Ortiz, H. (2013), 'Financial value: economic, moral, political, global', *Hau: Journal of Ethnographic Theory*, 3(1), 64–79.

Ortiz, H. (2014), 'The limits of financial imagination: free investors, efficient markets, and crisis', *American Anthropologist*, 116(1), 38–50.

7

How to use ANT in inventive ways so that its critique will not run out of steam?

Michael Guggenheim

'Critique behaves like blasé tourists who would like to reach the most virgin territories without difficulty, but only if they don't come across any other tourists' (Latour 2013, 85).

It is one of the recurring tropes of ANT to portray itself as nothing but an empiricist endeavour to understand the socio-material word, abstaining from the blasé practices of critique. When I was drawn to ANT as a student in the 1990s, I became quickly puzzled by the strong claims that ANT cannot be delineated, that it is supposed to be a method, or a 'sensibility' but not a theory, while at the same time the proponents of ANT voiced very strong critiques of other approaches (see e.g. Law 2008). Reading ANT texts, it is hard not to conclude that ANT itself is a critical theory of sorts, full of critical judgements about the world. In short, very often, ANT is like a blasé tourist itself.

In this chapter, I take up to question ANT's relationship to critique, not in order to point the finger at ANT, but to explore what ANT as a critical theory can accomplish. I ask with regard to *which* element of the world ANT is (not) critical, how and why it is so, and what the effects of this selective criticality are.

Taking up the quote by Latour cited in the beginning of this chapter, I use the metaphor of travel to do this. Travel is conceived as a mode of exploring the world (in this case the world of social science) and reporting about it, to help others to understand that world. In other words, to ask about critique means to reflexively think about the travel arrangements of ANT – how it relates to sites and practices of study, and how these travel arrangements relate to those of other travel groups, that is, other groups of social scientists. In short, I suggest to look, in good ANT fashion, at the practices (and the practice of rhetorics), not the rhetorics of practice of social research (this chapter deserves to begin with a Bourdieusian rhetorical formula).

The goals of this chapter are twofold. The first is a critique of ANT. ANT practitioners use the same travel arrangements that others use. I will thus suggest that ANT could be less blasé about its own travel arrangements. There is no basis for ANT scholars' tendency to insist on a category split between good ANT travellers and bad tourists for the sake of it; they should rather embrace the fact that all travel arrangements have complex socio-material effects that are difficult to predict and understand.

The second goal is an empirical analysis of the travel arrangements of ANT with the tools of ANT. I reconstruct the ways in which ANT practices critique and why. Given the brevity required, other versions of ANT exist that cannot be covered here.

I will first explain what I mean by critique. Then, I will detail why ANT has such a dim view of critique and finally I will introduce four different kinds of critique within ANT. ANT as a critique of natural science, ANT as a non-critique of design, ANT as a critique of theories of society and finally ANT as a speculative critique of social practices.[1]

What is critique: a primer on travel infrastructures

The main problem of ANT seems to be that its routinised critique of critique begins with a naïve understanding of the specific properties of various kinds of travel. Like critical theory itself, it assumes that we can read from the travel devices, whether something is critique or not.

To understand the problem of critique, let me first elaborate how to conceive of the practice of critique from an ANT perspective. The most important notion of critique is Marx' 'the self-clarification of the struggles and wishes of the age' (Marx 1975, 209). For ANT, this is problematic, as it does not locate any agency. Luc Boltanski has helpfully summarised the notion of critique in a more compatible language: Critique is based on 'critical judgements on the social order which the analyst assumes responsibility for in her own name, thus abandoning any pretention to neutrality' (Boltanski 2013, 4). The 'self-clarification' that Marx hints at is an actor called 'analyst.' As ANT scholars, we might add that it is not just an analyst, but a network that takes responsibility. We could thus say that critique is an arrangement that clarifies struggles by taking responsibility in the name of that arrangement, and thus abandons any pretension to neutrality.

For Boltanski, there are two additional elements that matter: First, because critique is rooted in an *empirical* description of a problem, it cannot make recourse to 'spiritual or moral resources of a local character' (Boltanski 2013, 5). Rather, it needs to engage in a critique of the social order, not just some local aspect of it (saying that the sea is dirty is not critique; saying that plastic in the sea is a symptom of industrialisation is). None of this hints at an abstract definition of critique as blasé tourism.

Second, critique is reflexive. It needs to bring critical practice to the actors. It must 'grasp the discontent of actors' and 'explicitly consider them the very labour of theorisation, in such a way as to alter their relationship to social reality' (Boltanski 2013, 5). From an ANT point of view, we could say that critique brings back the outcomes of scholarly research – the new world with its new connections – to those who are affected by a problem. Telling people to clean the sea up is not critique, making the consumption of plastics difficult so that no more plastic is produced is. Again, this does not sound like blasé tourism, but rather like a complex job that requires lining up many socio-material elements.

What confuses ANT and critical theorists alike are two things. First, some seem to think that only a specific set of terms such as 'capital accumulation, class, property relations, land rent, exploitation, ...' can amount to critique (Brenner, Madden, and Wachsmuth 2011, 230). Similarly, Fortun complains that the AIME platform (An Inquiry into the Modes of Existence (MoE)) has no entries for terms like 'asbestos,' 'disaster' or 'petrochemicals' and therefore is not up to the job of critique (Fortun 2014, 317). But as Boltanski's work with Thévenot has shown, there are many different linguistic and material repertoires for critique, and there is no reason to restrict critique to a particular set of terms (Boltanski and Thévenot 1991).

Second, the notion of critique is overtly intentional: Critique is when someone wants to engage in acts of critique. Yet, only a cursory glance at the world shows a more complex picture. Some actors engage in what they think are acts of critique, but these are roundly ignored and fail *as critique*, while other actors engage in practices they do not conceive of as critique, while they are understood and reacted to as if they were critique (Vikkelsø 2007). Critique, then, is a situated achievement rather than an intentional act. Equipped with these preliminaries, we can have a look at the practices of ANT and how they work as critique, and what their particular travel arrangements are.

Criticising lazy travel modes by taking a budget flight: the problem of loose translations

If we allow for critique to be a practice that can occur in different forms and vocabularies and if we try to see what is understood in the world as critique, then we can easily see how ANT sometimes is a form of critique. ANT, so to speak, is like a tourist on a package holiday who moans about how crowded the Costa Brava is.

ANT denounces the existing practice of all other social science as wrong, because it omits the socio-material constructedness of the world from social theory (see, for example, Callon and Latour 1992; Latour and Woolgar 1979). More specifically, the failure of critical theory is to misunderstand the hidden, powerful 'macro-actors' as hidden agents instead of as socio-material networks (see the quote at the beginning of this chapter and Callon and Latour 1981).

This critique of critique is obviously a form of critique itself: These texts do not empirically engage with the research practices of other critical social scientists. Similarly, the generic claim of ANT that *all* other social research does not account for material agency operates in a similar, Thomas-Cookish, way.

But why are many ANT practitioners routinely opposed to critique? Judging all travel for not being culturally and ecologically sensitive is in itself a facile move that misunderstands the cosmopolitics of travel. This obviously gets at the core of all problems of critique: How much should we denounce those in a particular situation for being in that situation, or how much should we understand their situation as being a necessity, a practice or a position in a field? In what sense is not trying to *understand* a particular position already a denunciation?

At its core is the following problem: ANT (rhetorically) dislikes what I have called loose translations (Guggenheim 2015). Loose translations are translation steps that make big jumps, for example, from one medium to another, or from one kind of argument to another. But for ANT, to strengthen a network means to minimise translation steps. The problem of critique is that it is based on a particularly challenging loose translation, namely the one from description to how the world could be changed, or from is to ought. For example, the notion of 'capitalism' operates as a loose translation that jumps from a complex socio-material arrangement to a single *problematic* whole. Latour identifies such loose translations with blaming 'powerful agents hidden in the dark acting always consistently, continuously, relentlessly' (Latour 2004, 229).

This loose translation is an easy target, because ANT can always point to the gap between capitalism as empirical multiplicity and problematic whole and claim that the gap is unaccounted for. But doing so is a result of its own empiricist image: It cannot imagine that a theory, qua being a theory, can make translations that do not start at the object but at the theory. ANT must always assume that critique is the worst form of laziness, of trying to insert hidden theoretical 'powerful agents' as the cause for the way the world is, and the reason that the analyst cannot change the world.

There is a theoretical problem here: For ANT (social) science is always translation and representation. Although it is constructivist in its *self-description*, the constructivism is ultimately representational. There is a gap between the empirical claim when observing scientists, that they 'construct' and thereby create a world, which always contains a normative and political element, and ANT's self-description as an empirical practice that is merely descriptive. The making of the world for ANT always starts with something out there, never with something a social scientist would like to achieve.

In AIME (Latour 2013), Latour offers other ways to deal with the world; yet, these are disconnected from (social) science: 'fiction' is modelled on an idea of art that is unhinged from problems of the world, or 'morals' is based on moral reasoning (Latour 2013). But social science as critique could be imagined as a mode that combines 'fiction,' 'representation' and 'morals,' though it is neither of the three (I will later call this 'speculation'). Such a social science would be precisely what focuses on the construction of the world in a triangle of what we want to achieve, a given world and the tools we have at our disposal to do so. Before we can explore this further, let me move to another basic versions of ANT as critique.

Ecotourism with unintended consequences: ANT as critique of natural science

Many ANT texts, beginning with 'Laboratory Life' (Latour and Woolgar 1979), manage to embark at the same time on a budget airline trip and an ecotouristic adventure (as indeed, many tourists routinely do). In the early laboratory studies, ANT tried to tread as carefully as possible to not damage the local ecosystem. These studies claim to move away from a normative philosophy of science towards an empirical analysis of science as practice. By doing so, they only want the best for the locals: A more truthful account of laboratory work. Yet, the whole situation descends into acrimony. The natural scientists see themselves misrepresented. They prefer the previous version in which they appear as heroes in search for eternal truth, rather than as muddling, politicking actors. They also claim that they better know what they are doing than their observers: Witness what came to be called the science wars (Gieryn 1999; Jurdant 1998; Sokal and Bricmont 2004).

In this version of critique, ANT operates in a mode of description. It claims that it only describes, that it is not a theory of the world, and that it does not critique natural science. Indeed, Latour always makes the point that he is a 'lover' of natural science.[2] Curiously, Latour misrecognises the situation when he writes:

> I have rarely heard critiques of the descriptions that 'science studies' has given of scientific networks (on the contrary, the veracity of these descriptions has always been recognized [...]). And yet the alternative versions [we] have proposed ... have been hotly contested by some of the very researchers whose values we were trying to make comprehensible ... . The very words 'network' and 'fabrication' are sometimes enough to shock our interlocutors [...]. What poor diplomats we have been!
>
> *(Latour 2013, 12)*

For Latour, it is absolutely crucial to keep apart 'the accounts the Moderns have invented [...]; the values they have held to during this same history [...]; [and] finally, my own formulation [...] of this same experience' (Latour 2013, 12).

How is it possible that a description by a lover operates as critique of his object of love? The description clashes with the self-descriptions of the objects because these self-descriptions are informed by *other theories* of what these actors are doing. The scientists themselves described their own practice with a realist and positivist theory of science.

The very idea that descriptions and critique can easily be kept apart ironically follows a logic of differentiation that ANT has set out to undo. Apart from the fact that many STS descriptions were indeed hotly contested by scientists (see e.g. Labinger 1995), the language of 'poor diplomacy' only shows how little Latour understands his own travel arrangements. He insists on keeping descriptions and values apart, because only doing so allows for the assumption that the objects of love understand that their own accounts are wrong and those of ANT are better.

Ironically, during the science wars, the opponents of ANT and STS did not bother to keep critical accounts, say the feminist versions of Donna Haraway and Evelyn Fox Keller, apart from those of the purported lovers. The opponents even failed to spot that feminist scholars explicitly criticised ANT for their supposedly uncritical accounts. What mattered for the scientists was that they felt *misrepresented*.

To sum up, what critique is, and whether it succeeds in the unavoidably reflexive encounters with its object, is demonstrably not a quality of the intentions, nor the labelling of a scientific theory or approach, but of the *encounter* with its object. ANT practitioners in this second version are like tourists who buy eco-friendly trips to animal reserves and are surprised if the locals who live everyday with the animals neither approve of the photographs that are taken of them, nor of the advice to let the animals roam free.

The spa holiday: ANT as a (non-)critique of architecture and design

While the relationship with science has been strained, ANT has had much better relationships with other objects of inquiry, most notably architecture and design (for example, Latour has written a regular column for Domus, one of the main architecture magazines, and architects such as Alejandro Zaero Polo, Andres Jaque or Nerea Calvillo have embraced it). The identical practice of describing socio-material practices has resulted in the case of science in the vigorous objection of those described and in the case of design in an enthusiastic embrace.

The travel destination matters to how the relations with the locals pan out. The problem is not necessarily the mode of travel, but how the travel mode fits local infrastructures. Spa holidays are increasingly popular (I like them too): You don't venture very far, but do it with the least amount of trouble: Everyone is nice to you. Afterwards, the spa hotel publishes your praise on their website and your words look very much like theirs.

Structurally, this second version operates very much like the first: It uses the description of a practice to formulate a critique of social science. The empirical description of building and design practices serves to critique other social science as ignoring the ongoing recomposition of socio-material arrangements at the expense of semiotic or structural analysis. The difference between science and design is that in the case of the latter, the attribution of agency to designers and the notion of 'fabrication' and 'construction' to describe their work perfectly fit their self-description. This is unlike scientists, who abhor the idea that objects in the world are connected to their own agency. When Latour alleges that 'when we try to reconnect scientific objects with […]their web of associations, […] we always appear to *weaken* them, not to *strengthen* their claim to reality' (Latour 2004, 237), then this only holds for *scientific* objects. For *design* objects, the opposite is true, as can be gathered from any architect's description of her project. There are thus no critical effects of ANT descriptions of architecture. The reason is not that ANT practitioners fail to be critical, but because the very same methods and theories that have critical effects in one field fail to do so in a different field. ANT scholars end up on spa holidays not because they chose that destination, but because their hosts like their travel accounts.

Executive business travel: ANT as modes of existence

> Our flight must take place above the clouds, and we must reckon with a rather thick cloud cover. We must rely on our instruments. Occasionally, we may catch glimpses below of a land … that remind us of something familiar, or glimpses … of landscape with the extinct volcanoes of Marxism. But no one should fall victim to the illusion that these few points of reference are sufficient to guide our flight.
>
> *(Luhmann 1995, l)*

Even ANT practitioners sometimes feel the need to fly above the clouds. Getting somewhere far away means that we cannot stop in every village and try to understand how it is different from the one next to it. Hence, our reports afterwards are more concerned with talking about the quality of different airlines and whether the connections at the airports worked, rather than the actual destinations. Such business travel needs to be taken seriously as a practice, even though most travellers are routinely dismissive of it as a second-hand experience of non-spaces. Bruno Latour, in particular, has, in his second career as 'philosopher' (Latour 2010),[3] made a business of exploring new airlines and rarely visited airports. Like Luhmann, he prefers to bypass the extinct volcanoes.

ANT as a theory of MoE is a full-on critique of 'modernity' (Latour 1993, 2013). Modernity here means the way we – which is everyone who uses a *vocabulary* that refers to items such as 'science,' 'politics,' the 'environment,' etc. – understand our world.

For Latour, the fundamental mistake of modernity is not that it has destroyed the world or brings injustice (as critical theorists would have it), but that the way we conceive of how anything comes into existence is misguided. As with Luhmann, the main point for Latour is that by flying high enough, we learn to avoid prioritising one mode, learn to give justice to each mode and 'learn to respect appearances,' based on the central notion of 'felicity condition': This is the idea that there are specific ways how each mode operates and that it is fundamentally futile to overextend this mode as long as we operate within it.

'Modes of existence' (MoE) represent thus a large-scale critique of all other ways of understanding and critiquing modernity that fail to understand the operation of each mode of existence. Hence, Latour's self-designed tests of whether the theory of MoE succeeds are based on how it respects each mode:

> does the detection of one mode allow us to respect the other modes? … Can the inquiry mutate into a diplomatic arrangement … while a new space is opened up for comparative anthropology by a series of negotiations over values?
>
> *(Latour 2013, 475)*

The enemies of all these tests are theories that are attached to a single mode, such as the existing philosophy of science, theology, sociology of culture, etc. According to Latour, these all fail, first, because they are not based on ethnographic inquiry and second because they reduce it to some element that does not correspond to the self-description of the field.

The notion of 'ideology' in critical theory describes something very similar namely, the fact that the self-description of the world by existing mode-specific theories is wrong.

69

For 'critical' theory, the wrongness of ideology derives from the fact that actors try to obscure the 'real' working of these modes in order to exert power over those who do not see behind ideology. Ideology justifies a practice that is wrong.

For Latour, there is nothing wrong with practices, only with descriptions of practices.[4] Thus, the descriptions of Marxists and economists, of philosophers of science and Marxist critics of science, etc., are equally wrong, because they do not match the empirical complexity of the world. One irony of this approach is that it forgets that philosophy (of science, the economy, etc.) is also a *practice*. It is not clear why it should not also be taken seriously as empirical phenomenon and thus given the status of being 'right' like any other practice.

One important question then is why practices should always be right, and never objects of critique, while theoretical descriptions should be critiqued and are almost all wrong. Crucially, here Latour's critique mirrors the Marxist critique: The *standard* for critique is pregiven by the theory: In the case of Marxists, ideology is always with the others, who are deemed to veil the capitalist system. In the case of MoE, it is always the theories that veil the logic of MoE. Another irony then is why a theory that purports to take local practices seriously does not allow itself to suggest how such local practices could be critiqued, changed and improved? Taking practices seriously in their empirical complexity surely entails to suggest that some practices are better than others, once we agree on what these practices are supposed to do? A more refined programme, and one probably far more in line with the intention of the author, would be to ask how and why descriptions match or do not match practices?

The exchange student: ANT as incubation and speculation

When you are young, you want to go abroad to learn, and you stay with a family whose internal dynamics will become equally perturbed as your own life by this experimental cohabitation in the name of cultural exchange. In ANT, there are a number of projects and practices that come variously under the rubric speculation (Wilkie, Rosengarten, and Savransky 2016), inventiveness (Lury and Wakeford 2011) or, our own, incubation (Guggenheim, Kräftner, and Kröll 2017).

There are important differences between these approaches, but they all begin with the typical focus of ANT approaches on socio-material entities, 'a criticality that is oriented towards tracing the complex messy entanglements of societies with all their strange, weird and wonderful hybrid objects' (Ward and Wilkie, 2009, 2). But this empirical outlook is not guided towards a critique of theories, but towards an experimental *modification* of practices with the help of various devices. Unlike in critical social science, speculative practitioners do not denounce practices, nor do they try to change practices according to a given theoretical standard. Rather, inventive or speculative *methods* are employed to change the world in unforeseen directions. They begin with the insight that in moments of controversy, the social world is performatively opened up. This 'moment of dispute' is also recognised by Boltanski as a possible moment of critique, 'when actors express their moral claims … they 'perform' the social in an innovative way' (Boltanski 2013, 12).

But rather than simply analysing such moments, inventive methods aim to *create* such situations. Like the student on a gap year, they jump head-on into an unforeseen situation. Many of these methods derive not only from social science, but take their leads from scientific experiments, design or arts practices. The reason for these choices is that the latter have a history of self-consciously interfering with socio-material devices into the world. ANT could have learned this from its own analysis of such practices, but too often this has not happened.

Such inventive methods become tools of critique, because they allow ANT practitioners to intervene in practices by giving actors in the field the resources at hand to clarify and change their situation. But at the same time, this does not happen with a participatory naivety, in which the social scientists already know what is wrong, what should be achieved or even how anything could be achieved. Rather, a methodological, theoretical and material repertoire is employed in unforeseen ways. Just like the effects of having an exchange student in your house, such work is highly risky, and may be contested by the locals. It does not necessarily play to easy forms of 'participation' in which local practices are taken as standards. Rather, it brings strategic devices that are strange to the locals to create new situations that have not existed before.

Returning home

After this extensive tour of travel modes, we can now answer the question 'What about critique?' in a rather straightforward way. First, as I have shown, ANT should be a little bit less blasé about its own travel arrangements. There are reasons why people take cheap flights and ANT practitioners sometimes do so too, for the same reasons. Sometimes, the goal is to quickly get somewhere, as we have seen in the version of ANT as critique of social science and MoE. Sometimes, ecotourism inadvertently turns into a mess, as we have seen in the attempts of ANT to be sympathetic, descriptive and non-critical towards science. Critique, in these cases, is not a matter of intention, but of reception. Finally, ANT can also be a practice of creating situations in which actors can open up and recompose the world. This mode though forces us to give up both the radical empiricism of ANT and the stance that we know how to change the world. We then need to engage in a radical encounter with the world in which we accept and embrace our practice as intervention, without knowing where it will end up.

Notes

1 This basic logic of this chapter shares many similarities with the chapters by Daniel López and José Ossandon in this volume, though each of us focuses on a different analytical plane.
2 Latour's first German book is subtitled 'investigations of a lover of the sciences' (Latour 1996).
3 It is notable that in Latour's self-understanding as an Anglo-French STS/sociologist/anthropologist, business travel as writing the 'Modes of Existence' equals philosophy. For Luhmann, writing in the provinces of Germany, flying above the clouds is easily compatible with sociology.
4 Muniesa follows this route when considering ANT as critique of capital in this volume.

References

Boltanski, Luc. 2013. *On Critique: A Sociology of Emancipation*. Oxford: Wiley.
Boltanski, Luc, and Laurent Thévenot. 1991. *De La Justification. Les Economies de La Grandeur*. Paris: Gallimard.
Brenner, Neil, David J. Madden, and David Wachsmuth. 2011. 'Assemblage Urbanism and the Challenges of Critical Urban Theory'. *City* 15 (2): 225–40. doi: 10.1080/13604813.2011.568717.
Callon, Michel, and Bruno Latour. 1981. 'Unscrewing the Big Leviathan: How Actors Macrostructure Reality and How Sociologists Help Them to Do So'. In *Advances in Social Theory and Methodology: Toward an Integration of Micro- and Macro-Sociologies*, edited by Karin Knorr-Cetina and Aron Cicourel, 277–303. Boston: Routledge.
———. 1992. 'Don't Throw the Baby Out With the Bath School! A Reply to Collins and Yearley'. In *Science as Practice and Culture*, edited by Andrew Pickering, 343–68. Chicago, IL: Chicago University Press.

Fortun, Kim. 2014. 'From Latour to Late Industrialism'. *HAU: Journal of Ethnographic Theory* 4 (1): 309–329.

Gieryn, Thomas. 1999. 'Home to Roost: Science Wars as Boundary Work'. In *Cultural Boundaries of Science. Credibility on the Line*, edited by Thomas Gieryn, 336–62. Chicago, IL: Chicago University Press.

Guggenheim, Michael. 2015. 'The Media of Sociology: Tight or Loose Translations?' *British Journal of Sociology* 67 (June): 345–72.

Guggenheim, Michael, Bernd Kräftner, and Judith Kröll. 2017. 'Incubations: Inventing Preventive Assemblages'. In *Inventing the Social*, edited by Noortje Marres, Michael Guggenheim, and Alex Wilkie, 65–92. Manchester: Mattering Press.

Jurdant, Baudouin, ed. 1998. *Impostures Scientifiques. Les Malentendus de l'Affaire Sokal*. Paris: La Découverte.

Labinger, Jay A. 1995. 'Science as Culture. A View from the Petri Dish'. *Social Studies of Science* 25: 285–306.

Latour, Bruno. 1993. *We Have Never Been Modern*. Cambridge, MA: Harvard University Press.

———. 1996. *Der Berliner Schlüssel. Erkundungen Eines Liebhabers Der Wissenschaften*. Berlin: Akademie.

———. 2004. 'Why Has Critique Run out of Steam? From Matters of Fact to Matters of Concern'. *Critical Inquiry* 30 (Winter): 226–48.

———. 2010. 'Coming Out as a Philosopher'. *Social Studies of Science*, 599–608. doi: 10.1177/0306312710367697.

———. 2013. *An Inquiry into Modes of Existence: An Anthropology of the Moderns*. Cambridge MA: Harvard University Press.

Latour, Bruno, and Steve Woolgar. 1979. *Laboratory Life. The Social Construction of Scientific Facts*. London/Thousand Oaks/New Delhi: Sage.

Law, John. 2008. 'On Sociology and STS'. *The Sociological Review* 56 (4): 623–49. doi: 10.1111/j.1467-954X.2008.00808.x.

Luhmann, Niklas. 1995. *Social Systems. Writing Science*. Stanford, CA: Stanford University Press.

Lury, Celia, and Nina Wakeford, eds. 2011. *Inventive Methods : The Happening of the Social*. London: Routledge.

Marx, Karl. 1975. 'Letter to A. Ruge, September 1843'. In *Early Writings*, translated by Rodney Lingstone and Gregor Benton, 209. New York: Vintage.

Sokal, Alan D, and J Bricmont. 2004. *Intellectual Impostures: Postmodern Philosophers' Abuse of Science*. London: Profile Books.

Vikkelsø, Signe. 2007. 'Description as Intervention: Engagement and Resistance in Actor-Network Analyses.' *Science as Culture* 16 (3): 297–309. doi: 10.1080/09505430701568701.

Ward, Matt, and Alex Wilkie. 2009. 'Made in Critical land: Designing Matters of Concern'. In Networks of Design: Proceedings of the 2008 Annual International Conference of the Design History Society (UK) University College Falmouth, 3–6 September. Universal-Publishers.

Wilkie, Alex, Marsha Rosengarten, and Martin Savransky, eds. 2016. *Speculative Research: The Lure of Possible Futures*. London: Routledge.

Section 2

Engaging dialogues with key intellectual companions

Anders Blok, Ignacio Farías and Celia Roberts

While authors affiliating themselves with actor-network theory are generally wary of origin stories, few would presumably question the notion that ANT itself is the product of many intellectual sources and inspirations brought and wrought creatively together. Overall, such flows tend to generate a pattern of repetition and difference, in which certain figures and tropes from 20th-century Euro-American intellectual history stand out more clearly than do others (and yes, for better or worse, they do tend to be Euro-American!). Eventually, any serious engagement with ANT will have to take note of such a patterned hinterland of ideas.

For this section, we invited authors to critically explore and discuss some of the key intellectual sources of ANT, by paying attention both to well-acknowledged (but possibly less well understood) and to silenced, forgotten or potential sources that can still enrich ANT as an intellectual practice. Quite evidently, the dialogues with key intellectual companions staged across the sections' chapters represent only a rough selection of such potential sources, and some well-known figures indeed suggest themselves as missing from the selection (one could mention, for instance, the two Michel's, Foucault and Serres). That said, we believe that the selection pertains to a number of encounters already wedded into core ANT concepts and dialogues, as well as points to roads less travelled, and hence further options for future collective exploration.

The section opens with what is arguably a well-known, yet surprisingly under-appreciated intellectual encounter: The one between first-wave ANTs, Bruno Latour in particular, and French poststructuralist philosopher Gilles Deleuze. As Casper Bruun Jensen details in his chapter, the famous quip from Michael Lynch that 'ANT' would have been better named 'Actant-Rhizome Ontology' (ARO!), and hence exhibited its debt to Deleuze's rhizomes up front, indeed contains more than a fleeting truth to it. ANT, like Deleuze, exhibits ontological ambitions. Contrary to Lynch's suggestion, however, Jensen seeks not only to affirm this generative connection, but also to suggest ways in which further doses of Deleuzian sensitivity might move ANT in more productive directions, not least when it comes to its own politics of inquiry.

Taking a clue from the remaining part of Lynch's suggestion, that of the actant, Alvise Mattozzi's chapter distils an answer to the deceptively simple question of what ANT can still learn from semiotics. As is well known, early ('French') ANT in the 1970s learned much from the toolbox of semiotics as the study of textual functions, including from Algirdas J. Greimas' so-called actant model of narration. The concept of actants, indeed, remains key to the signature ANT move of distributing agency ('symmetrically') amongst humans and non-humans. Yet, as Mattozzi is keenly aware, the deceptive bit lies in the 'still': To him, this small word serves to open up a wide-ranging inquiry that takes us not only backwards, into the history of semiotics, but also forward, into a potential future of a more semiotics-aware ANT.

This exploration is continued, with some importantly different inflections, in the chapter by Jérôme D. Pontille on what the ANT collective has forgotten about its roots in the anthropology of writing. Unlike (classical) semiotics, writing is taken here in its most material definition: As any act of inscribing, scribbling, tracing and drawing whereby not just science attains its cognitive stabilisation (in the 'immutable mobiles' studied by ANT), but which make possible all forms of state and other organised governing of sociotechnical assemblages. Placing ANT in the wider landscape of the anthropology of writing staked out, amongst others, by Jack Goody since the 1960s, Jérôme D. Pontille (the name itself, we note, a textual invention) argues the value of reasserting such a legacy in the present age of information automatisation, which threatens to render mundane documentary work invisible and strengthen neopositivist imaginaries.

Picking up the theme of conceptual un- and redoing, Noortje Marres' chapter explores overlaps and tensions between ANT and the American-pragmatist lineage in philosophy, by way of asking what the former might learn from the latter in attending to its present challenges. Foremost amongst these, Marres suggests, is a series of 'returns,' to themes of interpretation, society, epistemology and critique, which ANT scholars may have too smugly believed themselves to have already surpassed. Here, Marres takes inspiration from the pragmatist-inflected notion that ANT's intellectual project might be less about deflation, and more about the reconstruction of how to think about the relation between the problems of philosophy and the problems of the world we inhabit together. In saying this, however, she also behooves us to note all the various ways in which the intellectual climate of today's world differs from that of the pragmatists'.

In the following two chapters, authors stage their intellectual encounters on the specific territory of the body, thus marking out one important direction in which further work in the ANT vein lies ahead. First up, Ericka Johnson asks why ANT needs feminist scholar of technoscience Donna Haraway for thinking about (gendered) bodies; and she answers in part by drawing on her own research experiences in the (hyper-gendered) world of medical–gynaecological expertise, technologies and patient-directed interventions. Johnson details three conceptual gifts from Haraway, which have been taken up widely in the ANT collective: Those of the modest witness, the cyborg and the material-semiotic apparatus of bodily production. Relative to the otherwise kindred conceptual and methodological domains of ANT, she argues, Haraway allows for attending to a broader cultural–political arena of power structures writ large, including in her consistent problematisation of the very notions of gender, bodies and biologies.

Michael Schillmeier, in turn, proceeds to think with and about a different kind of bodies, those suffering from dementia, as well as with a different intellectual companion to ANT, the early 20th-century British philosopher Alfred N. Whitehead. Here as well, Whitehead's processual thinking is well acknowledged as a precursor in the ANT canon; yet, Schillmeier's core claim is that ANT scholars have far from exhausted Whiteheadian resources when it comes to thinking about processual and immaterial aspects of experience. Indeed, he claims, present-day ANT provides only limited conceptual tools in understanding the everyday realities, issues, concerns and problems of dementia. Conversely, he suggests, thinking 'cosmo-politically' about the event of dementia, in ways resonant with key Whiteheadian themes, provides an avenue for seriously extending and reassembling key ANT tenets on the questionability of the social.

In the section's final chapter, Martin Savransky brings us full circle, in a certain sense, by engaging with one of ANT's key intellectual interlocutors who is herself in equal measure a creative heir to the philosophies of Gilles Deleuze, pragmatism, and Alfred N. Whitehead: The contemporary Belgian philosopher of science Isabelle Stengers. Long recognised in the Anglophone world as Bruno Latour's close philosophical ally, Savransky is adamant to dramatise the relation of ANT to Stengers differently, in a way that points to hitherto-unacknowledged divergences. Such divergence, he suggests, pertains to the very heart of ANT and its principle of 'generalised symmetry' which, when read via Stengers' ecologies of practice, is finally allowed to become fully situated in its methodological and political import. As such, and fittingly for our overall ambition in this section, Savransky leaves us with a future-oriented challenge: How will ANT change and mutate in the future, once the divergent path held out by Stengers comes to be seriously explored – beyond any simplistic sense of intellectual 'enemies' and 'allies'?

8

Is actant-rhizome ontology a more appropriate term for ANT?

Casper Bruun Jensen

Latour mentions actant-rhizome ontology only once, in his contribution to *ANT and After* (Latour 1999a). There, he recalls actor-network theory by reminding readers about the concerns that led to its development. But he also plays with recalling ANT in the sense of taking it back, like a flawed product. Thus, he first recalls the actor, then the network and then theory. It is at this point that the strange term actant-rhizome ontology appears:

> The third nail in the coffin is the word theory. As Mike Lynch said some time ago, ANT should really be called 'actant-rhizome ontology'. But who would have cared for such a horrible mouthful of words - not to mention the acronym ARO? Yet, Lynch has a point. If it is a theory, of what it is a theory? It was never a theory of what the social is made of, contrary to the reading of many sociologists who believed it was one more school trying to explain the behaviour of social actors.
>
> *(1999a: 19)*

In a gesture often repeated, Latour insists that, 'faithful to the insights of ethnomethodology,' ANT aims only to learn from the actors what they do, how and why. And he seems to agree that actant-rhizome ontology is in some sense a more appropriate name than actor-network theory. Only the word is too ugly, a 'horrible mouthful,' which would never have caught on.[1]

Indeed, the ethnomethodologist Michael Lynch, who coined the notion of ARO did not intend it as a compliment. It appeared in his review of Geoffrey Bowker's (1994) *Science on the Run* that was in equal measure a series of complaints lodged against ANT. While Bowker was praised for his 'matter of fact presentations,' his ANT colleagues were said to 'engage in a kind of conceptual art long on interpretation and short on detail' (Lynch 1995: 167). ANT, Lynch continued, 'licenses vaguely Machiavellian stories of how innovative persons and agencies manage to establish global networks' (168). If these stories had been advertised fairly as '"actant-rhizome ontology," the theory would of course never have caught on in the English-speaking world.' But, alas, the non-event was not to be and ANT did catch on.

Beneath the cosmetic agreement that ARO is a more accurate descriptor than ANT, there is thus profound disagreement. Lynch finds silly the idea of extending agency to non-human actants. He evidently views the notion of the rhizome as loose and metaphoric. And talking of ontology is, for him, a way of relegating ANT to the realm of the metaphysical. The only real point of agreement is that calling ANT ARO would have been a PR disaster.

Of course, Lynch was right that ANT is not a conventional social theory. And if it is not a singular ontology either, it is at least a strategy for tracing, articulating and adding to the practical ontologies that make up the world (Gad and Jensen 2010). Yet, even though ANT is unlikely to be renamed, it seems worth exploring the points of intersection and divergence between the famous actor-networks and the Deleuzian rhizomes that inspired Lynch's ironic suggestion (see also Jensen and Rödje 2010). Central will be the question of how to extend ANT's analytical and political sensitivities beyond the immediacy of material connections. We will have to learn, with Deleuze (1994: 220), to move 'from science to dreams and back.'

Of the three terms that make up ARO, actant is hardly problematic (see also Mattozzi this volume) and Latour's intermittent use of ontology has already been subject to commentary (Jensen 2017). However, the relation between Deleuzian rhizomes and the network remains largely unexplored. Accordingly, much of the following will focus there.

Follow the references!

Follow the actors! Given the centrality of this slogan for ANT, it seems only fitting to start by tracing what Latour, the actor, has said about and done with Deleuze.[2]

Following the references, however, is a less than satisfying task because Latour's engagement with Deleuze, as with most of his intellectual inspirations, is quite sporadic. Even though Latour has referred to Deleuze as the greatest French philosopher (alongside Michel Serres) (in Crawford 1993: 262), references appear only intermittently and usually without much commentary (e.g. Latour 2014: 15; 2016a, 2016b: 312).

Though touched upon only casually, however, certain ideas are clearly resonant. Thus, for example, the invocation of a '"composite" that "has to hold together on its own," as Deleuze and Guattari would say,' though appearing only briefly (Latour 2013: 242) is not difficult to connect with the crucial notion of hybrid networks that gain reality through processes of stabilisation. Similarly, Latour's (2005: 95n119) approval of Deleuze's description of relativism as 'not the relativity of truth but the truth of relation' seems more than coincidental, since it aligns with his own effort to simultaneously avoid the connotation of constructivism with a kind of social or ideologically induced 'consensual hallucination' and the fundamentalism of believing in truths that escape the networks that articulate them (Latour 2000).

One of the topics to which Latour returns in his scattered Deleuzian remarks is the relation between difference and repetition (Deleuze 1994).[3] Thus, Latour and Lépinay (2009: 39) wrote that Gabriel Tarde's theory conforms to the fundamental principle of Deleuze's difference and repetition according to which 'invention produces differences; repetition allows for their diffusion.' Depicted as the fundamental Deleuzo-Tardean rhythm of social activity, this is also, of course, the fundamental rhythm of Latour's networks.

The network bore early imprints of the rhizome, too. In a mid-1990s article dedicated to 'clarifying' the many misunderstandings proliferating about actor-networks already then, Latour (1996b: 49) stated that:

> Literally, a network has no outside. It is not a foreground over a background, nor a crack onto a solid soil, it is like Deleuze's lightning rod that creates by the same stroke the background and the foreground... instead of surfaces one gets filaments (or rhizomes in Deleuze's parlance).

Ten years later, Latour (2005: 129) paused to comment that he was indebted to 'a very special brand of active and distributed materialism of which Deleuze, through Bergson, is the most recent representative.'

It would thus appear that networks and rhizomes are at least close relatives. But can we be so sure? In fact, does sifting through the brief Latourian comments on Deleuze make us any wiser as to their specific import? It seems that we will have to look further than Latour's own specifications.

A nearly total affinity?

The introduction to *A Thousand Plateaus* is organised around several principles of rhizomatic analysis. There is a principle of connection and heterogeneity: 'any point of a rhizome can be connected to any other, and must be' (Deleuze and Guattari 1987: 7). According to the principle of multiplicity, the multiple must be 'treated as a substantive,' (8) and decoupled from 'any relation to the One as subject or object, natural or spiritual reality, image and world.' The principle of asignifying rupture states that: 'a rhizome may be broken, shattered at a given point, but it will start up again on one of its old lines, or on new lines' (9). Each of these principles is readily discernible in actor-network theory, which, as Latour (1999a: 15) emphasised, was indeed intended to convey an image 'like Deleuze's and Guattari's term rhizome… a series of transformations – translations, transductions – which could not be captured by any of the traditional terms of social theory.'

Henning Schmidgen (2015: 23) notes that Deleuzian themes were important to Latour from early on. Working under the anthropologist Marc Augé in the Ivory Coast, he analysed development projects using concepts from *Anti-Oedipus* (Deleuze and Guattari 1983) and found inspiration in its transversal theory of bodies and technology, which saw people as component parts in machinic assemblages (Schmidgen 2015: 103), the latter later turning into heterogeneous networks (Latour 1987). Schmidgen (2015: 133) further observes that both Latour and Deleuze broach the question of what actors and society consist of and what holds them together with an attitude of 'radical openness.'[4]

There are further resonances. In *The Logic of Sense*, Deleuze (1990a: 118) wrote that individuals are 'infinite analytic propositions. But while they are infinite with respect to what they express, they are finite with respect to their clear expression, with respect to their corporeal zone of expression.' In this dense formulation, we find an antecedent of Latour's (1988) principle of irreduction according to which actors are indeed also 'infinite propositions,' though at any given moment they do have a specifiable 'zone of expression.' We also find a precursor to the ANT critique of analyses that assume given distinctions between micro- and macro-scales, putting in place instead an analytics of scale making. Similar to the Deleuzian formulation, this entails that one can consider an actor as either a relatively stabilised entity that relates to others, or as a network composed of changing relations. Moreover, Latour (1999b: 287) would write of the world's composition of 'many different practical ontologies' (see also Jensen 2017), much as Deleuze (1990a: 125) had depicted the 'frontier established' by the interactions of heterogeneous forces as a 'metaphysical surface.'

The network as a pattern of interacting forces: The image is distinctly Nietzschean (Lee and Brown 1994, Jensen and Selinger 2003, Schmidgen 2015: 6). Deleuze and Latour, however, grapple with their Nietzschean legacies in markedly different ways. While Deleuze (1983: 68–73) was adamant that Nietzsche's thought has nothing to do with the caricature that might make right, Latour – no doubt due to the recurrent critique of ANT for its interest in powerful actors, shrewd scientists and scheming politicians – is at pains to distance himself from Nietzsche. In *Pandora's Hope* (1999b: 216–235), he argued that Socrates and his opponent, the proto-Nietzschean sophist Callicles, agreed on almost everything, and specifically on an elitist conception of politics, which Latour aimed to bypass (Jensen and Selinger 2003: 202–9, cf. Deleuze 1983: 58–59).

Despite this bit of exorcism, Latour's analyses of bodies and their powers remain strikingly resonant with Deleuze's Nietzsche and Spinoza. Thus, Deleuze's (1990b) discussion of Spinoza revolves around bodily encounters just as Latour (2004) addresses how actors learn to be affected by others, and Latour's notion of well-articulated facts (1999b, 2000) is reminiscent of the evaluative principle of affirmative and reactive forces that Deleuze extracts from Nietzsche.

Only well-articulated if they are kept alive to many concerns and firmly embedded in networks, facts exhibit not the relativity of truth but the truth of relation, as Deleuze had written. Furthermore, the principle of links and knots that became central to Latour's (2004a) move from matters of facts to matters of concern also has a Deleuzian flavour, resonating with the idea that sense must be seen 'as the problem to which propositions correspond' (1990a: 121). Thus, we can establish a series leading from a radical empirical openness to an exploratory and interrogative disposition, to the gradual forming of links and knots and eventually to well-articulated propositions and matters of concern. As we can never fully know what actors consist of, what power is or how society is composed, the labour of redescribing relations and making propositions adequate to the metaphysical surfaces we inhabit continues indefinitely.

Obviously, then, there are significant overlaps and shared inclinations between Deleuze's rhizomes and Latour's networks. As long as our discussion remains situated on the terrain of similarity, however, the rhizome will remain unable to add something very *distinctive* to the network. Furthermore, comparison on the basis of similarity is contrary to the spirit of Deleuze and Latour for both of whom difference at once precedes and supersedes identity. It is thus time to consider divergence.

Divergence

If the affinity between rhizomes and networks is less than total, where does divergence occur? For some commentators, this question has little sense. The philosopher Graham Harman, for example, finds between the network and the rhizome not a series of bridgeable gaps but a vast chasm. The 'metaphysical surfaces' to which I previously alluded can illustrate the point, since Deleuze (1990a: 125) also wrote that they are an 'effect of deep mixtures – a physics which endlessly assembles the variations and pulsations of the entire universe.' But, Harman insists, the idea of an underlying flow of becoming, or a set of pulsations, has no analogue in Latour. Moreover, contrary to Deleuzian objects, which are continuously changing, Harman (2009: 6) depicts Latourian entities as 'so highly definite that they vanish instantly with the slightest change in their properties.' Thus, 'no off-the-cuff remarks by Latour about his fondness for Deleuze can outweigh the utterly … non-Deleuzian foundations of Latour's own metaphysics' (Harman 2016: 158). Case closed.

Or perhaps not. For while it is indeed difficult to find a Latourian version of a flow of becoming,[5] ANT objects do not vanish and reappear as instantaneously as all that. Part of the issue is that Harman's metaphysical focus tends to abstract from the objects and issues that Latour actually studies. Latour's philosophical interest, after all, is an empirical one, and as he notes (1999b: 287), we are always faced with 'many practical metaphysics, many different practical ontologies.'

If we turn to studies like *Aramis* (Latour 1996a) or the pédofil of Boa Vista (Latour 1995), we do not, in fact, encounter trains and soil samples instantly fading in and out of existence. Instead, we are faced with objects that, much like Deleuze's, are gradually modified as they enter into relational composites and begin to exchange properties with other entities. When Latour (1999b: 311) advances the notion of 'relative existence,' to make it 'possible to define existence not as an all-or-nothing concept but as a gradient,' thereby allowing 'for much finer differentiations than the demarcation between existence and non-existence,' we are obviously a far cry from an image of things disappearing with the slightest transformation (see also Jensen 2010: 19–31). To understand the points of divergence between the rhizome and the network, we are thus obliged to make finer discriminations. And no one offers better guidance for doing so than the philosopher of science and erstwhile student of Deleuze, Isabelle Stengers, who also happens to be one of Latour's closest intellectual companions (see Savransky this volume).

In *We Have Never Been Modern*, Latour (1993) was constantly on the attack: The philosophy of science got it wrong, postmodernists got it wrong, social constructivists got it wrong. Once their epistemological errors had been fixed, and the false modern dichotomies between nature and culture, science and politics bypassed, the path would open to his hybrid parliament of things in which humans and non-humans could be made to peacefully coexist.

Yet, Stengers (2000: 124) is doubtful: The problems of contemporary science are hardly reducible to the 'error of epistemologists.'[6] Moreover, she goes on:

> 'error' does not have to be any more denounced than power. It explains nothing, except insofar as it is a product of the network, characteristic of the style of the network that belongs to our epoch, and of the political problem it poses.

As we know, however, Latour has spent nowhere near the same energy denouncing power as epistemology (Haraway 1997: 33–35; Schaffer 1991; see also Johnson, this volume).

In an early discussion, Nick Lee and Steve Brown (1994: 587) turned Deleuze against ANT to make just this point. Rather than rhizomatic, they concluded, the network had a tree-structure: 'All lines (routes, connections) are subordinated to the point, and all points are made to resonate with the centre. This is the space of measurement and calculation – ordering space.' And if one follows the scientist-entrepreneur of *Science in Action* (Latour 1987), it does indeed seem that actor-network theory is infatuated with striation, with tree logic, and thus with the state, and power. Accordingly, the network appears rather powerless against power's seductions.

While Deleuze and Guattari's advocacy of minor sciences and lines of flight is often seen to epitomise radical politics, it seems that, for Latour, 'any revolutionary impulses recede firmly into the background' (Schmidgen 2015: 7).[7] It is not coincidental that *The Invention of Modern Science* (Stengers 2000) was dedicated to Latour and Guattari 'in memory of an encounter that never took place.'

Stengers, in fact, accepts Latour's (1999b: 106–108) depiction of scientific truths as the consequence of the making of 'links and knots.' However, she is quite circumspect about the *reach* of these truths. And the grounds for her concern are simultaneously political and pragmatic. The fact that scientists manage to create particular facts within the rarefied settings of the laboratory confers upon them neither the capacity nor the right to determine the implications of those facts among a heterogeneous ecology of practices. While the making of a fact is an event, since it creates a distinction between a before and an after, the *extension* of the fact depends on other events involving non-scientists that inhabit other practices. The scientist is therefore unable to dictate the scope of the event, and accordingly any attempt to extend 'facts' into other practices is always accompanied by risk.

If Stengers characterises (2000: 64) the 'parliament of things' (Latour 1993) as a 'difficult success,' it is thus not because it fails to conform to any standard critique of science. To be sure, the sciences become open to critique whenever they fail to respect the specificity and creativity of other practices, thus entering the mode of what Deleuze and Guattari called major, or state, science. Conversely, however, the creativity specific to the sciences, that which gives them the capacity to formulate their own problems and develop their own solutions, must also be acknowledged. The ability to do so, however, is threatened by what could be called the affinity for power of the early ANT (Latour 1987).[8]

The crucial issue is thus to separate the parliament of things from its potential majoritarian inclinations. As Stengers will suggest, this can be achieved by aligning it with Deleuze's notion of adventurous, minor sciences.

Experimental realignments

In an early critique, Nick Lee and Steve Brown (1994: 787) argued that the network's tendency to engulf everything meant the loss of any space for 'irreducible otherness.' And the statement that 'literally, a network has no outside' (Latour 1996b: 49) seems to vindicate this interpretation. We might, however, pause at the word 'literally.' Does the space of the other perhaps exceed the network in a *non-literal* manner? Perhaps, as Deleuze would say, *virtually*? This possibility is at least raised in *Reassembling the Social*, which introduces the notion of a fluid, formless plasma, which has not yet been brought into the collective (Latour 2005: 241ff).[9]

But then how does such 'bringing into' occur? What ties together the rhizome and the network on this point is the famous principle of irreduction (Latour 1988), according to which the 'coming together of heterogeneous components…is…the first and last word of existence' (Stengers 2018: 105). For scientists, the issue is how to invent the means to bring new entities into the collective, which also means protecting them against ever-vigilant sceptics standing ready to reduce them to fictions (think of climate change). It concerns learning how to confer on non-human entities the capacity to confer back on the scientists the power to speak in their name (Stengers 1997: 164).

In accordance with the principle of irreduction, this process of *reciprocal conferral* has no fixed procedure and entails no particular form of relation. Even so, Latour's career-long emphasis on tracing connection tempts a reductive imagination, always in search of material, or even better, directly observable, connections (Krarup and Blok 2011). In contrast, Deleuze (1994: 220) wrote that imagination 'crosses domains, orders and levels, knocking down the partitions.' What held his interest was thus a movement 'from science to dreams and back again': Between what is presently fiction yet may later become real, between plasma and network, or between the virtual and the actual.

Looking to one side, a young face represents science in action. Looking in the other direction, a mature, bearded face symbolises ready-made science. I am referring, obviously, to Latour's (1987) famous version of the Janus head. Unfortunately, the usage of this figure has become rather formulaic, its interpretation itself 'ready-made.' Almost all interest has focused on the manoeuvres of scientists as they make facts, while almost all critique has centred on the bearded scientists toasting in celebration of their achievements. As Stengers (2010: 42) remarks, however, the Janus head depicts a 'contrasted unity.' A one-sided focus on science in action loses sight of the fact that 'the dreams of the youth, his ambitions, are bearded ones.'

Science critics might well reply that this is precisely the problem. But, Stengers argues, Deleuze and Guattari (1994) help to understand why it is unsatisfying to engage scientists' dreams only in the mode of denunciation, for while they 'certainly did not agree with the old bearded-face explanation... they nevertheless asked us to relate science as creation with science's 'own specific means,' which are associated, one way or the other, with the possibility of a scientist getting a beard' (2005: 154).

At issue is thus a dissociation of two aspects of ready-made science. Learning to celebrate the risky achievements of science does not entail endorsement of scientific conquest or denunciation of other practices. The image, instead, is the one in which each science operates as its own

> minority adventure, as Deleuze and Guattari positively characterize a minority as what does not dream to become a majority. And it is precisely because a minority collectively produces a divergence without a dream of convergence, of representing a future majority or consensus, that some transversal connections are possible.
>
> *(Stengers 2005: 158)*

The notion of transversal connections – Deleuzian rhizomes – might nevertheless appear to be a far cry from Latour's parliament of things, which is often seen to embody a fundamentally pragmatic and reformist view.

Consider, for example, Graham Harman's (2014: 63) exemplification of Latourian politics. Although we may feel repulsed by Vladimir Putin's annexation of parts of Ukraine, Harman observes, the sense of outrage and moral superiority must be tempered by the practical realisation that 'certain chickens are left to the wolves because the collective itself cannot face the wolves, under penalty of disruption.' Had action been taken against Putin-the-wolf, the consequences might have been even worse.

One might object that the depiction of Latour as a promoter of pragmatic 'realpolitik' and conflict avoidance disregards his recent discussions of climate change disruptions as well as more programmatic interventions like *War of the Worlds* (Latour 2002). Adding to the problem, Harman's example leaves out consideration of the non-human agency and thing-politics so central to the parliament of things. To better grasp the rhizomatic potentials of Latour's material politics, it is fruitful to turn instead to Isabelle Stengers's (2000: 159) comments on the children's story 'The Three Little Pigs.'

If one reads 'The Three Little Pigs,' Stengers observes, no amount of social construction will obviate the distinction between the fictive protection of houses made of straw and the real protection offered by brick. The narrative thus locates us firmly within a world of material recalcitrance. Yet, before this reality is taken as given, and the safety of pigs entrusted to 'experts discussing bricks and cement,' it is necessary to probe what their solution to the problem of wolves will 'take as acquired.' Posing this question makes us see that houses made of brick are a solution to the problem of the wolf defined as a menace. If we were able to invent other relationships with the wolf, our imaginative horizon might extend beyond the necessity of reinforcing the walls. Indeed, we might be encouraged to explore a world in which chickens, pigs and wolves could coexist.

Unavoidably, Stengers (2000: 159) wryly notes, experts in 'protection against destructive wolves' will counter that this kind of proposal is idealist, dangerous or impossible. But just as scientists are not free to extend the event associated with their laboratory facts into other practices, experts in cement and safety are also unqualified 'to follow through all its consequences the logic of the story they are advocating.' After all, a principled refusal of, or disinterest in, any search for alternative solutions might lead to

> a story in which other wolves, even more threatening, will intervene, in which the bricks and cement will no longer suffice, in which we will be taken up in an endless move toward ever more costly and rigid modes of protection.

Clearly, coexistence is not a matter of mutual affection, and disruption always remains a possibility. Yet, Stengers' way of dealing with the threat of disruption relies neither on a return to any form of material determinism *nor* on any kind of quasi-pragmatic reformism. Instead, she guides us to a position from where it can be perceived that the distinction between Latourian thing-politics and Deleuzian 'dreams' of alternative futures has always been blurry.

Such futures would be shaped by actors learning to confer on other entities (like menacing wolves) the capacity to confer back on their own collectives the ability to imagine and materialise different forms of coexistence.

In a slightly more rhizomatic guise, the parliament of things thus emerges as an institution for experimenting with minoritarian practical ontologies.

Rhizome

It is evidently just as pointless to call for a renaming of ANT as it was for Latour to issue his original recall. But perhaps a new name is not required to allow the rhizome to infiltrate and complicate the network. For one thing, it would impose the obligation on actor-network theorists, beneficial in my view, to become less literal-minded about the connections they trace. Unable to hide behind actors who they claim to simply follow, they would also have to take more risks and be more explicit about why they take them.

What would be extended in this process would thus not be the network as such but rather *our curiosity* about the continuous composition and reconfiguration of the world multiple (Omura *et al*, 2018). The aim would not be to *capture* within the network all those entities – like earth-beings (de la Cadena 2015) or spirits (Ishii 2017) – whose forms do not easily fit, but rather to learn to relate to worlds composed by things and forces far more heterogeneous and unruly than those ANT has conventionally dealt with (see e.g. Jensen 2015, Jensen and Blok 2013, Jensen, Ishii and Swift 2016). At issue is an extension of ANT's already considerable imaginative ontological re-figurations.

In brief, then, parsing ANT through the rhizome offers a way to intensify its experimental disposition. Together, ANT and the rhizome continue to work the metaphysical surface.

Notes

1 As far as I am aware, only Latour has addressed Lynch's proposal to rename ANT as ARO. Since, among the progenitors of ANT, Latour is also the only to regularly invoke Deleuze, the following discussion centres on his work.

2 Henning Schmidgen (2015: 6) suggests that '"Even the famous 'Follow the actors!' the basic principle of the anthropology of science as conceived by Latour, derives from a concept of Deleuze and Guattari's: namely, the 'nomad, ambulatory sciences'" (see also Jensen 2012).

3 This theme particularly held Latour's attention during the Tardean interval (after the Serresian phase, but before Souriau, James and Whitehead took over: As we know Latour changes intellectual predecessors at a dizzying pace). More generally, *Difference and Repetition* is the Deleuzian work

to which Latour returns with the greatest regularity (e.g. 1993: 72, 1996: 49, 1997: 179, Latour and Lépinay 2009: 39).

4 Long before Latour became famous for insisting that *We Have Never Been Modern*, and Haraway (2008) proposed that we hadn't been human either, Deleuze (2004: 90–94 [1966]) had characterised humans as '"a dubious existence."'

5 As noted below, however, it appears in the form of 'plasma' in Latour (2005).

6 It is curious that Latour defines the problem in terms of error since this situates his analysis on the same terrain – of knowledge rather than networks – as the epistemologists whom he criticises. Deleuze's (1990a: 120) characterisation of error in *The Logic of Sense* as '"a very artificial notion, an abstract philosophical concept, because it affects only the truth of propositions which are supposed to be ready-made and isolated"' is very apt here. Here, too, Deleuze's formulation resonates with Latour's later movement from matters of fact to matters of concern.

7 On the one side, Latour constantly evokes the language of common sense. On the other, Deleuze is constantly on the attack against common sense and good sense ('"partial truth associated with the feeling of the absolute"' (Stengers 1997: 70)). While Latour likes to don the mantle of diplomat, Deleuze declared shameful Leibniz' dictum to never aim to overthrow established sentiments (Stengers 2000: 15, also Jensen 2006).

8 Latour has attempted to take such criticisms on board though with mixed success (e.g. Latour 2004b). Among friendly interpreters (like Henning Schmidgen) as well as staunch supporters (like Graham Harman), the image of Latourian politics as fundamentally 'reformist' rather than 'radical' remains strong.

9 Tellingly, Graham Harman (2009: 132) skips very quickly over this relation.

References

Bowker, Geoffrey. 1994. *Science on the Run: Information Management and Industrial Geophysics at Schlumberger, 1920–1940*. Cambridge, MA: MIT Press.

Crawford, T. Hugh. 1993. 'An Interview with Bruno Latour,' *Configurations* 1(1): 247–69.

de la Cadena, Marisol. 2015. *Earth-Beings: Ecologies of Practice Across Andean Worlds*. Durham, NC & London: Duke University Press.

Deleuze, Gilles. 1983. *Nietzsche & Philosophy*. London: Athlone.

Deleuze, Gilles. 1990a. *The Logic of Sense*. New York: Columbia University Press.

Deleuze, Gilles. 1990b. *Expressionism in Philosophy: Spinoza*. New York: Zone.

Deleuze, Gilles. 1994. *Difference and Repetition*. New York: Columbia University Press.

Deleuze, Gilles. 2004 [1966] 'Humans: A Dubious Existence', in *Desert Islands and Other Texts 1953–1974*. New York: Semiotext(e), pp. 90–94.

Deleuze, Gilles and Félix Guattari. 1983. *Anti-Oedipus: Capitalism and Schizophrenia*. London: Athlone.

Deleuze, Gilles and Félix Guattari. 1987. *A Thousand Plateaus: Capitalism and Schizophrenia*. Minneapolis & London: University of Minnesota Press.

Deleuze, Gilles and Félix Guattari. 1994. *What Is Philosophy?* New York: Columbia University Press.

Gad, Christopher and Casper Bruun Jensen. 2010. 'On the Consequences of Post-ANT,' *Science, Technology and Human Values* 35(1): 55–80.

Haraway, Donna. 1997. *Modest_Witness@Second_Millennium.Femaleman©_Meets_Oncomouse™ – Feminism and Technoscience*. New York: Routledge.

Haraway, Donna. 2008. *When Species Meet*. Minneapolis & London: University of Minnesota Press.

Harman, Graham. 2009. *Prince of Networks: Bruno Latour and Metaphysics*. Melbourne: re:press.

Harman, Graham. 2014. *Bruno Latour. Reassembling the Political*. London: Pluto Press.

Harman, Graham. 2016. 'Response to Altamirano and Ivakhiv,' *Global Discourse* 6(1/2): 157–160.

Ishii, Miho. 2017. 'Caring for Divine Infrastructures: Nature and Spirits in a Special Economic Zone in India,' *Ethnos* 82(4): 690–710.

Jensen, Casper Bruun. 2010. *Ontologies for Developing Things: Making Health Care Futures through Technology*. Rotterdam: Sense.

Jensen, Casper Bruun. 2006. 'Established Sentiments, Alternative Agendas, and Politics of Concretization,' *Configurations* 14(3): 217–44.

Jensen, Casper Bruun. 2012. 'Anthropology as a Following Science: Humanity and Sociality in Continuous Variation,' *NatureCulture* 1: 1–24.

Jensen, Casper Bruun. 2015. 'Experimenting with Political Materials: Environmental Infrastructures and Ontological Transformations,' *Distinktion: Scandinavian Journal of Social Theory* 16(1): 17–30.

Jensen, Casper Bruun. 2017. 'New Ontologies? Reflections on Some Recent 'Turns' in STS, Anthropology and Philosophy,' *Social Anthropology* 25(4): 525–545.

Jensen, Casper Bruun and Anders Blok. 2013. 'Techno-Animism in Japan: Shinto Cosmograms, Actor-Network Theory, and the Enabling Powers of Non-Human Agencies.' *Theory, Culture and Society* 30(2): 84–115.

Jensen, Casper Bruun and Evan Selinger. 2003. 'Distance and Alignment: Haraway and Latour's Nietzschean Legacies,' in Don Ihde and Evan M. Selinger (eds.), *Chasing Technoscience: Matrix for Materiality*. Indianapolis: Indiana University Press, pp. 195–212.

Jensen Casper Bruun and Kjetil Rödje. 2010. 'Introduction,' in Casper Bruun Jensen and Kjetil Rödje (eds.), *Deleuzian Intersections: Science, Technology, Anthropology*. Oxford & New York: Berghahn, pp. 1–37.

Jensen, Casper Bruun, Miho Ishii and Phil Swift. 2016. 'Attuning to the Webs of *En*: Ontography, Japanese Spirit Worlds, and the 'Tact' of Minakata Kumagusu,' *Hau: Journal of Ethnographic Theory* 6(2): 149–172.

Krarup, Troels M. and Anders Blok. 2011. 'Unfolding the Social: Quasi-Actants, Virtual Theory, and the New Empiricism of Bruno Latour,' *The Sociological Review* 59(1): 42–63.

Latour, Bruno. 1987. *Science in Action: How to Follow Scientists and Engineers Through Society*. Cambridge, MA: Harvard University Press.

Latour, Bruno. 1988. *The Pasteurization of France*. Cambridge: Cambridge University Press.

Latour, Bruno. 1993. *We Have Never Been Modern*. New York: Harvester-Wheatsheaf.

Latour, Bruno. 1995. 'The 'Pédofil' of Boa Vista: A Photo-Philosophical Montage,' *Common Knowledge* 4(1): 144–87.

Latour, Bruno. 1996a. *Aramis, or the Love of Technology*. Cambridge, MA: Harvard University Press.

Latour, Bruno. 1996b. 'Om aktør-netværkteori. Nogle få afklaringer and mere end nogle få forviklinger,' *Philosophia* 25(3/4): 47–64. Translated into English as 'The Trouble with Actor-Network Theory,' available at http://www.bruno-latour.fr/node/379.

Latour, Bruno. 1997. 'Trains of Thought: Piaget, Formalism and the Fifth Dimension,' *Common Knowledge* 6(3): 170–91.

Latour, Bruno. 1999a. 'On Recalling ANT,' in John Law and John Hassard (eds.), *Actor-Network Theory and After*. Oxford: Blackwell Publishers, pp. 15–25.

Latour, Bruno. 1999b. *Pandora's Hope: Essays on the Reality of Science Studies*. Cambridge, MA: Harvard University Press.

Latour, Bruno. 2000. 'A Well-Articulated Primatology: Reflections of a Fellow-Traveller,' in Shirley Strum and Linda Fedigan (eds.), *Primate Encounters: Models of Science, Gender and Society*. Chicago, IL & London: University of Chicago Press, pp. 358–81.

Latour, Bruno. 2002. *War of the Worlds: What About Peace?* Chicago, IL: Prickly Paradigm Press.

Latour, Bruno. 2004. 'How to Talk About the Body: The Normative Dimension of Science Studies,' *Body & Society* 10(2–3): 205–29.

Latour, Bruno. 2004a. 'Why Has Critique Run out of Steam? From Matters of Fact to Matters of Concern,' *Critical Inquiry* 30(2): 225–48.

Latour, Bruno. 2004b. *Politics of Nature: How to Bring the Sciences into Democracy*. Cambridge, MA & London: Harvard University Press.

Latour, Bruno. 2005. *Reassembling the Social: An Introduction to Actor-Network-Theory*. Oxford: Oxford University Press.

Latour, Bruno. 2013. *An Inquiry into Modes of Existence: An Anthropology of the Moderns*. Cambridge, MA & London: Harvard University Press.

Latour, Bruno. 2014. 'Agency at the Time of the Anthropocene,' *New Literary History* 45(1): 1–18.

Latour, Bruno. 2016a. 'Life among Conceptual Characters,' *New Literary History* 47(2–3): 463–76.

Latour, Bruno. 2016b. 'Onus Orbis Terrarum: About a Possible Shift in the Definition of Sovereignty,' *Millennium: Journal of International Studies* 44(3): 305–20.

Latour, Bruno, and Vincent Antonin Lépinay. 2009. *The Science of Passionate Interests: An Introduction to Gabriel Tarde's Economic Anthropology*. Chicago, IL: Prickly Paradigm Press.

Lee, Nick and Steven D. Brown. 1994. 'Otherness and the Actor-Network: The Undiscovered Continent.' *American Behavioral Scientist* 37(6): 772–90.

Lynch, Michael. 1995. 'Building a Global Infrastructure,' *Studies in History and Philosophy of Science* 26(1): 167–72.

Omura, Keiichi, Atsuro Morita, Grant Otsuki and Shiho Satsuka. 2018. *The World Multiple: The Quotidian Politics of Knowing and Generating Entangled Worlds*. London & New York: Routledge.

Schaffer, Simon. 1991. 'The Eighteenth Brumaire of Bruno Latour,' *Studies in the History and Philosophy of Science* 22(1): 174–92.

Schmidgen, Henning. 2015. *Bruno Latour in Pieces: An Intellectual Biography*. New York: Fordham University Press.

Stengers, Isabelle. 1997. *Power and Invention*. Minneapolis & London: University of Minnesota Press.

Stengers, Isabelle. 2000. *The Invention of Modern Science*. Minneapolis & London: University of Minnesota Press.

Stengers, Isabelle. 2005. 'Deleuze and Guattari's Last Enigmatic Message,' *Angelaki* 2: 151–167.

Stengers, Isabelle. 2010. 'Experimenting with *What is Philosophy?*,' in Casper Bruun Jensen and Kjetil Rödje (eds.) *Deleuzian Intersections: Science, Technology, Anthropology*. Oxford & New York: Berghahn, pp. 39–57.

Stengers, Isabelle. 2018. 'The Challenge of Political Ontology,' in Marisol de la Cadena and Mario Blaser (eds.), *A World of Many Worlds*. Durham, NC & London: Duke University Press, pp. 83–112.

9

What can ANT still learn from semiotics?

Alvise Mattozzi

Semiotics, still

'Still' is the pivot, around which the question I have been asked to answer turns. 'Still' connects the past with present and, from there, with the future: It suggests that ANT has previously learned from semiotics and that it can possibly keep learning from semiotics.

Therefore, 'still' acknowledges a long-term relationship between ANT and semiotics. Such relationship has been frequently explored, reenacted and recalled by Latour and it has been sealed by John Law's definition of ANT as 'material semiotics.' Despite all that, such relationship has been often overlooked, disregarded, forgotten.[1]

Therefore, in order to answer the question making up the title of this contribution, I first need to recover the history of such relationship.

Semiotics, then

Semiotics (or semiology) is, from an etymological point of view, 'the science of signs.' As such, semiotics studies signification as the outcome of sign processes: Something, material, present to perception – the sound of a word, the coloured cloth of a flag, the shape of an emoticon, the look of a car or the cut of a suit – *stands for* something else, more immaterial, abstract and absent – the meaning of a word, a nation, an emotion, a lifestyle, a social class.

The first founder of modern semiotics, the American pragmatist philosopher Charles Sanders Peirce, thought the sign as a threefold relation (Figure 9.1a) among:

- an *object* or referent;
- a *representamen*, i.e. the actual sign – the configuration, which represents the *object*;
- an *interpretant*, i.e. the further configuration elicited by the *representamen*, usually intended as the idea created in the mind, but which does not need to be a mental representation.

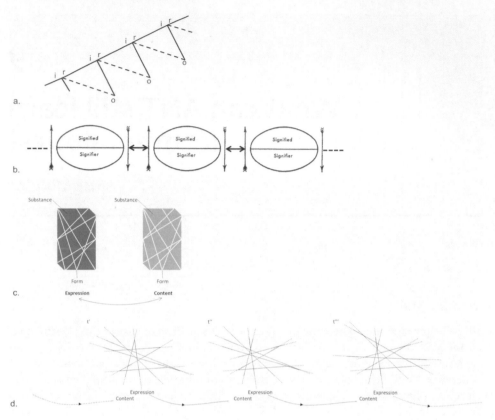

Figure 9.1   Signification models: a. Peirce, o = object, r = representamen, i = interpretant (my elaboration); b. de Saussure's ([1916] 1959: 115); c. Hjelmslev's (my elaboration); d. ANT's (my elaboration)

Peirce's (1868) semiotics is a complex classification of signs, developed by considering the types of relations the three elements can entertain. The most famous classification is the one based on the relation between the *representamen* and the *object*, which can produce:

- an *icon* or *likeness*, i.e. a resemblance, as with a figurative images;
- an *index*, i.e. physical or causal relations, like a pointing finger or smoke for fire;
- a *symbol*, i.e. conventional relations, like the word 'dog' for the domestic barking animal.

The Swiss linguist and originator of structuralism Ferdinand de Saussure is the second founder of modern semiotics. Being a linguist, Saussure focused only on verbal language, considering it grounded in conventional signs, akin to Peirce's *symbols*. However, Saussure thought the sign as binary: A relation between a *signifier* – the sound of a word as perceived – and a *signified* – the concept recalled by the perception (Figure 9.1b).

Despite the focus on verbal language, Saussure ([1916] 1959: 16) acknowledged that language is comparable to other sign systems like 'the alphabet of deaf-mutes' or 'symbolic rites.' All of them are studied by 'semiology,' i.e. 'the science that studies the life of signs within society.'

Peirce's one is a philosophical semiotics aimed at developing a theory of knowledge, which could provide the rules for scientific inquiries, by considering which kinds of signs are used by scientists within their deductions, inductions or abductions.

Saussure's one is a scientific semiotics, aimed at providing a method for studying languages or, more in general, sign systems. Following Saussure's approach, linguistics and semiotics have been developed as methodologies providing terms, categories and models to describe-analyse[2] sign systems, as well as their specific empirical manifestations.

Semiotics: from signs to relations

All along the 20th century, various semiotics have been elaborated assuming as ground the sign and developed through various related notions such as representation, symbol, language, code, communication, etc. Within such framework, signs have been often reduced to what they represent, according to given societal, cultural or mental structures. Therefore, signs have been often reduced to their most de-situated, disembodied and immaterial aspects. Such approach to signs is clearly at odds with ANT. No wonder that 'semiotics readings' pursuing such dualist – material/immaterial-ideal – and transcendent understanding of signifying processes have been considered 'incompatible' with ANT (Farías and Mützel 2015: 524).

Nevertheless, such incompatibility is the result of a partial view of semiotics, based on a simplified and isolated conception of the sign, put forth and adopted, first of all, by many semioticians. Such conception of the sign does not, however, pertain neither to Peirce nor to Saussure, nor to some of their heirs. Peirce and Saussure developed their semiotics by addressing issues that take place beneath and above the sign. They indeed intended signs as mediating entities and mediation as a process, taking place through relations. Examples of mediating entities considered by semiotics can be: The *interpretant*, between *object* and *representamen*, within the Peircian sign (Figure 9.1a); language, between sound and thought (Saussure [1915] 1959: 112); *forms* between *substances of expression* and *of content* (Hjelmslev [1943] 1961; Figure 9.1c, see below), *enunciation* between *language* (*langue*) and *speech* (*parole*) (see below). These mediating entities not only allow establishing relations, but are constituted by relations: For Peirce (1898), the basic categories from which signs arise are relational; for Saussure ([1916] 1959: 122), in language 'everything is based on relations,' so that signification is never reducible to the simple coupling of *signifier-signified*, but it has always to be considered in relation with other couplings (Figure 9.1b). The latter approach has been radicalised by the Danish linguist Louis Hjelmslev ([1943] 1961), for whom signification is solely based on relations and relations among relations, i.e. relations among configurations of relations or, using Hjelmslev's terms, relations among '*forms*': A *form* of *expression* (*signifier*) and a *form* of *content* (*signified*) (Figure 9.1c).

Signs are then just the 'tip of an iceberg' and semiotics is actually concerned with the 'complex [...] work' under the 'tip' (Marrone 2002: 14, my translation), carried out by relations and mediations.

While signs, considered as isolated entities reduced to the immaterial representations to which they refer, can be of no relevance for ANT (Farías and Mützel 2015: 524), relations and mediations certainly are.

Relations have indeed provided the common ground for the exchange between ANT and semiotics. It is not coincidence, then, that ANT considers semiotics, not the 'science of signs,' but the 'science of relations' (Law 2002: 49).

What ANT has learned from semiotics

The 'semiotic insight' of 'the relationality of entities' (Law 1999: 4), according to which 'everything in the social and natural worlds [is] a continuously generated effect of the webs of relations within which they are located' (Law 2008: 141), is what, in general, ANT has learned from semiotics. As Annemarie Mol (2010: 257) acknowledges

> In [d]e Saussure's version of semiotics, words do not point directly to a referent, but form part of a network of words. They acquire their meaning relationally, through their similarities with and differences from other words. Thus, the word "fish" is not a label that points with an arrow to the swimming creature itself. Instead, it achieves sense through its contrast with "meat", its association with "gills" or "scales" and its evocation of "water". In ANT this semiotic understanding of relatedness has been shifted on from language to the rest of reality. Thus it is not simply the term, but the very phenomenon of "fish" that is taken to exist thanks to its relations.

Because of this extension of relationality 'from language to the rest of reality,' of its 'ruthless' application 'to all materials,' ANT has been considered 'a *semiotics of materiality*' (Law 1999: 4). As such, it 'forget[s] about signs and signification, [...] only retain[ing] the stress on interdependence' (Mol and Mesman 1996: 420), thus producing a version of semiotics which 'is not about meaning' (Mol and Mesman 1996: 429).

And yet, the 'insight' of relationality has emerged to be the ground for the relationship between ANT and semiotics only at a later stage.

At first, ANT has learned from semiotics a method. Only through the use of such method, the 'insight of relationality' has emerged as a shared ground between ANT and semiotics.

In semiotics, indeed, Latour initially found a way to describe-analyse agency (Latour 2014a), regardless of the ontological status of agents, by considering relations among entities and how they are transformed. Latour, together with semiotician Paolo Fabbri, was thus able to write the first ANT science studies article, by using semiotics as a 'methodology' able to take 'sociology of science at the heart of [scientific] articles' (Latour and Fabbri 1977: 82, my translation). In the following years, thanks also to the collaboration with biologist and semiotician Françoise Bastide, Latour and other ANT scholars have drawn on semiotics as

- a "method" that allows describing the "interdefinition of actors and the chains of translations" (Latour [1984] 1988: 11) or that allows "following, along the design phase, the user as is inscribed, translated" in a technical object (Akrich 1990: 84, my translation)
- a set of "tools" "used to compare what Einstein says about the activity of building spaces and times with what sociologists of science can tell us" (Latour 1988: 3),
- a way to map "a common ground, a common vocabulary, that would be intermediary between [empirical descriptions] one hand and the ontological questions [...] on the other" (Latour 2000: 251).

ANT has then learned from semiotics an insight about relationality and a method based on relationality. While the insight is still relevant as a general theoretical framework, the method semiotics provides has been relevant for ANT in a more circumscribed way: Mainly during the 1980s and early 1990s, for a limited number of ANT scholars. Nevertheless, for Latour, semiotics has continued to play a relevant role as descriptive–analytical methodology, because it provides an 'organon' or 'toolkit' able 'to record important variations' (Latour 2014b: 265).

A semiotics of relational transformations

What Latour refers to as 'organon' for ANT is not semiotics in general, but a very specific strand of semiotics: The one developed by the French-Lithuanian semiotician Algirdas J. Greimas and his collaborators (Greimas and Courtés 1979).

Greimas has turned the linguistics elaborated by Saussure and Hjelmslev into a semiotics not only by extending beyond verbal language the signifying configurations to be described-analysed, but also dynamicising them, by taking into account signification as a transformation occurring among configurations. He has achieved such dynamicisation by integrating Saussure's and Hjelmslev's framework with narratology and the theory of enunciation.

As for narratology, Greimas drew on the analysis of folktales elaborated by Russian folklorist Vladimir Propp and revised it through the relational syntax elaborated by French linguist Lucien Tesnière – the actual coiner of the term 'actant.' Thus, Greimas was able to develop a narrative syntax – considered the syntactic ground of signification – which allows describing transformation of relations among actants.

As for enunciation, by drawing and operationalising the theory of enunciation proposed by the French linguist Émile Benveniste, Greimas was able to describe-analyse the discursive dynamics taking place among various frames of reference. Benveniste introduced the notion of 'enunciation' in order to account for the individual act of appropriation of *language* (*langue*) through which *speech* (*parole*) is produced. Such appropriation entails various tensions between the person, the time and the space from which the appropriation takes place and the person, the time and the space within the produced sentence (Figure 9.2). The descriptions-analyses of these tensions and of the related dynamics give way to accounts of the circulation of meanings through various frames of reference, as well as of the shiftings among points of view and of the positionings of utterers and recipients (Figures 9.2 and 9.4).

These are the features that Latour has found interesting in Greimassian semiotics, which cannot be found in other semiotics that tend to be more rigid and more bound to signs and verbal language, rather than relational transformation more in general.

ANT thus uses Greimassian semiotics to describe the relational transformations it is interested in. However, it has always used Greimassian tools in a limited way, without borrowing 'all of [semiotics'] argument and jargon' (Latour 2005: 55) and formalism, in order to prevent to efface actors' own language, instead of highlighting it (Akrich 1992a).

Nevertheless, the use of semiotics' tools has been extended from literary text, initially considered by Greimas, to 'settings, machines, bodies, and programming languages' (Akrich and Latour 1992: 259), thus overcoming the limits that Latour ([1984] 1988: 183) saw in the way semiotics had been enacted before ANT.[3]

Latour's *infralanguage*, used in order 'to help [analysts] become attentive to the actors' own fully developed metalanguage' (Latour 2005: 49), is indeed built on Greimassian semiotic tools and is used in order to account for

- narrative dynamics – related to actants' actions and transformations – through semiotic tools like: "actant/actor", "competence/performance", "dictum/modus", "do/make-do", "figurative/non-figurative", "prescription/proscription/affordances/allowances", "program/anti-program of action" (Figure 9.3)
- enunciational dynamics – related to the way actants' actions and transformations are framed and made to circulate: "enunciation", "delegation", "shifting in/out/down" (Figure 9.2 and 9.4).

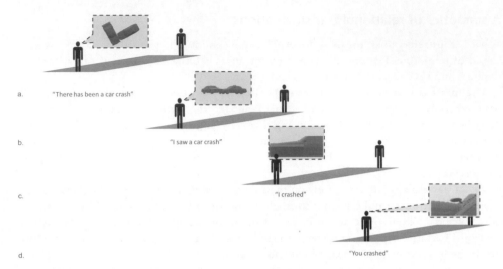

Figure 9.2 Examples of enunciational relations referring to the same narrative relations

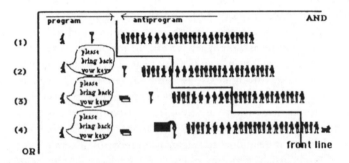

Figure 9.3 Example of the use of semiotics categories and models (Akrich and Latour, 1992: 263). Here, the following models are used: 'AND/OR relations,' 'program' and 'anti-program of action.' The 'program of action' is the goal an 'actor' wants to achieve – in this case the hotel manager wants to have the keys back; the 'anti-program of action' is the opposite programme, in this case carried out by the other 'actors,' the customers, and consists in keeping the key. The diagram shows the way in which, for each new association (AND), the entire set of relations is replaced (OR) by another set

The model built on the category 'association/substitution' or 'AND/OR relations,' able to map relations more in general, is also a semiotic model, coming originally from linguistics (Figure 9.3).

Coming out as a semiotics

As we have seen, ANT shares with semiotics – and especially with the semiotics emerged from the Saussurean tradition – an interest in describing-analysing relations and mediations through specific tools, however forgetting, apparently, about signification (Mol and Mesman 1996: 420). Despite the fact that signification, sense and meanings are not issues often directly addressed by ANT, they are not forgotten, but reframed, breaking away from a dualist and transcendent way of conceiving them, clearly at odds with ANT (Farías and Mützel 2015).

Akrich and Latour (1992: 259) assumed meaning as ANT's object of study by reframing it as 'how one privileged trajectory is built, out of an indefinite number of possibilities.' Akrich (1992a; 1992b) has made clear what 'trajectories' are through her research works on technical objects. For her, signification takes place through differences emerging through displacements, seen as passages from one moment to another of the configuration of relations characterising a specific technical object or as passages between the actions an artefact disposes and the actions a user unfolds. These displacements outline trajectories that can be then seen as 'concatenations of mediations,' in which each mediator 'transform[s] [...] the meaning or the elements it is supposed to carry' (Latour 2005: 39). More recently, Latour ([2012] 2013: 236) has clarified that 'sense' is 'the direction or trajectory [...] traced by a[ny] mode [of existence] and [...] defin[ing] both the predecessors and the successors of any course of action whatsoever,' whereas 'signs,' intended as figures undergoing the same general dynamic of sense, characterise one specific mode of existence, i.e. [FIC]tion. Therefore, for Latour, sense precedes signs and does not need them to unfold.

This way of conceiving sense (Figure 9.1d), signification and meanings is very far from the *signifier-signified* relation, to which these are often reduced by certain strands of semiotics. Yet, it is similar to the dynamics outlined by Peirce for signification, seen as chains of interpretants (Figure 9.1a). Moreover, it complies with signification as intended by Hjelmselv (Figure 9.1c), especially if considered, as Gilles Deleuze and Felix Guattari ([1972] 2003: 241) did, like 'flows of form and substance, content and expression,' i.e. not as a rigid relationship between a *signifier* determining a *signified*, but as encounters between different *agencements* (or configurations), for instance, the *agencement* of the movements of a technological object, with the *agencement* of the movements of users, referring to a situation considered by Akrich (1992a; 1992b).

By addressing signification and sense, ANT turns out to be a semiotics in itself. Moreover, it turns out to be a quite innovative semiotics, able to combine in a new way the two main semiotic traditions, the one derived from pragmatism, more philosophical, and the one from structuralism, more scientific. Through a pragmatist reading of the structuralist Greimas, homologous to the pragmatist reading of the structuralist Hjelmslev, carried out by Deleuze and Guattari ([1972] 2003), ANT is indeed able to provide Deleuze and Guattari's reflections on signification an actual descriptive–analytical methodology and empirical grounds on which to probe it.

What ANT can learn from other semiotics, still and again

Being ANT a semiotics, the initial question needs to be modified into 'what can ANT still learn from *other* semiotics?'

The answer to this question cannot but waive basic notions like signification, relationality and mediation, already appropriated and reformulated by ANT in the process of becoming a semiotics.

What is left to learn, as it happened at the beginning of the relation between semiotics and ANT, are then descriptive–analytical tools, i.e. terms, categories and models.

Greimassian semiotics, from which ANT has mainly drawn the already learned tools, has indeed kept refining old tools and elaborating new ones, of which ANT knows very little, since the relation between ANT and semiotics faded during the 1990s.

However, new and refined tools are not the only ones ANT can learn. Only few scholars have indeed followed Latour's example and used semiotic tools in a systematic and extended way, especially after the 1990s. Therefore, old tools are often unknown to many ANT scholars, so that they can also be learned again.

Given the limited space, I will provide just two examples of the tools ANT can learn anew and of their background that needs to be relearned. The two examples will refer to the main sets of relations, which these tools allow to describe-analyse.

Narrative relations

As Latour (2014b) has recently noted, the semiotic category of 'actant/actor' has been key for his work in order to describe the unfolding of agency, by accounting for actions and transformations of any entity. The terms constituting such category have been used by Latour and other scholars in at least three ways. Each of them gives relevance to different aspects of the terms and of the category.

First, 'actant' has been used to point to 'anything that acts' (Latour 1992), regardless of its ontological status, size, scale, features, etc., thus providing the principle of symmetry with a descriptive term.

Secondly, 'actant' has been used in tension with the term 'actor,' thus having the possibility to distinguish between actants. The fact that anything that acts is an actant does not mean that what acts is reducible to the performed action and then that it is 'just' an actant. Any actant can enjoy other relations and have other features that distinguish it from other actants. Thus, for example, the action of reminding to bring the hotel room's key back can be performed by a written note or by a weight attached to the key (Figure 9.3). They are both actants and, as for the action they perform, they are the same actant: They occupy the same position within the network manager-desk-key-customer, by providing the latter with a certain competence, namely a knowledge. However, on another level, with reference to other relations, they are different – they are different actors: A written note, a weight. The first is white, flat, made of paper bearing inscriptions; the second is brown, bulgy and made of metal. Because of that, notwithstanding they perform the same action related to providing a knowledge to the customer, they do that in different ways, so that the second results to be more effective in contributing to the success of the hotel manager's program of action (Figure 9.3).

Thirdly, 'actor' has been intended not just as an actant provided with its various features besides the action it accomplishes, but as 'what is made the source of action' (Latour 2014a). Thus, 'actor' has been intended as the specific entity to which agency is attributed, despite the fact that agency always unfolds through many actants. Akrich (1993) has called the 'actor' seen in this way 'author.'

As we can see, by deploying semiotic tools, 'variations,' differences, can be recorded and comparisons made, among entities, as well as among their features. Allowing the detailed description-analysis of actant-actors and of their actions was indeed the aim of Greimas' narrative grammar. In order to achieve such aim, Greimas elaborated many more terms and categories than those actually used by ANT. He, for instance, introduced other levels between the actant and the actor like actantial and thematic roles. Thus, for Greimas, any actor can cover one or more thematic roles, which, in turn, are performed by covering one or more actantial roles.

Akrich (1990; 1993) is one of the few[4] to have actually took advantage of a more stratified articulation of the 'actant/actor' category, by considering also 'positions' (Akrich 1990) or 'postures' (Akrich 1993) – akin to Greimas' 'thematic roles.' Analysing the design process of a pay-TV service, she has been able to show that the connection between the receiver and the VCR had been neglected, since it would have disrupted the superposition of 'actant,' 'position-posture' and 'actor-author' necessary for the service to work. With

- "actants", she referred to elements of the receiver related to a certain action, like the button "enter";
- "position-posture", she referred to roles assumed by entities, in relation to the milieu in which they would act as, for instance, "the subscriber", "the viewer", "the paying viewer";
- "actor-author" she referred to the entity to which the action can be attributed.

For the service to work properly, the three entities need to superpose, so that to the action of pushing 'enter' corresponds a 'viewer,' to which the action can be attributed and thus made responsible for payments.

ANT can thus learn again to further articulate the 'actant/actor' relation, in order describe and compare more in detail actants and actors.

Such opportunity is even higher today, thanks to the questioning that both ANT and Greimassian semiotics have carried out of action, in order to give relevance to passions and affects. This parallel reconsideration of action, acknowledged by ANT (Hennion, Maisonneve and Gomart 2000), has led both to thematise more and more the role of the body. Jacques Fontanille (2004), French semioticians collaborator of Greimas, has recently proposed to integrate the body or, better, a schematic version of it that works for humans as well for non-humans, into Greimas' narrative syntax. Fontanille has thus provided the actant with a body, allowing to account for actants' senses and passions or, more in general, affects, intended as 'change[s], or variation[s], that occur when bodies collide, or come into contact' (Colman 2005: 11).

Fontanille, in a way similar to the one attempted by Tim Ingold (2007) for the description of materials and their perceptions, shows how contacts and collisions among bodies can be described as pressures-penetrations and envelopments involving internal substances – like flesh – and their envelopes-surfaces – like skin.

I deem Fontanille's schematisation of the body a new tool ANT can learn from Greimassian semiotics, which allows ANT to address up-to-date issues like affects and to describe their unfoldings in detail.

For instance, only by considering the bodies of the actors accompanying the hotel's key (Figure 9.3), we can actually account for their different efficacy. While the written note acts only through its inscribed envelope-surface, which has to be taken into account by the customer's body taking a certain distance from it, the weight reminds of its presence continuously when in contact with the customers' body, by pressing the latter on its surface-envelope through its specific internal-substance and envelope-surface, which provide weight's consistency and shape.

Enunciational relations

The Greimassian enunciational model (Figure 9.2) has been key for Latour's work in order to distinguish fiction from science (Latour 1988; Figure 9.4), technology from fiction (Latour 1992), science from law, to describe religion and, eventually, to organise the various modes of existence (Latour [1998] 2017; [2012] 2013).

Indeed, through it, Latour has been able to account, distinguish and compare the ways in which entities circulate among different frames of reference. For instance, he has shown how

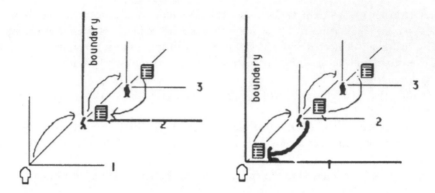

Figure 9.4   Enunciational dynamics in fiction and in science (Latour 1988, 14)

in fiction, figures tend to be shifted out within nested frames away from the original situation of enunciation, whereas in sciences, references tend not only to be shifted out within nested frames, like in fiction, but also to be shifted in, back towards the original situation of enunciation (Figure 9.4).

Despite the relevance the Greimassian enunciational model has had for Latour, it has been neglected by ANT scholars. Because of that, it is certainly a tool that ANT can learn again and, in a certain way, anew from semiotics.

As it should emerge from the references to frames, as well as from the examples provided here, the Greimassian enunciational model can contribute to provide tools to develop an ANT approach to media, which would be able to consider the relation between technology and what is displayed in and through technologies – what Latour ([2012] 2013) would call the intersection between fiction [FIC] and [TEC], or what media scholar Roger Silverstone called the 'double articulation' of media.

This is not such a novelty, given that enunciation, in general, and the Greimassian model more specifically have been widely used in the neo-Latin-speaking world to analyse various forms of communication, probing its efficacy especially for forms of visual communication (advertisements, paintings, cinema, television, etc.). These analyses have mainly focused on the dialogue images disposed with the recipient. The Greimassian model of enunciation could be further used, however, in order to account for the role of media technologies, and especially the role of interfaces, which dispose various shiftings in and out among frames of reference, as well as modes of existence.

Such an approach could lead to rearticulate the present debate around ANT and media. Indeed, the enunciational model, by fully considering mediation (Hennion [1993] 2015), allows accounting for 'the ways in which particular framings of entities are made to circulate,' as asked by Marres and Rogers (2008, 276). However, at the same time, by being akin to a 'diffusionist model[…], which conceive the dissemination of information as a movement from source to recipients,' criticised by Marres and Rogers (2008, 276), it also allows to account for the framed entities and thus account for dynamics that are usually filed under the notion of 'interpretation' – deemed key by Nick Couldry, but criticised by Marres and Rogers. However, such 'interpretation' would take place through dynamics akin to those studied by Akrich (1992) for technical objects (see above), related to the passage between the actions a certain frame and what framed dispose, and the action a recipient unfolds. Using semiotic tools relative to both narrative and enunciational relations would make such approach possible and, I deem, productive.

Conclusions: The conditions at which ANT can learn from semiotics still and again

From semiotics, and specifically from Greimassian semiotics, ANT has learned the use of various tools – terms, categories and models – enabling the description-analysis of relations. Through their use, ANT has learned the 'insight of relationality.' Furthermore, by unfolding these tools and this insight, ANT has turned out to be itself a semiotics.

Today, ANT can still learn new tools, as well as refined old ones. Indeed, since the 1990s, the exchange between ANT and Greimassian semiotics has faded and ANT has stopped being up-to-date about other semiotics. Given that the tools previously learned have been used only by a limited number of ANT scholars and, though with relevant exceptions, for a limited time, besides the new tools, ANT can also learn again the old ones – and actually this would be a necessary step in order to learn the new ones.

This chapter has then been less of a showcase of new Greimassian semiotic tools, than a refresher of old ones and of the grounds on which the previous learning has been possible.

At the end of this walkthrough, another question can thus be raised: 'Why has ANT stopped learning from semiotics?' Answering it would not only require another chapter, but a further research.

As for now, I intend to conclude by only touching upon the conditions at which I consider ANT can learn still and again from semiotics. Given that the failing of these conditions cannot but lead to a missed learning, their listing provides a first outline for a possible answer to the question emerged here in the conclusion.

ANT can learn from other semiotics still and again if

1 it is interested in description – if it is actually 'descriptive' (Law 2008: 141). If ANT is mainly interested in other activities like developing concepts, theorising, speculating, providing empirical examples for specific concepts, then, there is not much to re/learn, since semiotic tools are not meant to enable these activities – at least not directly;

2 it is interested in describing in detail relations, or, in Latour's terms, 'associations,' of any kind, included those 'out of which [actors are] made' (Latour 2005: 233);

3 it provides, within its architecture, relevant room and autonomy for the methodological level, a 'middle ground' (Latour 2000: 252), in between the theoretical-conceptual and the empirical ones, around which ANT, like other STS approaches, 'continuous[ly] variat[e]' (Jensen 2014). Indeed, the methodological level is the one on which semiotics mainly operates and from which it addresses the other levels (Figure 9.5);

4 it considers, within the methodological level, not only data-gathering methods, summarised by 'follow the actors themselves' and 'generalized symmetry,' which have always characterised ANT as more of a methodology than of a theory (Law 2008: 141; Mol 2010: 253; Sayes 2014: 144), but also the descriptive–analytical methods and tools. Basically, what Latour calls the *infralanguage*.

Indeed, this methodological aspect of semiotics has been the one that led Latour to learn it at first:

> the systematic study of texts in this French tradition became what was imported into an American context as 'Theory.' While on this side of the Atlantic, I took it as exactly the opposite of 'theory': as the chance to acquire an empirical method.
>
> (Latour 2016: 468)

This is still the aspect that can lead ANT to learn from other semiotics, still and again.

Conceptual-theoretical level

4. Epistemological level
Assessment of the consistency of
level 3 and of the descriptions and
discoveries taking place on level 2.

3. Methodological level
Notions and procedures giving way
to level 2 are here elaborated.

Infralanguage + Follow the
actors+Symmetry + principles for writing
good accounts

2. Descriptive level
Metalinguistic representation of
level 1.

Accounts - Analyses

Empirical level

**1. Material studied or Object-
language level**

Empirical level

a.   b.

Figure 9.5   Comparison among disciplinary architectures: a. Greimas' (Greimas and Courtés [1979] 1982: 107 and 171); b. Latour's (2005)

Notes

1 But see Muniesa (2015) for an outline of ANT where semiotics is considered; Beetz (2017) and Høstaker (2005) for an account of the Greimassian semiotics in Latour's work; Blok and Elgaard (2011) and Schmidgen ([2011] 2015) for a general account of Latour's work, where semiotics is considered; Lenoir (1994) for a critique of Latour's use of semiotics.
2 After Hjelmslev ([1943] 1961: 131), in semiotics, an analysis is considered a description of relations. In order to highlight such connection, I will use the two terms together.
3 Actually, Greimassian semiotics has always tended to describe-analyse configurations beyond language and literary texts, which also resulted in early forms of ANT (Greimas [1972] 1995: 73). Since the 1990s, such extended use of semiotics' tools has been systematic (Floch [1995] 2000; Marrone 2002).
4 Denis and Pontille (2010) recover Akrich's proposal.

References

Akrich, M. (1990) 'De la sociologie des techniques à une sociologie des usages', *Techniques et Culture*, 16: 83–110.
——— (1992a) 'Appendice. Sémiotique et sociologie des techniques: jusqu'où pousser le parallèle?', in Centre de Sociologie de l'Innovation (ed.), *Ces réseaux que la raison ignore*, Paris: L'harmattan, 24–30.
——— (1992b) 'The De-scription of Technical Objects', in W.E. Bijker and J. Law (eds.), *Shaping Technology/building Society: Studies in Sociotechnical Change*, Cambridge, MA: The MIT Press, 205–224.
——— (1993) 'Les objets techniques et leur utilisateurs : de la conception à l'action', in B. Conein, N. Dodier, et L. Thévenot (eds.), *Les objets dans l'action*, Paris: Editions de l'EHESS, 35–57.
Akrich, M. and Latour, B. (1992) 'A Summary of Convenient Vocabulary for the Semiotics of Human and Nonhuman Assemblies', in W.E. Bijker and J. Law (eds.), *Shaping Technology/building Society: Studies in Sociotechnical Change*, Cambridge, MA: The MIT Press, 259–264.
Blok, A. and Elgaard, T.J. (2011) *Bruno Latour: Hybrid Thoughts in a Hybrid World*, London: Routledge.

Colman, F.J. (2005) 'Affect', in A. Parr (ed.), *The Deleuze Dictionary*. Edinburgh: Edinburgh University Press, 11–12.

Deleuze, G. and Guattari F. ([1972] 2003) *Anti-Œdipus. Capitalism and schizophrenia*, transl. R. Hurley, M. Seem and H.R. Lane, Minneapolis: University of Minnesota Press.

Denis, J. and Pontille, D. (2010) *Petite sociologie de la signalétique. Les coulisses des panneaux du mètro*, Paris: Presses de l'École des mines.

Farías, I. and Mützel S. (2015) 'Culture and Actor Network Theory', in J.D. Wright (ed.), *International Encyclopedia of the Social & Behavioral Sciences*, Oxford: Elsevier, 523–527.

Floch J.M. ([1995] 2000) *Visual Identities*, transl. P. Van Osselar and A. McHoul, London: Continuum.

Fontanille, J. (2004) *Soma et séma: figures du corps*, Paris: Maisonneuve et Larose.

Greimas, A.J. ([1972] 1995) 'Toward a Topological Semiotics', transl. P. Perron and F.H. Collins, *Nordic Journal of Architectural Research*, 8(4): 65–81.

Greimas, A.J. and Courtés, J. ([1979] 1982) *Semiotics and Language. An Analytical Dictionary*, transl. L. Christ et al., Bloomington: Indiana University Press.

Hennion, A. ([1993] 2015) *The Passion for Music: A Sociology of Mediation*, transl. M. Rigaud, London: Routledge.

Hjelmslev, L. ([1943] 1961) *Prolegomena to a Theory of Language*, transl. F.J. Whitfield, Madison: The University of Wisconsin Press.

Høstaker, R. (2005) 'Latour – Semiotics and Science Studies', *Science & Technology Studies*, 18(2): 5–25.

Ingold, T. (2007) 'Materials against Materiality', *Archaeological Dialogues*, 14(1): 1–16.

Jensen, C.B. (2014) 'Continuos Variations. The Conceptual and the Empirical in STS', *Science, Technology & Human Values*, 39(2): 192–213.

Latour, Bruno ([1984] 1988) *The Pasteurization of France*, transl. A. Sheridan and J. Law, Cambridge, MA: Harvard University Press.

———— (1988) 'A Relativistic Account of Einstein's Relativity', *Social Studies of Science*, 18(1): 3–44.

———— ([1993] 2000) 'The Berlin Key or How to Do Words with Things', transl. P. Graves, in P. Graves (ed.), *Matter, Materiality and Modern Culture*, London: Routledge, 10–21.

———— ([1998] 2017) *Piccola filosofia dell'enunciazione*, Roma: Aracne.

———— (2000) 'On the Partial Existence of Existing and Nonexisting Objects', in L. Daston (ed.), *Biographies of Scientific Objects*, Chicago, IL: The University of Chicago Press, 247–269.

———— (2005) *Reassembling the Social: An Introduction to Actor-Network-Theory*, Oxford: Oxford University Press.

———— ([2012] 2013) *An Inquiry Into Modes of Existence*, transl. C. Porter, Cambridge, MA: Harvard University Press.

———— (2014a) 'How Better to Register the Agency of Things | Tanner One – Semiotics' *Tanner Lectures*, Yale University, 26th and 27th of March 2014 (www.bruno-latour.fr/node/562, visited on the 12 of December 2018).

———— (2014b) 'On Selves, Forms and Forces', *Hau: Journal of Ethnographic Theory*, 4(2): 261–266.

———— (2016) 'Life among Conceptual Characters', *New Literary History*, 47: 463–476.

Latour, B. and Fabbri, P. (1977) 'La rhétorique de la science [pouvoir et devoir dans un article de science exacte]', *Actes de la recherche en sciences sociales*, 13(1): 81–95.

Law, J. (1999) 'After ANT: Complexity, Naming and Topology', in J. Law and J. Hassard (eds.), *Actor Network Theory and After*, Oxford: Blackwell, 1–14.

———— (2002) *Aircraft Stories: Decentering the Object in Technoscience*, Durham, NC: Duke University Press.

———— (2008) 'Actor Network Theory and Material Semiotics', in B.S. Turner (ed.), *The New Blackwell Companion to Social Theory*, Oxford: Wiley-Blackwell, 141–158.

Lenoir, T. (1994) 'Was the Last Turn The Right Turn? The Semiotic Turn and A. J. Greimas.', *Configurations*, 2(1): 119–136.

Marres N. and Rogers, R. (2008) 'Subsuming the Ground: How Local Realities of the Fergana Valley, the Narmada Dams and the BTC Pipeline Are Put to Use on the Web', *Economy and Society*, 37(2): 251–281.

Marrone, G. (2002) *Corpi sociali. Processi comunicativi e semiotica del testo*, Torino: Einaudi.

Mol, A. (2010) 'Actor-Network Theory: Sensitive Terms and Enduring Tensions', *Kölner Zeitschrift Für Soziologie Und Sozialpsychologie*, 50: 253–269.

Mol, A. and Mesman J. (1996) 'Neonatal Food and the Politics of Theory: Some Questions of Method', *Social Studies of Science*, 2(2): 419–444.

Muniesa, F. (2015) 'Actor Network Theory', in J.D. Wright (ed.), *International Encyclopedia of the Social & Behavioral Sciences*, Oxford: Elsevier, 80–84.

Peirce C.S. (1868) 'On a New List of categories', *Proceedings of the American Academy of Arts and Sciences*, 7: 287–298.

de Saussure, F. ([1916] 1959) *Course in General Linguistics*, transl. W. Baskin. New York: Philosophical Library.

Sayes, E. (2014) 'Actor–Network Theory and Methodology: Just What Does It Mean to Say That Nonhumans Have Agency?', *Social Studies of Science*, 44(1): 134–149.

Schmidgen, H. ([2011] 2015) *Bruno Latour in Pieces. An Intellectual Biography*, transl. G. Cunstance, New York: Fordham University Press.

10

What did we forget about ANT's roots in anthropology of writing?

Jérôme D. Pontille

It is something that is often forgotten: ANT is partially born from an interest in written traces, and some of its main roots are deeply embedded in anthropology of writing, especially Jack Goody's work (Goody, 1977, 1986; Goody & Watt, 1963). Indeed, mapping sociotechnical networks largely amounts to following small traces of paperwork, marks on sheets of paper, specific words in documents (e.g. authors names, citations), files, minutes, reports, etc. Beyond the scientific literacy as the starting point of ANT's investigations, such a focus on written traces calls for unfolding the utterly graphic quality of modern societies.

Our world, namely writing societies (as anthropologists used to put it), is literally saturated with traces and written objects. Any human being is constantly defined with birth and death certificates, identity papers, school or professional degrees, employment contracts, fingerprints, property acts, marriage agreement... and of course handwritten signatures. Watches, calendars and diaries, clocks, rules and yardsticks, scales... and of course money are also crucial in the coordination and synchronisation of actions. Similarly, a contemporary city would not exist without its architectural plans and drawings, street plaques, directory signs, road markings, shop signs... and of course its map, on both printed and online versions. The State itself would be little, if anything at all, without administrative lists, regular population census, archives of many kinds, a lot of maps of different scales, resource inventories... and of course national statistics.

In this chapter, I want to acknowledge the basic place of writing in our world by going back to ANT's roots in anthropology of writing. Such a move, I argue, is helpful in better understanding, at least, two crucial aspects of the contemporary life that are constantly performed and enacted by written traces: Forms of reasoning and modes of governing. These practices are commonplace in ANT studies, and the best-known vocabulary used by scholars to describe them is one of the 'immutable mobiles' (Latour, 1986) and their stabilising properties. However, following the developments in anthropology of writing is key in unfolding and (re)discovering how far the multiplicity of written traces goes beyond immutable mobiles, and gives access to hitherto neglected practices in producing knowledge and performing politics.

Jérôme D. Pontille

The materials of writing and cognition

If the borrowings from the anthropology of writing may be various in ANT-inspired studies, one gesture is particularly significant: Departing from the usual conception of both writing and cognition as a solely intellectual activity. Such a plea for a more open, widened definition of reasoning oriented ANT's early investigations towards unexpected and overlooked practices.

Inscribing, scribbling, tracing

Latour's laboratory ethnography (Latour & Woolgar, 1986) shared with Goody's early work (Goody, 1977, 1986; Goody & Watt, 1963) a very specific approach to writing. Driven by the will to withdraw from traditional 'phonetic' approaches that treat writing only as a linguistic phenomenon that mirrors orality, they apprehended writing and, more generally, any graphic operation 'for themselves,' especially foregrounding properties that are independent of speech. In a similar vein, Derrida (1967), another crucial inspiration for this analytical and empirical gesture (Lenoir, 1998), constantly argued that the forms and materiality of texts are active, fully fledged elements of meaning itself, which cannot be detached from written artefacts.

These assumptions lay foundations for a materialist approach that dramatically reorients investigations of writing. Not only could texts and documents be examined from a different angle, but writing practices themselves became a worthwhile object to be scrutinised more carefully. Among them, Goody famously insisted on the importance of studying the manufacture of lists and tables. Making a written list (of goods, of guests, of administrative entities, etc.) is both commonplace and a very distinct activity from enumerating items orally. First, the list 'reduces oral complexity to graphic simplicity' (Goody, 1977, p. 70). Second, as writing gives a relatively permanent form to words, the listed items are materialised (e.g. traces on clay tablets, pigments on papyrus, ink on paper, etc.) and, consequently, can 'be inspected, manipulated and re-ordered in a variety of ways' (Goody, 1977, p. 76). The shift from the auditory evanescence to the visual permanence significantly opens up the ways in which written words, and subsequently sentences, can be separated, aggregated and hierarchised. Unexpected combinations are thus made possible, sometimes giving rise to the production of new categories and distinct classification systems.

More generally, such a materialist approach to writing radically broadens the scope of components that can be explored. Beyond the numerous spatial arrangements of sentences and words on a scribbled or printed page, it also calls attention for the distinct surfaces and the multiple materials involved in writing practices. For instance, after some excursions into the bench space of a scientific laboratory at the Salk Institute, Latour and Woolgar insist on the diversity: 'a large leatherbound book... blank sheets with long lists of figures... writing on pieces of paper... numbers on the sides of hundreds of tubes... pencilling large numbers on the fur of rats... coloured papertape to mark beakers' (Latour & Woolgar, 1986, p. 48). Taking the materials of writing into consideration brings the heterogeneity of written objects to the fore, even within an otherwise circumscribed space like a scientific laboratory.

In this regard, the radical ANT gesture, especially in laboratory studies, resided in breaking with the focus on the sole 'content' of scientific texts. The attention towards theoretical statements and discovery claims, cherished by the previous science studies tradition, was considerably redirected to a wide range of scribbling and writing practices. The main notion that drove this movement was that of 'inscription,' directly borrowed from Derrida. A very generic term, inscriptions can be written words, of course, but also 'all traces, spots, points,

102

histograms, recorded numbers, spectra, peaks, and so on' (Latour & Woolgar, 1986, p. 88). Comprising a multiplicity of graphic expression, the notion is ideally suited for describing the detailed process of fact production in science, from the very first inscriptions meticulously traced at the workbench to articles published in scientific journals. More extensively, it is particularly fruitful to investigate how far a workplace – being a laboratory, a legal court or a department in any organisation – is itself 'a hive of writing activity' (1986, p. 51) in which numerous written objects proliferate.

The careful attention to inscriptions driven by a materialist approach calls for expanding the investigation of writing practices even further. As soon as the aim is to repopulate the manufacture of the written world, going beyond inscribed marks, traces and words becomes obvious. Therefore, machines and technologies naturally come into the picture. The printing press, and the reproduction of documents at a large scale it introduced, is one of the more manifest examples (Eisenstein, 1983). This is also the case of the various 'inscription devices' (Latour & Woolgar, 1986, p. 51), such as a mass spectrometer or a bioassay that transforms pieces of matter into written traces of all kinds (dots, curves, figures, etc.). Likewise, staplers, paperclips, post-it notes, coloured folders with rubber band, computers, printers and other office furniture are the key in the daily production of written objects in many work settings (Gardey, 2008). Widening the inquiry to these literary technologies is an important gesture to get free from the usual image attributed to writing. Rather than a single, individualist act performed by a few exceptional minds, writing appears as a collective performance and an equipped practice.

The heterogeneous components of cognition

Notably inspired by developmental psychology, Goody's central concern was cognition. Writing, Goody (1977, 1986) claimed, has facilitated the development of particular cognitive skills due to its different properties from orality, especially its fixity and sturdiness. Materially inscribed, signs are made durable and acquire more or less permanency over time. Simultaneously, writing goes with a spatial arrangement of information that spurs on diverse cognitive activities such as memory, comparison, computation and ranking. This groundbreaking theoretical standpoint confronted a long tradition in social sciences: Instead of presupposing essentialist or naturalistic differences between populations' minds, scholars should consider the 'intellectual technologies' these populations used.

The materialist exploration of writing as a peculiar 'intellectual technology' considerably renewed our understanding of the institutionalisation of scientific objectivity. Shapin (1984) notably showed how what he called a 'literary technology' (the practice of enabling the reader to become a witness of the experimental scene) was designed as part of the early experimental programme so as to enlarge the public involved in the generation of authentic matters of fact. Similarly, Bazerman (1988) underlined the role of specific writing technologies (e.g. text formats, reference styles and specialised journals) in the shaping of scientific knowledge. This equipment appeared essential to secure some ways of reporting and recording, but also to multiply memory practices and to stabilise modes of reasoning. Eisenstein (1983) has famously demonstrated the cognitive role of the printing press in what she called an 'unacknowledged revolution' in religion and in science. The identical reproduction of documents actually revealed the existence of simultaneous distinct versions of a text, leading to comparison with extreme scrutiny for the sake of precision that sometimes resulted in the production of unexpected, new knowledge. In a similar vein, by drawing on both Goody's and Eisentein's contributions, Latour (1986) famously characterised inscriptions in these terms: the more immutable, the more powerful.

Following scientists and technicians in the field and at the workbench, ethnographers also described in minute detail the gestures at the core of fact production. Not only were written objects found diverse and numerous in workplaces, but they were also handled in many ways. Manipulated, checked, compared, sorted out, compiled, assembled, partially copied from one another, such a dance of inscriptions and documents actually made manifest the manufacture of cognition with and through written objects. In this step-by-step process, modes of visualisation – such as lists, images, field guides, tables, numbers and curves – proved to be particularly important (Lynch & Woolgar, 1990). Cognitive innovation in the realm of scientific reason was thus investigated as a collective and embodied activity, both deflating the central place of individual genius and exalting the significance of 'thinking with eyes and hands' (Latour, 1986).

The material aspects of cognition and the role of inscriptions in the concrete forms of reasoning are also at the centre of another stream of works that cultivated a constant conversation with ANT and anthropology of writing, and go by the labels of *distributed cognition* or *situated action*. The way Hutchins (1995), Kirsh (1995), Lave (1988) and Suchman (1987) accounted for inscriptions and their materials did not oppose to the structuralist linguistic view on writing, though, but to cognitive sciences and, more generally, experimental psychology. Written artefacts here, among a larger set of 'cognitive artifacts' (Norman, 1991), are taken into consideration to demonstrate the externality and the diversity of cognition processes which, these authors claimed, should not be studied from a strictly mentalist perspective. When observed 'in the wild' (Hutchins, 1995), cognition appears to lie on complex and hybrid 'systems' in which different kinds of material representations are put into circulation. Complementing the early laboratory studies, these works paid attention to the diverse material properties of these 'external' representations, their spatiality and the various ways they equip memory, calculation or computation. Doing so, they set the basis for an ecological approach to workplaces in which cognition and knowledge are never fixed and internal phenomena, but heterogeneous and sociomaterial processes.

A case study: updating a biomedical database

Shadowing Kelly, I learned how far a hospital is a great setting to illustrate the role of the materials of writing and cognition. Kelly was a data management technician involved in the updating of a medical database dedicated to the predictive factors for a joint disease, based on the monitoring of a cohort of 880 individuals who were willing to be examined once a year over the period 1992–2002 (Pontille, 2010). This database assembled different kinds of traces: Annual health check X-rays of all painful joints, biological samples (blood serum itself and DNA) and a clinical examination conducted through a detailed questionnaire on patients' physical state (e.g. feelings of fatigue, severe pain and ability to move around) and mental state (e.g. feelings of being a burden to others and insomnia), as well as their daily activities (e.g. work activity, social life, family support and nervous tension).

These various inscriptions had been stacked in a particular room of the hospital department where the medical record of each patient was stored in some box files. Although these documents seemed well ordered at first sight, I discovered the mess resulting from their liveliness. Every time a patient attended an annual health check, the nurses and doctors would fetch their respective files in order to fill them in. The files, thus, were in a daily circulation between the storage room and the rheumatology unit, and over the ten years of patient follow-up, a growing disorder gradually set in.

The first thing Kelly did struck me: She did not carefully read the various written materials composing the database to be updated. Rather, she sorted each patient's documents in chronological order in box files, and arranged them in alphabetical order on the shelves. Carried around, moved, put in piles, sorted and arranged, the patients' files were ordered in a systematic manner that performed the stabilisation of the working environment. By handling and manipulating these files and box, Kelly actually imposed a particular spatial organisation on the documents. In doing so, the shelves and the box files constituted a visual memory, enabling Kelly to prepare and anticipate future activities: The documents were spatially arranged in a way that facilitates routines in perception and action (Kirsh, 1995).

Therefore, I understood that the handling of the written objects was not purely a manual task. Simultaneously, Kelly noticed the sort of files that are part of the database, the dates of the health checks or the patients' names. As a consequence, she became increasingly familiar with the range of data that composed the material infrastructure of the medical database. In other words, Kelly acquired an intimate knowledge of the informational ecology she had to deal with in the next steps: Gradually updating the medical database mainly consisted in reading, cross-checking some data, copying a piece of information from one document to another, converting some inscriptions into figures, storing specific documents in a cardboard folder and so on. Through such a processual work, Kelly verified the general validity of data and their medical coherence. In other words, she made sure that these data became reliable scientific information, and could be taken as the starting point for articles to be published in specialised journals.

Governing with words and figures

Anthropologists and ANT scholars have not questioned the relationship between writing practices, written artefacts and cognition in isolation, however. In their accounts, the manipulation of lists, the circulation of inscriptions and the reproduction of texts are always tied to concrete modes of organising and governing. This is very clear in *Logics of writing* in which Goody (1986) associated specific writing practices with the emergence of modern institutions. It is also explicit in *Science in Action* where Latour (1987) recalled that in the 18th century, the French explorer Lapérouse and his team privileged the map in a notebook instead of a map drawn in the sand, because the King (Louis XIV) asked them to bring geographical knowledge back to Versailles in order to anticipate new commercial journeys. Numerous scholars have pursued these reflections, also following Michel Foucault's invitation to investigate knowledge practices and the exercise of power altogether (Foucault, 1978). The role of writing in organisations, and notably within governments, has been discussed, essentially, in two apparently opposite directions. On the one hand, privileging a focus on sturdiness and immutability, scholars described the ordering powers of standardised writing. On the other, researchers insisted on the diversity and ambiguity of written objects that can be found in organisations and administrations, ordering devices and fragile components of lively sociomaterial ecologies.

Standardisation

Numerous scholars studied organisations, and especially bureaucracies, through the lens of writing practices and their progressive standardisation. This is notably the case of Chandler (1977), who famously foregrounded the role of the emergence of innovative accounting and management devices in the dramatic growth of companies during the second half of the 19th century. In his view, the 'managerial revolution' that made possible such unprecedented development was mostly an 'information revolution' (Chandler & Cortada, 2003). Firms, especially railroads and telegraph companies, faced new challenges and grew new 'needs' for precision, notably in their internal processes of coordination. This led to a huge amount of innovation that considerably reorganised the forms of communication in organisations, and especially writing. All functionalism left aside, a lot of subsequent studies, informed and inspired by ANT, confirmed and complemented this narrative, showing how, at the turn of the 20th century, companies and administrations progressively adopted numerous normalised written artefacts (Yates, 1989) and invested in the standardisation of writing practices (Gardey, 1999; 2001), structuring in the same trend organisational forms, information flows and ordinary work (Agar, 2003; Beniger, 1986).

Scientific management crystallised this formalisation process (Thévenot, 1984). On the one side, it expanded the scope and reinforced the role of accounting devices, and more generally quantification, in the daily organisation of work (Miller & O'Leary, 1987). On the other, it reconfigured organisations as 'systems,' rationalising not only every physical gesture in factories, but also clerical work within offices, simultaneously drawing on a systematic struggle against idiosyncrasies and small talks, considered as 'informal,' and thus useless, communication (Jelinek, 1980). Above all, the managerial control of written communication gave birth to an important movement of mechanisation. Once formalised, clerical work could be seen as a trivial matter that did not require real intellectual skills. Some of it (like copying) could be taken care of by devices, some other (like typing, or sorting files out) could be handled by unqualified employees, who happened to be almost exclusively women (Gardey, 2001). These projects of mechanisation were reinforced by the first attempts to build computers and the birth of digitisation, during which engineers identified writing and reading tasks they considered as already 'mechanical' (routines), in order to delegate them to machines (Agar, 2003).

In the vocabulary of ANT, these accounts show that writing practices and devices have become one of the most powerful 'tools for managing complexity' (Callon, 2002). Embodying a 'political economy of representations' (Law, 1994, p. 27), they perform a variety of 'modes of ordering' that is at the core of modernity. These studies highlight a general shift in this political economy: Since the end of the 19th century, the massive use of standardised representations and the circulation of a growing number of immutable mobiles have turned administrations and companies into huge centres of calculation, which have been governing more and more aspects of our lives. Modern societies have thrived on a pervasive normalised writing infrastructure that has grown steadily and yet remained largely unnoticed.

Beyond formality

Though their contributions to the understanding of the history of organisations and governing practices are decisive, these narratives may lead to a univocal and partial comprehension of writing practices in organisations. Inscriptions, indeed, are not limited to immutable mobiles, and writing practices go beyond formalisation and standardisation. Three aspects are particularly missing in the accounts which stick to the ordering properties of writing devices: The fragmentation of writing infrastructures, the dynamic and uncertain processes that concretely ensure the circulation of written artefacts and, more generally, the disorder that comes with writing, even when it is used as an ordering technology.

First of all, governing practices themselves never take the form of a perfect panopticon, the all-seeing surveillance device imagined by Bentham and discussed by Foucault. Far from being totalising enterprises that draw on manifold standardised inscriptions to build an exhaustive representation of the world, they are always situated, partial and selective. For instance, governing a city such as Paris on a daily basis goes through the multiplication of what Latour & Hermant (1998) called 'oligopticons,' that is, narrow windows giving an always fragmented and incomplete view of urban reality. Each view is fuelled with specialised inscriptions (concerning water, electricity, telephony, traffic, meteorology, town planning, etc.) that are never fully interoperable and do not mechanically allow to zoom in or zoom out the city as a whole.[1]

Beyond the usual discourses about the seamless and transparent circulation of information, various studies emphasised the processual and dynamic nature of writing practices dedicated to organising and governing. The procedures shaping the information dissemination, as formalised and standardised as they may be, are never straightforward. Rather, a meticulous work made of verification, control and validation is constantly performed. The case of law, as studied by Latour (2009) and other in ANT-inspired studies, is particularly telling in this regard: Far from a mere and fluid flow of documents from one step to another in a legal procedure, making a case goes with numerous minute writing acts that articulate, in practical terms, a personal situation into a petition to the judge, a claim into a file, an opinion into a legal decision. Following the dynamics of inscriptions actually revealed the crucial place they take in the concrete passage and adjustable constitution of law (Fraenkel et al., 2010; Latour, 2009). Similarly, the daily performance of regulations goes through constant adjustments and realignments that concretely make possible the mobility of artefacts that are never as immutable as they may appear from farther away (Weller, 2008).

As soon as the activities are carefully described, writing practices no longer appear as operations of formalisation mostly, if not exclusively, directed towards a univocal process of rationalisation. Their extreme variety comes to the fore, bringing to light the messy side of the writing production process and the disorder it unavoidably provokes (Hull, 2012; Kafka, 2012). This is also the case with numbers, which are generally considered as intrinsic vehicles for accurate representations. Instead, the role of false numbers in diverse documents and professional settings is the key, as temporary or conditional devices enabling to proceed by trial and errors, to give room to indecision for a little while, or to make forecasts (Lampland, 2010).

Surely, city landscapes are perfect sites for the appearance of unexpected urban inscriptions, alongside the official public lettering displayed to organise distinct places (e.g. street names plates and road markings) and to manage distinct flux (e.g. directory signs and real-time information). Among these alternative graphic manifestations, the multiplicity of inscriptions and sometimes disruptive written objects (such as banners, stamps, posters and stickers) accompanying public demonstrations is commonplace (Artières & Rodak, 2008), even though these demonstrations may be, to some extent, authorised and supervised by municipal officials. By contrast, graffiti is particularly interesting in that its sudden appearance, in terms of sites, size and frequency, cannot be anticipated and controlled. On the other hand, it simultaneously gives rise to graffiti removal policies and dedicated daily practices in large cities around the world (Austin, 2001; Shobe & Banis, 2014).

As Law (1994) showed early on, these two sides of writing, ordering technology and fragile and messy practices, are not as paradoxical as they may seem. The situated writing operations that take place in organisations and administrations show how much governing goes through the circulation of swarming inscriptions and draws on disordered workplaces.

A case study: verifying files in a bank

Observing as ordinary and arid, a process as files verification is a great way to apprehend the role of writing infrastructures in organisations, while steering clear of all formal reductionism. I had the opportunity to conduct an ethnography in a workplace dedicated to such activity in the banking sector (Denis, 2011). There, employees were mandated to scrupulously verify the absence and presence of pieces of information in documents that had been filled in by commercial operators and their customers in order to open a bank account or activate new services.

Their offices were surprisingly messy, populated by all kinds of written artefacts: Sticky notes, loose sheets of all size, copies, folders, binders, corporate flyers and so on. This overwhelming presence of inscriptions (on desks, shelves and even the floor) was also manifest on the screens before which every employee was sitting: Various kinds of software were constantly running, and numerous 'windows' were open simultaneously. The movement that was constantly transforming this bewildering landscape was itself particularly intriguing. Files were displaced every day, notes circulated from one desk to another before ending down in the trash bin, new ones were created every five minutes or so, binders were open, filled in or emptied out and so on.

What also struck me was the volatility of files themselves, which proved way less stable and rigid than I first imagined. In fact, once on the desk of an employee, the first thing a customer file was subjected to was disassembly. Folders were open, paperclips and staples were removed, sheets were displaced and the pieces were reorganised all over the desk. Above all, after being verified, the pieces were reassembled in a new file that could sensibly differ from the previous one. Some documents might have been pulled out, whilst new ones (printed versions of screenshots, for instance) might have been added. Throughout its verification, the content and shape of the file changed more or less dramatically.

Parts of the verification process also consisted in providing graphical additions to some documents. For instance, when the employees discovered dubious blanks in files, instead of simply dismissing the guilty file, they would generally manage to fill in the form by themselves. Interestingly, the documents were also modified during the mere action of reading, which, I understood, was not a matter of eyes only, and implied the use of pencils and highlighting pens. During the verification process, the bank employees were adding marks in the margins, underlining or highlighting words and figures, progressively incorporating new inscriptions to every form, loose sheet or copy they handled. These transformations were not only ways of easing and securing the reading of the files. They changed the very intelligibility of every document, which would bear markers for their next reading and traces of their own verification.

Instead of going against data processing and files verification, I rapidly understood that such heterogeneous and volatile writings, overcrowded screens and desks drowned under papers were an essential part of the ecology from which data emerged and in which they circulated. So were the physical components of files, such as staples, folders, paperclips and tabs, as well as the colour of ink, the paper formats and weights, and so on. They seemed to be the very conditions under which files could pursue their journey through the company. Mess and labour are therefore crucial dimensions of the daily functioning of centres of calculation, preventing the circulation of immutable mobiles from seizing up.

Conclusion: writing infrastructures and the politics of accounts

From mundane traces to official documents, from marks on paper to data: By recalling the relationships between ANT and anthropology of writing, this chapter aimed at redirecting attention towards the manifold writing infrastructures that ground modern societies. Several aspects call for this 'rediscovery' of writing practices, as we witness the so-called data revolution and the neopositivism that seems to go with it. The invisiblisation of work is one of them. Openness and transparency put to the fore certain properties of data that neglect the frictions their circulation generates (Edwards et al., 2011), and deny the need for most of data work and workers (Denis & Goëta, 2017). Anthropology of writing invites us to pay attention to what this new era of information automatisation owes to what happened in firms and administrations at the turn of the 20th century. Identifying these continuities will help to question the moral economy of work contemporary centres of calculations carry.

Another aspect concerns the politics of facts. Early on, Dorothy Smith called attention to the 'documentary reality' that characterises contemporary societies, and urged to question the apparent neutrality of written accounts that populates our lives. While taking the appearance of universal facts, she explained, these formal accounts, as mundane as they may appear, are always the product of specific organisations and tailored for their specific needs (Smith, 1974). Since then, a lot of scholars raised these concerns, investigating, for instance, the politics of categories (Bowker & Star, 1999), highlighting the role of quantification and accounting devices in the construction of reality and its corollary: The silencing of certain entities and problems (Carruthers & Espeland, 1991; Espeland, 1993; Miller & O'Leary, 1987; Porter, 1996), or more recently discussing the generalisation of audit practices (Strathern, 2000). Thick descriptions of the situated production and circulation of written accounts of all kinds, including data, directly contribute to such a gesture of de-naturalisation, notably by surfacing the most mundane and processual dimensions of the documentary construction of reality. Such de-naturalisation should not be understood as an overall critique or deconstruction of that reality, though. As it has been witnessed in the case of the virulent debates around climate change, surfacing the concrete conditions though which written accounts are generated may aim at reinforcing trust in institutional procedures (Edwards, 2010; Latour, 2004, 2017).

To end this chapter and illustrate this last point, let us go back to the manufacture of scientific texts. Attribution technologies are crucial in the shaping of scholarly accounts, even though largely under-problematised. In academia, evaluation is firmly linked to authorship that is regularly referred to as the primary 'currency' for hiring and advancement (Biagioli, 1998; Pontille, 2004, 2016). And the myth of individual excellence is mostly made effective by attaching one's name to scientific publications. Furthermore, with the use of citation counts and author-level metrics in various research assessment exercises, scientific names have been progressively turned into 'units' to be counted. This trend considerably masks the institutional, economic and political contexts in which research is actually performed, by totally disregarding and belittling the collective work done. Against the never-ending competition for the sake of individual excellence, one could promote the systematic use of collective names in publishing. Alongside the shared custom in high-energy physics (Galison, 2003), some initiatives are already available in the literature (e.g. The SIGJ2 Writing Collective, 2012; Collectif Onze, 2016).

As part of this politics of accounts, the present chapter itself suggests another option. The author-in-the-text is 'Jérôme D. Pontille' whereas the author-in-the-flesh is twofold: Jérôme Denis and David Pontille.[2] Since the end of the 1990s, we have used the attachment of our two civil names as a way 'to diffract' our own authorship, following Haraway (1996), and to disrupt the emphasis of single individuals performed by most of research assessment frameworks. As the editors of the ANT Companion were concerned with two different chapters bearing the same name(s), we created a fictional author as a way to pursue the diffraction process. Such a gesture is perfectly in line with the semiotic foundation of anthropology of writing and ANT (see the second footnote in Latour, 1988), which both insisted on the generative process of writing practices that bring new entities to existence and participate in their maintenance. The different fieldworks mobilised in this chapter come from either Denis' or Pontille's, or Denis and Pontille's investigations.

Notes

1 See Färber, this volume.
2 See also David J. Denis's chapter in this volume.

References

Agar, J. 2003. *The Government Machine: A Revolutionary History of the Computer.* Cambridge, MA: MIT Press.

Artières, P., & Rodak, P. 2008. Écriture et soulèvement. Résistances graphiques pendant l'état de guerre en Pologne (13 décembre 1981-13 décembre 1985). *Genèses,* 70: 120–139.

Austin, J. 2001. *Taking the Train: How Graffiti Art Became an Urban Crisis in New York City.* New York: Columbia University Press.

Bazerman, C. 1988. *Shaping Written Knowledge: The Genre and Activity of the Experimental Article in Science.* Madison: The University of Wisconsin Press.

Beniger, J R. 1986. *The Control Revolution: Technological and Economic Origins of the Information Society.* Cambridge: Harvester.

Biagioli, M. 1998. The Instability of Authorship: Credit and Responsibility in Contemporary Biomedicine. *The FASEB Journal* 12(1): 3–16.

Bowker, G. C., & Star, S. L. 1999. *Sorting Things Out: Classification and its Consequences.* Cambridge, MA: MIT Press.

Callon, M. 2002. Writing and (Re)writing Devices as Tools for Managing Complexity. In J. Law & A. Mol (Eds.), *Complexities. Social Studies of Knowledge Practices* (pp. 191–217). Durham, NC and London: Duke University Press.

Carruthers, B., & Espeland, W. N. 1991. Accounting for Rationality: Double-Entry Bookkeeping and the Emergence of Economic Rationality. *American Journal of Sociology,* 97(1), 31–69.

Chandler, A. D. 1977. *The Visible Hand: The Managerial Revolution in American Business.* Cambridge, MA: Harvard University Press.

Chandler, A. D., & Cortada, J. W. 2003. *A Nation Transformed by Information: How Information Has Shaped the United States from Colonial Times to the Present.* Oxford: Oxford University Press.

Collectif Onze, 2016. Enquêter, écrire et publier en collectif. *ethnographiques.org,* 32, www.ethno graphiques.org/2016/Onze.

Denis, J. 2011. Le travail de l'écrit en coulisses de la relation de service. *Activités,* 8(2), 32–52.

Denis, J., & Goëta, S. 2017. Rawification and the Careful Generation of Open Government Data. Social. *Studies of Science* 47(5): 604–629.

Derrida, J. 1967. *De la grammatologie.* Paris: Les Éditions de Minuit.

Edwards, P. 2010. *A Vast Machine. Computer Models, Climate Data, and the Politics of Global Warming.* Cambridge, MA: MIT Press.

Edwards, P. N., Mayernik, M. S., Batcheller, A. L., Bowker, G. C., & Borgman, C. L. 2011. Science Friction: Data, Metadata, and Collaboration. *Social Studies of Science* 41(5):667–90.

Eisenstein, E. L. 1983. *The Printing Revolution in Early Modern Europe.* Cambridge: Cambridge University Press.

Espeland, W. 1993. Power, Policy and Paperwork: The Bureaucratic Representation of Interests. *Qualitative Sociology,* 16(3): 297–317.

Foucault, M. 1978. *The History of Sexuality: The Will to Knowledge.* London: Allen Lane.

Fraenkel, B., Pontille, D., Collard, D., & Deharo, G. 2010. *Le Travail des Huissiers: Transformations d'un métier de l'écrit.* Toulouse: Octares.

Galison, P. 2003. The Collective Author. In M. Biagioli & P. Galison (Eds.), *Scientific Authorship. Credit and Intellectual Property in Science* (pp. 325–55). New York: Routledge.

Gardey, D. 1999. The Standardization of a Technical Practice: Typing (1883–1930). *History of Technology,* 15(4), 313–343.

Gardey, D. 2001. Mechanizing Writing and Photographing the Word: Utopias, Office Work, and Histories of Gender and Technology. *History of Technology,* 17(4), 319–335.

Gardey, D. 2008. *Écrire, calculer, classer. Comment une révolution de papier a transformé les sociétés contemporaines (1800–1940).* Paris: La Découverte.

Goody, J. 1977. *The Domestication of the Savage Mind.* Cambridge: Cambridge University Press.

Goody, J. 1986. *The Logic of Writing and the Organization of Society.* Cambridge: Cambridge University Press.

Goody, J., & Watt, I. 1963. The Consequences of Literacy. *Comparative Studies in Society and History,* 5(3), 304–345.

Haraway, D. J. 1996. Modest Witness: Feminist Diffractions in Science Studies. In P. Galison & D. J. Stump (Eds.), *The Disunity of Science: Boundaries, Contexts, and Power* (pp. 428–441). Stanford, CA: Stanford University Press.

Hull, M. S. 2012. *Government of Paper. The Materiality of Bureaucracy in Urban Pakistan.* Berkeley: University of California Press.

Hutchins, E. 1995. *Cognition in the Wild.* Cambridge, MA: MIT Press.

Jelinek, M. 1980. Towards a Systematic Management: Alexander Hamilton Church. *The Business History Review,* 54(1): 63–79.

Kafka, B. 2012. *The Demon of Writing. Powers and Failures of Paperwork.* Brooklyn: Zone Books.

Kirsh, D. 1995. The Intelligent Use of Space. *Artificial Intelligence,* 73(1–2), 31–68.

Lampland, M. 2010. False Numbers as Formalizing Practices. *Social Studies of Science,* 40(3): 377–404.

Latour, B. 1986. Visualisation and Cognition: Drawing Things Together. *Knowledge and Society,* 6: 1–40.

Latour, B. 1987. *Science in Action. How to Follow Scientists and Engineers through Society.* Cambridge, MA: Harvard University Press.

Latour, B. 1988. Mixing Humans and Non-Humans Together. *Social Problems,* 35(3), 298–310.

Latour, B. 2004. *Politics of Nature: How to Bring the Sciences into Democracy.* Cambridge, MA: Harvard University Press.

Latour, B. 2009. *The Making of Law: An Ethnography of the Conseil d'État.* Cambridge: Polity Press.

Latour, B. 2017. *Facing Gaia: Eight Lectures on the New Climatic Regime.* Cambridge: Polity Press.

Latour, B., & Hermant, E. 1998. *Paris ville invisible.* Paris: La Découverte / Les Empêcheurs de penser en rond.

Latour, B., & Woolgar, S. 1986. *Laboratory Life: The Construction of Scientific Facts.* Princeton, NJ: Princeton University Press.

Lave, J. 1988. *Cognition in Practice: Mind, Mathematics, and Culture in Everyday Life.* Cambridge: Cambridge University Press.

Law, J. 1994. *Organising Modernity. Social Ordering and Social Theory.* Oxford: Blackwell.

Lenoir, T. 1998. Inscription Practices and Materialities of Communication, in *Inscribing Science Scientific Texts and the Materiality of Communication* (pp. 1–19). Stanford, CA: Stanford University Press.

Lynch, M., & Woolgar, S. (Eds.). 1990. *Representation in Scientific Practice.* Cambridge, MA: The MIT Press.

Miller, P., & O'Leary, T. 1987. Accounting and the Construction of the Governable Person. *Accounting, Organizations and Society,* 12(3): 235–265.

Norman, D. 1991. Cognitive Artifacts. In J.M. Carroll (Ed.), *Designing Interaction* (pp. 17–38). Cambridge: Cambridge University Press.

Pontille, D. 2004. *La Signature scientifique : Une sociologie pragmatique de l'attribution.* Paris: CNRS Éditions, "CNRS sociologie" series.

Pontille, D. 2010. Updating a Biomedical Database: Writing, Reading and Invisible Contribution. In D. Barton & U. Papen (Eds.), *Anthropology of Writing: Understanding Textually-Mediated Worlds* (pp. 47–66). London: Continuum.

Pontille, D. 2016. *Signer ensemble. Contribution et évaluation en sciences.* Paris: Economica, "Etudes sociologiques" series.

Porter, T. 1996. *Trust in Numbers: The Pursuit of Objectivity in Science and Public Life.* Princeton, NJ: Princeton University Press.

Shapin, S. 1984. Pump and Circumstance: Robert Boyle's Literary Technology. *Social Studies of Science,* 14(4), 481–520.

Shobe, H., & Banis, D. 2014. Zero Graffiti for a Beautiful City: The Cultural Politics of Urban Space in San Francisco. *Urban Geography* 35(4): 586–607.

Smith, D. E. 1974. The Social Construction of Documentary Reality. *Sociological Inquiry* 44(4): 257–68.

Strathern, M. (ed.). 2000. *Audit Cultures. Anthropological Studies in Accountability, Ethics and the Academy.* London and New York: Routledge.

Suchman, L. 1987. *Plans and Situated Actions: the Problem of Human-Machine Communication.* Cambridge: Cambridge University Press.

The SIGJ2 Writing Collective, 2012, What Can We Do? The Challenge of Being New Academics in Neoliberal Universities, *Antipode,* 44(4): 1055–1058.

Thévenot, L. 1984. Rules and Implements: Investment in Forms. *Social Science Information,* 23(1): 1–45.

Weller, J.-W. 2008. La disparition des deux bœufs du Père Verdon. Travail administratif et statut de la qualification. *Droit et Société,* 67: 713–755.

Yates, J. 1989. *Control through Communication: The Rise of System in American Management.* Baltimore, MD and London: The Johns Hopkins University Press.

11

As ANT is getting undone, can Pragmatism help us re-do it?

Noortje Marres

Actor–network theory (ANT) has been a giver of many gifts. It has produced new analytic vocabulary, research protocols, many memorable slogans and dictums, and a distinctive intellectual style. It has identified future directions of travel for thought. It has shown philosophers and ethnographers alike that there is a way of being in a field while not being *of* the field (Kelly, 2012). ANT has taken the famous British talent for self-depreciation a step further, by turning it into an embodied research practice, as in the case of the celebrated sociologist of technology describing his meeting with a company director in a deserted reception area while slouching uncomfortably in a chair that was too low (Law, 1994). However, as the German social and media theorist Erhard Schuttpelz put it rather brutally, 'all concepts are going downhill all the time.' I'm afraid that this extends to methods, and that the concepts and methods of ANT are no exception, and in these notes, I would like to sketch out briefly why I think ANT is not escaping this harsh but possibly fair fate, and what we might do about the inevitably challenging repercussions. However, it is also appropriate to note here that, in a variation on ANT's many slogans, there can be 'no re-composition without de-composition!' My charting of the undoing of ANT in this chapter is undertaken purposefully, with the express aim of identifying tasks for a redoing of ANT, which I will list in the conclusion.

Un- and redoing in ANT and Pragmatism

Indeed, one of the most generative, and possibly enduring, set of terms offered by ANT is that of 'composition' and 're-composition.' The idea, briefly put, is that something must get undone in order for it to be done. This does not only extend to the associative coupling which today can surely be called Latour (2005) speak – to assemble is to reassemble – but also to specific modalities of enquiry – or knowledge practice: To research is to re-search (Rogers, 2013), and to invent is to reinvent (Lury and Wakeford, 2012). Crucially, furthermore, the undoing of things in order for them to be redone, ANT style, does not only happen in spectacular moments, in periods of the so-called great transformations; composition and recomposition is the way of many practices, institutions and formations most of the time. The double movement of un- and redoing then does not only signal that an epoch-defining rupture may be unfolding but can equally indicate that a mundane process is underway.

This latter insight is central to the intellectual challenge that ANT poses to reifications and naturalisations of all sorts, to engrained habits of thought and action across the domains of science, politics, religion, economy and society, where it is still the default to presume the autonomous existence of entities (reality, the people, god, society, etc.), and that their legitimacy – their right to endure – is predicated on a pregiven autonomous existence. (This, in turn, has the consequence that many institutions today continue to be heavily invested in bracketing or even denying the ways in which the aforementioned entities are marked or even informed by practices, technologies or environments.) For ANT, by contrast, the autonomy of any entity is best treated as an accomplishment, one likely to depend on supporting arrangements and support acts, which are not just 'social' but include environmental entities previously categorised as nature, and which, when they cease to be operative, herald the undoing of the autonomy of said entity, and possibly of said entity itself, which will then require a redoing of some sort.[1]

In insisting in this way on de- and recomposition as a formative process that occurs across domains, ANT places itself in direct continuity with Pragmatism, the American philosophical and sociological tradition[2]: Like ANT, early 20th-century philosophers like Charles Peirce, William James and John Dewey had as one of their principal objectives to challenge naturalism and reification, and it was to this end that they developed a relational ontology – briefly put, the proposition that processes of association are critical to existence and the coming into existence of things. Like ANT, moreover, John Dewey (1938) identified 'problematisation' as a defining operator of these processes, not only in knowledge, but also in politics, morality and art. The undoing of entities – the loosening of associations – in order for them to be redone – for *different* associations to enter into the making of said entities – is not a mechanical process, but a profoundly risky one, one that puts entities and their relations *at stake*. As John Dewey (1908) put it, in the problematic situation, a 'conflict of tendencies' makes itself felt, one that can only be resolved by 'changing conditions.' To which ANT added: ...*and* by bringing different entities, and the possibility of their existence, into the mix. For both Pragmatism and ANT, problematisation is not some clean, authored process, in which to define or frame a problem is always already to render it solvable, but an unsettling occasion, on which the appearance on the scene of an entity that was previously disregarded unsettles the existing roles, relations, environments and capacities of agents and/or categories (they are *not* acting, *not* 'in control,' *not* having it covered), and necessitates their requalification. While most of this was already contained in Dewey's concept of the problematic situation, ANT added the multiplication of entities as a key dimension of this process. In studies of technoscience, the invention of biological, military and moral entities – the vaccine, the bomb and the pregnancy test – problematisation was shown to operate not only on roles, relations, environments and capacities of actors, but to enable the coming into being of new entities.

Even if I got to know ANT before I started reading American Pragmatism, the former for me is in many ways a redoing of the latter, with some important twists. Much can be said at this point, but what makes ANT most decisively different from Pragmatism, in my view, above and beyond its aforementioned commitment to variable ontology – the possibility of new entities coming into existence – is its commitment to heterogeneity, to heterogeneous composition, and the heterogenisation of domains. This can be loosely summed up as the insight that most things that matter arise out of association of entities deriving from diverse domains – science and politics, medicine and religion, engineering and culture, the home and the street. Some of ANT's most well-known slogans point in this direction: Science is politics by other means (Latour, 1993); engineering is sociology by other means (Latour, 1996). In ANT, it becomes clear that an associationist ontology, one which posits ongoing de- and recomposition, if consistently pursued, does not only unsettle institutional investments in naturalism and reification *within* different domains, but equally challenges institutional boundaries *between* domains, such as between science and politics, society and technology. If we pursue processes of association, including dis- and reassociation, we end up, not only with recognition of the hybrid character of much that exists in this world, but also with an insight into the ongoing reality and generative capacities of boundary-crossings.

To be sure, the notion that adopting an associationist ontology means pursuing or even affirming boundary-crossing is not unfamiliar to Pragmatism: Dewey (1991/1927) famously wrote that industry gives rise to new types of associations 'which break existing forms.' Still, for Pragmatists like Dewey, situations are ultimately containable within bounded conditions, as problem-solving follows problematisation as surely as night follows day. The moment in which the endurance of things and relations are put at stake – the moment of problematisation – in Pragmatism is a passing moment, and one, moreover, that can be assigned a designated place in a cycle of enquiry: It is at the beginning, as the problematicness of a problematic situation first becomes apparent, that 'a conflict of tendencies' makes itself felt, or 'something seems to be missing' (Dewey, 1908). In the Deweyian schema presented in the Logic of Enquiry, however, at the end of the cycle of enquiry – or mobilisation, or creation – humpty dumpty *will* have been put back together again.[3] Which is to say, Dewey's problematic situation had inscribed into it the expectation that decomposition will and should be followed by recomposition. ANT's notion of hybridisation points towards a dynamic that is more complex, one where entanglement renders the imposition of boundaries ever more challenging, and the ambivalence of some of ANT's proponents about precisely this notion of hybridisation suggests that they do not have Dewey's confidence on this point.

Even if ANT introduced the proposition of hybrid networks, and later, heterogeneous assemblages, much work conducted under this banner is very much like Pragmatism in its commitment to contain the moment of problematisation. ANT adopted from Dewey the idea of enquiry as cycle, and ANT, too, boxed in problematisation in a 'first phase' – the moment of the opening up of an existing proposition, say the combustion engine (Callon, 1980) – to be superseded by a 'next phase' of translation and then stabilisation of the new network. It's one reason why Latour's 'We've never been modern' remains an important book. Here, problematisation is shown to exceed efforts to contain it, not just on the institutional level – which Dewey recognised with his point noted earlier about the need to break existing forms – but also on the ontological level: In Latour's account, problematisation is the operator of a process of entanglement, as connections between distributed entities turn out to proliferate from the moment of problematisation onwards ('the climate is changing'; or in a contemporary example, 'AI is changing how institutions think'), gaining the capacity to disrupt established categories (there is nature, here is society; here is technology; there is democracy and the rule of law, etc.). Hybrids prove ever harder to contain. The recomposition or renewal of categories will leave these very categories transformed, as they are undeniably and indelibly marked by the boundary-crossings of their constituent entities.

However, it seems to me that here we reach a fork in the road and a path not taken, at least by ANT. Most of its proponents are not *that* interested in pursuing notions of boundary-crossing, the unsettling of presumptions of agency or the renewal of categories as entailing their transformation. Some of the most well-known inheritors of ANT, today, are located in the social studies of markets, digital media studies and in urban and architectural research, and in each of these cases, ANT is arguably deployed in order to elucidate a bounded domain. More precisely, this inheritance of ANT – that its successful uptake takes the form of the renewal of sub-disciplines – suggests that domains and methodologies must be clearly bounded in order for them to receive the ANT treatment). The further elaboration of ANT as a methodology of heterogeneous mapping, and as an agenda of heterogenisation, seems destined to be undertaken from a located, and to an extended 'domain specific' vantage point.[4] As the authors of the 1999s volume on the aftermath of ANT were clearly aware, ANT, as a more general approach or 'logic of enquiry,' is unlikely to escape the logic of de- and recomposition. Without further ado, I therefore would like to briefly discuss how I think ANT is being undone today and how it could, perhaps, be redone. In doing so, I will also touch on the few points of similarity and difference between ANT and Pragmatism noted earlier. For the sake of convenience, I will concentrate on three specific undoings of ANT as a general approach, each of which has to do with the resurfacing of categories, both empirically – in the world – and methodologically – as problems for research – that ANT claimed to have put to rest: (1) Interpretation is back. (2) Society is back. (3) Epistemology is back.

ANT's undoings: interpretation, society and epistemology

One of the legacies of ANT is its appetite for neologism, its commitment to replace familiar terms from the sociological and philosophical lexicon with new ones: Collective for society, trace for data, and actant for actor and object. These attempts at substitution entailed the banishment of certain key categories, including the three listed in this section title. I would like to propose here that these terms are ready to be reactivated, or have already been reactivated. This clearly puts me at risk of making a megalomaniac argument.[5] If I think this is worth it, it is because I am strongly, and increasingly stubbornly, convinced that some core assumptions of ANT – not least the notion that it is possible for researchers to 'follow traces,' and that this does not involve or require buying into any given 'theory,' at least not initially – have lost some of their efficacy, and even, are ceasing to be tenable. And, more importantly, that facing this challenge – that our propositions are going downhill, to speak with Schuttpelz, today possibly even more so than yesterday – is critical if we, or others, are to get somewhere with the task of redoing ANT in the face of its undoings. The challenge or danger is as follows: Each of the three key words identified in my short but heavy list earlier has been shown, by ANT *and* Pragmatism, to lead away from associationist ontology into a split world (dualisms!), where it is all too easy to succumb to reification. Keeping this danger in the front of our minds should then also help us to keep what is to be gained from the redoing of ANT firmly on the horizon.

Interpretation is back

Earlier, I defined ANT as an associationist ontology, but the approach is probably best known for its proposal to extend agency to non-humans, and its commitment to include objects, infrastructures and environments as active elements in social life. One important consequence of this, and one with which many anthropologists and some sociologists have taken issue, was the bracketing, in ANT, of the role of *interpretation* in social life, as well as in social research. ANT qualifies humans and non-humans as actants: They leave their trace in the world not through expression ('what they say'), but by way of their actions ('what they do'). And the role of ANT itself was not to offer interpretation, but to assemble traces into descriptions or accounts. However, one of the defining issues in the area of science, technology and society, today, precisely concerns interpretation, or more precisely, the reworking of the boundary between interpretation and action. As computation has emerged as a defining element in technological and scientific infrastructures, issues of intelligibility are once again in the foreground. Contemporary debates about bias are a case in point. In recent years, algorithms have been exposed as discriminatory, as they amplify the racist and sexist assumptions latent in the data from which they learn to detect patterns, such as databases recording police arrests (Eubanks, 2018), while in science the allegation of bias is treated by some as a 'cheap' trick deployed by subalterns who are trying to get in by policitising science, and in so doing debase the value of knowledge.[6]

In some respects, today's debates about bias seem highly susceptible to ANT treatment:

'bias' tends to be defined as an attribute of an actor (be it human or non-human), or an effect of an analytic operation, and both these conceptions of bias are highly reductive, as they tempt us to disregard the *process* by which bias emerges from relations *between* machines and humans, and many other legal, environmental, moral entities, besides. However, at the same time, an understanding of these relations in terms of 'actants,' as ANT would propose, does not enable us to grasp the problematic denoted as 'bias' either: It is rather the *labelling* of actors that is at issue here (Becker, 1963; Hacking, 1986). The deliberately thin categories of actant and trace are too ephemeral if the aim is to express the normative effect that comes with the negation of difference/the other.[7] *To get at the latter, we need to follow through the production of bias as an interpretative process.*

In the wake of controversies about bias, a seemingly ANT-like challenge surfaces: Do we blame society, or science and technology for these discriminatory effects? Surely, these controversies provide opportunities to point out that it must be both, that associations between humans and machines produce at least some of these discriminatory effects? ANT certainly has something important to contribute to these debates insofar as it compels us to widen the range of possible answers beyond the binary, to say either both or neither. It can show us the limits of an understanding of the politics of technology that treats bias as either subjective attribute or an objective effect. The problem is that the ANT manoeuvre of letting the hybrid emerge – it's both! – does not appear sufficient to account for this type of problematic situation: It is not only on the level of *association between humans and non-humans*, but from a dynamic of *interaction between entities and categories* that problematic effects arise here. *To say that bias emerges from hybrid networks doesn't articulate the problem.* Pointing to network formation between entities, or even, the situated enactment of entities, does not quite get us there. It does not really allow us to specify the process by which actors qualify their relations to others through the application of *categories*. To uncover that kind of operation on alterity – and by extension diversity – as something that is negated or affirmed, amplified or disavowed, we need to recover the qualities of interpretative process, the process by which a category gains a hold on an entity. This relative blind spot of ANT for interpretative effects has long been highlighted in feminist science studies (Leigh Star, 1990) but today's digital forms of circulation enable significantly more interactive and amplificatory interpretative circuits, requiring renewed engagement with this process and their effects.

The social is back

This brings us to the vexed but unescapable issue of the social. ANT is closely associated with what others with a taste for ceremonial declarations called the 'Death of the Social' (Rose, Baudrillard) and for which more constructively and/or creatively minded scholars minted other terms: the post-social (Knorr-Cetina), the more-than-human (Braun and Whatmore). In Latour's provocative appropriation of Tatcher, 'there is no such thing as society.' Today, however, at least from where I am writing in the UK, the social seems to be very much back on the agenda. As Will Davies (2015) noted, the last years have seen a proliferation of new 'socials,' as the label social is being slapped onto a variety of technical or commercial sounding propositions, from social media, to social innovation, social technology, social marketing, social bonds, social enterprise, social design, social listening and so on. To be sure, many of these new socials prove to be all too thin labels covering over a proposition that has much more to do with technology than with society: To add 'social' in practice often means to displace services away from public institutions and into alternative and – oh irony – hybrid architectures, realised with the aid of a mix of public and private funding, and more often than not involving digital platforms (for a more detailed discussion, see Marres and Gerlitz, 2018). In other words, the social in this vocabulary becomes an near-synonym of the displacement of the facilitation of 'togetherness' from a public responsibility to a profit-making activity, and as such pretty much means the opposite of what social policy, social housing, etc., used to mean in the 1960s. However, if we probe deeper, it becomes clear that 'sociality' is not only a public-facing label but equally denotes a methodology. As Kieron Healy, Bernard Rieder and Wendy Chun have forcefully reminded us, sociological methodology – from Simmel's triadic closure and Merton's homofily – has crucially informed the methods materialised in the 'new' social innovation paradigm (for a discussion, see Marres, 2017).

What does this mean for ANT? At the very least, it means that rejection of the very term ('social') has become unviable as an *empirical* strategy: The rise to prominence of social technologies clearly falls within the domain of study that ANT was designed, lest we forget, to elucidate: Science, technology and innovation. However, arguably more challenging is that the proposition of social technology – whether or not we find it credible or acceptable – is threatening to undo some of the propositions of early ANT, not least the notion of the 'laboratization of society.' In early work by Latour, Callon and Law, the extension of scientific infrastructures beyond the laboratory was assumed to make society more like science. However, when we consider how distinctively social methodologies are being operationalised in technological and analytical architectures across society today, it becomes clear that the opposite effects may equally arise. As platforms operate on the basis of liking and friending, social life becomes more like 'society' as imagined by sociologists: Preoccupied with reputation, pursuing the accumulation of social capital. Laboratisation does not make social life resemble a laboratory, but is resulting in the reformatting of activities in social sciences' image (reputation as currency, anyone?). However, at the point where we can see how the laboratisation thesis is being unsettled today, we can learn something important from Pragmatism: One of the lasting contributions of John Dewey has been to insist that experimentalism does not just denote a scientific paradigm, that it does not just belong to the sciences (Marres, and Stark, forthcoming). It is equally an approach in politics, morality and aesthetics. As the British novelist Joanna Kavenna puts it: 'If you're uncertain you will test all the parameters. If you're certain you will protect the citadel. The former is where things can shift because you have tested all the possibilities.'[8] The proliferation of experimental propositions across society and culture indicates not simply that the realm of science is being extended. They may equally denote the growing resonance of experimental traditions from literature, the arts and politics.

Epistemology is back

ANT has been closely associated with the ontological turn, the move – very briefly put – of locating entities and processes first on the plane of existence, and only then on the plane of knowledge. But clever readers and authors of ANT were quick to point out that it's not that straightforward. ANT is not existentialism. When you try doing so in relation to an actual example, it turns out to be really hard to separate knowing from existing, and that, indeed, became ANT's point (as indeed, it was Pragmatism's before that). Nevertheless, ANT joined fellow travellers like the philosopher Richard Rorty in embracing a deflationary approach to theories of knowledge (Lynch, 2013). Actor-network theorists have been very unforgiving about epistemology; it was and long remained one of their primary targets, coming right after Durkheimian sociology, and for a younger generation of actor-network theorists seemed mostly irrelevant, as other settings, from parliament to architecture studios, took the place of research and development laboratories as favoured research sites. Early ANT had sought to undo epistemology by dissolving its foundational problems, according to the following recipe: (a) Identify a general problem of epistemology (how do we access reality?); (b) empiricise it – in other words, turn it into a research-eable problem – by translating it into a phenomenon that could be observed ethnographically, in a field site and then, (c) make the philosophical problem disappear by respecifying it in practical terms. Following this recipe, in its more adventurous moments, ANT sought to dissolve the great foundational problems of epistemology: There is no 'gap between subject and object.'

I'm afraid this has reasons that would require a book of its own, but today it would be a mistake to attempt to make such foundational problems simply disappear. How do we know? How does the public know? Instead, these questions must be asked again and again, and can't be asked often enough. The rise of fake news and growing evidence of attempts at opinion manipulation online has put the normative question of the validity of knowledge claims, above and beyond the empirici-sable question of relevance, back on the agenda: Deception and duplicity abound in today's digital architectures, demonstrating the need for the identification not just of sources that are active, topical and engaged (the measurement of relevance in which search engines excelled), but the determination of validity conditions for knowledge claims (Bounegru, Marres and Gray, 2018). Campaigns against knowledge, in particular public knowledge, have too many allies today, and it seems unlikely that these can be satisfactorily dealt with by a strategy of deflation.

But here, too, Pragmatism can come to the rescue: For intellectuals like Dewey, the objective was never deflation. While his principal target was indeed epistemology, and most of all, the intolerance of doubt that epistemology was used to justify, his strategy was never to abandon the valuation of knowledge as a normative project: Pragmatisms' project was to re-construct the problems of philosophy. They used methods of re-specification, and highlighted the relational and situational character of even the most seemingly abstract phenomena and processes, as a way not of collapsing but of requalifying problems of knowledge, not in order to make epistemology disappear (deflate it), but in order to relocate them in relation to life, and, thus, to transform our understanding of the relation between the problems of knowledge and the problems of the world, and of society. They insisted that the distance between these different types of problems is often smaller than philosophers and sociologists, and scholars and 'people' – those committed to abstract knowledge and those committed to 'keeping it real' – like to believe. To re-specify problems of epistemology in field settings does not mean that empirical research can take the place of epistemology. Rather, the pragmatists were the precursors of empirical philosophy (Mol, 2000). Composition and recomposition, in Pragmatism, denote not only a process of recombinant association between humans and non-humans, but equally a methodology for the valuation of abstract entities, like truth.

Conclusion

These notes are brief and sketchy, but the point I am edging towards is that of an experimental approach to social and cultural enquiry.[9] Both Pragmatism and ANT sensitised us to new ways in which representing and intervening (Hacking, 1983) are combined in societies 'marked by the invention of science and technology' to use Isabelle Stengers's helpful phrase. The process of getting to know the world and changes in existing conditions in the world are intimately connected, this is what the concept 'the problematic situation' tells us, and is signalled by the twin notions of de- and recomposition. However, as Pragmatism recognised much earlier on, experimentality extends across domains, it operates not just in the domain of science, or innovation, but society too must be invented (Marres, Guggenheim and Wilkie, 2018), and so must democracy, and the arts, and culture. Commitment to these objectives does *not* necessarily mean that 'the world must become more like science.' The question rather is how different the recomposed collective is, can and is allowed to be from its previous iteration, and what tests can be organised and/or designed in order to establish whether and how this affirmation of diversity 'holds,' and can be made to hold, in empirical reality. Pursuing this question is likely to require engagement with experimental traditions beyond Pragmatism or ANT and indeed beyond the university narrowly defined, bound up as it must be with the diversification of the very collective we still like to label 'research community.' As Bruno Latour brought out his hammer and nails to set to work on ANT's coffin 20 years ago, it may be tempting for the rest of us to leave the burial ritual to him and his close collaborators. However, as time passes, I'm more and more convinced that if we are to get somewhere in the redoing of ANT, then the job of its undoing must be taken up not just by its inventors, but by many more of us.

Notes

1 In a conversation with Graham Harman (Latour, Harman and Erdélyi, 2011), keen to identify a feminist aspiration in early ANT, I referred to this as a 'metaphysics of the crutch.'
2 Several other contemporary approaches in the social sciences and humanities claim to be inspired by Pragmatism, and even use it to name their approach, like the French sociologists of the Epreuve, Boltanski and Thevenot. Another case in point is post-Habermasian Frankfurter school. It is certainly striking that very different schools of thought claim this common ancestry, and I do think that, on this point at least, ANT *is* 'different' insofar as it is one of the few, perhaps the only approach that remained committed to the ontological register in which a pragmatist like Dewey wrote. I'm not sure, however, that there is that much to gain from debating the differences between the various versions of pragmatist sociology claimed today, at least not, if the aim is to evaluate ANT's viability as a contemporary intellectual programme, as is my aim here.
3 This is not to say that Dewey did not acknowledge the possibility of enduring failure. But his conception of the problematic situation as an occasion for enquiry (knowledge formation) in the Logic of Enquiry supposes problem-solving as the horizon (in contrast to continuously and endlessly unfolding problematisations). With thanks to the editors for their comments.
4 In other work, I have discussed how the concept of problem-solving of Pragmatism must be replaced with the concept of issue formation, a process marked not only by the formation of cross-boundary associations but the expression of antagonism – I would now call it diversity – throughout this process. This has various consequences, including for how we understand the relation between empiricism – which problem-solving prides itself on – and critique – which is a necessary ingredient of issue formation. However, on this point too, I think we move away from ANT, insofar as ANTers themselves tend to avoid either critique or the discussion of it, putting their substantive commitments – be it their feminism, ecologism or postcolonialism – into the foreground instead.
5 Even worse, I have a number of further possible undoings of ANT on my list, with which I will not bother the reader in further detail, but I include it here: 4. Border-crossings can no longer be contained in the sub-political domain if they ever could. 5. Engineering is not, in fact, sociology

by other means. 6. The Internet is not a hybrid forum. 7. The experimental paradigms by which we're increasingly governed today do not facilitate hybrid forums either.

6 The latest attempt to pull a Sokal on 'grievance studies' is a relevant example here. https://en.wiki pedia.org/wiki/Grievance_Studies_affair. See also www.wired.com/story/why-men-dont-believe-the-data-on-gender-bias-in-science/.

7 Some ANTs are very aware of this problem and one way they have addressed this is by claiming not traceability but storytelling as their register of choice. However, if we, the scholars, define our craft as storytelling, why can't our actors? It seems to me that this proposed solution threatens to undo the ontological turn, or to render it less consequential insofar as it threatens to replace empiricist trace-following ANT with interpretation, without re-negotiating their relation.

8 Joanna Kavenna, HowtheLightGetsIn 2018, Philosophy Festival, Hay-on-Wye, 26 May 2018.

9 Whereas others proposed that addressing ANT's shortcomings requires its transformation into a (object-oriented) metaphysics, I propose that ANT's radical empiricism– it's insistence on on-going re-composition – must be translated into a experimentalist methodology (for a discussion, see Latour, Harman and Erdélyi, 2011).

References

Becker, H. (1963). *Outsiders - Studies in the Sociology of Deviance*, New York: Free Press.

Bounegru, L., Marres, N. and Gray, J. (2018) *A Digital Test of the News: Checking the Web for Public Facts, University of Warwick*, Workshop report, December.

Callon, M. (1980). 'Struggles and negotiations to define what is problematic and what is not.' In *The Social Process of Scientific Investigation*, edited by W.R. Knorr, R. Krohn, and R. Whitley, Dordrecht: Springer, pp. 197–219.

Davies, W. (2015). 'The return of social government: from 'Socialist Calculation' to 'Social Analytics'.' *European Journal of Social Theory*, 18(4), 431–50.

Dewey, J. (2007 [1938]). *Logic: The Theory of Enquiry*, New York: Saerchinger Press.

Dewey, J. (1991 [1927]). *The Public and Its Problems*, Athens, OH: Swallow Press/Ohio University Press.

Dewey, J. (1955 [1908]). 'Theory of valuation. Repr.' In *International Encyclopedia of Unified Science* 2 (4), edited by O. Neurath, R. Carnap, and Ch. Morris, Chicago, IL: University of Chicago Press.

Eubanks, V. (2018). *Automating Inequality: How High-Tech Tools Profile, Police, and Punish the Poor*, New York: St. Martin's Press.

Hacking, I. (1983). *Representing and Intervening* (Vol. 279). Cambridge: Cambridge University Press.

Hacking, I. (1986). *Making up People*, Cambridge, MA: Harvard University Press.

Kelly, A. H. (2012). 'The experimental hut: Hosting vectors.' *Journal of the Royal Anthropological Institute*, 18, S145–S160.

Latour, B. (1993). *The Pasteurization of France*, Cambridge, MA: Harvard University Press.

Latour, B. (1996). *Aramis ou L'amour des techniques*, Paris: La Découverte.

Latour, B. (2005). *Reassembling the social: An Introduction to Actor-Network-Theory*, Oxford: Oxford University Press.

Latour, B., Harman, G., and Erdélyi, P. (2011). *The Prince and the Wolf: Latour and Harman at the LSE: The Latour and Harman at the LSE*, Alresford: John Hunt Publishing.

Law, J. (1994). *Organizing Modernity*, Oxford: Blackwell.

Lury, C. and Wakeford, N. (Eds.). (2012). *Inventive Methods: The Happening of the Social*, London and New York: Routledge.

Lynch, M. (2013). 'Ontography: investigating the production of things, deflating ontology.' *Social Studies of Science*, 43(3), 444–462.

Marres, N. (2017). *Digital Sociology: The Reinvention of Social Research*, Cambridge: Polity.

Marres, N. and Gerlitz, C. (2018). 'Social media as experiments in sociality. Inventing the social.' In *Inventing the Social*, edited by N. Marres, M. Guggenheim and A. Wilkie, Manchester: Mattering Press, pp. 253–286.

Marres, N., Guggenheim, M. and Wilkie, A. (Eds.) (2018). *Inventing the Social*, Manchester: Mattering Press.

Marres, N. and Stark, D. (forthcoming) 'Put to the Test: Critical Evaluations of Testing.' *British Journal of Sociology*.

Mol, A. (2000). 'Dit is geen programma: over empirische filosofie.' *Krisis: Tijdschrift Voor Filosofie*, 1(1): 6–26.

Star, S. L. (1990). 'Power, technology and the phenomenology of conventions: on being allergic to onions.' *The Sociological Review*, 38(S1), 26–56.

12

Why does ANT need Haraway for thinking about (gendered) bodies?

Ericka Johnson

This chapter discusses three conceptual gifts that Donna Haraway has given to those of us who use theory to inform critiques of the present, to engage with and, hopefully, transform the world, as she may have put it (Haraway 1991: 23, 187; 1997: 269). These gifts have been inspiring to many of us who work with ANT, as we become snarled into material-discursive entanglements in our studies, in particular, with entanglements sometimes called 'techno-science.' These include a critical understanding of the modest witness; imploding binaries, which troubles our relationship to gender and technology and gives us the cyborg; and the material-semiotic, with its apparatus of bodily production. Throughout, I will discuss how Haraway and ANT have been in conversation and many of the close affinities between them, in particular around these concepts. At the same time, I will reflect on how these three conceptual gifts from Haraway have enriched ANT and challenged those working within ANT to address political concerns around power and identity – and how they may continue to do so further in the future.

Much of this essay draws on the work that Haraway did in the 1980s and 1990s. I choose to focus on this period because it is here that many of the Haraway concepts most frequently used within ANT come from, and here one finds her connections to ANT most clearly. It is also the period in which she is most focused on the (human, gendered) body, though she problematises both of those categories fundamentally. Analytic concern with the body is a domain in which Haraway pushes ANT, and bodies as both subjects and objects of knowledge production (a binary she implodes) are more integral to her intellectual project than they are to much of ANT (one can see a response to this push in Latour 2004). In the work from this period, Haraway both engages with ideas within ANT, acknowledging the agency of 'objects' (Haraway 1991: 198) by employing and expanding upon the semiotic (Haraway 1997: 22, 135–138), and critiques ANT (especially in Haraway 1992), by emphasising the importance of both the critical theory (Haraway 1991: 23) and the science studies researcher's situatedness (Haraway 1991: 191; 1997: 267, 270). She does so in a way that now, when one returns to it 20–30 years later, reminds one that ANT today is not what ANT was *circa* 1990.

Hence, in one of his essays reflecting on ANT, John Law writes about how early ANT caused a flood of responses from many different horizons, a flood in which Haraway's response can be included (cf. Star 1990; Amsterdamska 1990; Lee & Brown 1994). Haraway's critique, Law reflects, reminded ANT of its non-innocence, and forced those listening to become more explicitly political in their work, not least by her particular development of the powerful concept of the material-semiotic, but also through the trope of cyborg (Law 2007: 16). But perhaps the first step Haraway forced ANT to take, both in its approach to bodies, and in its approach to *everything* besides, was the analysis of semantics[1] as broad, power-infused patterns of cultural meaning-making, as 'contents and meanings, tropes and topics' (Haraway 1997: 22) of technoscience. This is in contrast to a more technical use of semiotics as methodology found in much of ANT (see Matozzi this volume and Lenoir 1994). Her use of semiotics in this way has garnered critique from within ANT but has also inspired many in the second generation of ANT to engage patterns of cultural meaning-making through analysis of the material-semiotic of technoscience, writ large.

Haraway's work has been in conversations in ANT, but it has also been engaged in fields as diverse as science and technology studies (STS) and philosophy of science, feminist technoscience studies, gender studies, cultural studies, somatechnics, environmental humanities and posthumanism. She has inspired analysis of technoscientific contingencies in many fields, and comes with ideas that inform researchers and others, including many artists.

Haraway's ideas have impacted my own work, and inspired me to think with some of her theoretical terms and challenges. To show how, in this chapter, I will reflect on a study of medical simulators I did many years ago as part of a project that involved gynaecologists, cultural studies researchers, gender studies researchers and an ANT component – the latter brought to the table by me. We were all looking at the introduction of a pelvic simulator to a Swedish teaching hospital. The simulator – invented, validated, used and marketed in the USA – was a model of the 'female' reproductive tract. The 'body' was connected by pressure sensors to an external computer which measured if, and how hard, a student touched the 'right' places during an internal gynaecological exam. It was a machine that would teach medical students how to know the gynaecological body; it was an artefact that could help the medical students 'learn to be affected,' to use a concept proposed for speaking about the body (Despret 2004; Latour 2004). The simulator was to be used to supplement or possibly replace an existing practice at the hospital which relied on live, volunteer, 'professional' patients who let the students examine them in a classroom-like situation.

The simulator and its move from the USA to Sweden offered up many 'knots' that could be untied, inspired by Haraway's assertion that technoscience is 'a knot of knowledge-making practices, industry and commerce, popular culture, social struggles, psychoanalytic formations, bodily histories, human and nonhuman actions, local and global flows, inherited narratives, new stories, syncretic technical/cultural processes, and more' (Haraway 1997: 129). It had moved from one culture to another, was part of a global flow of medical knowledge and was going to be integrated into local practices. There were tangled commercial and medical interests in its production, validation, sale and usage. Its integration into the Swedish context was awakening concerns about the patient body and subjectivity which had been addressed in very local practices of professional patient programmes, programmes with activist histories.

But some of Haraway's concepts, for example, her use of the material-semiotic and discussion of the god-trick and situated knowledge, coloured even more the specific thoughts and questions I and my colleagues asked to the simulator body during the research project. Her interest in tropes and models (Haraway 1997) was with us from the first group meeting, even when we were writing the research proposal. The simulator modelled the reproductive tract but did not model a pregnant body. Most female simulators at the time were birthing simulators, as have been models of the female body, historically (Sundén 2010; cf. Jordanova 1999; Waldby 2000). 'Our' simulator presented a model of the body which was not a reproducing body, merely a 'female' body, and told an entirely different narrative than models of the pregnant body. It gave the impression of a female adult subject, even if rather disembodied, which was not merely a receptacle for reproduction. And yet, the simulator also hinted that a female body was the reproductive tract – we asked how this little silicon model was supposed to be 'female' in contrast to the sea of unmarked male bodies that other simulators modelled? Why was the rest of the 'female' invisible, unmaterialised? Already in the planning stage, Haraway inspired us to ask about the tropes and narratives the model expressed.

Additionally, her work encouraged us to look for the 'god-tricks' the medical experts, engineers and designers were playing as they created this model of the female body, how the knowledge they were making was situated, even if it wasn't articulated as such. The simulator was supposed to help teach a particular exam and anatomical knowledge of a part of the body. We wanted to ask how the experts designing the simulator were mobilising knowledge that was first created and then reified in the models that were supposed to be inscriptions of that knowledge, inscriptions which could transfer or delegate the teaching of that knowledge to new doctors and medical students. ANT, and in particular laboratory studies that were being done within it, was and is also attuned to situating technoscience, particularly with the ethnomethodology-inspired approaches employed (see Latour & Woolgar 1979; Lynch 1985). Yet somehow, Haraway was bringing more to the table with her term 'situated,' reminding us of the sticky politics entangled in technoscience and the asymmetrical, non-innocent power of knowledge production practices. So, inspired by Haraway, we wanted to situate a simulator.

Modest witnesses and god-tricks

The concept of the god-trick and a reinterpretation of a modest witness are two gifts Haraway has given to ANT. God-tricks are those views from nowhere and everywhere, seen by no one and everyone, articulated in the passive form and representing truth as discovered in nature. It's a powerful concept that Haraway calls out, in the process giving science studies a politically charged term for the 'credible, civil man of science.' The ideas of witnessing and testifying were already in use, thanks to Robert Boyle and his air pumps, and thanks to Shapin and Schaffer (1985) and their history of how Boyle created the identity of a witness to science, of a disembodied, objective observer of public scientific experiments who could witness the production of knowledge. Thus, science studies already had a word for the man of the unmarked category who performs self-invisibility in the name of publicly witnessed science. But Haraway proposed a new, alternative, practice of witnessing, a new subject position: the modest witness. For her:

> Witnessing is seeing; attesting, standing publically accountable for, and psychically vulnerable to, one's visions and representations. Witnessing is a collective, limited practice that depends on the constructed and never finished credibility of those who do it, all of whom are mortal, fallible, and fraught with the consequences of the unconscious and disowned desires and fears.
>
> *(Haraway 1997: 267)*

Haraway shows how the civil man of science's voice can be identified, challenged and contested. As she writes in her work on situated knowledges, the god-trick produces knowledge that is 'unlocatable, and so irresponsible [...] unable to be called into account' (Haraway 1991 (1988): 191). She writes of natural sciences like physics and biology and their god-tricks of alleged clarity and referentiality, produced in the 'culture of no culture' (Traweek 1988), that is science, with the 'language of no language, the trope of no trope, the one self-referential world' (Haraway 1997: 138). ANT was also concerned with similar issues, as Latour writes, for example, on purification. Yet while Latour eloquently claims that the purification system is already clogged, the hybrids of nature and culture already overwhelming us (Latour 1993 [1991]: 50), Haraway imperatively shows us the god-tricks and (im)modest witnessing still being done in the tropes of technoscience (and medicine) and demands we out them. Her observations are similar to those made by Latour, but her call to arms is more explicit.

Haraway's concept of god-tricks is relevant for general science studies and the philosophy of science, even though she specifically articulates its theoretical inspirations and home in feminist theory, within a framework which avoids 'metaphors of organic and technological vision in order to foreground specific positioning, multiple mediation, partial perspective, and therefore a possible allegory for feminist scientific and political knowledge' (Haraway 1991: 3).

But rather than arguing for an alternative, feminist science, an idea that had traction in the 1980s,[2] Haraway called for 'a doctrine of embodied objectivity that accommodates para-doxical and critical feminist science projects: Feminist objectivity means quite simply *situated knowledges*' (Haraway 1991 (1988): 188). She coined the term situated knowledges and put words to how we and others were making truth claims. The practice of situating ones' (and others') knowledges involved not relativism, but 'a doctrine and practice of objectivity that privileges contestation, deconstruction, passionate construction, webbed connections, and hope for transformation of systems of knowledge and ways of seeing' (Haraway 1991 (1988): 191). Scholars in science studies, including within ANT, responded, contesting, deconstruct-ing, articulating webs and hoping and caring for transformations as if – because – the shape of our lives depended on it (see Berg & Mol 1998; Clark & Montini 1993; Dugdale 2000a; 2000b; Mol 2002; Moser 2006; 2009; Pols 2013; Puig de la Bellacasa 2017; Singleton 1998; Star 1990; Thompson 2005).[3]

Within a Haraway-enriched ANT, the job became one of describing how scientists per-form their god-tricks and old-school modest witnessing, of attending to narratives of 'ob-jectivity,' and using them as our material. But it also became one of reflecting over our own witnessing practices, our own god-tricks, of situating our partial perspectives and paying attention to the narratives and tropes we employ in the construction of knowledges. It also meant reflecting on how others, *and how we*, give voice to our material, and the distancing operations, relocating practices and political implications of representation. Haraway re-minded us that, 'the represented is reduced to the permanent status of the recipient of action, never to be a co-actor in an articulated practice among unlike, but joined, social partners' (Haraway 1992: 311–12). This is, at first glance, somewhat at odds with the way some ANT has considered the material 'objects' of technoscience as active and the ANT challenge to conduct symmetrical analysis. ANT responded to her critique that such an approach flattens power structures, paying more attention to the relationship of politics, representation and science (Latour 2004; Star 1991).

Heeding her warning, we in the gynaecological simulator project were also trying to situate our own knowledge-making practices – to try to figure out how we were impli-cated in what truths we were telling about the simulator and gynaecological practices. We all thought we knew that within a medical-professional setting, male patients have neutral bodies and female ones have (embodied) partial perspectives. Haraway, and other inspirational theorists working at the time, often in conversation with Haraway (Butler 1990; Fausto-Sterling 1992; 2000; Moi 1985; Keller 1996; Sedgwick 1990), gave us terms to make this more explicit in our empirical findings. We were six women in the project, bringing to it experiences of being feminist researchers and practitioners, but in very different professional contexts. We all had bodies that were identified as female by ourselves and others, and were looking at a simulator that was identified as female, too. For some of us in the project, especially those who daily practised gynaecology in the hospital, there was an ease with which they approached the modelled body and its understanding of biological sex, an ease which some of the others in the group did not share. The simulator contributed to this binary understanding of sex/gender – it was called female, sexed by having the internal organs to be touched during the exam, and gendered (given a social identity), materially. For example, the simulator was packaged together with a small blue blanket. In the US context, this blue blanket was an essential part of the gynaecological examination. It was placed over the abdomen to offer discretion to the patient, and when it slipped to the side during a simulated exam, the students were told immediately to pay attention to the blanket, to put it back over the woman's body. When the blue blanket was pulled out of the box as the simulator was unpacked in Sweden, it was met with mild confusion and then put to the side. Small blankets are not used to cover one's sex during a gynaecological exam in Sweden. The simulated body was given shame, embarrassment, discretion in the US, a gendered (social) identity that was articulated by the culturally expected nakedness in Sweden.

The difference between the anatomical sex of the simulator and the social aspects of its gender (the blanket) was a starting point of some conversations about the simulator within our group. However, those of us in the project who were inspired by Haraway were a little less comfortable with this easy distinction. This is because Haraway (and ANT) has succeeded in showing how binaries, and especially the sex/gender binary in Haraway's work, implode.

Cyborgs and imploded binaries

There is a reason '(gender)' is in parenthesis in the title of this chapter. I think Haraway would have been very keen to help us with the task of thinking about 'gendered' bodies sometime in the mid-1980s, but her work – and the flora of other theorists she was engaging with at the time – led to the demise of gender as a concept to denote the binary opposite of biological sex. As such, it also led to the demise of 'gender' as a legitimate category of difference. Obviously, this isn't true in every context. Gender is still a word used widely and even usefully, particularly outside of gender studies departments. But I'm going to let Haraway's work explain why the sex/gender binary imploded and, more importantly, why the category of gender is fraught with problems.

Much of this section draws from Haraway's wonderfully fun discussion in 'Gender for a Marxist Dictionary: The Sexual Politics of a Word,' which can be most easily found in *Simians, Cyborgs and Women* (1991). In this piece, she is asked to write about gender, but contends that gender is a problem in itself. She posits that gender is not a word or a concept, it is a semiotic account of complicated struggles (Haraway 1991: 3): '"Gender" was developed as a category to explore what counts as a "woman," to problematize the previously taken-for-granted' (Haraway 1991: 147).

At the time of her essay on sex/gender, the terms and their combination as a binary pair had been floating around as a way of differentiating the biological (sex) and the social (gender) when categorising bodies and people as man or woman. Or, to be honest, at that time, one was mostly categorised as a woman. Men had still not really been allowed a sex or a gender in much of the discourse. They were still the unmarked category. Haraway points this out later (1997), when she articulates gender as,

> always a relationship, not a performed category of beings or a possession that one can have. Gender does not pertain more to women than to men. Gender is the relation between variously constituted categories of men and women (and variously arrayed tropes), differentiated by nation, generation, class, lineage, color and much else.
>
> *(Haraway 1997: 28)*

One of the critiques of gender that Haraway makes is that gender fails to represent the complexity of subject positions and the violence that categorisation can do. Echoing the beginnings of an intersectional discourse, she reminds her reader that

> the positionings of African-American women are not the same as those of other women of colour; each condition of oppression requires specific analysis that refuses the separations but insists on the non-identities of race, sex, and class. These matters make starkly clear why an adequate feminist theory of gender must *simultaneously* be a theory of racial difference in specific historical conditions of production and reproduction.
>
> *(Haraway 1991: 146)*

She was writing in California in the 1980s – her knowledges are situated there, but she and the academics and activists she was in conversation with were nurturing seedlings of a critique of social theory's simplified categories that would spread around the world. In many places, this critique has cracked the foundations of analysis built on individual categories of the self like 'gender' but also 'class,' 'race,' 'ethnicity' and 'age.' Yet, interestingly, much of the foundational work within ANT rarely engaged with these categories at all, either individually or intersectionally. This is a point at which ANT needs Haraway, and not just for thinking about bodies.

But even if many theorists today are reluctant to use the concept of gender, many of us live with gendered bodies in contexts which categorise us that way and treat us differentially because of that category. And, as Haraway notes, there was a strongly liberatory and political push to use these categories by feminists, too (Haraway 1991: 130). But, as she notes, the sex/gender distinction was just as dependent on other categorising practices as on the distinction between a social and a biological that it claimed to articulate. Producing sex and gender was also producing (and produced by) equally dubious distinctions in other realms:

> [T]he political and explanatory power of the 'social' category of gender depends upon historicizing the categories of sex, flesh, body, biology, race, and nature in such a way that the binary, universalizing opposition that spawned the concept of the sex/gender system at a particular time and place in feminist theory implodes into articulated, differentiated, accountable, located, and consequential theories of embodiment, where nature is no longer imagined and enacted as a resource to culture or sex to gender.
>
> *(Haraway 1991: 148)*

ANT engages with idea of imploding binaries, particular Latour's ideas about hybrids and the nature/culture binary in *We Have Never Been Modern* (Latour 1993[1991]). But the nature/culture–sex/gender binary is not the only imploded binary brought forth by Haraway which is relevant to how ANT can think about bodies. Haraway expands these binaries in what is perhaps the most renowned of her early works, the *Cyborg Manifesto*. While she, in her work for the Marxist dictionary, notes that 'The difference between sex and gender depends on a related system of meanings clustered around a family of binary pairs: nature/culture, nature/history, natural/human, resource/product' (Haraway 1991: 130), in the *Cyborg Manifesto*, she attacks a plethora of binaries, jolting the reader out of the narrative and forcing one to grapple with the categories she is making and problematising. She names many pairs, but the most clearly discussed are the three binary categories that the figure of the cyborg implodes: The human/animal; the animal/machine and the physical/non-physical (Haraway 1991 [1984]: 151–153).

This figure of the cyborg forces us to acknowledge our partial, impure identities as humans, in an early (remember, this is 1984) encouragement to embrace partial perspectives.

> A cyborg world might be about lived social and bodily realities in which people are not afraid of their joint kinship with animals and machines, not afraid of permanently partial identities and contradictory standpoints. The political struggle is to see from both perspectives [of the animal and the machine] at once because each reveals both dominations and possibilities unimaginable from the other vantage point. Single vision produces worse illusions than double vision or many–headed monsters.
>
> *(Haraway 1991 (1984): 154)*

Haraway's more detailed work on modest witnessing and situated knowledges is yet to come, but already the cyborg is facilitating a new take on objectivity, scientific and otherwise. She pulls on discussions of the body and reproduction as a medical site of contention, and reminds the reader of body claiming politics of the (white, North American) movements to retake knowledge production about the body. The speculum shows up, and self-help, too (Haraway 1991 (1984): 169). But her cyborg strides beyond the body as a site of reproductive politics, bringing the discussion to a matter of how to 'find political direction in the 1980s in the face of the hybrids "we" seemed to have become world-wide' (Haraway 1991: 3). This use of the term 'hybrids' indicates that she and ANT are in conversation. Latour, for example, uses the concept of hybrid to connote the way nature and culture are mixed and churned up together in 'modern' explanations of scientific phenomena (Latour 1993 [1991]). But her cyborg has a more aggressive and liberational character. It became a political figure behind which academics, artists and activists rallied, in a way the ANT hybrid never did.

Working with the gynaecological simulator, we, too, saw the political power of the cyborg and the analytical relevance of imploding binaries. One of the gynaecologists we were working with had come to the question of this simulator and professional patients after a career of trying to improve the teaching of the gynaecological exam. Many decades earlier, as a young medical student studying abroad in a different European country, she had been shocked to find out that they would be expected to learn and practise the gynaecological exam on non-consenting female bodies which had been anaesthetised and prepped for surgery. When she protested, it was suggested that she, herself, could volunteer to let her fellow students train on her.

Carrying this experience back with her to Sweden (and suspecting that similar 'teaching opportunities' were employed in Sweden at the time, too), she eventually made her way into a professional position that allowed her to create a programme of volunteer bodies which allowed medical students to train the gynaecological exam on them in respectful ways, consciously and with consent. This was the programme which the simulator was going to be introduced into, perhaps to replace the professional patients, perhaps to supplement them, depending on how it worked. (It didn't work. It has been put back in its box.)

127

Inspired by the cyborg, we could see the professional patient example and the simulator demonstrate two imploding binaries – unique human/representative human bodies and living bodies/silicon models: In her experiences of the anaesthetised bodies being used as teaching material for young medical students, the practices of the hospital were already erasing a distinction between what is a human patient and what is an anatomical model upon which one could be trained. But this shift occurred again, decades later, when the same gynaecologist considered replacing or at least supplementing the professional patients with the newly acquired simulator. The professional patients implode a distinction between an individual, human body and a generic model of the anatomy. They are both anatomical models and individual patients, standard bodies and subjects. Like the imploded machine/human binary of the cyborg, the bodies of the professional patients are both representative, teaching models and individual patients. Then, years later, the distinction between the human body and the mechanical, silicon representation of that body was also erased. The simulator imploded the machine/human binary. These two examples show how binaries around bodies can implode in practice. Haraway gave us words for that and a theoretical concept, the cyborg, to articulate it.

The material-semiotic and an apparatus of bodily production

Haraway uses the term *material-semiotic*

> to highlight the object of knowledge as an active part of the apparatus of bodily production, without ever implying immediate presence of such objects or, what is the same thing, their final or unique determination of what can count as objective knowledge of a biological body at a particular historical juncture.
>
> *(Haraway 1992: 298)*

To do the work of articulating the material-semiotic, she gives us the *apparatus of bodily production*, a concept to understand 'the generation – the actual production and reproduction – of bodies and other objects of value in scientific knowledge projects' (Haraway 1991 (1988): 200–201).

With the *material-semiotic* and the *apparatus of bodily production*,[4] Haraway pointed those following her work[5] down the path of thinking about how the material world and semiotics play together – or *intra-act*, to appropriate a term from Barad's ontoepistemological agential realism (which Haraway does (1997: 116)) – with a concern for the subject, and the lived, embodied, emotional strength of subjectivities analytically inseparable from bodies and technologies. Mattozzi (this volume) notes that ANT shares similar concerns, especially with the material-semiotic, though Haraway is perhaps more attuned with bodies. The *material-semiotic* and an *apparatus of bodily production* are particularly tenable in thinking about bodies because they allow us to see how bodies and their 'boundaries materialize in social interaction among humans and non-humans, including the machines and other instruments that mediate exchanges at crucial interfaces and that function as delegates for other actors' functions and purposes. '"Objects" like bodies do not pre-exist as such. Similarly, "nature" cannot pre-exist as such, but neither is its existence ideological' (Haraway 1992: 298).

Haraway's concepts of the material-semiotic and the apparatus of bodily production speak to how the world – later spoken of in terms of materiality – pushes back at the knower, how it influences (or intra-acts with, Barad again) the way we make knowledge about it. This, too, I was able to see in the study of the simulator, particularly through the inclusion of its fat pad.

The simulator had a removable fat pad, placed above the internal organs, under the abdominal skin, to simulate an obese patient. It was the shape and thickness of one of those old mouse pads for computers back in the 1990s. I wondered how that centimetre of silicon was supposed to represent the very thick abdominal fat of an obese patient. When I asked the model designer about this, she explained to me that fat in the body is warm – body temperature – and therefore not very solid. It is bound in small capsules, but these have a tendency to move around almost (but not quite) like a liquid in parts of the body. Thus, when a patient is lying down on their back during the pelvic exam, the fat in their abdomen tends to slide downwards, off the peak of the stomach. During an exam, the fat is gently pushed out of the way. Not all of it, of course, but quite a bit of it. Therefore, the 'thin' fat pad gives the feeling that a doctor would have when examining a much 'thicker' patient.

Here, the 'fat' becomes a material-semiotic actor – it pushes back and produces an experience of 'fat' – and the production of knowledge about the fat becomes very specific, or contingent, as Haraway would say. It becomes through a practice (the bimanual examination) in a certain way (with the patient on their back) and under certain circumstances (at body temperature). The students are learning to be affected by the fat, but that knowledge is situated, and the model made of it (the fat pad) is also specific and situated. It is not representing 'obesity' in general, but a very specific way of knowing obesity. Significantly, the fat pad simulates not the actual body of the patient but the material-semiotic 'obese patient,' as known through an apparatus of bodily production, as a phenomenon of knowledge (cf. Barad 2007, see also Johnson 2008).

Why was this relevant? It showed that the anatomy of the simulator was an anatomy of both medical practices and the materiality they encountered – the fat inside a living body being known through a specific examination. Haraway insists on the importance of tracing constructions of material-semiotic objects of knowledge (which she does with the foetus, gene and race in her 1997 book) while at the same time reminding us that 'constructed' does not mean 'made up.' Her work encourages researchers to show how objects of knowledge are knots of knowledge-making practices, with bodily histories (see above and Haraway 1997: 129). In a way, the ANT idea of a circulatory model of science is related, but Haraway insists on the political imperative of tracing the sticky strings of the knot, articulating the politics stuck on them, as she does with the foetus, the gene and race. There is an affinity between ANT and Haraway here, but an insistence and activist undertone in Haraway's work that is not (always) as clear in ANT.

ANT and Haraway in conversation

The question posed in the title of this chapter is why ANT needs Haraway. Haraway was engaging ANT in her early work, both indirectly (through the concept of situated knowledges, for example, and her insistence that the materiality of the world (including bodies) pushes back. There are many affinities between ANT and Haraway's work, but she also troubles a couple of things that ANT was having some trouble with, at least in its early incarnations.

As one of my colleagues has pointed out, with Haraway, we as researchers can never get away from the fact that everything we do, no matter how hard we try, is going to be at least a little 'bad.' No matter how 'good' we try to be, we are always excluding someone, something. There is always a backside to our political choices, and every choice we make is political. But we must continuously try to do our best, anyway.[6]

> Objectivity is not about dis-engagement, but about mutual and usually unequal structuring, about taking risks in a world where 'we' are permanently mortal, that is, not in

'final' control. [...] Perhaps our hopes for accountability, for politics, for ecofeminism, turn on revisioning the world as coding trickster with whom we must learn to converse.

(Haraway 1991 (1988): 200–201)

This political conversation and imperative is a theme Haraway has emphasised even more in her later work (Haraway 2016). In the intervening years, her work has explored the (non) human relations and our ways of theoretically critiquing and engaging with these. It is powerful, political and potent, as one would expect. Yet, it is not as interested in gender nor as engaged explicitly with ANT.

So, why does ANT need Haraway to think about the body? She points us to the semiotic, and allows us to approach bodies and tropes, science and semantics, in a way that conflates all our presumed binaries that stem from nature and culture. There is an affinity with ANT here, but there is a tendency for Haraway to engage with a wider set of actors, to engage with structures and systems beyond the laboratory and scientific discourse. Like ANT, Haraway conflates nature/culture, but in a way that breathes life into analytical work by accounting for the situatedness, the contingency of the material-discursive while contextualising it in the larger culture. Her gifts help one rethink some of the fundamental assumptions about science, technology, culture and specifically gender and other relations of categorising. Her work gives us permission, and more importantly the theoretical tools, to assert situated knowledges and pervert the god-tricks of self-invisible modest witnesses. She gave us the cyborg to think about how technoscientific bodies are powerful tools to write our way into futures we would want to inhabit. And, perhaps most importantly for thinking about the body, she gave us the material-semiotic and the apparatus of bodily production to think critically about how meanings and bodies get made. Why is this important? Because, 'we need the power of modern critical theories of how meanings and bodies get made, not in order to deny meaning and bodies, but in order to live in meanings and bodies that have a chance for a future' (Haraway 1991 (1988): 187).

Acknowledgements

I am grateful for the comments and suggestions I've received on this text, in particular from Steve Woolgar, Casper Bruun Jensen, Anders Blok, Celia Roberts, Corinna Kruse, Lisa Guntram, Jenny Gleisner, Jelmer Brüggemann, Jeff Christensen, Katherine Harrison, Lotta Björklund Larsen, Nimmo Elmi and Else Vogel.

Notes

1 For a discussion of the semiotic turn in science studies, particularly in the work of Latour, Akrich, Hayles and Haraway, see Lenoir (1994).
2 See the special issue of Hypatia (2004).
3 Haraway's more recent work also stresses the urgency of engaging with the world and its inhabitants of all life forms in these contested *cenes (Haraway 2016).
4 Haraway speaks of the apparatus of bodily production in conjunction with Katie King's objects called 'poems' (Haraway 1991 (1988): 200–201; 1992: 289) and Barad's agential Realism (Haraway 1997: 116, 151–3).
5 As I mentioned earlier, Haraway's concept of the material-semiotic and an apparatus of bodily production spread widely outside of feminist science studies, catching fire in gender studies departments without strong science studies components and bringing Haraway's concepts into other fields like cultural studies and philosophy (Rouse 2004).
6 Thank you, Marianne Winther Jörgensen, 1966–2017.

References

Amsterdamska, O. (1990) Surely You are Joking, Monsieur Latour! *Science Technology & Human Values* 15(4): 495–504.

Barad, K. (2007) *Meeting the Universe Half-Way.* Durham, NC: Duke University Press.

Berg, M., & A. Mol (Eds.) (1998) *Differences in Medicine: Unraveling Practices, Techniques, and Bodies.* Durham, NC: Duke University Press.

Butler, J. (1990). *Gender Trouble: Feminism and the Subversion of Identity.* New York: Routledge.

Clarke, A. E. & T. Montini. (1993) The Many Faces of RU 486: Tales of Situated Knowledges and Technological Contestations. *Science, Technology & Human Values* 18(1): 42–78.

Despret, V. (2004) The Body We Care For: Figures of Anthropo-zoo-genesis. *Body & Society* 10(2/3): 111–134.

Dugdalem, A. (2000a) Materiality: Juggling Sameness and Difference. In J. Law & J. Hassard (eds.), *Actor Network Theory and After.* Oxford: Blackwell, pp. 113–135.

Dugdale, A. (2000b) Intrauterine Contraceptive Devices, Situated Knowledges, and the Making of Women's Bodies. *Australian Feminist Studies* 15(32): 165–176.

Fausto-Sterling, A. (1992) *Myths of Gender: Biological Theories about Women and Men.* New York: Basic Books.

Fausto-Sterling, A. (2000) *Sexing the Body: Gender Politics and the Construction of Sexuality.* New York: Basic Books.

Hankinson Nelson, L. & A. Wylie (Eds.) (2004) Special Issue on Feminist Science Studies *Hypatia.* 19(1).

Haraway, D. (1991) *Simians, Cyborgs, and Women. The Reinvention of Nature.* Free Association Books: London.

Haraway, D. (1992) The Promises of Monsters. A Regenerative Politics of Inappropriate/d Others. In L. Grossberg, C. Nelson, P. A. Treichler (eds.), *Cultural Studies.* New York: Routledge, pp. 295–337.

Haraway, D. (1997) *Modest_Witness@Second_Millennium.FemaleMan©_Meets:OncoMouse^{TM}* New York: Routledge.

Haraway, D. (2016) *Staying with the Trouble. Making Kin in the Chthulucene.* Durham, NC: Duke University Press.

Johnson, E. (2008) Simulating Medical Patients and Practices: Bodies and the Construction of Valid Medical Simulators. *Body and Society* 14(3): 105–128.

Keller, Fox E. (1996) *Reflections on Gender and Science.* New Haven, CT: Yale University Press.

Latour, B. (1993 [1991]) *We Have Never Been Modern* (original title: *Nous n'avons jamais été modernes: Essais d'anthropolgie symétrique*). Translated by Catherine Porter. Cambridge, MA: Harvard University Press.

Latour, B. (2004) How to Talk About the Body. *Body & Society* 10: 205–229.

Latour, B. & S. Woolgar (1979) *Laboratory Life. The Construction of Scientific Facts.* Sage: Beverly Hills.

Law, J. (2007) *Actor Network Theory and Material Semiotics*, version of 25th April 2007, available at www.heterogeneities.net/publications/Law2007ANTandMaterialSemiotics.pdf downloaded August 30, 2017.

Lee, N. & S. Brown (1994) Otherness and the Actor Network. The Undiscovered Continent. *American Behavioral Scientist* 37(6): 772–790.

Lenoir, T. (1994) Was That Last Turn A Right Turn. The Semiotic Turn and A.J. Greimas. *Configurations* 2(1994): 119–136.

Lynch, M. (1985). *Art and Artifact in Laboratory Science: A Study of Shop Work and Shop Talk in a Research Laboratory.* London; Boston: Routledge & Kegan Pau.

Moi, T. (1985) *Sexual/Textual Politics: Feminist Literary Theory* (1985; 2nd edition 2002). New York: Routledge.

Mol, A. (2002) *The Body Multiple: Ontology in Medical Practice.* Durham, NC: Duke University Press.

Moser, I. (2006) "Sociotechnical Practices and Difference: On the Interferences between Disability, Gender and Class", *Science, Technology & Human Values* 31(5): 537–565.

Moser, I. (June 2009) "A Body that Matters? The Role of Embodiment in the Recomposition of Life after a Road Traffic Accident", *Scandinavian Journal of Disability Research*, 11(2): 83–99.

Pols, J. (2013) Knowing Patients. Turning Patient Knowledge into Science. *Science, Technology & Human Values* 39(1): 73–79.

Puig de la Bellacasa, M. (2017) *Matters of Care: Speculative Ethics in More Than Human Worlds.* Minneapolis: University of Minnesota Press.

Rouse, J. (2004) Barad's Feminist Naturalism. *Hypatia* 19(1): 142–161.

Sedgwick, Kosofsky E. (1990) *Epistemology of the Closet*. Berkeley: University of California Press.

Singleton, V. (1998) *Stabilising Instabilities: The Role of the Laboratory in the UK Cervical Screening Programme in Differences in Medicine*. Durham, NC: Duke University Press.

Star, S.L. (1991) Power, Technology, and the Phenomenology of Conventions: On Being Allergic to Onions. In J. Law (ed.), *A Sociology of Monsters: Essays on Power, Technology and Domination*. London: Routledge, pp. 22–56.

Sundén, J. (2010) 'Blonde Birth Machines: Medical Simulation, Techno-Corporeality and Posthuman Feminism', in E. Johnson and B. Berner (eds.), *Technology and Medical Practice. Blood, Guts and Machines*. Farnham, Surrey: Ashgate.

Thompson, C. (2005) *Making Parents. The Ontological Choreography of Reproductive Technologies*. Boston, MA: MIT Press.

Traweek, S. (1988) *Beamtimes and Lifetimes. The World of High Energy Physicists*. Boston, MA: Harvard University Press.

Waldby, C. (2000) *The Visible Human Project: Informatic Bodies and Posthuman Medicine*. New York: Routledge.

13

How does thinking with dementing bodies and A. N. Whitehead reassemble central propositions of ANT?

Michael Schillmeier

Introducing ANT, the questionability of the social and dementia

Despite its non-cohesive character, it seems fairly safe to say that the research efforts of Actor-Network Theory (ANT) gained considerable importance in sparking controversies about a variety of central ideas, concepts and methods of modern thought including the questions of agency, subject and object, society and nature, human and the non-human, science and technology, body and materiality, just to name a few.

Since the early Laboratory Studies in the late 1970s, ANT scholars have been experimenting with the outrageous claim that all entities – human or not – are societies, thereby contributing to speculative thought that can be found in Gabriel Tarde (2012) and most vibrantly in Alfred North Whitehead's ontology (Whitehead [1929]1978). Embracing a process-oriented (ethno-) methodology, ANT has been interested in tracing how different actors, facts, positions, perspectives or thoughts come into being as an emerging, collective and distributed accomplishment. The latter arise from controversies, issues, uncertainties, hesitations and ambiguities of what counts as a shared problem, who and what is involved in the questions that arise and how it is dealt with in the course of engagement (Callon 1986, Hassard & Law 1999, Latour 2005, Mol 2002, Passoth, Peuker & Schillmeier 2012).

Rather than granting explanatory power to 'the social' in understanding the formation of societal beings, for ANT, it is the very *questionability of the social* that leads to the complexities of mediating uncertainties and differences and thereby fabricating the emergence, maintenance, alteration and perishing of actor-worlds (Latour 2005). These actor-worlds, in turn, provide the potentiality for novel social processes and related societal accomplishments. Importantly, then, in ANT, societal variances of in/stability and agentic dis/abilities are due to the situated ways of how different bodies, thoughts, feelings and things are related.

What makes ANT a good ethnographic research heuristic is its well-known obligation to learn from the specificities of everyday life in order to redescribe existing accounts by 'adding a new agent to the description' (Latour 2005: 147), which, in turn, may allow to rethink taken-for-granted ideas about them (Callon & Latour 1981). In that sense, ANT does not only talk about 'translation' viz. 'mediation' as one of its core ideas (Callon 1981) but becomes a *technique of translation* between the empirical and the conceptual and how both sides are made real, configured and reconfigured by the very process of relating itself (cf. Jensen 2014).

It was ANT's insistence on the questionability and transformability of the social that made it attractive for my own research on *cosmopolitical events* in the context of health, illness and dis/ability (Schillmeier 2010, 2011, 2014). Like Latour's controversies (Latour 2005), or Garfinkel's idea of breaching (Garfinkel 1967), yet without the researchers' directive, bodily events like Alzheimer's disease, a stroke or bacterial/viral infections unbutton the taken-for-granted social relations and societal accomplishments, thus unfolding what I have called the 'cosmopolitics of illness' (Schillmeier 2014). The latter refers to all embodied practices, processes and events that disrupt, question and alter, and thereby politicise, the 'cosmos' of the seemingly given and expected modes of ordering everyday life.

ANT, I believe, offers a valuable research technique to engage with dementia's cosmopolitical breach of conventions in everyday life. However, ANT turned out to be a limited conceptual tool in understanding everyday realities, issues, concerns and problems of dementia. Writ large, ANT has not been very strong in addressing the temporal and immaterial aspects of experience as part of the material, embodied process of world-making, which nevertheless proved vital in understanding dementia. Conversely, this is precisely what ANT may learn from dementia: To open its concerns for the temporality and immateriality of experience as essential parts of embodied living, of *actor-world making,* of the *worlding* process of the actor-worlds in place (Tsing 2010).

The inclusion of the temporality and immateriality also complexifies ANT's conceptual reflections. Following Latour's quest for an 'empirical metaphysics' of world-making (Latour 2005: 51pp), ANT cannot abstain from philosophical questions and thought. More radically put, it is precisely the philosophical questions which become apparent in thinking with the lives of people with dementia that offer most important insights into speculating with the metaphysics of how actor-networks come into being, maintain and change. For this purpose, my thinking with dementing bodies is accompanied by ideas from Alfred North Whitehead, which have proven to be most helpful for a better understanding of (dementing) bodies as *affected/ing societies in action*.

Thinking with Mrs M and A. N. Whitehead

For more than ten years, I have been researching the lives of people who are experiencing Alzheimer's Disease (Schillmeier 2009, 2014, 2017). My readings of dementia are meant to contribute to accounts in the fields of care studies, sociology of health and illness and dis/ability studies that offer alternative readings to economic, individualistic, biomedical and deficit models of illness and disability (Kitwood 1997, Kohn & McKechnie 1999, Mol 2008, Moreira et al. 2014). As a 'matter of care' (Puig de la Bellacasa 2017), it is essential to draw attention to how people with dementia live their everyday life, how they perceive and engage with the world and how it changes their relations with others and *vice versa*. This includes addressing the agency of people with dementia. To be sure, this does not mean to romanticise their experiences of dementia, but to free them from the stigma of passivity and/or negativity. Only then, we may grasp the requirements of what it could mean to live with dementia, to *live well* with it and to experiment with possibilities to improve people's life with dementia (Moser 2008, 2010).

The brief data-events, which I analyse in this chapter, are part of a documentary film called *The Day That Got Lost in a Handbag* (2000) by Marion Kainz, in which Mrs M's experiences of dementia and care are portrayed.[1] Mrs M, diagnosed with Alzheimer's disease, lives in a nursing home. The documentary begins with Mrs M in her nightgown, walking from the dark end of the corridor of the ward towards the filmmaker, Mrs Kainz. Mrs M is nervous, agitated.

MRS M: I am all over the place [Ich bin ganz ausser mir] – Good day! With whom do I have the pleasure?

KAINZ: It's me. Marion!

MRS M: Marion, what is on hand (t)here [was liegt da vor]?

KAINZ: Whereabouts?

MRS M: Here, with me and the surrounding [bei mir und der Umgebung]!

KAINZ: You are here at the […] nursing home. It is a home for the elderly and a nursing home.

MRS M: A home for the elderly and a nursing home? Why haven't they told me anything about that?

KAINZ: I am sure that you must have just forgotten it in this moment since you are nervous.

MRS M: And now [nun]? What is going to happen now [jetzt]?

KAINZ: You can continue to live here!

MRS M: Continue to live here? What does this mean?

KAINZ: You have already been here for a while.

MRS M: Already for a while? And nobody has told me so! How is this possible?

In the situation when bodies experience dementia, parts or the whole external environment turn into isolated givens, mere *matters of facts* and become *matters of felt unknowns*. In effect, the ensemble of embodied social relations, knowledge practices and orderings in place becomes unfamiliar, vague and messy, and they frequently lack meaning. Being trapped in the dementing event, Mrs M stumbles into the unknown and the questionable. In that moment, she herself and her environment turn alien. Her self, other people and things around her become important precisely because they are abstract, problematic and questionable. The uncanniness of the situation seems to force her to think about herself and the environment, and makes her enrol others to think with her and investigate about *what is on hand*.

Undoubtedly though, Mrs M lucidly addresses her problem. She says that she is *ausser sich*, which not only articulates her negatively affected situation, but literally translates as *being outside of herself*. Mrs M does not know *was da vorliegt*, what *lies in front of*, of '*what is on hand.*' We may speculate that to be *ausser sich* not only refers to being highly affected and agitated not knowing what the situation is all about, but also signifies that she is looking at herself from the outside. She looks at herself as she looks at her surrounding; both veil *what is on hand*. In short, and altogether, Mrs M expresses a fivefold problem: (1) Being negatively affected; (2) being *ausser sich*, and a threefold incomprehensibility of *what is on hand*; (3) with herself; (4) with the surrounding and (5) as the word 'and' suggests, with the *relation* between her and the environment.

The scene unfolds a rather human endeavour that clearly defines and selects events, which are crucial for understanding the specific and individuated situation, a situation that nevertheless appears massively abstract, vague and terrifying. Mrs M seems to miss out on what might make a situation a 'good' experience. For Whitehead, 'the good' should not be conflated with morality as its sole aspect ([1938]1968: 76). Rather, if we follow Whitehead, it has to do with the human ability to 'understand structure' in the way we 'abstract its dominating principle from the welter of concrete realized facts, confused and intermixed' detail (ibid.). As Whitehead notes:

> This clarity of human vision both enhances the uniqueness of each individual occasion, and at the same time discloses its essential relationships to occasions other than itself. It emphasizes both finite individuality and also the relationship to other individuals.
>
> *(ibid. 77)*

Still, for Mrs M, the matters of fact are on hand as merely individual occurrences. Her body and the environment are intensively felt since they do not contribute to her gaining clarity of what is perceived and how they interrelate. Although clearly perceived, her body and her environment do not allow to abstract from the presence of *mere matters of embodied and environmental facts*. Paradoxically speaking, these matters of fact are too abstract to enable Mrs M to *abstract from* their individual presence and understand how they relate to other matters of fact, and what makes these relations meaningful, familiar and traceable.

We might thus say that Mrs M's situation, and the different ways her and other bodies, things and practices are related in and through that situation, remain unspecified and lack structure – both in space and time. The dementing process veils their *situated relevance* otherwise necessary to understand how she and her environment relate in that situation, how her situation came about and what can be anticipated or expected from it. Under the intense impression of highly abstract bodies, things and practices, Mrs M wishes to get answers from Mrs Kainz that might help her to reassemble a clarity of vision, which would, in turn, contribute to decipher what is on hand.

Having said this, Mrs M does indeed have a vague understanding of the structure of her situation: It is about herself and the environment. Both are important to pay attention to; both matter and are of concern to her. Mrs M forgets that she forgets and hence she is not familiar with such a situation. However, in looking at this, we must presume that she somehow remembers – although consciously unaware of this – that she has been feeling good in the past, where situations were differently perceived, in which her body, her environment and how they relate had gained meaning. Why else should Mrs M be so distressed by the situation, if not for the way she felt *differently* with her body and her environment in past situations?

Mrs M feels *ausser sich*. *Being outside or beside herself,* Mrs M seems to express her inability to abstract from her merely given body and environment. She can only look at her body and environment without being able *to be* with her body and her environment in ways that could offer traceable social relations and societal orderings to possibly enable her to see more than just negatively felt, given matters of fact. The question of *what is on hand*, of *was liegt da vor,* is a question about given, static facts. To be with herself would require to be part of a process of abstraction from matters of fact that would enable her to trace their concrete social relations and practices. This would then allow to get an understanding about *what is going on*, which would address a process rather than mere facts. Abstracting from the mere presence of things would help Mrs M to resituate her embodied life and how it has evolved in relation to others. Her questions in the scene address her longing to gain a feeling for the situation and how she *belongs to it, partakes in it and becomes with it.*

We can speculate with Mrs M being *ausser sich* and argue that her body is protesting against the devastating effects of being outside of the social practices and processes that assemble societal accomplishments like bodies and things and how they relate in more or less ordered and meaningful ways (including, for instance, living in a nursing home setting). Mrs M's feelings iterate one of ANT's central claims: *No actor becomes, is and acts in isolation to its environment and other actors.* No actor just exists on its own; actors are in the ways they become with others. As societies, Whitehead already reminded us, bodies and things cannot be/come by themselves: 'there is no society in isolation' (Whitehead [1929]1978: 98). Every 'society must be set in a wider environment permissive of its continuance,' as Whitehead wrote (ibid. 99). Mrs M's body though, merely 'lies in front of,' is simply present, like the environment, which lacks a social background, exiled from its societal accomplishments and thus does not allow Mrs M to translate its importance into a 'conceptual event,' of grasping an understanding of it. As societies, bodies and things are enduring, that is temporal, entities. Feeling of the mere presence of things, as Mrs M does, veils their trajectories and biographies, i.e. their social becoming and societal being.

To be sure, Mrs M's affective state shows clearly that what she experiences is not nothing. Mrs M is forced to feel and think with the mere presence of her body and environment and how these presences affect her being. Mrs M is confronted with the lure of matters of fact, with the unknown potentiality of 'time-less' or 'eternal objects' as Whitehead calls them (Whitehead [1929]1978). These objects have the potentiality to socialise Mrs M's unsocial world into conceptual experiences of enduring and related bodies and things. However, with the dementing event, their potentialities remain to be discovered, traced or invented, and require others to help to do so.

This is the paradoxical situation that Mrs M reveals to and for us: As 'time-less' objects of potentiality, the scene reminds us of our processual existence, and hence also that objects are more than mere matters of fact. Instead, they are actors of unrealised possibilities. They provoke Mrs M's 'appetite' for translating their uncanny importance into canny relations she can live well *with and in*. Mrs M experiences a situation that has perceptual and affective life, is full of objects, full of intensities, but it remains – as purely affective relations do – 'massive and vague' (Whitehead [1929]1978). It is rather an unpleasant situation and Mrs M is longing for conceptual enjoyment to ease the suffering.

In accordance with my speculative interpretation so far, Mrs M's feeling of being *ausser sich* may allude to the experience of being outside of known and understand-*able* societal relations. Her existence is thrown into a cosmopolitical situation that disrupts, questions and alters the social processes that previously made the personal ordering of Mrs M both societally stable (enduring) and lively (changing and changeable). It is precisely 'the social,' understood as the productive processual relation between heterogeneous entities, between contrasting differences, between the past, present and future, the material and immaterial, the (purportedly) social and (purportedly) non-social, and so on, that seems to be missing in the dementing situation. In an uncanny way, Mrs M is experiencing the questionability of the social along the mere presence of bodies and things!

Actor-networks of affective perception

This vital *social* complexity cannot be emphasised enough: Every entity is more than it-self. Every actor-network unfolds temporalities of affective relations that relate subjects and objects. And every actor-network *is* experience affectively charged (cf. Whitehead [1933]1967: 176).[2]

Here, Mrs M's situation brings to the fore a very specific relation between two different modes of experience which Whitehead calls 'causal efficacy' and 'presentational immediacy.' While the former is about bodily, non-sensuous, affective experiences, the latter is about sensory perception (cf. Whitehead [1929]1978). Presentational immediacy reveals the spatio-temporal coexistence of different beings of concern, i.e. perception as the 'here, now, immediate, and discrete' (Whitehead [1993]1967: 180) which assembles the relational potentialities (also) of Mrs M's contemporary world. Mrs M's contemporary world is objectified 'under the aspect of passive potentiality' of sense-data 'such as colours, sounds, bodily feelings, tastes, smells together with the perspectives introduced by extensive relationships' (Whitehead [1929]1978: 61). These entities only endure in the now; they do not reveal if they had a past or what their future will look like. Thus, by themselves, the sense-data do not provide any material for interpretation. To be clear, with Mrs M, we become aware of a *general* aspect of perception that we all share, given that we are creatures who perceive with our senses. It reveals our life with contemporaries, the present as a *slice of the world*, as mere potentiality. Sensory perception in such a reading is not the mind's screening of an outer world, but the embodied practice and process which 'provides the lively, changing aspect of such experience, the passing shapes, colours and clouds which punctuate our engagement with the world' (Halewood 2011: 52).

Still, sensory perception appears as a rather limited mode of activity for interpreting the worlding of the world, if we consider it as the sole foundation of perception. Whitehead ([1933]1967: 181p) stresses that every sense perception is fused with non-sensory perception, termed 'causal efficacy,' and hence cannot be understood without the latter. Following Whitehead, our embodied being can be considered crucially as a matter of memory, whereby non-sensuous perception is central to understand how this process of remembrance makes bodies endure and change:

> It [causal efficacy] is that portion of our past lying between a tenth of a second and a half second ago. It is gone, and yet it is here. It is our indubitable self, the foundation of our present existence. Yet the present occasion while claiming self-identity, while sharing the very nature of the bygone occasion in all its living activities, nevertheless is engaged in modifying it, in adjusting it to other purposes. The present moment is constituted by the influx of the other in that self-identity which is continued life of the immediate past within the immediacy of the present.
>
> *(Whitehead [1933]1967: 181)*

Whitehead's reading of perception brings to the fore what might be termed the 'socialness' of perception, which links sensory and non-sensuous experiences with objects and subjects, past, present and future. This social process enables the constitution and enduring of orderings, including all sorts of bodies, which deserves in this precise sense to be called 'societies' (cf. Debaise 2017, Halewood 2011). As such, Mrs M's objects *on hand* (her body, the environment), her concerns for them and the affective tone which does arise from feeling them are important sources with which to rethink the very subject/object relation (Debaise 2017). Whitehead states

> [t]hat an occasion is a subject in respect to its special activity concerning an object; and anything is an object in respect to its provocation of some special activity within a subject. Such a mode of activity is termed a 'prehension'. (...) There is the occasion of experience within which the prehension is a detail of activity; there is the datum whose relevance provokes the origination of this prehension; this datum is the prehended object; there is the subjective form, which is the affective tone determining the effectiveness of that prehension in that occasion of experience. How the experience constitutes itself depends on its complex subjective forms.
>
> *(Whitehead [1933]1967: 176)*

Such a reading of experience as a social process suggests that subject and object are 'relative terms' (Whitehead [1933]1967: 176). With the notion of prehension, Whitehead refers to a double-faced occasion of the becoming of subject and object. It is a complex interrelated 'mode of activity' whereby objects provoke 'special activity within a subject' (ibid.) which reshapes the object(s) prehended.

Following such a reading, we may conclude that Mrs M is clearly looking for clues to gain knowledge, in the sense of a conceptual feeling for and about the situation in which she finds herself. Still, the prehension of the subject/object-relation should not be conflated with the knower/known relation of a modernist bifurcating conception of subject and object. As Whitehead (ibid. 177) stresses, knowledge is 'nothing more than an additional [consciously discriminated, MS] factor in the subjective form of the interplay of subject with object.' And he adds that '[t]his interplay is the stuff constituting those individual things, which make up the sole reality of the Universe. These individual things are the individual occasions of experience, the actual entities' (ibid.).

Speculating with Mrs M's experience in this direction, I suggest that it assembles the problem of the dementing situation as a question of prehension that reveals subjective forms of perceptual concernedness with certain objects in place. Being *objects* of her concern, Mrs M is concerned about herself and the environment as givens that are received without knowing where they came from and how they will unfold. It unravels an *actor-network of non-knowing*, a mere relation of affect provoked by objects felt in a massive but vague sense, lacking specific assemblages of knowledge. One may say that Mrs M's actor-networks are temporally too short to achieve the possibilities of knowledge formation.

In fact, Whitehead has put it almost exactly as Mrs M did: An object 'must be experienced in virtue of its antecedence' (ibid.178), objects are 'the "data"' for the occasion for experience, they are '"lying in the way of"' (ibid. 179). Entities as objects seem to be precisely of Mrs M's concern, since she names their literal meaning by asking *was liegt da vor, what is on hand,* which literally translates *as what is lying in the way of her.* As objects, they are not passive, they affect Mrs M and she feels them rather negatively due to a lack of actor-networks of memories that extend the temporal span of causal efficacy so much so that different and more concrete and familiar actor-worlds of feelings, practices and knowledge could have emerged.

While the prehension of herself and the environment as massive but vague objects makes her feel rather agitated, nervous and anxious, the encounter of Mrs M with the filmmaker is perceived in a welcoming and thus differently valued manner. *With whom do I have the pleasure?,* Mrs M asks, seeking for a positively valued mode of relation. Mrs M seems to spot a similarity between herself and the filmmaker; something they possibly share (e.g. outlook, language), or in short, a societal relation that links similar-looking bodies and that may be of benefit in her situation. She welcomes an association, which may help her to look out for more mediators, to expand her temporally sparse network. It is likely a way of gaining time in order to meet more objects, mediated with the help of Mrs Kainz, objects that may affect her differently, actor-networks she may come to enjoy.

This seems to be Mrs M's main concern: Look here, she implies, objects matter, my reality matters, my mode of being, my body and the environment matter in the way they incorporate valued facticity that happens to be present, massive and vague. But she also wishes to extend her actor-network into the one that gains knowledge about the situation, knowledge that may help to tame the experience of the vivid presence of rather abstract bodies and uncanny things. Mrs M enacts and is enacted by the differences of value-laden relations that need to be cared for. Or, as Whitehead has put it: 'Our enjoyment of actuality is a realization of worth, good or bad. It is a value-experience. Its basic expression – Have a care, here is something that matters!' (Whitehead [1938]1968: 116).

Actor-networks of symbolic references

Sensory perception as presentational immediacy alone does not give Mrs M a secure hold; it does not give her a feeling of being in place, of being part of a personal order, since it only offers a rather partial and meaningless slice of existence that makes her body and the environment *contemporaries*. As such, it does not allow Mrs M to trace their histories and connectedness, to specify their importance. Mrs M's question, *Was liegt da vor? Bei mir und der Umgebung*? [What is on hand, with me and the environment], addresses very precisely her dilemma and invites us to speculate about it, to experiment with it and to care about it. In doing so, we get the sense that Mrs M relies too much on her sensory perception, situated in the presence and co-presence of objects.

Mrs M's worlding world, her being and becoming with and for others, has temporally petered out, since her embodied experience that unfolds the process of her existence by linking the past with the present and onwards to the future is troubled. With the dementing event, the presentational immediacy does not connect with the embodied experiences that link the different pasts with the present. In the dementing situation, Mrs M seems to be affected by 'temporally thin' objects, which do neither reveal their past nor whether they have a future. Mrs M is caught in the present, and trapped in the now that is felt massive and vague. Furthermore, Mrs M seems to rely strongly on the very short-spanned embodied memories ('causal efficacy'). We do not know how long the present, the 'now,' is for Mrs M. What we know is that these temporally thin objects affect her rather badly in the way she perceives them.

Within the dementing situation, her world thins out so much so that the occasions of experience, and the actual entities of Mrs M's world (including herself) do not gain enough endurance to become part of a personal ordering. They just come and go and do not socialise in the way that they may allow societal relations of togetherness assembling the past, present and future, the material and immaterial, of Mrs M's eventful personal order. In the dementing moment, Mrs M is stuck in the event of existence, shackled within the no longer and not yet. *And nobody has told me so… How is this possible?*, she agonises and is eager to enlarge her networks and look for social possibilities in order to free herself from an impossible situation.

Whitehead calls 'symbolic reference' the relation between causal efficacy and presentational immediacy. As a bodily event, this mixed mode of perception *interprets* experience (Whitehead [1929]1978). It seems that in the dementing moment, the interrelation between symbol and meaning is troubled and enacts her affective state as well as her eagerness to interrogate the 'massive and vague' feeling to possibly enjoy the conceptual, imaginative and thus interpretive power that derives from how the felt present becomes actualised through the felt past.

We can only speculate about it, but asking Mrs Kainz about the situation, Mrs M tries to enlarge her actor-network with words, which possibly turn into events and give meaning to symbols or *vice versa*. Through the use of spoken language, Mrs M tries to re-reference the embodied and materialised symbols and meanings that make up her situation. Mrs M's effort is to find out how the spoken language may symbolise, i.e. translate bodies, people and things in a way that they provide possibilities of understanding the situation. Mrs M tries to multiply symbolic references, which may enable to trace meaningful references that may be/come part of her personal ordering in space and time. In this sense, we may infer that she wishes to regain the *pragmatic relevance* of conceptual freedom that may *justify* her situation, to herself and to others, and might help to reduce her anxieties, insecurities and agitated state.[3]

Conclusion: bodies, perceptions and affects

This all-too-brief thinking with Mrs M and aspects of Whitehead's process philosophy highlighted the importance of different modes of embodied perception that make up the complexity, specificities, limits and potentialities of different actor-networks. What we can learn from dementia is how we may understand how good and/or bad things come into being. As we have seen, the good and bad as understood in processual philosophy is not just a moral question but extends to the ontological fabric of (human) experience and gains pragmatic value about *what* experience is in the different ways of *how* it is experienced (cf. Whitehead [1938]1968). To come closer in understanding the latter, people with dementia can be considered as essential partners to do so. People with dementia make us aware of the ontological importance of the affective in understanding the *value-laden becoming of things* – human and non-human alike. The dementing experience draws our attention not only to the entanglements of the social and the societal, the material and the temporal, but also to the importance of *affective* relations.

The discussion has allowed a deep understanding of one of ANT's central but all-too-abstract claim – that all things are societies – by addressing their variable 'temporal thickness' (Debaise 2013: 104) and their variable affective/ing aspects. *How* things gain temporal thickness is a core question of the social. I have tried to show that within the dementing experience, the temporal thickness thins out and is experienced along the questionability of the social as a source for *negatively felt* situations governed by more or less meaningless things. This analysis arguably draws attention to the fragility of all actor-networks. However, my analysis also highlights how objects may offer potentials to turn bodies and environments into positively felt relations of good situations. We have seen that Mrs M plays a crucial role in tracing possible ways to do so and how people with dementia may contribute to the conduct of caring practices.

Notes

1 On documentary as a 'cosmopolitical technique of affect/ion,' see Schillmeier (2018).
2 Actor-Networks as experience allude to a pragmatist reading of 'experience' shared by James, Dewey and Whitehead. As Dewey ([1929] 1971: 4) argues: 'Things interacting in certain ways *are* experience; they are what is experienced.' See also James ([1912]2003).
3 See Whitehead ([1929]1978: 180pp) on the pragmatic justification of symbolism.

References

Callon, M. (1981) Struggles and Negotiations to Define What Is Problematic and What Is Not: The Socio-logic of Translation. In K. Knorr, R. Krohn & R. Whitley (eds.) *The Social Process of Scientific Investigation*. Dordecht, Holland: D. Reidel Publishing Co, 197–220.

Callon, M. (1986) Some Elements of a Sociology of Translation: Domestication of the Scallops and the Fishermen of St. Brieuc Bay. In J. Law (ed.) *Power, Action and Belief: A New Sociology of Knowledge?* London: Routledge & Kegan Paul, 196–233.

Callon, M. & Latour, B. (1981). Unscrewing the Big Leviathan: How Actors Macro-Structure Reality and How Sociologists Help Them Do So. In K. Knorr Cetina & A.V. Cicourel (eds.) *Advances in Social Theory and Methodology: Toward an Integration of Micro- and Macro-Sociologies*. Boston: Routledge & Kegan Paul, 277–303.

Debaise, D. (2013) A Philosophy of Interstices: Thinking Subjects and Societies from Whitehead's Philosophy. *Subjectivity* 6(1): 101–111.

Debaise, D. (2017) *Speculative Empiricism: Revisiting Whitehead*. Edinburgh: EUP.

Dewey, J. ([1929] 1971) *Experience and Nature*. Second Edition. La Salle: Open Court.

Garfinkel, H. (1967) *Studies in Ethnomethodology*. Englewood Cliffs, NJ: Prentice Hall.

Halewood, M. (2011) *A.N. Whitehead and Social Theory*. London/New York/Delhi: Anthem Press.

James, W. ([1912] 2003) *Essays in Radical Empiricism*. Mineola, New York: Dover Publications.

Jensen, C.B. (2014) The Conceptual and the Empirical in STS. *Science, Technology, & Human Values* 39(2): 192–213.

Kitwood, T. (1997) *Dementia Reconsidered. The Person Comes First*. Buckingham: Open University Press.

Kohn, T. & R. McKechnie (1999) (eds.) *Extending the Boundaries of Care: Medical Ethics & Caring Practices*. Oxford/New York: Berg.

Latour, B. (2005) *Reassembling the Social. An Introduction to Actor-Network-Theory*. Oxford: Oxford University Press.

Law, J. & J. Hassard (1999) (eds.) *Actor Network Theory and After*. Sociological Review Monographs/ Oxford: Blackwell.

Mol, A. (2002) *The Body Multiple. Ontology in Medical Practice*. Durham, NC: Duke University Press.

Mol, A. (2008) *The Logic of Care: Health and the Problem of Patient Choice*. London/New York: Routledge.

Moreira, T., O'Donovan, O. & E. Howlett (2014) Assembling Dementia Care: Patient Organisations and Social Research. *BioSocieties* 9(2): 173–193.

Moser, I. (2008) Making Alzheimer's Disease Matter. Enacting, Interfering and Doing Politics of Nature. *Geoforum* 39: 98–110.

Moser, I. (2010) 'Perhaps Tears Should Not Be Counted But Wiped Away. On Quality and Improvement in Dementia Care', in A. Mol et al. (eds.) *Care in Practice: On Tinkering in Clinics, Homes and Farms*. Bielefeld: transcript, 277–300.

Passoth, J.H., B. Peuker & M. Schillmeier (2012) (eds.) *Agency without Actors. Rethinking Collective Action*. London/New York: Routledge.

Puig de la Bellacasa, M. (2017) *Matters of Care. Speculative Ethics in More Than Human Worlds*. Minneapolis/London: University of Minnesota Press.

Schillmeier, M. (2009) Actor Networks of Dementia. In J. Latimer & M. Schillmeier (eds.) Un/ knowing Bodies. The Sociological Review Monograph Series. Oxford: Blackwell, 141–160.

Schillmeier, M. (2010) *Rethinking Disability: Bodies, Senses, and Things*. London/New York: Routledge.

Schillmeier, M. (2011) Unbuttoning Normalcy – On Cosmopolitical Events. *The Sociological Review* 59(2): 514–534.

Schillmeier, M. (2014) *Eventful Bodies – Cosmopolitics of Illness*. London/ New York: Routledge.

Schillmeier, M. (2017) The Cosmopolitics of Situated Care. In: Gill, N., Singleton, V. & C. Waterton (eds.) *Care and Policy Practices*. The Sociological Review Monograph Series. London: Sage, 55–70.

Schillmeier, M. (2018) Foreword. In Brylla, C. & H. Hughes (eds.) *Documentary and Disability*. London: Palgrave Macmillan, v–vii.

Tarde, G. (2012) *Monadology and Sociology*. Melbourne: re.press.

Tsing, A. (2010) Worlding a Matsutake Diaspora: Or, Can Actor-Network Theory Experiment with Holism? In T. Otto & N. Bubandt (eds.) *Experiments in Holism. Theory and Practice in Contemporary Anthropology*. Oxford: Wiley-Blackwell, 47–66.

Whitehead, A.N. ([1933]1967) *Adventures of Ideas*. New York: The Free Press.

Whitehead, A.N. ([1938]1968) *Modes of Thought*. New York: Free Press.

Whitehead, A.N. ([1929]1978) *Process and Reality. An Essay in Cosmology*. Corrected Edition. New York/ London: The Free Press.

14

What is the relevance of Isabelle Stengers' philosophy to ANT?

Martin Savransky

Stengers and ANT: dramatising a divergent alliance

It is a difficult situation – that which the question posed to me by the editors of this *Companion* puts me in – of pondering on the relevance of the philosophy of Isabelle Stengers to this other style of work, however we may wish to define it, that has become condensed under the name of Actor-Network Theory (ANT). This is because the question of relevance is always specific, belonging to an event of a coming to matter, and does not lend itself to whole-sale judgements or categorical pronouncements. Without some careful consideration, then, the question might incite quick, uninspired reactions that betray one's own trajectories and predilections – either the relevance of her philosophy to ANT is a matter of course, and what follows needs not be written; or it is the product of a profound misunderstanding, and what follows requires the effort of a restorative critique.

Such reactions are nothing, however, if not the prolongation of what this eventful relationship, that which I insist – despite the relentless dishonouring of this word by capitalist knowledge economies – on calling *relevance* (Savransky, 2016), demands we refuse. The prolongation, that is, of the questionable Western philosophical tradition that will only set up infernal alternatives between what are divergent yet occasionally resonant paths. A tradition that will only admit distinctions between 'for' and 'against,' either/or, and will understand the battles and conflicts as symptoms of a more transcendent reason to be unveiled, one finally capable of realising the modern dream of an ecumenical, perpetual peace– ecumenical, that is, except for those that will be subsequently cast as irrational.

Believing with William James (1956) that rationality is more of a sentiment than a faculty, this is a tradition I seek to refuse. As I see it, what is at stake instead is to explore some of the divergent connections *through* which these two projects have woven a certain intellectual kinship. Indeed, because insights from ANT and Stengers' philosophy, respectively, have been gaining reception in the Anglophone world at a somewhat uneven pace, scholars noting or drawing on the work of Stengers in relation to the former have so far tended to emphasise the resemblances and affinities that connect both oeuvres: some shared philosophical influences (Watson, 2014), some resonant materialist and empiricist attitudes (Nimmo, 2011), a shared distaste for critical operations (Jensen, 2014) and some connected terminology (Blok & Farías, 2016).

One cannot deny that partial connections do exist, that attention to them may be productive whenever –as with the sources cited earlier – what is at stake is not a mere exegesis but the retooling of diverse concepts and propositions to develop new questions and problems. And they are especially pertinent when entertained in direct relation to one of ANT's most prominent figures – namely, Bruno Latour, whose work both founds and *exceeds* ANT.[1] For Latour and Stengers have confounded the philosophical tradition which dictates that one cannot become a worthy philosopher until one has succeeded in waging war on every other worthy philosopher. Instead, they have developed a fertile friendship as well as a very public *alliance*, generating many occasions for reciprocal learnings and borrowings across and between their respective paths. And like many alliances, this one is based not on a general convergence of purposes, but on a certain affinity at the level of the problematic itself. An affinity for the need to create words that would cease prolonging that Great Divide that separates 'us,' the moderns, who can distinguish facts from values, the natural from the cultural, the scientific from the political, the rational from the passionate; from all the others, who mess all those neat distinctions up.

But while stories of connection can be productive, stories of convergence are rather dangerous. For they confuse a problematic alliance with the question of relevance itself, of how their respective responses to this problematic come to matter to each other (Savransky, 2016). Which is to say that, by pacifying their differences, other stories prevent readers from discerning the equally fertile borders through which these paths may be connected through difference. The result is that when the attempt is made not to craft creative connections but to *locate* Stengers' philosophy in relation to ANT, a curious articulation is formed: on the one hand, ANT is regarded as indebted to Stengers' philosophy, even as being in certain respects 'a methodological expression' of it (Whatmore, 2003: 83). On the other, the latter is classed as a branch of an ANT tree, under what has been referred to as 'post-ANT' (Michael, 2016: 126). The making of partial connections becomes here a blurring of borders, and any reciprocal learning is turned into a form of convergence rather than an opportunity to learn how these projects *diverge, that is, to appreciate the singularity of the respective dramas and trajectories that force each form of thought to be set in motion* (on this, see Savransky, 2018).

The challenge in addressing the relevance of Stengers' philosophy to ANT, of refusing the aspiration of a transcendent truth capable of accommodating everything and everyone, is therefore that of dramatising the fact that such an *alliance* neither requires, nor presupposes, a shared project or a general consensus. As Stengers (2015:16) wrote in acknowledgement of her alliance with Latour at the beginning of her *In Catastrophic Times,* the forging of alliances is a process that testifies to the fact 'that agreements between sometimes diverging paths are created thanks to, and not in spite of, divergence.' It is thus in response to the question of the *relevance* of Stengers' philosophy to ANT that what I seek to do in what follows is not to emphasise affinities or to highlight differences, but to dramatise their *divergence*.

Of course, a dramatisation is never a neutral operation, and the attempt made here cannot be reduced to a didactic 'compare and contrast'– instead, it is always partial and pragmatic, responding to the insistence of a possibility. What follows is no exception. I seek to dramatise a divergence not least because it is this very concept, 'divergence' itself, and the crucial role it plays in Stengers' speculative proposition of an 'ecology of practices,' that may perhaps constitute one of the elements capable of interesting ANT allies who choose to engage her philosophy. And if it is of interest, it is because it is susceptible of asking questions that ANT scholars may consider relevant, even if they are not their own questions. The divergence of an ecology of practices is liable, in other words, to situate one of ANT's political and methodological footholds: the principle of generalised symmetry (Callon, 1986; Latour, 1993; Latour, 2005). As it becomes perceptible that the relationship between symmetry and asymmetry is not itself a symmetrical relationship, attending to divergences in an ecology of practices, I suggest, prompts us to experience a space of political relations irreducible either to modern asymmetries or to generalised symmetry – an inappropriable space where angels fear to tread.

The angels' demand: situating ANT

'For fools rush in where angels fear to tread,' wrote Alexander Pope (2013: 80, §625) in his famous poem from 1711, *An Essay on Criticism,* reminding the nascent literary critics of the time that bad critique is much more harmful than the possible going-astray of a literary adventure. This already seems like an apt refrain for what distinguishes an alliance from the infernal alternatives of the modern tradition. For it reminds us that the angels' fear needs to be honoured, it demands that we slow down, and proceed with care; but it does not in itself constitute a divine prohibition. In this case, moreover, 'where angels fear to tread' has yet another, entangled, but more circumscribed sense. This is that, as an *ally* of ANT myself, it has for a long time seemed to me that what animates the concerns of the actor-network theorist, when imagined as a conceptual persona, is intimately related to a certain figure of the angel, in the sense in which this figure has been characterised by someone who is arguably one of the most direct philosophical influences of ANT, namely, Michel Serres.

Serres' (1995: 7) angels, heirs of Hermes – the God of communication, of circulations and of translations – are messengers, quasi-objects, mediators, bearers of heterogeneous signals and hybrid information: 'Angels are *messengers.* [...] They appear and then disappear. It is said that they move through space at the speed of their own thoughts.' As Serres' conversational character, Pia, goes on to suggest:

> In the oldest traditions, angels do not necessarily take on human appearance; they may also inhabit the universe of things, whether natural or artificial [...] The light that comes from the sun brings messages, which are decoded by optical or astrophysical instruments; a radio aerial emits, transmits and receives; humans do not need to intervene here. [...] We exchange information with objects that appear more as relations, token, codes, and transmitters.
>
> *(Serres, 1995: 52)*

Making angels (or what ANT scholars might call 'hybrids') and their networks visible – in travelling letters, in human migrations, in scientific labs, through airports, in storms and climate patterns, in DNA ribbons, in the rain and the wind, in material designs, in the art of moving bodies, human and nonhuman – is no doubt the aim of Serres' rather swashbuckling book. But is this not also, less romantically, the aim of actor-network theorists – might we call them *angel-network theorists?* – as travellers following ancient and novel 'angelic paths' (or, hybrid networks) that the modern tradition of the Great Divide has insisted on keeping apart? Is ANT, in some sense, not an attempt to retrace those paths that the modern tradition centrifugally expels into all kinds of hierarchies and classifications, into little circumspections and confines subsequently baptised as 'social,' or 'natural,' or 'political,' or 'technological,' or 'scientific' or what have you, continuously enthroning and dethroning God, yet banishing these hybrid angels at every turn?

The banishing of angels is indeed the effect of the 'crisscrossed schema' of the moderns wherein Latour (1993: 34), in his landmark *We Have Never Been Modern,* had located their power:

> They have not made Nature; they make Society; they make Nature; they have not made Society; they have not made either, God has made everything; God has made nothing, they have made everything. There is no way we can understand the moderns if we do not see that the four guarantees serve as checks and balances for one another. The first two make possible to alternate the sources of power by moving directly from our natural force to pure political force, and vice versa. The third guarantee rules out any contamination between what belongs to Nature and what belongs to politics, even though the first two guarantees allow a rapid alternation between the two. Might the contradiction between the third, which separates, and the first two, which alternate, be too obvious? No, because the fourth constitutional guarantee establishes as arbiter an infinitely remote God who is simultaneously totally impotent and the sovereign judge.

This modern constitution, with its checks and balances, was *powerful* because it, in turn, located everyone – natural scientists, social scientists, philosophers, theologians, politicians, technicians – in their own little contradictory confines, on their respective trenches of the Great Divide. From there, they would appeal to their own transcendences and purified causalities, and in so doing insist on the banishing of angles and their hybrid modes of existence – neither natural nor social, neither scientific nor political, neither godly nor earthly, and yet at once real and meaningful, relational and collective, divine and mundane, human and other-than-human. But this modern operation was not almighty either – as both Serres and ANT scholars argue, angels nevertheless persisted in their proliferation. What it created, instead, was a particular kind of asymmetry, where nothing relates to anything else, unless it is *entirely* explained (away) by, and thus reduced to, something else. Start with the modern concept of Nature, and everything else becomes but particles in motion. Start with the modern concept of Society, and everything else becomes but discourse and ideology. It is in response to this particularly *modern* kind of asymmetry that ANT is born – as a counter-modern, empirical angelology concerned with the retracing of the tehcnoscientific networks through which angels arise thanks to, and in spite of, the trenches put up by modernity.

One can thus appreciate the very dear place that the so-called *principle of generalised symmetry* holds in ANT's development. The principle, that is, to inhabit the middle kingdom, not to deploy the categories of the moderns to explain (away) the world, 'not to impose a priori some spurious *asymmetry* among human intentional action and a material world of causal relations' (Latour, 2005: 76). The metaphor of 'generalised symmetry' originally alluded to the attempt to correct the ultimately asymmetrical operation of the sociology of scientific knowledge (Latour 1993: 94–96), but we might in retrospect say that it was not the happiest of metaphors. Whenever it has been taken as a theoretical prescription, an ontological claim that would assert that 'nothing is natural or cultural, everything is a relation of forces,' or worse, a global political project animated by a moral command demanding that we 'make the world symmetrical again,' the principle of generalised symmetry has suffered the same sad fate as those modern concepts that have all too often been used as weapons: it transforms ANT into a general doctrine, capable of quashing any practice that diverges from its path. This has, of course, been an early and enduring criticism, that as Nick Lee and Steve Brown

(1994: 774) have put it, ANT is 'so liberal and so democratic that it has no Other [...] it has made itself into a "final" final vocabulary.' As they go on to expand:

> Having converted the world into a play of forces, [ANT] has no way of circumventing the formulaic circle of expansion, domination, and collapse. ANT has achieved a metalinguistic formulation– inscribed as problematization, interessement, enrolment, mobilisation, and dissidence (Callon 1986)– into which any sequence of human or nonhuman actions can be encoded. This amounts to a foreclosure of all alternative descriptions of the world through the assertion of total democracy and complete ontological monadism.
>
> *(Lee & Brown, 1994: 781)*

One can hardly imagine a better description of what happens when a proposition, with its singular interests and constraints, is turned into a programme that abstracts and legislates at everyone else's peril. And while this critique has been duly 'noted' (Law, 1999, Latour, 1999), the degree to which it has infected the habits of ANT scholars more generally remains in my view an open question. It is echoed, for instance, in Latour's (2005: 76, n. 86) footnoted concern about the perils of taking symmetry as a *theoretical* principle, in a book that *nevertheless* presents ANT as a particular theory of associations, and thus as a peculiar approach to sociology. Namely, the concern

> that readers concluded from it [the principle] that nature and society had to be 'maintained together' so as to study 'symmetrically' 'objects' *and* 'subjects', 'non-humans' *and* 'humans'. But what I had in mind was not *and*, but *neither:* a joint *dissolution of both collectors.*

Central as it was, the principle of generalised symmetry appears to have fuelled considerable theoretical hubris when it was a very *specific* kind of operation that was sought, one defined primarily in negative terms, as a rejection of the particular vectors of asymmetry that characterise modernity. Alas, general principles are never especially apposite to the crafting of specific operations. Following Latour's (2005) footnote, by contrast, I would thus propose to understand 'symmetry' as a *technical constraint:* rejecting modern asymmetry is what the hybrid mode of existence of angels itself demands if one is to become sensitive to them, to follow their paths and networks, to translate their messages, to have them matter. Start from Nature, from Society, from Politics, from God, and you will miss them as they pass.

Understanding ANT's rejection of modern vectors of asymmetry as a technical constraint perhaps allows us, in turn, to situate the dreams of this empirical angelology otherwise. For it makes present that ANT may be best described not as a theory of the social, or of science; not as a theory of 'material-semiotics,' or of performativity; and not even as an ontology of action. Indeed, we might even go as far as to suggest that, despite its best efforts, ANT may not be another way of doing 'social science' at all (and that may be one of its strengths)! By contrast, I would tentatively propose to characterise ANT (angel-network tracing?) as its own singular *practice,* not in the ordinary acceptation of the term, but in the speculative sense given to this notion by Isabelle Stengers – a singular vector of divergence, of value–creation, situated by constraints imposed by the requirements and uncertainties it introduces, by the mode of existence of what makes it think, hesitate and hope, and its corresponding obligations. Perhaps we can risk characterising ANT thus, as a singular practice animated by, and concerned with, the existence of a multiplicity of angels and the associations they make relevant.[2] One that *requires* that hybrid angels not be banished by modern asymmetries, and one that is, in turn, *obligated* to follow their paths and networks wherever they may lead. Perhaps that is ANT's own *mode of divergence* – for it is angels that make ANT practitioners think, hesitate and hope.

To be sure, at a time when capitalism continues to expand the territory of its cemetery of practices, the invention of a new practice is no minor achievement. What is more, to characterise and celebrate ANT as a *practice* rather than as a theory, an approach or an ontology allows us to shift our mode of attention. Indeed, I must agree with Latour (1999) that the 'T' ('theory') in ANT deserves a nail in the coffin (see also Jensen, this volume). Because no matter how immanent, how democratic, how empiricist, how *symmetrical* it may claim to be, a theory (not to mention an 'ontology') can hardly resist the temptation to legislate for everyone and everything, just as a different *approach* to the social sciences can hardly resist the temptation to compete with other approaches for who might be best equipped to define the nature of the social. Yet as Latour (2013) has more recently suggested, the networks with which ANT operates do not designate a general theory of being, but correspond to one mode of existence among others.

What the characterisation of ANT as a practice makes present, therefore, is that the other side of modern asymmetrical relations is not 'generalised symmetry' but a manifold otherwise of possible relations to be dramatised. Indeed, as mathematicians would tell us, the relationship between symmetry and asymmetry is not itself a symmetrical relationship. And unlike a theory, which strives for the Archimedean point – indeed, the symmetrical mean – between everything it lays claims to, a practice has no such privileged foothold: from its singular situation, it must proffer its divergence into the world, among a multiplicity of other practices characterised by different modes of divergence, and is thus unable to singlehandedly determine *how* it may interest others, how it may matter to them. In other words, ANT's rejection of modern asymmetry helps us address the problem of the ways in which the moderns present themselves in a manner that disqualifies others, but as Stengers (2011: 360) herself suggests, '[s]ymmetry is not itself a solution to this problem.'

It is by appreciating ANT as a singular practice that I propose to further explore the relevance of Stengers' philosophy in relation to it. Not because the latter may provide the solutions to the problem that the rejection of modern asymmetry enables us to explore, but because what it may make felt among ANT practitioners is precisely the importance of *allowing themselves to be situated* by their own practice – of affirming that while the world is populated by angels, it is not thoroughly *made of* them; that beyond angels and their paths there are other beings and connections whose differences cannot be symmetrised; that what makes ANT practitioners think, hesitate and hope is what defines and conveys significance to *their* practice, one among others.

Our symmetry, and theirs? Divergences of an ecology of practices

If Isabelle Stengers' philosophy can be regarded as a possible ally to ANT, it is doubtless because some of the problems that animate both of them enjoy a certain affinity and resonance. But Stengers is no actor-network theorist. Her experimentation with this imperative not to continue ploughing the Great Divide does not, in and of itself, *found* a new practice. It is instead concerned with the radical mutation of what that ancient practice called 'philosophy' might become capable of were it to refuse participation in the anonymous game of proclaiming yet another transcendental truth, yet another promise of eternal salvation (see Savransky, 2018). Which is to say that it is animated by an exploration of the question of what a philosophy *of* practices might become, when this 'of' designates not simply an object of thought but an entire technique of habitation – of attending to those modes of thinking, doing and feeling *amongst which* philosophy is forced to think; *with which* it may forge alliances against their ongoing capitalist devastation; *in the presence of which* it may risk propositions that gently shift the ways in which they may be capable of presenting themselves (Savransky, 2016).

This task must, from the outset, come to terms with a profoundly asymmetrical space, in the modern sense of the term. For indeed, one of the possibles to which it must invent a response, one of the risks it is required to take, is that of characterising the singularity of divergent practices so as to wrest the dreams, hopes and fears that animate them, from the powers to authorise and disqualify by which some practices have deservedly acquired their 'modern' badge. Crucially, Stengers undertook this task in response to a specific event, that is, the poisonous wrath of the Science Wars, which had physicists and social scientists, respectively, defending against each other the primacy of a 'physical' and a 'social' reality, angrily claiming for themselves an 'exclusive position of judgement over and against all other "realities," including those of all other sciences' (Stengers 2005a: 183). And it is in response to the speculative possibility of learning to produce another mode of astonishment, another way of presenting what makes a practice diverge, that Stengers articulates her proposition of an 'ecology of practices': 'How can we make it possible for a modern practitioner to present herself, justify her practice, draw attention to what interests her, without that interest coinciding with a disqualification?' (Stengers, 2010: 49).

It might be tempting to address this question itself as a matter of orchestrating a symmetry between modern and non-modern practices. But the aforementioned discussion may have already made present that such a response will not do, for '[t]he "we" who symmetrizes doesn't speak "for everyone," using words that also apply to us and to those we will encounter.' (Stengers 2011: 362). A profoundly political shift is at stake here. Symmetry may be a relevant requirement for some practitioners, but it is not for that reason a solution capable of becoming a general means, a tool susceptible of securing a perfect communication among practices. To extend symmetry as a general principle would amount to confronting other practices, situated by different constraints, with the need to abandon their hopes and doubts, their dreams and their fears, to leave behind what matters to them, what makes them hesitate, whether these be the detection of neutrinos, the understanding of human emotions or the immaterial reality of ghosts. As such, symmetry itself would become the tool that thwarts other adventures in divergence. By contrast, Stengers' ecology of practices has no recourse to such general harmonisations, and it cannot hope to represent the dreams of others in ways that would gain overall consent:

> We, who are not angels but think in political terms, must therefore create obstacles that prevent us from rushing toward others while requiring that they resemble what we might become, obstacles that prepare us to wonder about their conditions, the conditions they might establish for eventual exchange. This is *our* problem. Its construction in no way ensures that the [other] will meet with us (any more than the construction of an experimental device ensures that the being we wish to mobilize will show up). Our words are relative to our practices and we now ask that they will tell us which obligations will guide us where angels fear to tread.
>
> *(Stengers, 2011: 362)*

In contrast to the tendency to read Stengers' philosophy *via* ANT, the proposition of an ecology of practices should make anyone tempted to carry 'symmetrical anthropology' as a rallying flag hesitate. If in erecting itself as a vocabulary capacious enough to provide the 'uniquely adequate description of a given situation,' where every actor would be assigned a 'part in the plot' (Latour 2005: 130), ANT occasionally ran the risk of allowing nothing to escape from its symmetrical webs, the proposition of an ecology of practices means nothing if it does not actively affirm an outside, always relative to its specific wager, but always inappropriable and beyond its grasp.[3]

As such, by reminding us that 'our words are relative to our practices,' the notion of an ecology of practices introduces the possibility of a political revaluation of non-symmetrical relationships. It introduces this possible revaluation by making present that 'symmetry' is relative to this *practice* called ANT – it is a constraint relevant to those who are made to think, hope and dream by the passing of angels. But their passing is not everyone's problem, and it does not turn 'symmetry' into a general mean, relevant to every practice. Beyond it, where angels fear to tread, multiple and always singular forms of divergence insist and persist, bringing a multiplicity of other practices to life. What the proposition of an ecology of practices seeks to create, thus, is not a dissolution of these borders, the achievement of a final arrangement of relations among practices that could enable them to live together happily ever after. It corresponds instead to the very challenge of approaching each practice

> as it diverges, that is, feeling its borders, experimenting with questions which practitioners may accept as relevant, even if they're not their own questions, rather than posing insulting questions that would lead them to mobilise and transform the border into a defence against their outside.
>
> *(Stengers, 2005a: 184)*

This requires a new mode of appreciation. An appreciation of practices that is concerned with the possibility of disentangling that which may function as a *force*, luring practitioners into their own risky adventures of learning, from the *power* with which modern practices may turn their own situated lessons into wide-ranging generalities, capable of judgements that thwart others' trajectories of apprenticeship. For after all, 'no unifying body of knowledge will ever demonstrate that the neutrino of physics can coexist with the multiple worlds mobilized by ethnopsychiatry' (Stengers 2011: vii). That, in fact, is the test of the speculative proposition of an ecology of practices: cultivating, on a case-by-case basis, the possibility of turning polemical contradictions into immanent contrasts, where practices can describe their singularity, the hopes, dreams and fears that animate them, the values they create, without at the same time turning those attachments into rights, their values into claims, their dreams into theories.

And if this proposition is speculative, this is not least because it offers no guarantees and no methodological prescriptions – a speculative proposition does not have the power to bring about that which it calls for (Savransky, 2017). Indeed, the proposition of an ecology of practices is not a programme that would have every practice consenting to a new way of describing what matters to it, and it excludes 'the possibility that it might become a source of values everyone would be required to submit to, and in whose name everyone's place and relationships could be determined' (Stengers 2011: 367). It involves political transformations but not in the sense that it would create representatives of practices in a new Parliament of Things, which for its part *requires* practitioners interested in the challenge of collectively reinventing the dreams and fears that shape the territory of their practices, but in no way prepares them 'to meet someone who refuses to cooperate, to play the game, to take an interest in the challenge associated with those values' (ibid.: 366).

The ecology of practices involves political transformations in the sense that it establishes non-symmetrical relationships. Unlike the asymmetries of the Modern Constitution, which imply hierarchisation, infernal alternatives, judgements and disqualifications, these *non-symmetrical passages* authorise nothing – they belong to an experience of metamorphosis that, in the presence of those who nevertheless remain other, enables one to feel one's own attachments in a different way. And metamorphoses, while they may be shared, are hardly ever symmetrical, for '[w]e can never fully understand another's dreams, hopes, doubts and fears, in the sense that an exact translation could be provided, but we are still transformed as they pass into our experience' (ibid.: 371). Indeed, the non-symmetrical metamorphoses on which an ecology of practices wagers open a host of other political relations belonging to a world without symmetrical mean yet crossed by tangential passages, infinitely close contacts along a curve of divergence, and oblique angles of exchange involving transductions and transitions across borders. Relations that establish no new settlement, no new agreed definition of the common world, but 'an experience of deterritorialization,' conferring on the singular manner in which a practice inhabits its situation the power to make it think. There is no achievement of symmetry here. What such deterritorialisation accomplishes, instead, is to make the practitioner brush against the borders of her practice – to provoke

> an experience of "feeling" one's own territory, and as result a practitioner, *modern or not*, may […] discover that her practice situates her, resulting in a nonequivalence that she is meant to actualize, that makes her hope, doubt, dream, and fear.
>
> *(Stengers, 2011: 372)*

Where angels fear to tread: an experience of capture

To suggest that an ecology of practices involves a multiplicity of non-symmetrical deterritorialisations and metamorphic political relations means that it can be neither produced at will, nor by discursive argumentation. No single vocabulary or argumentation, however capacious, will be able to bring it about:

> such an argumentation is ruled by the fiction of the everybody or the anyone – 'everybody should agree that …'. 'anyone should accept this or that consequence …' – a fiction which downgrades to good will and enlightenment the creation of the possibility of a conjunction, 'this and that' where the disjunction 'this or that', leading to war, ruled before.
>
> *(Stengers, 2005a: 194)*

The possible actualisation of an ecology of practices, therefore, is never guaranteed – it may happen, but it is never deserved, and does not correspond to any right. There is thus no final answer to the question that animates Stengers' philosophy: it postulates no ideal horizon, and no set of concepts that will give theorists the power to roam the world identifying and recognising their analytical pertinence in spite of specific and always singularly demanding situations.

As I have intimated earlier, between symmetry and non-symmetry lurks a marked political *option* – a divergent path that concerns the question of what the always unstable possibility of a problematic coexistence may require before it itself becomes perceptible. And it is notably around more openly political, or 'cosmopolitical,' questions, that the divergences that partially connect these two paths of thought and inquiry (Stengers' and ANT's) are conspicuously expressed. While Latour (2004) would on occasion characterise 'comsopolitics' as the vector and collective achievement of a new distribution of powers, the provisional construction of a new, angel-inclusive, democratic assembly oriented towards progressive composition of a good common world, in Stengers the term is deeply aligned to the experience of reciprocal reterritorialisation that accompanies the ecology of practices. A profoundly non-symmetrical experience indeed, because the cosmopolitical corresponds precisely to the requirement 'of imbuing political voices with the feeling that they do not master the situation they discuss.' As such, it makes resonate the ongoing problematic of a political arena that is 'peopled with shadows of that which does not have a political voice, cannot have or *does not want to have one*' (Stengers 2005b: 996). Thus, cosmopolitics designates not the realisation of maximal democratic inclusion, where an entire 'cosmos' may be assembled, but a generative 'unknown' constituted by what a multiplicity of divergent practices may eventually become capable of:

> Cosmopolitics introduces what is neither an activity, nor a practice, but *the mode in which the problematic copresence of practices may be actualized*: the experience, always in the present, of the one into whom the other's dreams, doubts, hopes, and fears pass. It is a form of *asymmetrical reciprocal capture* that guarantees nothing, authorises nothing, and cannot be stabilized by any constraint, but through which the two poles of the exchange undergo a transformation that cannot be appropriated by any objective definition.
>
> *(Stengers 2011: 372. emphasis added)*

We thus come full circle – to discover that the circle is no longer perfect. It has broken loose. ANT, angel-network theory, angel-network tracing: the problem has changed. Neither a matter of course, nor a mere misunderstanding, addressing the relevance of Stengers' philosophy itself involves accepting a possible experience of capture, requiring a metamorphosis of one's own. Unlike so many of the theories populating the contemporary academic landscape, Stengers' concepts do not provide anyone with ready-made tools. They cannot be passed from hand to hand without transforming the habitual gestures of the hand that holds them. This is because these tools provoke thought to respond to a possible whose coming about will have the character of a pluralistic event – a response to a held out hand, by another hand stretched out from the other side, from an immanent but inappropriable beyond. For this reason, when addressing her philosophy, one should move carefully, and remember: one might find oneself there, where angels fear to tread.

Notes

1 One of the ways by which ANT continues to recreate itself amongst its practitioners is no doubt through an ongoing interrogation about what ANT is. Relatedly, there is, of course, an equally open question about where it begins and where it ends vis-à-vis the work of its most prominent representatives. Because of the singular alliance between Stengers and Latour, I shall focus here exclusively on some of the latter's direct contributions to ANT, while leaving for another occasion an examination of Latour's (2013) most dramatic departure from it in his AIME project, where networks become one mode of existence among others.

2 As mentioned in the previous footnote, it is in his *Inquiry* that Latour (2013) dramatises a movement beyond ANT that simultaneously situates the latter as concerned with only one mode of existence among others, with their own specific "felicity conditions."

3 And in this sense, while Latour's (2013) AIME project constitutes a stepping out of ANT that, in turn, situates the latter, one would be forgiven from wondering whether, in some significant respects, rather than an attempt to affirm an outside, AIME might not rather be an attempt to expand the power of our grasp. But this is a question for another exploration.

References

Blok, A. & Farías, I. (2016), *Urban Cosmopolitics: Agencements, Assemblies, Atmospheres*. London: Routledge.

Callon, M. (1986), 'Some Elements of a Sociology of Translation: Domestication of the Scallops and the Fishermen of St Brieux Bay'. In J. Law (ed.), *Power, Action and Belief: A New Sociology of Knowledge* (pp. 196–229). London: Routledge & Kegan Paul.

James, W. (1956), *The Will to Believe and Other Essays in Popular Philosophy*. Minneola: Dover Publications.

Jensen, C. (2014), 'Experiments in Good Faith and Hopefulness: Towards a Postcritical Social Science', *Common Knowledge*, 20(2), 337–362.

Latour, B. (1993), *We Have Never Been Modern*. Cambridge, MA: Harvard University Press.

Latour, B. (1999), 'On Recalling ANT'. In J. Law & J. Hassard (eds.), *Actor-Network Theory and After* (pp. 15–25). Oxford: Blackwell Publishing.

Latour, B. (2004), *Politics of Nature*. Cambridge, MA: Harvard University Press.

Latour, B. (2005), *Reassembling the Social*. Oxford: Oxford University Press.

Latour, B, (2013), *An Inquiry into Modes of Existence*. Cambridge, MA: Harvard University Press.

Law, J. (1999), 'After ANT: Complexity, Naming and Topology'. In J. Law & J. Hassard (eds.), *Actor-Network Theory and After* (pp. 1–14). Oxford: Blackwell Publishing.

Lee, N. & Brown, S. (1994), 'Otherness and the Actor Network: The Undiscovered Continent'. *American Behavioral Scientist*, 37(6), 772–790.

Michael, M. (2016), *Actor-Network Theory: Trials, Trails and Translations*. London: Sage.

Nimmo, R. (2011), 'Actor-Network Theory and Methodology'. *Methodological Innovations Online*, 6(3), 108–119.

Pope, A. (2013), *An Essay on Criticism*. Cambridge: Cambridge University Press.

Savransky, M. (2016), *The Adventure of Relevance*. Basingstoke & New York: Palgrave Macmillan.

Savransky, M. (2017), 'The Wager of an Unfinished Present: Notes on Speculative Pragmatism'. In Wilkie, A; Savransky, M. & Rosengarten, M. (eds.), *Speculative Research: The Lure of Possible Futures* (pp. 25–38). London & New York: Routledge.

Savransky, M. (2018), 'Isabelle Stengers and The Dramatization of Philosophy'. *SubStance*, 47(1), 3–16.

Serres, M. (1995), *Angels: A Modern Myth*. Paris: Flammarion.

Stengers, I. (2005a). 'Introductory Notes to an Ecology of Practice'. *Cultural Studies Review*, 11(1), 183–196.

Stengers, I. (2005b). 'The Cosmopolitical Proposal'. In Latour, B. (ed.), *Making Things Public: Atmospheres of Democracy* (pp. 994–1003). Cambridge, MA: MIT Press.

Stengers, I. (2010), *Cosmopolitics I*. Minneapolis: University of Minnesota Press.

Stengers, I. (2011), *Cosmopolitics II*. Minneapolis: University of Minnesota Press.

Stengers, I. (2015), *In Catastrophic Times*. London & Luneburg: Open Humanities Press and Meson Press.

Watson, M. (2014), 'Derrida, Stengers, Latour, and Subalternist Cosmopolitics'. *Theory, Culture & Society*, 31(1), 75–98.

Whatmore, S. (2003), 'Generating Materials'. In Pryke, M. et al. (eds.), *Using Social Theory: Thinking Through Research* (pp. 89–104). London: Sage.

Section 3

Trading zones of ANT

Problematisations and ambivalences

Ignacio Farías, Anders Blok and Celia Roberts

Trading zones in science are frictional arenas for exchange processes among academic disciplines and traditions. Here, concepts, methods, questions and sensibilities come to be appropriated, translated, transformed and repurposed in processes that are not exempt from problematisations and ambivalences concerning the things exchanged. Trading zones are thus open spaces, in the sense that they involve seeking inspiration and learning from a certain discipline, while at the same time posing critical questions, highlighting absences and adding supplements. The figure of the trading zone is thus particularly helpful to grasp the spirit of the contributions gathered in the following section, as they involve engagements with ANT aimed at both problematising specific absences and tensions generated by ANT intellectual practices and retooling existing versions of ANT for other projects and purposes. What unites all these contributions, and makes them good companions to think with, is that they involve friendly forms of critique and reinvention, in the sense that they are dedicated to making propositions on how to continue thinking and inquiring near or alongside ANT.

A first problematisation comes from Nigel Clark's concern with what can go wrong when people get interested in the non-human. Clark makes a powerful argument about how the major strength of ANT, namely, the radically relational understanding of all existents, especially non-humans, can turn into its Achilles' heel, if, and this is crucial, ANT scholars would stick to only one form of relationality epitomised by the notion of entanglement. Such term would entail the symmetrical co-production of parts entangled, thus overlooking what Clark describes as radically asymmetrical forms of relationality shaping how human collectives unilaterally depend on and are exposed to non-human processes and forces. What can go wrong, Clark argues, is not just the knowledge that is thus produced, but most importantly the politics that derives from it. A proof for this may be found in the new vocabularies of asymmetry that Latour has recently developed, as he has turned to the problem of how to face Gaia in the current climate regime.

Something like a reversal of Clark's problematisation can be found in Kane Race's exploration of the unfortunate and somewhat strange non-relationship between ANT and queer theory. The reversal of the problem might be formulated as what goes wrong when some versions of ANT overlook not just queer bodies, but bodily becomings and affectivities in general. The problem Race suggests for ANT is at least double. On the one hand, ANT becomes unilaterally concerned with the emergence of 'molar assemblages,' that is, with large and stabilised sociotechnical arrangements or infrastructures, thus losing sight of twisted, rhizomatic and disruptive molecular encounters. On the other hand, ANT sticks to the problem of agency and actorhood, even if it is now relationally distributed, thus missing the eventfulness of the social – or what Race describes as a move from actor-networks to event-networks. But, instead of then dismissing ANT, what Race's contribution does is to read queer theory as an actor-network theory of queer bodies and politics. We have always been actor-network theorists, Race implies, when demonstrating the central role played by human–non-human articulations in the work of the most influential queer theorists.

Concerned with the extent to which ANT's relational ontology is capable of tracing the virtual, affective and atmospheric capacities of both human and non-human bodies, Derek P. McCormack's contribution points also to the problem of actualism in ANT's conceptual languages. However, instead of just seconding Nigel Thrift, who once described ANT as 'champagne without the fizz,' McCormack shows how various authors have attempted to perform an affective turn with the help of ANT, as it allows to pay attention to the objectual, technical and infrastructural arrangements of affective experiences and, more accurately, affective atmospheres. And, yet, McCormack argues, ANT accounts need to be careful not to treat the atmospheric as an object or an actor-network. The challenge, he observes, is to experiment more with affective modes of writing, producing accounts that are less about describing, as ANT scholars famously proclaimed, than about becoming attuned to unformed and moving objects. Interestingly, McCormack's own contribution might offer an example of such writing, as it begins with a story of his own encounter with ANT, while at the same time refusing to produce a bounded version of ANT.

How to care for our accounts is also the question that Sonja Jerak-Zuiderent explores in her conceptual-ethnographic piece on two very different sets of practices of prostate cancer screening and the affective and ethico-political implications of how medical doctors and medical researchers engage with patients. At stake, she argues, is the problem of agency and what human actors do with the agency that is gifted to them in such socio-medical assemblages. In two ethnographic vignettes, Jerak-Zuiderent contraposes her own disconcertment as an ethnographer with healthcare practices based on friendly, but solid and authoritative knowing from the side of the doctors vis-à-vis her own identification as a scholar with the more speculative and motile practices of researchers in a large cancer prostate study. Caring for others, while caring for our accounts, requires then a commitment to more-than-knowing practices, that is, a commitment for speculation as a mode of engaging with otherwise neglected things. Caring also for ANT, Jerak-Zuiderent makes the effort of exploring the least mainstream ANT contributions that fit well with the feminist tradition in science and technology studies concerned with questions of care.

Probably, the most transformative reading of ANT at work in this sections' companionate trading zone can be found in Francis Halsall's discussion of whether ANT is equivalent to contemporary artistic practices. In his reading, central tenets of ANT, especially those concerning the decentring of subjects and objects, are indeed shared with contemporary artistic practices and need to be understood as effects of late capitalism and the proliferation of multiple systems of communication and control. The consequences of such reading are significant and rather critical. The first one, Halsall observes, is that the claims to validity of ANT accounts cannot refer to epistemological criteria, but only to aesthetic choices. What does that mean for ANT, he asks. The second one, Halsall claims, is that ANT would be subject to the same criticism that many contemporary art practices are confronted with: That by celebrating the dispersion and fractality of subjects and objects, it reproduces the cultural semantics and cultural figures of late capitalist regimes.

Staying closer to the challenge of ANT to the sociologies of the social and with that to the constitution of Euro-American modernity, Marcelo C. Rosa explores the convergences between ANT and Southern sociologies. The article provides a detailed discussion of key mechanisms through which sociologists from the South have been historically participating in a global division of academic labour, whereby theory is developed in Euro-America and applied in the South, producing dynamics of internal colonialism and reproducing a liberal ontology of the social. In this context, Rosa sees in the ANT/Southern sociology encounter an opportunity to strengthen a knowledge politics that is not bounded to a fixed modern liberal ontology, but committed to visibilising variable and emergent ontologies. More specifically, Rosa postulates that the work of Eduardo Viveiros de Castro and Marisol de la Cadena could be seen as exemplars of a crossing between ANT and Southern sociologies, a crossing that results in bringing 'the geos' into ANT's preoccupation with ontological politics.

Finally, in yet another turn of the screw, Wen-Yuan Lin is concerned with the extent to which ANT's infra-language to describe articulations of heterogeneous actors might be reproducing a specific Euro-American naturalistic ontology of differences in kind that result in a language of purification and hybridisation. The problem, he observes, is the extent to which ANT might end up concealing other ontologies by redescribing them in terms of its naturalistic infra-language. Inspired by Viveiros de Castro's development of a perspectival anthropology resulting from its encounter with Amerindian ontologies, Lin discusses how researching with ANT into the ontology of Chinese medicine (CM) should have a recursive import for the articulation of a different, CM-inflected ANT. Based on CM philosophy and practices in Taiwan, Lin provides an empirical account of the ontological groundings of three key CM notions, those of correlativity, propensity and patterning. CM, Lin argues, involves different modes of doing difference in different locations according to different schools of practices. What would it mean for ANT to do difference this way, always open to subversions and equivocations? How to reinvent ANT as an intellectual practice not defined by a specific infra-language? And where in the process does ANT (need to) end? These are precisely the kind of questions that are posed in ANT's trading zones.

15

What can go wrong when people become interested in the non-human?

Nigel Clark

My account of actor-network theory and its tussling with the non-human might surprise many of its adherents for I will be suggesting that ANT has significant limits precisely where others discern its greatest strengths.

ANT counsels us to challenge assumptions that scientific or technological specialists have the right to superscribe our conduct in the world based on their own privileged access to the workings of 'nature.' At the same time, by affording non-humans a vital role in the constitution of our collective existence, ANT resists the fallback of conventional social thought that the stuff of physical existence boils down to whatever sociocultural actors choose to make of it.

More than simply inducting the non-human into social thought, ANT foregrounds relationality. Or rather, particular *modes* of relating. As the conceptual and titular centrality of the 'network' indicates, ANT prompts us to conceive of humans and non-humans as interconnected, consubstantial, co-constituted: Which is to say, not simply meeting in some middle ground as pregiven entities – but actually shaping each other in and through their encounters.

Social scientists are no strangers to this kind of relationality, especially in the era of globalisation. Neither are natural scientists – in a time when studies of complex systems abound. What ANT has done so effectively, however, is to merge these two sets of concerns into a kind of meta-relationality. It has achieved this by generating forms of inquiry that permit us to track back and forth and through all manner of heterogeneous entities without their *a priori* placement in categories of the social or the natural.

But in its very success in elaborating upon a particular kind of relationship, I suggest, ANT has diverted attention away from other sorts of relating – and from the ways non-humans might be implicated in these other relational modes. This is a claim that I expect to be controversial, both because ANT scholars have good reason to be deeply invested in notions of entanglement or inter-implication, and because – if we look closely – it is not hard to find traces of other kinds of relating in ANT.

Keeping these reservations in mind, I make a case for supplementing the concern with reticulated, co-constitutive relations with some of these other modes of relating – with no claim to being exhaustive. One alternative kind of relating I discuss is the situation in which some things or processes serve as the condition of possibility of other things – what has been termed antecedent or subtending relations (Clark and Hird, 2014). In such cases, relationships – if we can still call them that – are not so much 'mutual' or 'inter' – but tend to be one-way or at least profoundly uneven (Clark, 2011: 46–50). Radically asymmetrical or subtending relations are especially important in the context of the non-human, I propose, if we take into account that the earth and cosmos came into existence long before our own species appeared.

Another kind of relating – perhaps even less 'relational' – characterises those situations where there are deep, constitutive rifts between or within entities. While ANT scholars tend to assume it is networking 'all the way down,' there are other thinkers who attest that fracturing or separation is every bit as 'fundamental' as connectivity or entwinement.

Finally, we might also consider a sort of relating/non-relating that occurs in the interiority of things – human or non-human. That is to say, all the inner qualities or goings-on of an object that are at any moment not available to enter into interactions with other objects – for which some theorists dare to use the unfashionable term 'essence.'

This is not a contest. I am not setting up these other forms of relating as superior to or more momentous than the relationalities favoured by ANT. So why make a fuss about different modes of relationing? Surely, at a time in which human imbrication with non-human world seems to reaching unprecedented degrees of intensity and extensiveness, when our entwinings stretch from the innermost recesses of the genetic code to the operating state of the entire planet, it is entanglement that is our most pressing concern? At a moment when the politicisation of nature is so urgent but so precarious – why flirt with relational modalities that seem to come with dubious or risky political implications? And why pick on ANT when social thought is still populated by schools and approaches that barely attend to the non-human?

What my 'other' types of relating/non-relating bring into relief, in their various ways, is a kind of excess – a before, a beyond/beneath, a within – to the networking of human and non-human. This excess, in its different manifestations, is a reminder that there are limits to our capacity to collectively reshape our worlds. So while I emphatically agree that we ought to be constantly advancing 'the political' into new contexts, I also want to insist that not everything is open to negotiation, and that there is value in engaging with those domains or regions of existence that resist our overtures. If too much attention is focused on the inter of interconnectivity – or in some registers, the intra of intraconnectivity – there is a risk that we underestimate the power of the non-human to take us by surprise, to undermine or overwhelm our collective endeavours. To put it another way, privileging the mutuality of human–non-humans relations will encourage us to direct our inquiries to objects, events and processes where 'we' and our non-human 'others' are discernibly co-present – in the process discouraging us from dwelling sufficiently on the more distant, withdrawn or obdurately inhuman.

And this is where things can go wrong, sometimes badly. For if non-human forces or phenomena can both give and take away the conditions we've come to depend upon, then we will be dangerously unprepared if our eyes are only on the knot or node of *inter*connectivity. By the time we have become *entangled* with the surging wave, the firefront, the predaceous organism, it may already be too late.

Focusing on the canonical work of Bruno Latour, and using two examples that already feature in ANT research – microbial life and climate change – I show how following things themselves to their logical conclusion draws us more deeply into the excess of the non-human than most ANT has been willing to venture.

There is, however, a twist to my tale. Just as I have suggested that ANT afficionados might be surprised by my critique, so too have I been taken by surprise. For in his recent work, especially his encounters with Gaia, Latour has plunged into the very before, beneath, beyond that I have in the past insisted ANT would never take seriously. This 'metamorphosis,' I propose, potentially transforms the relationships between the political, the epistemological and the ontological that have until now set the contours of ANT – raising some intriguing possibilities of reimagining the inhuman.

Networking with non-humans

Let's start with what goes right when ANT interests itself in the non-human. Thinking in terms of entangled, co-constituted humans and non-humans encourages us to see how certain kinds of problems emerge from treating these as distinct realms – even as we go about generating evermore dense and bristling mixtures.

Immersing themselves in the knots, nodes and networks that configure ordinary existence, ANT researchers remind us that heterogeneous gatherings are the rule not the exception, and that objects or identities do not precede the jostle of relationality. 'Things are everywhere mixed with people, they always have been,' insists Latour (2003: 37). Or in the words of John Law 'Contra appearances, nature is always entangled with culture and society. To negotiate the structure of one is to negotiate the structure of the others' (2004: 121).

A methodology and guide for living as much as a conceptual framework, ANT offers procedures for spotting, tracking and grappling with a rich bestiary of significant actors. It seeks to make our worlds fairer, more livable, less crisis-prone by demonstrating how the arrangements we affirm as 'modern' too often entail things slung together crudely and incautiously. Eschewing templates and decrees, ANT advocates a careful (re)ordering of our sociomaterial networks so they will be more likely to do the things we want, hold up under pressure and leave room for the emergence of the new. Networks will have a much better chance of performing in these ways, it is claimed, if we all take responsibility for their composition and maintenance rather leaving the crucial decisions to a select contingent of experts. And so Latour calls on 'all of us, scientists, activists, politicians alike, to compose the common world' (2010: 12).

One of the great appeals of this vision is that it is at once political, epistemological and ontological. Classic ANT is political in the way that it multiplies and redistributes capacities to speak and act, epistemological in its meticulous attention to the processes through which knowledge is composed, ontological in that it shows how truth claims and political mobilisations transform the fabric of existence. It is onto-political-epistemological in that it sutures together these different aspects of world-building, and fuses them in a blazing ambition to democratise the very composition of the cosmos.

But what is perhaps ANT's most beautiful aspiration, I propose, is also a source of tension and foreclosure. Tension because the core commitment to the question of how scientists and other knowledge claimants access reality does not necessarily sit comfortably with the idea that there are worlds of non-humans interacting amongst themselves. And foreclosure because the intent on demonstrating how material existence itself bears the imprint of our efforts to engage with it has meant that there is a preference in ANT to focus on those aspects of reality where these traces are most discernible – at the expense of other fields or domains where their significance is more questionable. In short, some modes of relating, types of object and forms of experience are more conducive to the running together of the epistemological, the ontological and the political than others. And indeed, as I will try to show using the examples of microbial life and climate change, some of ANT's own most audacious claims about non-human 'things-in-themselves' tend to get sidelined in the process.

Microbial underworlds, planetary climates

Latour's (1988) account of the performances through which Louis Pasteur elevated microbes into active participants in 19th-century French society is justifiably regarded as an ANT classic. The story demonstrates how Pasteur and his allies enterprisingly assemble novel networks that render germs visible – in the process enabling both scientists and microbes to emerge as consequential actors. In this way, *epistemology* – Pasteur's procedures for studying and disclosing microbial worlds – is presented as inseparably bound up with the *ontological* positing of microorganisms as a specific category of being and with the *political* process of constructing effective sociotechnical networks.

But how does this sit alongside Latour's equally celebrated assertion in the *Irreductions* section of the *Pasteurization* book that non-human things-in-themselves 'lack nothing,' that they 'get by very well without any help from us' (1988: 193), and consequent on this condition of autonomy, that '(e)very actant makes a whole world for itself'? (1988: 192). It is this 'symmetrical' treatment of non-humans and humans that helps us to see how microbes share the agency in Pasteur's networks and to grasp why we cannot expect them to always bow to our will. But if we consider the centrality of epistemology in the Pasteur tale – what becomes of this symmetry? What's going on in Latour's canny but slippery assertion that 'after 1864 airborne germs were there all along' (1999: 173).

By the time Latour had assembled the Pasteur narrative, evolutionary theorist Lynn Margulis was already deep in conversation with chemist James Lovelock about the role of microbial life in the Gaia hypothesis – of which more later. Margulis, whose early accounts of the symbiotic composition of bacteria faced formidable disciplinary resistance, knows all about the work that needs to be done to build a scientific network. But at the same time she would also insist that she and fellow scientists are themselves one of the outcomes of bacterial interactivity. For as Margulis liked to remind us, not only did microbes generate the unthinkably complex web of interconnectivity scientists call the biosphere, they also serve as the building blocks and ongoing life support systems of all the composite beings that are 'big like us' (1998; see also Hird, 2009).

In the 'later Latour' and indeed in the 'earlier Latour' of *Irreductions*, one can imagine a deep fidelity to the world-building and body-building achievements of microbial life, a deference that might well run to acknowledging that vast associations of microbes helped bring Pasteur into being well before Pasteur brought microbes into modern social life. But it is the pre-eminence of epistemology in the *Pasteurization* story, I want to suggest, that holds these other more 'ontological' possibilities at bay, for it is the prioritisation of the apparatuses, procedures and performances through which Pasteur lends existence to the microbial underworld that leaves the strongest legacy in ANT. To put it another way, what things do amongst themselves or how they might incidentally provide conditions of possibility for human life take a back seat, while the question of how 'we' reveal, translate and orchestrate these things moves firmly into the driver's seat.

From our perspective, microbes are very small and climate change is very big, though as the Lovelock–Margulis Gaia thesis makes clear, climate and microorganismic life have been shaping each other throughout most of earth history (a reminder, as if you needed it, that *interactivity* or co-constitutive relations are indeed important!). Climate change isn't conspicuous in *We Have Never Been Modern*, but Latour opens the book with the problem of the thinning ozone layer (1993: 1). At a time when natural science held a near monopoly over the presentation and interpretation of evidence about global environmental change, Latour offers a timely reminder that other kinds of inquiry and storytelling play a constitutive role in the 'event' of the ozone hole – a message that characteristically does not permit critical thought the consolation of old school social constructionism.

As climate change establishes itself as a vital theme of Latour's work, it joins the 'proliferation of hybrids' announced by *We Have Never Been Modern*. As one of the 'collective experiments mixing humans and non-humans together' (Latour, 2003: 5), global warming is a controversy that cannot be left to the authority of the natural sciences and their claim to speak of a single unified nature, but must instead be taken as an incitement for new political experiments in which the very constitution of the cosmos is opened to painstaking collective negotiation.

Much of the subsequent debate in and around successive UN climate change conferences would seem to bear this out. But what we also need to keep in mind is that human-triggered climate change only makes sense in the context of a planet upon which climate is changeable, in the bigger picture of an earth with a great many climatic transformations in its turbulent history. Global climate, in this regard, always involves a multitude of interacting components. But only once out of an exceedingly long series of climatic fluctuations and shifts does it appear that anthropogenic ingredients have been a significant part of this mix.

Eloquent in its engagement with contemporary climate change, classic Latourian ANT, I am suggesting, offers few guidelines as to how we might account for the deep history of earthly climate, for the way climate shaped our species long before we shaped climate, or for how past climatic oscillations might have carved deep rifts in human experience – including being implicated in the loss of whole branches of the hominid lineage. Though such questions *do* resonate with Michel Serres' stunning evocation of an upheaving earth at the close of *The Natural Contract*: 'A thousand useless ties come undone, liquidated, while out of the shadows beneath unbalanced feet arises essential being, background noise, the rumbling world: the hull, the beam, the keel, the powerful skeleton...' (1995: 124).

What Serres seems to be returning us to is not simply a planet that is indifferent to our entreaties, but an earth whose very indifference is also the condition of possibility of our existence. Which brings us back to the question of how the same non-humans who ANT credits with making whole worlds quite apart from any human summoning seem somehow to end up depending upon us to work them into a meaningful, well-constructed cosmos.

In excess of networking

In any consideration of ANT's prevailing framing of the non-human, we should be mindful of the centrality of overhauling the 'modern' human/non-human or society/nature dualism. Nothing is more disparaged in ANT than a natural science that claims to speak for a 'nature' from which all traces of society have been evacuated, and its counterpart, social thought that conjures a human subject insulated from its physico-material context. In order to 'combat the epistemologists who are trying to purify science of any contamination by the social' (1999: 19), Latour takes aim at 'the strange invention of an "outside world"' while simultaneously swiping at the old school humanists' equally phantasmic 'brain in the vat' (1999: 3–7).

Along these lines, any epistemology that does not acknowledge its own implication in the admixture of the human and non-human, any mode of inquiry that invests in a reality that is 'independent, prior, single and determinant' (Law, 2004: 137) is charged with circumventing and undermining the due processes of the onto-political. 'Nature' purged of its social entanglements and co-constituents 'appears as what it always was, namely, the most comprehensive political process ever to gather into one superpower everything that must escape the vagaries of the society "down there"' (Latour, 1999: 297).

Classic ANT prohibits direct access to the 'out there' on the grounds that all inquiry is a translation – and thus a transformation – rather than a mirroring of its object. Or rather, every epistemological intervention to some degree alters the expansive field or network through which 'subjects' and 'objects' are constituted. In accordance with this alternative 'relational' onto-epistemology, then, any non-naïve study of the non-human will likely home in on the knots of connectivity, the zones of mutual entanglement, the practices of co-constitutions of the human and non-human. And in this way, each carefully framed study, each cautious intervention or each well-constructed experiment makes its own modest addition to the democratisation of the cosmos.

But what happens in this scenario to those celebrated legions of non-humans who persist in making worlds of their own – impervious to our inquiries and imaginings? What is their fate if we cannot speak of them without immediately foregrounding our entanglement with them? In this regard, it is revealing that philosopher Graham Harman, one of Latour's most outspoken fans, concedes that across the entire Latourian corpus, there are to be found 'only flickering hints of networks devoid of human involvement' (2009: 124). And the same might be said of ANT scholarship across the board.

If in theory ANT posits world upon world of resoundingly non-human actors, in practice, it would appear that the realm of the inhuman has shrunk to the point where it overlaps with 'our' world with almost no remainder. By the time Latour asserts that 'the very extension of science, technologies, markets, etc. has become almost coextensive with material existence' (2007: 7), it seems the best non-humans can do is to periodically resist our enrolments or to subsist in the shadowy interstices of our own far-reaching sociotechnical networks.

What has happened here, I would suggest, has a lot to do with the quest to bring ontology – our beliefs, however speculative and provisional, about what exists – into line with epistemology – how we access the world, and politics – how we might compose our worlds otherwise. As Harman has it: 'Latour conflates (the) issue of the *knowability* of the real with the entirely different question of whether the real *exists* beneath the current relational networks of actors' (2014: 59). While Latour-as-ontologist (Harman would even say metaphysician) joyfully liberates the inhuman from our probing and scribing, Latour-as-epistemologist reframes the 'out there' in terms of how it is accessed, which almost always seems to mean *human* access. And in this way, the political task of democratising the cosmos comes to rest ultimately on our entanglements with non-humans – at the expense of their entanglement with each other, or their asymmetrical relations, or our exposure to the fallout of their encounters.

While ANT most often couches its epistemological and onto-political advances in counterpoint to the modern natural and social sciences, there is a telling contrast – in this regard – with post-war continental philosophy.

Like ANT, much recent philosophy has sought to make sense of the deep and momentous malfunctional streak in modern thought and practice – and especially with the events that punctuate 'a century of unutterable suffering' (Levinas, 1998: 80). What Michel Foucault – in conversation with Maurice Blanchot – refers to as 'thought from the outside' (1987, 16) might be defined, amongst other things, by its positing, probing and speculating about those forms of 'extremity' that exceed our capacity for knowing or even imagining. The term often used for the experience of events that resist sense-making, resist scientific capture, resist any form of reappropriation, is 'the impossible' (Lynes, 2018: xxxviii).

The excess with which this enigmatic 'impossibility' or 'experience of the outside' grapples has been construed in different ways: As a great roiling inhuman exteriority, as a bubbling over of potentiality from within, as an insurmountable rift gouged by catastrophe and trauma. But where this mode of philosophical inquiry tends to diverge from the main thrust of ANT has been its willingness to unhinge ontology from epistemology. While ANT, I have been suggesting, tends to downsize the realm of existence until it is more-or-less congruent with the reach of epistemology, philosophical 'thought from the outside' has tended to contract the domain of epistemology such that it nestles precariously amidst the vastness of being and becoming.

By no means am I proposing that these currents of continental philosophy do all the work we need to confront the inhuman. Far from it, they and their inheritors have much to learn from the rigour and meticulousness with which ANT approaches the non-human – especially in its more mundane manifestations. But what philosophy's tussling with exteriority and extremity *does* is to remind us of the range of modes of relating other than the *inter*actions of reciprocity, entanglement and co-enactment.

Like Serres' confrontation with the earthquake – which indeed channels philosophical 'thought from the outside' – we are prompted to consider not only the abyssal reaches beneath human–non-human networking, but the inescapable exposure of every human endeavour to events that exceed our capacities to know or negotiate. And this tells us something about the fragility of the networks 'we' would assemble and maintain that goes far beyond the question of injudicious construction.

But here is the twist in the tale. Over recent years, sparked by increasingly profound concern with the planetary predicament, ANT has begun to take its own plunge into excess and extremity. In the process, things-in-themselves have started to break free of their epistemological holding pattern and reassert something of their brute and boisterous early promise.

ANT's metamorphosis?

Not long ago, in keeping with my concerns about ANT's predilection for a certain species of relationality, I grumbled: 'It is difficult to imagine Bruno Latour, or those in his orbit, speaking of nature as ground' (2011: 45). But as Latour's engagement with climate change, Earth system science and the Gaia hypothesis has deepened, I've had to repeal this verdict (see Clark, 2015). In the paper 'Agency at the Time of the Anthropocene,' Latour announces: 'The prefix "geo" in geostory does not stand for the return to nature, but for the return of object and subject back to *the ground* —the "metamorphic zone"' (2014a: 16, my italics; see also Latour, 2017: 57–58). In more detail:

> Why does it seem so important to shift our attention away from the domains of nature and society toward the common source of agency, this "metamorphic zone" where we are able to detect actants before they become actors ... where "metamorphosis" is taken as a phenomenon that is antecedent to all the shapes that will be given to agents?
>
> *(Latour, 2014a: 13)*

The *Modes of Existence* project too, despite its explicit focus on 'moderns,' entails a delving into forms of relating that exceed human–non-human interaction. In the modes of 'reproduction' and 'metamorphosis,' we are introduced to a protean domain of beings that '*precede* the human infinitely,' in what Latour goes on to describe as 'a sort of matrix or kneading process from which the "human" can later take nourishment, perhaps, can in any case branch out, accelerate, be energized, but that it will never be able to replace, engender, or produce' (Latour, 2013: 203; see also Clark, 2019; Conway 2016, 49– 50).

With this move, I suggest, we have gone beyond non-humans lurking in the gaps of our networks or even knocking at the door of the collective. Not only does the realm of metamorphosis not depend upon us in any way, but as the condition of our own emergence and sustenance, it is also beyond our powers of reconstitution.

In order to embrace this notion of an abyssal and generative ground, Latour – like the philosophers 'of the outside' – must free himself of the obligation to show exactly how it is that we access or know about this 'exteriority.' To put it another way, he needs to unleash ontology from epistemology, to unhinge the question of 'the *knowability* of the real' from the issue of 'whether the real *exists* beneath the current relational networks of actors.'

What does this mean for the political, which until now – I have been arguing – has also been bundled together with the ontological and the epistemological? In a generous and open exchange between Harman and Latour in 2008, both thinkers reconsidered their own previous conflation of the political and the ontological (Clark, 2013; Latour, Harman and Erdélyi, 2011). Harman conceded that he may have been 'careless in the manuscript [of *Prince of Networks*] in equating ontological democracy with political democracy.' Latour responded by agreeing 'that there is actually no connection at all with the idea of the multiplicity of beings and any sort of democratic position,' before graciously pondering whether it 'might be the weakness of *Politics of Nature* that it gives this impression' (Latour, Harman and Erdélyi, 2011: 96–97).

In the same conversation, Latour also clarifies his more general position with regard to relationality – attesting that his own philosophy does not simply privilege interactive relations but also affirms the 'irreducible singularity' of each participating object. For without this integrity, he insists, networking and communicative processes would have nothing tangible to work with (2011, 43–44, 49, 63; Clark, 2013). With this acknowledgement of a kind of pre-relationality, we catch an echo of the thematic of withdrawnness and inaccessibility that concerns theorists of 'extremity' or 'radical otherness.' Here, it is worth recalling how, for Emmanuel Levinas and others in his wake, the incitement to ethical and political relating is not interconnectedness so much as a desire to reach out across a disconnect or a difference, this being the very summons for the coming-into-relation that is compassion, care or the quest for justice (Clark, 2011: 70–74; Levinas, 1969: 21–30). And in this regard, it might be said that the 'irreducible singularity' of the encounter with otherness is a kind of 'metamorphic zone' in which actants are lured or urged to become actors.

But as Latour likes to remind us, his own priority is the worldly problems or challenges he refers to as 'empirical matters' (Latour, Harman and Erdélyi, 2011: 41). In his own words:

> When faced with vast philosophical concepts like mutations in space, time and agency, my research strategy has always been the same: let's try to find a neat empirical site where it is possible through fieldwork to obtain precise answers to speculative questions.
>
> *(2016: 4)*

We shouldn't be surprised, then, to find that in his ongoing engagement with climate, Gaia and the Earth system, Latour's prime concern is to find his way to a substantive, scaled-down framing of the planetary predicament. Consequently, rather than engaging with the mass of the Earth or the cosmos to which our planet opens, his focus is increasingly on the 'tiny, fragile, slim' region where living and nonliving processes are in ceaseless interchange (2016: 7; see also 2017: 140). Neither should we be surprised to see that such a circumscribing of Earth-sized crises enables a certain regrouping of the ontological, the epistemological and the political.

Latour's currently favoured 'neat empirical site' is the 'critical zone' – a relatively recent scientific term for a parcelled cross section of the Earth system of varying sizes that runs from treetops, through the soil, and down as far as the rocky substrate (2014b, 2016, 2017: 93). For him, critical zone science – research done by interdisciplinary teams made up of soil scientists, hydrologists, ecologists, biogeochemists and others – offers a localised point of entry to planetary challenges that does not sacrifice the complexity of the Earth system (2016: 4). As Latour explains: 'critical zones define a set of interconnected entities in which the human multiform actions are everywhere intertwined' (2014b: 3).

This decision to alight on a defined segment of the planetary body – as disclosed by critical zone science – enables the knowability of the real, what actually exists, and what can realistically be contested or renegotiated to once more coming into alignment. Rather like Pasteurians of old, 'critical zoners' show us how it is possible to conjoin empirical inquiry with political process – as they assemble concepts, instruments and a slice of the 'living planet' in a way that brings Latour back to his preferential task – the *progressive composition of the common world*' (Latour, 2014b: 1; 2004).

But if much is to be gained from this circumscription of planetary issues, we might also detect a return to ANT's familiar trade-off between the restricted realm of human–non-human co-enactment and the radical outsideness of the fully inhuman – and all the issues this raises. Meanwhile, contemporary thinkers of 'extremity' and the 'experience of the outside' are themselves engaging more explicitly with the planetary crisis. Literary theorist Gayatri Chakravorty Spivak draws our attention to 'the necessary impossibility of a "grounding" in planetarity' (2003: 82), while philosopher Philippe Lynes calls on us to 'confront the creation, the rediscovery of the world in a restricted sense with the *invention* of the earth as the experience of the impossible' (2018: xlvii). While Latour's approach – in continuity with much of the tradition of ANT – sets out to render earthly problems thinkable, tangible, doable, the theorists of 'planetary impossibility' remind us that whatever sense we and our scientist colleagues can make of our terrestrial environment is inevitably provoked, contaminated and haunted by our inescapable vulnerability to the excesses of the Earth and cosmos. And that the unfolding crisis of the Earth involves losses that cannot ever be recuperated, suffering that is inconsolable, changes so overwhelming that they threaten the very logics through which we apprehend the world.

The 'possible' and the 'impossible,' however, are not opposed conditions or options. For as 'thought from the outside' would insist, it is the very confrontation with 'impossible' otherness that ultimately inspires our attempts to make sense of the world, our ongoing efforts to cast bridges or networks across the rifts that open, from time to time, in the body of the Earth. The task of trying to compose common worlds, we might say, being the very necessity to which a sense of 'necessary impossibility' summons us.

In the broader and as-yet-unfinished story of ANT, what Latour's moving between critical and metamorphic zones seems to acknowledge is that there is indeed an outside, a considerable remainder to 'tiny, fragile, slim' domain where humans and non-humans are intertwined. If the axiom of classic ANT seemed to be that it was human–non-human entanglement 'all the way down,' it now looks increasingly clear that prioritising the zone of co-enactment is a choice, a decision, a deliberate cut made in an infinitely vaster field of existence. While Latour insists upon the onto-epistemological-political necessity of carving out 'critical zones,' so too does he confront us with the unfathomable, immeasurable and impolitic 'metamorphic zone.' Just how these irreducible domains might play out will not only be a question of how people interest themselves in the non-human – but a matter of what the inhuman makes of *us*.

References

Clark, N. (2011) *Inhuman Nature: Sociable Life on a Dynamic Planet*. London: Sage.

Clark, N. (2013) 'Review: *Prince of Networks*, Harman, G and *The Prince and the Wolf*, Latour, B, Harman, G and Erdelyi, P,' *Contemporary Political Theory* 12; e15–e19.

Clark, N. (2015) 'Metamorphoses: On Philip Conway's Geopolitical Latour', *Global Discourse* 6(1–2): 72–75.

Clark, N. (2019) 'Political Geologies of Magma', in Bobbette, A. and Donovan, A. (eds.) *Political Geology: Active Stratigraphies and the Making of Life*. Cham: Palgrave Macmillan. Pp 263–292.

Clark, N. and Hird, M. (2014) 'Deep Shit', *O-Zone: A Journal of Object-Oriented Studies*, online at: http://o-zone-journal.org/issue/ 1: 1: 44–52.

Conway, P. (2016) 'Back Down to Earth: Reassembling Latour's Anthropocenic Geopolitics', *Global Discourse* 6(1–2): 43–71.

Foucault, M. (1987) 'Maurice Blanchot: The Thought from Outside', in Foucault, M. and Blanchot, M. (eds.) *Foucault/Blanchot*. New York: Zone Books. Pp 9–58.

Harman, G. (2009) *Prince of Networks: Bruno Latour and Metaphysics*. Melbourne: Re.press.

Harman, G. (2014) *Bruno Latour: Reassembling the Political*. London: Pluto Press.

Hird, M. (2009) *The Origins of Sociable Life: Evolution after Science Studies*. New York: Palgrave Macmillan.

Latour, B. (1988) *The Pasteurization of France*. Cambridge, MA: Harvard University Press.

Latour, B. (1999) *Pandora's Hope: Essays on the Reality of Science Studies*. Cambridge, MA: Harvard University Press.

Latour, B. (2003) 'Atmosphère, Atmosphere', in May, S. (ed.) *Olafur Eliasson: The Weather Project*. London: Tate Publishing. Pp 29–41.

Latour, B. (2004) *Politics of Nature: How to Bring the Sciences into Democracy*. Cambridge, MA: Harvard University Press.

Latour, B. (2007) 'A Plea for Earthly Sciences', *Annual Meeting of the British Sociological Association*, East London, April 2007. Available at: www.bruno-latour.fr/articles/index.html

Latour, B. (2013) *An Inquiry into Modes of Existence: An Anthropology of the Moderns*. Cambridge, MA: Harvard University Press.

Latour, B. (2014a) 'Agency at the Time of the Anthropocene', *New Literary History* 45(1): 1–18.

Latour, B. (2014b) 'Some Advantages of the Notion of 'Critical Zone' for Geopolitics', *Procedia: Earth and Planetary Science* 10: 3–6.

Latour, B. (2016) 'Is Geo-logy the New Umbrella for All the Sciences? Hints for a Neo-Humboldtian University', Cornell University, 25th October 2016. Available at: www.bruno-latour.fr/sites/default/files/150-CORNELL-2016-.pdf

Latour, B. (2017) *Facing Gaia: Eight Lectures on the New Climatic Regime*. Cambridge: Polity Press.

Latour, B., Harman, G. and Erdélyi, P. (2011) *The Prince and the Wolf*. Winchester: Zero Books.

Law, J. (2004) *After Method: Mess in Social Science Research*. London and New York: Routledge.

Levinas, E. (1969) *Totality and Infinity: An Essay on Exteriority*. Pittsburgh: Duquesne University Press.

Levinas, E. (1998) *Entre Nous*. London: Continuum.

Lynes, P. (2018) *Futures of Life Death on Earth: Derrida's General Ecology*. London: Rowman & Littlefield International.

Margulis, L. (1998) *The Symbiotic Planet: A New Look at Evolution*. London: Phoenix.

Serres, M. (1995) *The Natural Contract*. Ann Arbor: University of Michigan Press.

Spivak, G. C. (2003) *Death of a Discipline*. New York: Columbia University Press.

16

What possibilities would a queer ANT generate?

Kane Race

Actor-network theory and queer theory might appear to be incommensurable pursuits. Sexual bodies – their practices, pleasures, vulnerabilities and constraints – have never been an abiding interest or explicit focus within the branch of Science and Technology Studies most closely associated with ANT. And while both fields are thoroughly indebted to the work of Michel Foucault, the ethnographic approach to the making of science, technology, facts and realities that is a hallmark of ANT produces different emphases than those typically found within queer theory. Where ANT draws attention to the material associations and practical relations that transform contingent relations into durable, concrete realities, queer theorists tend to stress the overarching power of the ideological, normative, disciplinary, discursive and/or psychic structures that produce queer lives as abject, vulnerable and suspect within prevailing social worlds. ANT is distinguished by the precision with which it attends to the sociomaterial arrangements that Foucault called *dispositifs* and gives them consequential form. Queer theory has been characterised by its disciplinary fixation with 'hidden patterns of violence and their exposure' (Sedgwick 1997, p. 143) and the 'overdetermined relationalities' of the normative, disciplinary structures it takes as objects of critical attention (Wiegman & Wilson 2015, p. 10). Indeed, queer theory's favoured strategies of dismantling, countering, subverting, exposing or otherwise disrupting hegemonic norms emerge from a sense (keenly shared among proponents) that established institutions, truths, norms and realities damage the social standing, material prospects and overall lives of those subject to regimes of gender and sexuality (among other social categories). Given these differences, the view from ANT that critique has 'run out of steam' (Latour 2004a) risks propelling queer theorists towards the chorus of voices that habitually (if reductively) dismiss ANT as a politics-free zone lacking any coherent analysis of systemic oppression and/or structural violence.

These differences in motivation, orientation and approach can, however, obscure certain synergies between the two fields. Both fields reject essentialism and determinism; conceive identities and realities as enacted, practised or performed (Butler 1990; Butler 2010; Callon 2010; Mol 1999; Mol & Law 2004); investigate experimental practices and publics (Marres 2012; Michael 2002, 2018; Race 2018; Warner 2002); and adopt constructivism as a guiding premise and analytic approach. On this point, queer theory does not necessarily hold that the things said to be constructed are 'easily deconstructible' (Latour 1999, p. 115) or can be done away with at a whim (a charge Latour has levelled at social constructionists). Indeed, at the first academic conference I ever attended, I distinctly recall one of queer theory's founding figures, David Halperin, rejecting that idea, observing that the Sydney Harbour Bridge is as constructed as the truths and discourses queer critics find worrisome. This doesn't mean it is an illusion or can be dismantled overnight. On the contrary, it is a massive edifice that has a materiality and significance we must acknowledge, account for and grapple with.[1]

This example is suggestive on a number of fronts: That bridges might be built to enable further traffic to flow between promontories conventionally assumed to harbour radically different ecologies of practice, inclination and thought, and that the tenders such a proposal calls forth might well enrich each of these terrains rather than operate as a presumed source of environmental damage and critical havoc. Since queer theory first hit its stride in North American literature departments during the heyday of post-structuralist theory, it would be easy to reduce its critical scope to language, representations, texts and discourses. But the field's concern with the arrangement and experience of sexual and gendered life has given it a sensitivity to certain non-human actants, the material significance of which ANT has helped many fields bring more sharply into view.

In this chapter, I will argue that ANT's insistence on the contingency of relations among human and non-human actants might help break the stranglehold of totalisation sometimes said to characterise queer theorists' take on the 'ideologies and institutions of intimacy' they dub heteronormativity (Berlant & Warner, 1998). This could generate more constructive forms of attention to projects of 'queer world-making' as well as the publics (or 'counterpublics') such projects call forth and/or are deemed to depend upon. In the first part of this chapter, I conduct a close reading of Lauren Berlant and Michael Warner's influential (1998) essay 'Sex in Public' – perhaps the most robust and compelling early attempt to bring queer theory into articulation with key traditions of social theory – alongside congruent work by Sara Ahmed (2014). My reading aims to bring out the activity of non-human actants in the 'buffer zones' and 'elaborate support systems' (Ahmed 2014) these essays rally against, asking whether they might be understood more constructively, following ANT, as *infrastructures*.

Worlds capable of sustaining queer lives are commonly considered to be composed of cultures, communities, arrangements and/or publics that this literature characterises as 'fragile and ephemeral' (Berlant & Warner 1998, p. 561; cf. Ahmed 2014). Arguably, ANT could help remedy such fragility by providing insights into how stronger worlds or realities are built and sustained. But in the second part of this chapter, I suggest that ANT could benefit just as significantly from a better appreciation of the critical necessity and/or disruptive effects of these worlds 'assembled out of the experience of being shattered,' as Ahmed puts it (2014), the 'alter-ontologies' (Papadopoulos 2018) wrought from queer experiments in intimacy.

Intimacy, institutions and infrastructures

More than any other, the concept of *heteronormativity* makes queer theory cohere as a field today, providing a theoretical basis for the field's extension beyond sexuality studies (and attempts to 'carve out a buffer zone for a minoritized and protected subculture' (Warner 1993, p. 3) to devote critical attention to 'any consequential social difference that contributes to regimes of sexual normalization' (Hall & Jagose 2012, p. xvi). The concept is most rigorously theorised in Berlant and Warner's (1998) essay 'Sex in Public,' which sets out to promote 'the radical aspirations of queer culture building' and 'unsettle the garbled but powerful norms…implicated in the hierarchies of property and propriety that we will describe as heteronormative' (p. 548). They define the latter term in a lengthy footnote:

> By heteronormative we mean the institutions, structures of understanding and practical orientations that make heterosexuality seem not only coherent – that is, organized as a sexuality – but also privileged. Its coherence is always provisional, and its privilege can take several (sometimes contradictory) forms…[consisting] less of norms that could be summarized as a body of doctrine than a sense of rightness produced in contradictory manifestations – often unconscious, immanent to practice or to institutions.
>
> *(p. 548)*

As this excerpt suggests, heteronormativity does not refer to 'anything like a simple monoculture' for the authors. Rather, it achieves 'much of its metacultural intelligibility through the ideologies and institutions of intimacy' (p. 553).[2] 'Heterosexuality is not a thing,' they claim, because heterosexual culture 'never has more than a provisional unity' (p. 552). In this respect, they are closer to some of the central premises of ANT here than anyone has yet cared to admit. For example, Latour has linked the process of substantiation to the 'stability of the assemblages' that subtend it, claiming 'the best word to designate a substance is "institution"' (1999, p. 151; see also p. 307). For all of these theorists, 'thingness' (or substantiality) rests on the stability (or otherwise) of the institutions (or assemblages) that make a given proposition durable and sustainable.

At first glance, the terminology of 'ideologies and institutions' evokes traditional sociology's preoccupation with ideas, beliefs and customs. But when we dig down into the arrangements that give heterosexual culture its 'metacultural intelligibility,' we begin to encounter things of a much more material texture. The 'sense of rightness' heteronormativity confers is 'more than ideology, or prejudice, or phobia against gays and lesbians,' the authors explain; it is 'embedded in things and not just in sex' and 'produced in almost every aspect of the forms and arrangements of social life' (p. 554). Alongside the domains of the state, law, medicine, commerce, nationality and the 'conventions and affects of narrativity,' a host of acts and mundane objects are said to 'support and extend' heterosexual culture even if they are 'less commonly recognised as part of sexual culture' (p. 555) to produce a 'constellation of practices that everywhere disperse heterosexual privilege as a tacit but central organising principle of social membership' (p. 555). These include 'paying taxes, being disgusted, philandering, bequeathing, celebrating a holiday, investing for the future, teaching, disposing of a corpse, carrying wallet photos, buying economy size, being nepotistic, running for president, divorcing, or owning anything "His" and "Hers"' – a list whose elaboration is marked as 'a project for further study' (p. 555). That is to say, Berlant and Warner's analysis encompasses more than 'garbled but powerful norms' (p. 548): It extends to 'those material practices that, though not explicitly sexual, are implicated in the hierarchies of property and propriety' that appear in their description of heteronormativity (p. 548). These practices include mundane objects such as wallet photos, burial procedures, joint checking and packaged commodities; the regimes of 'normal life' the authors rail against are far less monolithic than some critics have asserted (Wiegman & Wilson 2015, p. 10).

In her blogpost 'Selfcare as Warfare' (2014), Sara Ahmed similarly considers how the object world is arranged to promote the care and survival of some bodies and not others. Riffing on a sentence penned by black lesbian feminist Audre Lorde when diagnosed with liver cancer (1980), Ahmed conveys the materiality of the forms of racial and heterosexual privileges that turn self-care into a war for certain subjects: 'caring for myself is not self-indulgence, it is self-preservation, and that is an act of political warfare' (Lorde 1988). 'Privilege is a buffer zone, how much you have to fall back on when you lose something,' Ahmed writes in a passage that recalls Berlant and Warner's conceptualisation of heteronormativity:

> I have in my own work been thinking of social privilege as a support system: compulsory heterosexuality, for instance, is an elaborate support system. It is how some relationships are valued and nurtured... Racial capitalism is a health system: a drastically unequal distribution of bodily vulnerabilities

Her illustration of this point is populated with human and non-human assemblages that make some lives easier to sustain than others:

> When a whole world is organised to promote your survival, from health to education, from the walls designed to keep your residence safe, from the paths that ease your travel, you do not have become so inventive to survive. You do not have to be seen as the recipient of welfare because the world has promoted your welfare ...Racial capitalism is a health system: a drastically unequal distribution of bodily vulnerabilities

While Ahmed frequently invokes 'macro' structures such as 'structural inequalities' and 'systemic' factors (a practice avoided within ANT), these references can be taken to denote the very real obduracy and material effects of those monumental assemblages that promote the health and welfare of white heterosexual men and their families at the expense of so many others.

Ahmed closes 'Selfcare as Warfare' by underlining the inventiveness she associates with queer (self-)care and connecting it to a collective politics of mattering:

> In directing our care towards ourselves we are redirecting care away from its proper objects [...] we are not caring for the bodies deemed worth caring about. And that is why in queer, feminist and anti-racist work self-care is about the creation of community, fragile communities, assembled out of the experiences of being shattered. We reassemble ourselves through the ordinary, everyday and often painstaking work of looking after ourselves; looking after each other.

Ahmed refers to 'communities' in this passage (a term steeped in sentimental humanism), but her use of terms such as assembling and reassembling is suggestive: Building effective systems of support across diverse constituencies is difficult: It requires a lot of labour and construction work. Actor-network theorists understand assemblages to be composed of heterogeneous elements that enter into relations with one another. The project Latour sets out in his well-known (2005) introduction to ANT, *Reassembling the Social*, involves tracing how heterogeneous elements become associated in more or less durable formations to produce particular kinds of agency, for example. From this perspective, the resonance of the term assemblage makes sense when one considers that the French term *agencement* can mean both *arrangements* and/or a kind of *acting* or *doing* (i.e. *agenc-ement*). Following ANT, one could ask, does the process of 'reassembling ourselves' encompass more than humans?

Elsewhere, Ahmed attends to how the arrangement of mundane objects directs attention towards heteronormative imperatives – or tacitly 'configures it users' (Woolgar 1992), to use the terminology of ANT (which Ahmed does not). In a memorable passage of *Queer*

Phenomenology (2006), she considers how the arrangement of her family's dining room puts certain objects on display (tables, seats, photographs, a fondue set, etc.) in a manner that directs attention towards the family line and the purported gift of heterosexual sociality (pp. 88–92). Ahmed emphasises throughout the disorienting and destabilising effects such arrangements can have for queer bodies, discussing the difficulty of making queer objects 'come into view' as possible objects to be directed towards (p. 91). While Ahmed raises the question of queer furnishings – i.e. what object-arrangements might furnish and support queer lives – she hesitates to offer concrete advice for queer interior design, posing instead another question: 'how we are orientated towards queer moments when objects slip. Do we retain our hold of those objects by bringing them back "in line"? Or do we let them go, allowing them to acquire new shapes and directions?' (pp. 171–172).

> Queer tables are not simply tables around which, or on which we [queers] gather. Rather, queer tables and other queer objects support proximity between those who are supposed to live on parallel lines, *as points that should not meet.* A queer object hence makes contact possible. The contact is bodily, and it unsettles the lines that divides spaces as worlds, thereby creating other kinds of connections where unexpected things can happen.
>
> *(p. 169)*

Queer furnishings and object-arrangements involve something more or other than assembling a more stable and durable world for certain minoritised subjects, in other words. They make connections 'where unexpected things can happen' (see Race 2018, pp. 171–187). This may involve disassembling or interfering with certain established protocols, arrangements, functions and designs: Those techniques devised to keep people (and the non-humans they are connected to) 'in line.' Ahmed does not deny that the identity and function of material entities may change in different settings, but wants to keep 'queer' open as a verb or doing-word rather than converge and settle down on any particular positive identity.

If we turn to the practices of queer world-making theorised in 'Sex in Public,' we find 'Berlant and Warner at their most Deleuzean' as Lee Wallace remarks (2011, p. 127):

> The queer world is a space of entrances, exits, unsystematised lines of acquaintance, projected horizons, typifying examples, alternate routes, blockages, incommensurate geographies (p. 558) – sites whose mobility makes them possible but also renders than hard to recognise as world making because they are so fragile and ephemeral.
>
> *(1998, p. 561)*

Queer world-making depends here on makeshift arrangements, perverse architectures and inventive tactics, and on 'parasitic and fugitive elaboration' in unlikely settings such as 'gossip, dance clubs, softball leagues and phone-sex ads' (p. 561) The authors argue that the culture developed within these spaces has 'almost no institutional matrix for its counterintimacies' (p. 562). The fragility and ephemerality of queer worlds are said to flow from the instability of the assemblages that make them possible: The tactical (rather than strategic) use of objects and inhabitations of settings that support it (see de Certeau 1984). A key example is the once-vibrant queer subculture of Manhattan's Greenwich Village:

> The gay bars on Christopher Street draw customers from people who come there because of its sex trade. The street is cruisier because of the sex shops. The boutiques that sell freedom rings and 'Don't Panic' T-shirts do more business for the same reasons. Not

all of the thousands who migrate or make pilgrimages to Christopher Street use the porn shops, but all benefit from the fact that some do. After a certain point, a quantitative change is a qualitative change. A critical mass develops. The street becomes queer.

(p. 562)

The reference to material practices and urban spaces in this passage resonates with Deleuze's theorisation of assemblages which maintains that extensive relations give rise to particular intensive qualities. But Deleuze and Guattari make a distinction between different kinds of assemblages that may help specify what is distinctive about *queer* ANT. Specifically, they distinguish between *molar* (or arborescent) assemblages which they conceive as organised, hierarchised, consolidated, unified, binarised and divisible (i.e. more or less stable institutions) – and *rhizomatic (or molecular)* assemblages which are, by contrast, unpredictable, unsystematic, flexible, flighty, horizontal, non-binary and indivisible without changing their nature (i.e. ephemeral). Where molar assemblages stratify, territorialise, hierarchise, binarise and subjectify, molecular assemblages involve processes of de-stratification, de-territorialisation, de-subjectification, molecularisation and scrambling. Furthermore, these rhizomatic structures are characterised by unpredictable 'lines of flight' that can be generative or perilous depending on what they connect up with in the course of such trajectories (Deleuze & Guattari 1987). Given the radical, anti-establishment commitments of Deleuze and Guattari's politics, readers tend to invest rhizomatic and molecular processes with greater critical value, though on closer inspection all assemblages have stratifying and de-stratifying tendencies. I would contend at any rate that ANT has largely focused on the emergence of *molar assemblages*: The construction and elaboration of actor-networks that become stable enough to produce durable technologies, effective mediations and scientific 'facts' – or else fail to do so, with this failure emerging as a missed opportunity or instructive example of how things go wrong (see Callon 1984; Latour 1996).

Queer theory has a more conflictual and contradictory set of critical investments. It's hankering for sociomaterial arrangements that better support queer lives is complicated by its sensitivity to the fact that any process of stratification or stabilisation (any molar assemblage) runs the risk of producing its own misfits and shutting them out – i.e. other queers whose needs and lives are not serviced by what is set up. Queer theory's critical relation to institutions and established structures that embattle queer forms of life means that it is just as interested in *disassembling* the social (as it is currently constituted) as *reassembling* it or tracing the associations that make, reproduce and sustain it.[3] Disruption, deviation, destabilisation and failure tend to be invested with some political value in this literature, rather than being framed as problems or missed opportunities as we see in ANT.

Some queer critics object to what they regard as the territorial nature of Berlant and Warner's defence of queer counterpublics (Castiglia 2000). But I think it is just as plausible to focus on the de-territorialising tendencies and ephemeral qualities of the counter-realities they discuss, and the processes of de-stratification and lines of flight they entail. Queer worlds have a rhizomatic character that is characteristic of much queer theory – hence its sensitivity to disorientation (Ahmed 2006), unexpected associations (Ahmed 2014; Delany 1999; Race 2018), 'alternate routes, incommensurate geographies' and 'unsystematized lines of acquaintance' (Berlant & Warner 1998, p. 561). The 'unlikely proximities' and unexpected encounters queer theorists care for are valued precisely for their 'unsystematized' character, the 'diagonal lines' they lay out in the social fabric, the 'slantwise' positions their constituent elements normally occupy in relation to each other (Foucault 1997, p. 137). In other words, it is the *disruption* of those lines that demarcate territories, hierarchise and stratify worlds, and hold things in place generally that confers them with their critical value and performative appeal. What would an ANT that explores such twisted, rhizomatic, disruptive encounters look like? How would it feel to become sensitised to what is serendipitous about their outcomes?

A second point is that the heterogeneous and elastic nature of the hegemonic structures queer theorists describe might be underlined with the kind of attentiveness to relational contingency found in ANT (Berlant & Warner 1998, p. 553). Ara Wilson discusses the emergence of what she calls a 'baggy notion of infrastructure' in this literature, often drawing on ANT, that 'combines material and symbolic domains, eschews technological determinism, and recognizes both systematicity and failure' (2016, p. 249). Brian Larkin defines infrastructures as 'objects that create the grounds on which other objects operate' before making the suggestive claim, 'their peculiar ontology lies in the fact that they are things *and also the relation between things*' (2013, p. 329). Could the material arrangements, mundane objects and stubborn technologies that prop up Berlant and Warner's 'ideologies and institutions of intimacy' be conceived more constructively as infrastructures?[4] At once social *and* material in composition and effects, infrastructures are said to disappear into the background when working properly or fade from view, much like the 'conventional furniture' Ahmed discusses in *Queer Phenomenology*:

> a queer furnishing might be about making what is in the background, what is behind us, more available as "things" to "do" things with. Is the queer table simply one we notice, rather than simply the table that we do things "on"? … As soon as we notice the background, objects come to life, which already makes things rather queer.
>
> *(2006, p. 168)*

Ahmed situates her discussion within phenomenology (a very different tradition of thought from ANT) and I do not want to draw too long a bow. But this passage evokes the unsettling insights and odd (queer?) effects that ANT occasionally generates in its accounts of the (normally tacit) sociomaterial arrangements that organise relational life.

Queer theory's emergent interest in infrastructures and relational ontologies seems to me a highly promising development for the field. It will surely help dispel the charge of 'overdetermined relationalities' that recent critiques of queer antinormativity hinge upon and assert (Wiegman & Wilson 2015). In my own work, I have found the concept of infrastructures indispensable for exploring transformations in sexual and drug-taking practices associated with the widespread use of digital devices to arrange sex (Race 2015a; 2015b). In more recent work, I have conceived these practices as *intimate experiments* (2017, 2018).

Bodies/pleasures/publics: experimentality and play

STS have produced an extensive literature on the conduct of experiments: Indeed, experimentation might be regarded as a key theme in the emergence of the field. Historical and ethnographic studies of scientific practices in laboratories and beyond have investigated the trials scientists engage in and the technical arrangements, practices and publics that are collectively assembled to produce knowledge regarded as legitimate, authoritative and valid (Latour & Woolgar, 1979; Lynch, 1985). One of its key contributions has been to *situate* scientific experiments (Haraway 1991), revealing the practical negotiations, forms of mediation and techniques of reduction involved in diverse instances of putting 'nature' to the test. By drawing attention to the contingencies entailed in the making of scientific facts, this literature has challenged the formalised claims of traditional epistemology while distancing itself from social constructionism in certain respects, as I have mentioned. Not only has ANT demonstrated how non-human actants play a far more active role in the making of scientific knowledge than theorists of discourse and deconstruction allow: Leading proponents characteristically refuse to debunk what is said to be constructed and express their commitment to taking seriously the procedures, criteria and arrangements used to conduct trials and evaluate and substantiate their findings (see Latour, 1999, 2004b: Stengers 2000, 2010). This position gives rise to notions such as 'ontological multiplicity' (Mol 2002), 'the multiverse' (Latour 2004b) and 'cosmopolitics' (Stengers 2010) – terms that affirm the immense diversity of experimental procedures while underlining the provisional nature of the realities they produce. It is not a matter of legitimating or delegitimating these procedures but defining and negotiating their parameters, while tracing the arrangements and associations that hold the realities they enact in place (Callon & Rabeharisoa 2004).

Given this extensive literature on experimentation, one might consider what sort of contribution greater attention to sexual bodies – their habits, pleasures, transformations and failures – could make to this field of scholarship. Queer theorists generally understand queer world-making to involve forms of inventiveness and experimentation with bodies, pleasures and intimate relations that recall Foucault's proposal in the *History of Sexuality* that 'bodies, pleasures and knowledges, in their multiplicity and their possibility of resistance' ought to form the 'rallying point against the regime of sexuality' (1978, p. 157). Foucault conceived these activities as both creative and destabilising, capable at once of dismantling certain relations while creating and assembling new ones. These are relations of 'differentiation, of creation, of innovation' rather than self-same identity. This is what interests Foucault and why they matter, as one gathers from the offhand remark he uses to round off this claim 'To be the same is really boring' (1997, p. 166; see Halperin 1995; Race 2018).

In recent years, ANT has extended its consideration of experimental practices and their publics well beyond the realm of scientific practices, to encompass aesthetic, political and ethical forms of experimentation and demonstration (Latour 2011; Latour & Weibel 2005; Marres 2012; Michael 2018; Papadopoulos 2018) But with the exception of an important body of feminist scholarship, sexual bodies rarely (if ever) make an appearance in the canonical pages of ANT, except perhaps as an object of various branches of the biomedical sciences.[5] Indeed, Latour's first foray into bodily matters seems rare enough within his field that he sees no problem naming it 'How to talk about the body' (2004b). 'Biopower should have a bio-counterpower,' he proclaims here somewhat grandly (p. 237) – an unusual moment of convergence with some of queer theory's express objectives.

A small but intriguing body of work within ANT does explore certain embodied passions: For music (Hennion 2007), drugs (Gomart & Hennion 1999) and consumer goods (Cochoy 2007). Though they have little to say about sex, these studies develop highly original pragmatist approaches to the *dispositifs* or 'event-networks' (Gomart & Hennion 1999) that amateurs among others establish to captivate subjects and/or enable their passions to be exercised and/or transformed. Hennion approaches taste as an experimental and 'pragmatic activity involving amateurs turned towards their object in a perplexed mode' (2007, p. 104), a 'concrete activity whose modes, practices and dispositifs can be described …[an event] in which the relevant subjects, objects and social groupings are co-produced' (Gomart & Hennion 1999, p. 228). Frank Cochoy (2007) investigates the material arrangements and forms of publicity that market actors devise to captivate shoppers in their capacity as sensory, embodied beings. Though drawing extensively on ANT, these scholars supplement the term *actor* with the term *event*. Where *actor* connotes heroic activity, sovereign mastery and control of a situation, *event* is neither active nor passive but refers to a provisional arrangement that enables something new or stimulating to happen – hence *event-networks*. The pragmatist approach elaborated here might sharpen queer theorists' attention to the sociotechnical settings and material arrangements that condition and effectuate bodily passions and enable certain transformations to occur (Race 2017, 2018). This would extend to the material components and pragmatic dimensions of the publics such passions enlist, engage or otherwise depend upon *and* the concrete effects of these mediations (see Marres 2012, pp. 82–105).

A considerable body of work has, in fact, emerged within queer theory that considers how experiments in sex, pleasure and intimacy give rise to genres of publicity. Significantly, this work includes experiments that engage and/or are said to depend upon contexts of public elaboration to produce their transformations. Michael Warner provides the most comprehensive and influential theorisation of publics within the field in his book *Publics and Counterpublics* (2002). He discusses how queer and feminist counterpublics 'can work to elaborate new worlds of culture and social relations in which gender and sexuality can be lived, including forms of intimate association, vocabularies of affect, styles of embodiment, erotic practices, and relations of care and pedagogy' (2002, p. 57). 'Sex in Public' makes a case for erotic experiments that make sex 'the consequence of public mediations and collective self-activity in a way that [makes] for unpredicted pleasures' (1998, p. 565). Both Warner and Berlant conceive the public and collective dimensions of such experiments to be critical for the effectiveness of broader projects of queer world-making (see also Berlant 2008; Warner 2000).

Much of the queer scholarship on publics is embroiled in a dialectics of oppositionality that actor-network theorists may query, complicate or refuse. Indeed, it is difficult to imagine how the oppositional conception of power/resistance evoked by this binary (publics/counterpublics) can be reconciled with the ontological multiplicity and heterogeneity actor-network theorists have trained themselves to find in public worlds (see Latour 2004b; Marres 2012; Mol 2002; Michael 2009). But attention to the discursive–pragmatic characteristics of counter/publics and the different forms of mediation they entail can be found within queer scholarship, not to mention the different forms of agency they engender. Michael Warner (2002, p. 118) underlines the 'variegated array' of spaces and venues Nancy Fraser associates with the emergence of feminist counterpublics before discussing how the forms of agency usually attributed to publics usually involve a 'direct transposition from private reading acts to the sovereignty of public opinion' (publics scrutinise, ask, reject, decide, etc.). Observing that the 'ideology of reading does not have the same privilege' in queer counterpublics, Warner writes:

> It might be that embodied sociability is too important to them; [or they are not] organised by the hierarchy of faculties that elevates rational-critical reflection …[or that they] depend more heavily on performance spaces than print; [or their] creative-expressive function. How then will we imagine their agency? …A queer public might be one that throws shade, prances, disses, acts up, carries on, longs, fantasizes, throws fits, mourns, "reads"….
>
> *(pp. 123–124)*

This delightful passage attends not only to the mediating devices implicated in the practice of 'making things public' familiar within ANT (see Latour & Weibel 2005), but also the diverse manners and modes of embodied agency, the possibilities of which queers are carefully attuned and sensitive to (perhaps especially so).

Meanwhile, the experimental activities of sexual bodies are emerging as a central focus of recent work in sexuality studies. Susanna Paasonen argues that constituting sex as play makes it possible to 'highlight improvisation driven by curiosity, desire for variation and openness towards surprise' as 'things that greatly matter in sexual lives and scholarly attentions' (2018, p. 3). For Paasonen, playfulness 'intermeshes and overlaps with those of improvisation, exploration, curiosity and experimentation, yet it also stands apart from them in foregrounding pleasure and bodily intensity as key motivations for sexual activity' (p. 2). As sexual bodies 'move towards different scenes, fantasies, bodies and objects, and become attached to them,' the forms, identities and subjectivities of participants are reorganised (p. 3). In my work on emergent sex and drug practices commonly referred to as chemsex, or Party 'n' Play, I draw on Simmel and Hughes' (1949) discussion of the play-form in 'The Sociology of Sociability' to explore the generativity of this framing (Race 2015a, 2018). Play is conceived here as a non-instrumental form of social activity governed by the exchange of stimuli that generates new connections through its exploration of the pleasures of association. Such connections frequently extend to non-human entities, as new uses, attachments, relations and forms of intimacy develop among the various human and non-human participants that get drawn into and reconfigured by the play-assemblage. Among these are domestic objects, consumable substances, a range of communication devices and home media, music, sex and drug equipment, furnishings, decorations, medications, clothes or costumes, lighting rigs, carpets, neighbours, friends, strangers and other odds and ends.

Latour (2005) proposes the term 'associology' to describe the project of tracing how different elements get associated in specific circumstances to produce what traditional sociologists take to be pre-existing, self-explanatory facts or entities. Erotic experimentation and sexual play might be regarded as particularly prolific scenes of associative events and activity, not only because of their promiscuous and labile nature and the sheer multiplicity of the associations they are capable of generating, but also their capacity to disorganise and reconfigure attachments, identities, relations and ontologies ('their bodies have become disorganised and exciting to them,' Berlant & Warner 1998, p. 564). This unsettling capacity of erotic bodies is of critical relevance – not only for private individuals, but for intervening in the social, political, aesthetic, material, semiotic, technical, mediating and institutional arrangements participants find themselves in, which privilege some forms of life while making others difficult or impossible, as queer theorists have demonstrated. It is time for ANT to get a grip on (or at least take seriously) those playful experiments in bodies and pleasures whose multiplicity, material relevance and critical significance Foucault makes such a compelling case for.

In his work on odiferous bodies, Latour concedes that *biopower* is 'the great question of this century' (2004b, p. 237) – if mainly to frame his intervention as the way to break science's 'imperial' grip on the body and the definition of its primary qualities. Conceiving the body as a dynamic interface that is always entangled with and articulated by various prostheses, mediators and sociotechnical arrangements through which 'we learn to register and become sensitive to what the world is made of,' he writes:

> to have a body *is to learn to be affected*, meaning 'effectuated', moved, put into motion by other entities, humans or non-humans. If you are not engaged in this learning you become insensitive, dumb, you drop dead.
>
> *(p. 206)*

Not only does this prevent easy recourse to 'authentic' or unmediated experience, it also generates and sustains critical interest in the diverse sociotechnical arrangements that articulate the worlds we are in: *how they matter, the dynamics of their construction, their associative power.* Most sensitive queers will have me leave the last word with Foucault but in the end, their points are similar: To be the same is really boring.

Notes

1 This remark occurred, from memory, during an energetic discussion among participants in the *Regimes of Sexuality* conference at the Australian National University, 1993.
2 They state early on '[h]egemonies are nothing if not elastic alliances, involving dispersed and contradictory strategies for self-maintenance and reproduction' (1998, p. 553).
3 Some take queer critiques of the future (Edelman 2004) – and it's 'antisocial thesis' more generally – to be the most radical and effective way of doing this (See Caserio et al. 2006). I am less convinced you can have one without the other (disassembling/reassembling) – as the wide-ranging impacts of Foucault's genealogies bear out.
4 Infrastructures constitute an explicit focus of recent work by Berlant, who highlights their 'movement or patterning of social form … the living mediation of what organizes life' (2016, p. 393). Referring to the world-sustaining relations of roads, bridges, schools, food chains, etc., she remarks 'even ordinary failure opens up the potential for new organisations of life' (p. 393).
5 An obvious exception is the extraordinary work of Donna Haraway, which positively teems with queer fabulations and attachments (Haraway 1991). For other critically important works that address sexual bodies using ANT, see de la Bellacasa 2017; Dugdale 1995; Keane 2004; Keane & Rosengarten 2002; Pienaar 2016; Roberts 2007; Rosengarten 2008; Sofia 1984; Stengers 1997; Waldby 1996.

References

Ahmed, S., 2006. *Queer Phenomenology*. Duke University Press.
Ahmed, S., 2014. Selfcare as Warfare. https://feministkilljoys.com/2014/08/25/selfcare-as-warfare/
Berlant, L., 2008. *The Female Complaint*. Duke University Press.
Berlant, L., 2016. The commons: Infrastructures for troubling times. *Environment & Planning D* 34(3), pp. 393–419.
Berlant, L. and Warner, M., 1998. Sex in public. *Critical Inquiry*, 24(2), pp. 547–566.
Butler, J., 1990. *Gender Trouble*. Routledge.
Butler, J., 2010. Performative agency. *Journal of Cultural Economy*, 3(2), pp. 147–161.
Callon, M., 1984. Some elements of a sociology of translation: Domestication of the scallops and the fishermen of St Brieuc Bay. *The Sociological Review*, 32(1), pp. 196–233.
Callon, M., 2010. Performativity, misfires and politics. *Journal of Cultural Economy*, 3(2), pp. 163–169.
Callon, M. and Rabeharisoa, V., 2004. Gino's lesson on humanity: Genetics, mutual entanglements and the sociologist's role. *Economy and society*, 33(1), pp. 1–27.
Caserio, R.L., Edelman, L., Halberstam, J., Muñoz, J.E. and Dean, T., 2006. The antisocial thesis in queer theory. *PMLA*, 121(3), pp. 819–828.
Castiglia, C. 2000. Sex panics, sex publics, sex memories. *boundary 2*, 27(2), pp. 149–175.
Certeau, M., 1984. *The Practice of Everyday Life*. Berkeley University Press.
Cochoy, F., 2007. A brief theory of the 'Captation' of publics: Understanding the market with little red riding hood. *Theory, Culture & Society*, 24(7–8), pp. 203–223.
de La Bellacasa, M.P., 2017. *Matters of Care*. University of Minnesota Press.
Delany, S. 1999. *Times Square Red, Times Square Blue*. New York University Press.
Deleuze, G. and Guattari, F., 1987. *A Thousand Plateaus*. University of Minnesota Press.
Dugdale, A., 1995. Devices and desires: Constructing the intrauterine device, 1908–1988. PhD thesis, University of Wollongong.
Edelman, L., 2004. *No Future: Queer Theory and the Death Drive*. Duke University Press.
Foucault, M., 1978. *The History of Sexuality: An Introduction. Vol. 1.* Vintage.
Foucault, M., 1997. Friendship as a way of life. In P. Rabinow (ed.) *Ethics: Subjectivity and Truth*. London: Penguin, pp. 135–140.

Gomart, E. and Hennion, A., 1999. A sociology of attachment: Music amateurs, drug users. *The Sociological Review*, 47(1), pp. 220–247.

Hall, D. and Jagose, A. eds., 2012. *The Routledge Queer Studies Reader*. Routledge.

Halperin, D.M., 1995. *Saint Foucault*. Oxford Paperbacks.

Haraway, D., 1991. *Simians, Cyborgs, and Women*. Routledge.

Hennion, A., 2007. Those things that hold us together: Taste and sociology. *Cultural Sociology*, 1(1), pp. 97–114.

Keane, H., 2004. Disorders of desire: Addiction and problems of intimacy. *Journal of Medical Humanities*, 25(3), pp. 189–204.

Keane, H. and Rosengarten, M., 2002. On the biology of sexed subjects. *Australian Feminist Studies*, 17(39), pp. 261–277.

Larkin, B., 2013. The politics and poetics of infrastructure. *Annual review of anthropology*, 42, pp. 327–343.

Latour, B., 1996. *Aramis*. Harvard University Press.

Latour, B., 1999. *Pandora's Hope*. Harvard University Press.

Latour, B., 2004a. Why has critique run out of steam? From matters of fact to matters of concern. *Critical Inquiry*, 30(2), pp. 225–248.

Latour, B., 2004b. How to talk about the body? The normative dimension of science studies. *Body & Society*, 10(2–3), pp. 205–229.

Latour, B. 2005. *Reassembling the Social*. Oxford University Press.

Latour, B. and Weibel, P. (eds.), 2005. *Making Things Public*. ZKM, Center for Art and Media Karlsruhe.

Latour, B., 2011. Some experiments in art and politics. *e-flux*, 23, pp. 1–7.

Latour, B. and Woolgar, S., 1979. *Laboratory Life*. Sage.

Lorde, A. 1988. *A Burst of Light, Essays*. Sheba Feminist Publishers.

Lynch, M. 1985. *Art and Artifact in Laboratory Science*. Routledge & Kegan Paul.

Marres, N., 2012. *Material Participation*. Springer.

Michael, M., 2002. Comprehension, apprehension, prehension: Heterogeneity and the public understanding of science. *Science, Technology, & Human Values*, 27(3), pp. 357–378.

Michael, M., 2009. Publics performing publics: of PiGs, PiPs and politics. *Public Understanding of Science*, 18(5), pp. 617–631.

Michael, M., 2018. On "Aesthetic Publics" The Case of VANTAblack®. *Science, Technology, & Human Values*, 43(6), pp. 1098–1121.

Mol, A., 1999. Ontological politics. A word and some questions. *The Sociological Review*, 47(1), pp. 74–89.

Mol, A., 2002. *The Body Multiple*. Duke University Press.

Mol, A. and Law, J., 2004. Embodied action, enacted bodies: The example of hypoglycaemia. *Body & society*, 10(2–3), pp. 43–62.

Paasonen, S., 2018. *Many Splendored Things: Thinking Sex and Play*. Goldmsiths.

Papadopoulos, D., 2018. *Experimental Practice*. Duke University Press.

Pienaar, K., 2016. *Politics in the Making of HIV/AIDS in South Africa*. Palgrave.

Race, K., 2015a. 'Party and Play': Online hook-up devices and the emergence of PNP practices among gay men. *Sexualities*, 18(3), pp. 253–275.

Race, K., 2015b. Speculative pragmatism and intimate arrangements: Online hook-up devices in gay life. *Culture, Health & Sexuality*, 17(4), pp. 496–511.

Race, K., 2017. Thinking with pleasure: Experimenting with drugs and drug research. *International Journal of Drug Policy*, 49, pp. 144–149.

Race, K., 2018. *The Gay Science: Intimate Experiments with the Problem of HIV*. Routledge.

Sedgwick, E.K., 1997. Paranoid reading and reparative reading. In E. K. Sedgwick (ed.) *Novel Gazing: Queer Readings in Fiction*. Duke University Press, pp. 1–37.

Simmel, G. and Hughes, E.C., 1949. The sociology of sociability. *American Journal of Sociology*, 55(3), pp. 254–261.

Sofia, Z., 1984. Exterminating fetuses: Abortion, disarmament, and the sexo-semiotics of extraterrestrialism. *Diacritics*, 14(2), pp. 47–59.

Stengers, I., 1997. *Power and Invention: Situating Science*. University of Minnesota Press.

Stengers, I., 2000. *The Invention of Modern Science*. University of Minnesota Press.

Stengers, I. 2010. *Cosmopolitics I*. University of Minnesota Press.

Waldby, C., 1996. *AIDS and the Body Politic*. Routledge.

Wallace, L., 2011. *Lesbianism, Cinema, Space*. Routledge.

Warner, M. (ed.), 1993. *Fear of a Queer Planet*. University of Minnesota Press.

Warner, M., 2000. *The Trouble with Normal*. Harvard University Press.

Warner, M., 2002. *Publics and Counterpublics*. Zone Books.

Wiegman, R. and Wilson, E.A., 2015. Introduction: Antinormativity's queer conventions. *Differences*, 26(1), pp. 1–25.

Wilson, A., 2016. The infrastructure of intimacy. *Signs*, 41(2), pp. 247–280.

Woolgar, S., 1992. Configuring the user. In J. Law (ed.) *A Sociology of Monsters*. Routledge, pp. 57–102.

17

Is ANT capable of tracing spaces of affect?

Derek P. McCormack

ANT is often presented as deliberately and definitely anti-autobiographical, concerned as it is with tracing agencies and arrangements beyond human life. While I am drawn towards ANT as a cluster of invitations for becoming oriented to the object world, I confess, however, that I have never warmed to its antipathy towards autobiography. In some respects, this is because it seems to me to miss the point. Rather than becoming more or less than human, it is just as interesting to think about the human as a fragile achievement sustained by all kinds of circumstantial encounters, relations and forces. To think biographically and/or autobiographically is not therefore to reaffirm the sovereignty of the subject; it is about tracing some of the ways in which human life is, in some sense, shaped by circumstantial encounters, relations and forces. Ideas, concepts or propositions are no less part of this process than anything else. To trace their impact is to trace the conditions under which certain capacities to affect and be affected emerge. It is also, perhaps, to think about how these ideas, concepts or propositions become part of a wider 'structure of feeling' (Williams, 1977) and the spaces of affect through which this structure of feeling is sensed.

The sites in which concepts and propositions circulate are spaces of affect. For me, then, it began during a graduate seminar on alternative political theory taught by Timothy Luke at Virginia Tech. Over the course of that term, we read many books, including Haraway's *Simians, Cyborgs, and Women* (1991), Guattari's *Chaosmosis* (1995) and Latour's *We have Never Been Modern* (1993). Informed by Tim's own emerging interests in cyborg environments and liberal subjectivity (1996; 1997), running through these books was the now familiar idea that agency, politics, ethics, etc., might in some way be more-than-human. However, what I recall about my first encounter with this cloud of ideas is that they baffled me: I found them absolutely foreign. But they were still alluring: They seemed to point not so much to a heightened role for the technical capacity of objects but to the possibility of re-enchanting the world with a sense that there were more things animating it than the modern constitution allowed. This is a reminder that an encounter with ANT, as with other sets of ideas, is always an affective encounter. Always as much about something sensed rather than something recognised. It shows up as an unsettling feeling before it ever resolves itself conceptually or empirically. The affective qualities of its propositions are part of the circumstances of the world impacting upon bodies in ways that can show up as feeling or emotion.

And yet, such attention to the differentiated dimensions of affect was absent from much earlier work associated with ANT, even if this work obviously went far beyond the writing of Latour. Nigel Thrift made this point in an important early articulation of non-representational theory (2000; see also Thrift 2008). Describing ANT as 'Champagne without fizz,' Thrift welcomed ANT's treatment of agency, its emphasis on invention and how it conceives of personhood as distributed performance. Conversely, it was more attentive to the slow and often drawn-out process of assembly than the 'flash of the unexpected' (2008: 214). It had too little sense of the intensity of the event. And, most importantly, it had an impoverished sense of the human, its capacities and powers. As a staging post on the way to developing a non-representational style of work, ANT had its place, but offered little when it came to tracing the affective aspects of the world.

There were exceptions, of course. Emilie Gomart and Antoine Hennion's remarkable discussion of the sociology of attachment (1999) was exemplary here. It was one of the few ANT-inspired pieces that attended to the affective dimensions of relations between humans and non-humans and also to the forms of embodied conduct conditioned by capacities to affect and be affected. As Hennion remarks in a recent reflection upon ANT at the time:

> The surprise that peels away from the flux of things is the most ordinary of experiences, for an audience member, a painter, a footballer, a drinker. It is an experience shared by professionals and amateurs alike. Out of the fabric of familiar things, a small but decisive deviation has effects that can be enormous, but they arise from the things themselves as they present themselves. It is the jazz improviser who plays the same piece a hundred times, and yet... wait, this time it's going this way, insisting on a quite new pathway. He follows it, tries it again... comes back, it has opened up a space.
>
> *(Hennion and Muecke, 2016: 296)*

Such work had the potential to allow for the tracing of spaces of affect: This potential was not realised, however, beyond a few studies. The forensic attention paid by ANT-inspired work to the tracing of objects and their capacities tended not to be accompanied by a tracing of the affective capacities of objects, spaces or bodies. Indeed, its style and emphasis have often mirrored that of the areas of expertise it has sought to understand, through the production of accounts that are descriptively and analytically dazzling but which give little attention to the role of affective, emotion and feeling in the engineering and experience of a range of devices, contexts and arrangements.

★★★★★

In some sense then, some of the concerns of ANT diverged from those of the 'affective turn' (Clough and Halley, 2007). This turn is, of course, characterised by different approaches in (sometimes) generative tension (Anderson, 2014). These differences matter because the more affect is grasped as a force irreducible to a quality of human experience then the closer some versions of affect theory seem to be aligned, at least potentially, with elements of ANT. In some ways, the focus of ANT on the capacities of objects to affect and be affected by each other resonates with a Spinozist-inspired account of affect in which bodies are also defined in terms of those capacities (Deleuze, 1988). Both seem to be about foregrounding the affective capacities of non-humans. But there are important qualifications. Spinozist accounts of affect tend to be allied with the philosophies of becoming and process found in Bergson and Deleuze (See Massumi, 2002). This emphasis tends to be less evident in ANT-inspired accounts of objects: The focus on scientific and technical objects means that for the most part, ANT is silent about the sensible excess of objects, although there are some exceptions (see Lee and Brown, 2002). There is little about how the force of things, excessive of any individual entity, has the capacity to enchant, perturb and perplex (Bennett, 2010). Equally, ANT places more emphasis on the actual than the virtual. As Martin Müller and Caroline Schurr put it in relation to the sociology of attachment developed by Gomart and Hennion, 'this notion of attachment exhibits more than a touch of residual actualism, for it takes attachment to arise out of networks as a mediated effect' (2016: 224). Müller and Schurr also observe this actualism in Latour (2004) in whose work the body is theorised in ways that 'mean it is circumscribed by and never exceeds the relations that describe it' (2016: 224). Moreover, in some versions of the kind of speculative realism that emerges in part from readings of ANT, the emphasis on process is taken as a symptom of the undermining of the distinctive existence of entities and objects (Harman, 2010).

While versions of affect theory and ANT foreground the affective capacities of the non-human, they do so with very different ontological emphases. It might be more accurate to suggest that ANT has the potential to help us think about how non-humans have affective capacities, while particular versions of affect theory, conceive of affect in terms of the non-human forces from which forms of human experience emerge. Both are open to the ongoing surprise of the world that always exceeds human agency, but the latter is perhaps more attuned to how this surprise can be sensed through the capacities of different bodies to affect and be affected by other bodies (Bennett, 2010; Thrift, 2008). Equally, certain versions of affect theory are much more attentive to how affective orientations around objects are implicated in the politics and experience of different kinds of bodies (Ahmed, 2010). This would suggest, however, that ANT and affect theory are separate and distinct. It might be better to think of both as providing different sets of object-orientations towards and within spaces of affect. Taken together they offer different ways of foregrounding the role of objects in generating attachments to worlds while also examining the conditions that produce the affective backgrounds of these worlds. Such backgrounds are always excessive of efforts to capture them while also being the target of a range of technologies for engineering and modulating the affective capacities of objects and bodies.

These capacities are increasingly distributed across devices, technologies and practices of different kinds. They are also engineered to generate particular forms of value. There are openings here for various forms of association between ANT and theories of affect. For instance, Müller and Schurr (2016: 224) have argued recently that ANT has much to gain from a closer engagement with assemblage thinking, not least because the latter can encourage the former to do more to conceptualise 'the capacities of bodies, both human and non-human, to affect and be affected.' They also suggest that thinking in terms of assemblage not only encourages attention to affect, but can also temper the tendency to think of affect as the outcome of relations through focusing attention instead on how desire becomes part of the generative processuality of relations.

It might be worth noting here that ANT is sometimes invoked in work about affect on the basis that the former is more empirical in orientation. For instance, Ratner and Pors (2013: 212) note that 'whereas affect theory is difficult to operationalise for empirical analysis, given how affect is always-already outside of conscious perception, actor-network theory (...), or ANT, lends itself rather generously to such endeavour' (see also Deville, 2015; Grossberg, 2010). And yet, this critique ignores how attention to the affective expands and multiplies the sense and scope of the empirical. If, as Lauren Berlant (2011) has noted, affect theory in all its manifestations is just another way of thinking about how the world impacts bodies, then the many forms this takes are only beginning to be understood. The important question should not be how ANT might allow affect theory to become more empirical, but how affect theory might complicate the senses of the empirical enacted through versions of ANT.

This expanded affective sense of the empirical might allow us to mobilise ANT to trace spaces of affect. For instance, Yael Navaro-Yashin (2009) combines ANT and theories of affect in an ethnography of conflict in Cyprus. She welcomes the attention paid by ANT to objects but qualifies this with the claim that 'if we were to limit ourselves to an ANT framework or methodology, we would have to call off any query into "affect" as referring to "human" factors too' (ibid: 10). At the same time, she draws support from Deleuze, Guattari, Massumi and Thrift to grasp affect as the 'non-discursive sensation which a space or environment generates' (ibid: 13). This allows Navaro-Yashin to (a) temper the horizontality of the rhizome with what she suggests is the multiple temporality of ruination, and (b) hold onto the expressive quality of language.

Attention to affect might also allow ANT-inspired work to revisit some of the technoscientific spaces and practices through which different entities become visible and stable. Such attention amplifies the ethnographies of laboratory life that were so central to the emergence of ANT through foregrounding the sensible and felt arrangements of bodies and technologies at these sites. For instance, Natasha Myers has explored the 'affective entanglements of inquiry in the laborator[ies]' where protein modellers render explicit different forms of molecular life by becoming affectively and kinaesthetically involved in their work (2015: xi). In the process, Myers argues that these models are more than immutable mobiles but are part of animated styles of affective labour that investigate forms of 'excitable matter.' Such work does not need to be restricted to the lab but can be extended to a range of sites at which science learns to be affected by the world (Lorimer, 2008).

As Myers' own work suggests, theories of affect can make ANT more attentive to how difference shapes and is shaped by spaces of affect. This is important in a couple of senses. On one level, it makes ANT more attentive to how affective arrangements and encounters are differentiated via objects through the lived experiences of gender, class and race (Ahmed, 2010). On another level, it can also complicate the affective ontologies of ANT and post-ANT works, not least through foregrounding the way in which diverse indigenous ontologies mobilise a range of affective capacities around objects and entities (De la Cadena, 2010).

A further set of interesting working arrangements between ANT and affect is via research into increasingly networked space-times of affective experience. To some extent, this involves examining how different forms of media technology and infrastructures participate in contemporary affective life. Paasonen, Hillis and Petit (2015) reflect upon the generation and circulation of 'networked affects' through a myriad of screens, devices and channels (see also Lindgren, 2017). Noting that 'ANT is not always included in the genealogies of the affective turn,' they point to the possibility of drawing on ANT-inspired theorisations of distributed, networked and emergent agency in order to examine how affects emerge through 'networked exchanges and encounters' (2015: 10). In their view, 'networked affect' refers to the work of generating affect across infrastructures, devices and bodies. Through such networks, the very materiality of media is being questioned (Peters, 2015). Recent work by Mark Hansen (2014), for instance, also draws attention to the fact that the temporality of contemporary media means that the operational present of affective life takes place prior to consciousness. As a consequence, the question of what counts as a media object is up for grabs. As work by James Ash (2017) illustrates, there is significant scope here for developing new conceptual vocabularies for rethinking the affective ontologies of contemporary media in ways that go beyond the kinds of objects that were the study of earlier versions of ANT. In the process, argues Ash, it might be possible to rethink the affective life of objects by developing a sophisticated grasp of their 'phased' space-times.

★★★★★

One way of grasping these spaces is via the concept of atmospheres. Atmospheres have recently become an important focus of efforts within the social sciences and humanities to grasp the affective materiality of space-times (see, for example, Adey, 2014; Anderson, 2009; Blok and Farias, 2016; Böhme, 1993; Ingold, 2012; Stephens, 2016; Stewart, 2011). Atmosphere is alluring because it is vague, and resists easy specification. It suggests something that, while potentially palpable in bodies, always subsists beyond, across and between those bodies. It provides a way of thinking between the materiality of physical processes and the sensory capacities and, yes, experiences, of different forms of life. And while they become what Ben Anderson (2014) calls an 'object-target' of various forms of intervention and experiment operating at a range of scales, atmospheres have a processual materiality that exceeds such efforts.

An interesting question that emerges from this work concerns whether atmosphere can be grasped as an object or entity. This might seem like a minor question but it has important implications. If atmosphere is grasped as an object, then it means that the relation between atmospheres and anything else is a relation between objects. It also means, perhaps more controversially, that nothing is excessive of the category of object, including whatever we might understand as experience. ANT is arguably helpful for thinking about the technical apparatus and practices via which atmospheres (as space-times that are simultaneously meteorological and affective) are measured, generated, targeted and modified as objects. It offers less, however, when it comes to apprehending the vague yet palpable force of atmospheres in different human and non-human bodies.

Indeed, in some important strands of post-ANT thinking, the notion of affective and atmospheric experience is dismissed rather too easily as the residue of a nostalgic humanism. For Timothy Morton (2013), notably, any appeal to atmosphere marks a return to a form of phenomenological romanticism. Clearly, the atmosphere around the earth can be grasped in terms of what Morton calls a hyper-object. It seems more problematic, however, to think of all the differentiations that take place in this atmosphere, and all the ways in which these differentiations are sensed by bodies of different kinds, also as objects. One of the questions that shapes efforts to think about and with affective atmospheres in the wake of the influence of ANT therefore concerns the degree to which they can be understood in terms of objects without losing the excessive qualities that define them. Or, indeed, if any affective account of objects can take excess seriously.

In some ways, this question has informed some of my recent thinking and writing about a particular device – the balloon (see McCormack, 2018). In many ways, the emergence of this object as a technoscientific device in the late 18th century was a key moment in the process through which the composition, properties and behaviours of the atmosphere became explicit as the focus of a range of scientific and philosophical concerns. While no longer so novel, this device remains central to the ways in which the dynamics of atmospheres are sounded and made explicit at a range of scales. A story about the balloon might be told in ways that are parallel to accounts of the air pump (Shapin and Schaffer, 1985). But those early experiments with the balloon were also important as affective events: They attracted enormous crowds in anticipation of witnessing the spectacle of a launch. And yet, it would be too easy to imagine a distinction here between an object (the balloon) and something excessive of an object (atmosphere). This is where ANT-inspired approaches can help: They can sensitise us to the ways in which a simple device such as the balloon begins to take shape as a set of processes that are given a degree of intensive and extensive stability.

In a rather different vein, thinking about atmosphere can also help us reflect upon how much of the ANT-inspired work on economic life can be supplemented. Unsurprisingly, much of this work has focused on the practices and technologies of calculation through which different forms of economic life, particularly markets, are organised and reproduced (Calıskan and Callon 2010; Callon et al. 2002; Callon and Muniesa 2005). The emphasis here is on how practices and devices actively contribute to the historically specific genera- tion of the economies they purport to study from a distance. To be sure, some of the works emerging in the wake of ANT do focus on the economy as passionate interests, particularly following the work of Gabriel Tarde on imitation and invention (see Barry and Thrift, 2007; Latour and Lépinay 2009; McFall, 2014). Equally, in more recent work, Joe Deville (2012; 2015) has argued that attention to the affective relations through which debt obligations and attachments are sustained provides an important bridge between the pragmatist orientations of ANT and affective turn.

Attention to atmospheres can be extended further in a way that makes such approaches to the economic more attentive to the affective structures of forms of economic life. In recent work, for instance, Ben Anderson (2016) has analysed the affective dimensions of neo-liberalism in part through thinking about the conditions of their emergence. For Anderson, atmospheres 'are part of the conditions of emergence of formation for particular neoliberalisms' (734). By this he means that the tone and style of particular meetings and gatherings were part of the milieu from which specific articulations of and investments in shared sets of ideas emerged, stabilised to some degree and travelled across and between dif- ferent sites. This is not specific to the formation of neo-liberalism, of course. In some ways, we might say that what Anderson is writing about is characteristic of any process by which gatherings of particular groups at particular spaces generate atmospheres that may be articu- lated through wider structures of feeling. In that sense, scientific laboratories, financial insti- tutions and educational establishments – including graduate seminars – are also characterised by such atmospheres and can become spaces through which different forms of attachment and association can emerge.

★★★★

My answer to the question posed by the chapter is a qualified yes. ANT can trace spaces of affect in useful ways by helping us think about the devices and practices through which these spaces are constituted, through which they become visible and through which they are governed. In this sense its emphasis on the technical is an advantage. Indeed, the tech- nical capacity to trace affective space-times has been central to contemporary forms of life: Stock markets, for instance, can be understood at least in part as ongoing tracings of the affective intensity of this life that modulate that intensity as soon as they are rendered visible. A range of other economic measurements performs the same function. Equally, there are an increasingly complex array of technologies and devices for tracking the various forms of behaviour exhibited by consumers as they respond to commodities and prices. In that sense, then, it is possible to trace affective space-times in a manner analogous to how various web-based technologies have been used by those influenced by ANT in order to map online controversies.

However, the risk here is that by tracing things in this way we turn processes and variations into particular kinds of stable objects when they are badly served by that move. Any attempt to trace spaces of affect with ANT is less about translation as it is about a certain kind of working arrangement that never stabilises as an object or entity or approach. This is particularly important if we understand the act of tracing in terms of the 'productive intersection of a form of content (actions, bodies, things) and a form of expression (affects, words, ideas) (Buchanan, 2014: 132–133).' Tracing, therefore, becomes a process through which the style in which an account is produced is affected by an encounter with an object, event or form of life. To trace is to be affected by the circumstances of that tracing.

This has implications for how spaces of affect are traced in writing. We might say that such writing begins in a certain capacity to become affected by the world. This is the kind of capacity exemplified, for instance, in the work of anthropologist Kathleen Stewart (2014a, 2014b). In this work, attention to the slow accretion of things is coupled with a sense of the unexpected whose relative neglect within ANT was lamented by Thrift. To be fair, of course, ANT, as a methodological approach, has tended to emphasise the importance of capacities to describe, and has valued writing in ways that capture the details of arrangements in practice. However, with some exceptions, such accounts have not tended to be characterised by what we might call an affective poetics. For Stewart, the challenge of writing affectively involves the task of finding ways of becoming attuned to always unformed objects: To ways of sensing, being moved, being arrested, etc.

As an exemplar of work that traces spaces of affect, Stewart's work resonates in one final way with the work of some proponents of ANT: This involves tiredness with critique as the dominant mode of engagement with and expression of the world. More generally, the emphasis on composition (2014b), albeit different, found in thinkers as diverse as Latour and Stewart points to the value of tracing affect as a generative, if not quite post-critical enterprise. Here, the point is not just to describe things better, but to think of description in terms of how it might contribute to the generation of space-times, however modest, in which different forms of association might be gathered, sensed and modulated.

Acknowledgments

My thanks to the editors of this volume for their patience with me during this process, and to Thomas Jellis for pointing me to some helpful literature.

References

Adey, P. (2014). Security atmospheres or the crystallisation of worlds. *Environment and Planning D*, 32(5): 834–851.

Ahmed, S. (2010). Happy Objects. In M. Gregg and G. J. Seigworth (eds.) *The Affect Theory Reader*. Durham, NC: Duke University Press. 29–51

Anderson, B. (2016). Neoliberal affects. *Progress in Human Geography*, 40(6): 734–753.

Anderson, B. (2014). *Encountering Affect*. Farnham: Ashgate.

Anderson, B. (2009). Affective atmospheres. *Emotion, Space and Society*, 2(2): 77–81.

Ash, J. (2017, forthcoming). *Phase Media: Space, Time, and the Politics of Smart Objects*. London: Bloomsbury.

Barry, A. and Thrift, N. (2007). Gabriel Tarde: imitation, invention and economy. *Economy and Society*, 36(4): 509–525.

Bennett, J. (2010). *Vibrant Matter: A Political Ecology of Things*. Durham, NC: Duke University Press.

Berlant, L. G. (2011). *Cruel Optimism*. Durham, NC: Duke University Press.

Blok, A. and Farias, I. eds. (2016). *Urban Cosmopolitics: Agencements, Assemblies, Atmospheres*. London: Routledge.

Böhme, G. (1993). Atmosphere as the fundamental concept of a new aesthetics. *Thesis Eleven*, 36: 113–126.

Buchanan, Ian M. (2014). Assemblage theory and schizoanalysis. *Panoptikum*, 13(20): 115–125.

Caliskan, K. and Callon, M. (2010). Economization, part 2: a research programme for the study of markets. *Economy and Society*, 39(1): 1–32.

Callon, M., Meadel, C. and Rabeharisoa, V. (2002). The economy of qualities. *Economy and Society*, 31(2): 194–217.

Callon, M. and Muniesa, F. (2005). Peripheral vision: economic markets as calculative collective devices. *Organization Studies*, 26(8): 1229–1250.

Clough, P. and Halley J. eds. (2007). *The Affective Turn: Theorizing the Social*. Durham NC: Duke University Press.

De la Cadena, M. (2010). Indigenous cosmopolitics in the Andes: conceptual reflections beyond "politics". *Cultural Anthropology*, 25(2): 334–370.

Deleuze, G. (1988). *Spinoza: Practical Philosophy*. Tr. Robert Hurley. San Francisco: City Lights Book.

Deville, J. (2015). *Lived Economies of Default: Consumer Credit, Debt Collection and the Capture of Affect*. London: Routledge.

Deville, J. (2012). Regenerating market attachments. *Journal of Cultural Economy*, 5(4): 423–439.

Gomart, E. and Hennion, A. (1999). A sociology of attachment: music amateurs and drug addicts. In J. Law and J. Hassard (eds.) *Actor Network Theory and After*. Oxford: Blackwell, 220–247.

Grossberg, L. (2010). *Cultural Studies in the Future Tense*. Durham NC: Duke University Press.

Guattari, F. (1995). *Chaosmosis: An Ethico-aesthetic Paradigm*. Translated by P. Bains and K. Pefanis. Sydney: Power.

Hansen, M. (2014). *Feed Forward: On the Future of Twenty-First Century Media*. Chicago, IL: University of Chicago Press.

Haraway, D. (1991). *Simians, Cyborgs, and Women: The Reinvention of Nature*. London: Routledge.

Harman, G. (2010). *Towards Speculative Realism: Essays and Lectures*. Roply, Hants: Zero Books.

Hennion, A. and Muecke, S. (2016). From ANT to pragmatism: a journey with Bruno Latour at the CSI. *New Literary History*, 47(2–3): 289–308.

Ingold, T. (2012). The Atmosphere. *Chiasmi International*, 14: 75–87.

Latour, B. (2004). How to talk about the body? The normative dimension of science studies. *Body & Society*, 10(2–3): 205–229.

Latour, B. (1993). *We Have Never Been Modern*. Cambridge MA: MIT Press.

Latour, B. and Lépinay, V. A. (2009). *The Science of Passionate Interests: An Introduction to Gabriel Tarde's Economic Anthropology*. Chicago, IL: Prickly Paradigm Press.

Lee, N. and Brown, S. (2002). The disposal of fear: childhood, trauma, and complexity. In J. Law, and A. Mol (eds.) *Complexities: Social Studies of Knowledge Practices*. Durham, NC: Duke University Press, 258–280.

Lindgren, S. (2017). *Digital Media and Society*. London: Sage.

Lorimer, J. (2008). Counting corncrakes: the affective science of the UK corncrake census. *Social Studies of Science*, 38(3): 377–405.

Luke, T. (1997). At the end of nature: cyborgs, 'humachines', and environments in postmodernity. *Environment and Planning A*, 29(8): 1367–1380.

Luke, T. (1996). Liberal society and cyborg subjectivity: the politics of environments, bodies, and nature. *Alternatives*, 21(1): 1–30.

Massumi, B. (2002) *Parables For the Virtual: Movement, Affect, Sensation*. Durham, NC: Duke University Press.

McCormack, D. (2018). *Atmospheric Things: On the Allure of Elemental Envelopment*. Durham, NC: Duke University Press.

McFall, L. (2014). The problem of cultural intermediaries in the economy of qualities. In J. S. Maguire and J. Matthews (eds.) *The Cultural Intermediaries Reader*. Sage: London, 42–51.

Morton, T. (2013). *Hyperobjects: Philosophy and Ecology after the End of the World*. Minneapolis: University of Minnesota Press.

Müller, M. and Schurr, C. (2016). Assemblage thinking and actor-network theory: conjunctions, disjunctions, cross-fertilisations. *Transactions of the Institute of British Geographers*, 41(3): 217–229.

Myers, N. (2015). *Rendering Life Molecular: Models, Modelers, and Excitable Matter*. Durham, NC: Duke University Press.

Navaro-Yashin, Y. (2009). Affective spaces, melancholic objects: ruination and the production of anthropological knowledge. *Journal of the Royal Anthropological Institute*, 15(1): 1–18.

Paasonen, S., Hillis, K. and Michael P. (2015). *Networked Affect*. Cambridge MA: MIT Press.

Peters, J. D. (2015). *The Marvelous Clouds: Toward a Philosophy of Elemental Media*. Chicago, IL: University of Chicago Press.

Ratner, H. and Pors, J. (2013). Making invisible forces visible. Managing employees' values and attitudes through transient emotions. *Int. J. Management Concepts and Philosophy*, 7(3/4), 208–223.

Shapin, S. and Schaffer, S. (1985). *Leviathan and the Air-Pump: Hobbes, Boyle, and the Experimental Life*. Princeton, NJ: Princeton University Press.

Stephens A. C. (2016). The affective atmospheres of nationalism. *Cultural Geographies*, 23(2): 181–198.

Stewart, K. (2014a). Road registers. *Cultural Geographies*, 21(4): 549–563.

Stewart, K. (2014b). Tactile compositions. In P. Harvey and E. Casella (eds.) *Objects and Materials*. London: Routledge, 775–810.

Stewart, K. (2011). Atmospheric attunements. *Environment and Planning D*, 29(3): 445–453.

Thrift, N. (2008). *Non-Representational Theory: Spaces, Politics, Affects*. London: Routledge.

Thrift, N. (2000). Afterwords. *Environment and Planning D: Society and Space*, 18(2): 213–255.

Williams, R. (1977). *Marxism and Literature*. Oxford: Oxford University Press.

18

How to care for our accounts?

Sonja Jerak-Zuiderent

What does the figure of the knower do with the gift of agency?

This chapter explores the question of how to care for our scholarly accounts by paying attention to the 'figure of the knower.' Attending to the figure of the knower provides a nuanced
complication of where and how we want to study agency and how this relates to knowledge
politics – one of the central concerns that shaped ANT as an intellectual practice. Starting
from the radicalisation of situated action by Gomart and Hennion (1999) that blurs dichotomies between humans and non-humans and between activity and passivity, attending to the
figure of the knower once again moves the question centre stage *how* those endowed with
'action [as] an unanticipated gift from the "dispositif"' (ibid.: 222) are equally 'involved in
the making of the world' (Puig de la Bellacasa, 2017: 39). One way of addressing agency
as an unanticipated gift from the dispositif is to keep on asking 'how to care.' And this, as
María Puig de la Bellacasa argues, implies a reorientation to something that mainly humans
do. 'Care is a human trouble' (ibid.: 2), she states; a 'condition that circulates through the
stuff and substance of the world as agencies without which nothing that has any relation to
humans would live well' (ibid.: 122), and at the same time it is something that is only possible in 'more than human worlds' (Puig de la Bellacasa, 2017). 'Affirming the absurdity of
disentangling human and nonhuman relations of care and the ethicalities involved requires
decentring human agencies, as well as remaining close to the predicaments and inheritances
of situated human doings' (ibid.: 2). Attending to the figure of the knower when asking the
question 'how to care' (ibid.) is therefore indebted to both: ANT concerns about decentring
human agency, while also focussing on what human people do, once agency has become a
gift from the dispositif. Or, as Lucy Suchman put it, 'the price in recognizing the agency of
artefacts need not be at the denial of our own' (Suchman, 2007: 285).[1]

Attending to agency as a gift from the dispositif and to the question of what is done with
it once given, I explore how the figure of the knower can be rendered differently based on
different ways of figuring (not) knowing and (not) moving/being moved by an 'other' and
how this matters for the question of 'how to care' for scholarly accounts. In the following,
I elaborate on two such figurings: 'solid knowing' and 'motile not-knowing.'[2] 'Solid' refers
to 'clear or fixed' entities, firm boundaries, imposing themselves (de Laet and Mol, 2000) and

shaping knowing as something that constantly tries to divide, fix and abstract by predicting, rationalising and ordering. 'Motile' instead refers to moving/being moved by an 'other' (Munro, 2012); not like mobile in the sense of crossing boundaries. It rather refers to a flickering, a shifting back and forth "between definition and deferral: between processes of objectifications and processes of personification, between material evidence (…) and processes of interpretation and judgement in which everything becomes tentative again" (Latimer 2013: 169). And I propose that for continuing to ask the question on how to extend ANT as an intellectual practice, attending to the figurings of (not) knowing and (not) moving/being moved by an 'other' matters.

To explore the importance of attending to how the figure of the knower is rendered, I introduce observations from my fieldwork on prostate cancer screening practices. During one observation, where Peter, a urologist, carries out a biopsy, I focus on the practice of solid knowing that is not moved by a frustrated and angry man. During a second observation, I focus on how Carol, a project back-office research nurse on a prostate screening study, does not know why she does certain things that allow her to be moved by the men in the study. To be clear, both of them care. Peter and Carol seem extremely polite, friendly, kind and attentive in what they do. But these elements of apparent care are not the conceptual repertoire of care I wish to mobilise in this chapter. To address the question of how to care, I attend to figurings of knowing as 'embedded in the ongoing remaking of the world' (Puig de la Bellacasa 2017: 28) and as 'fostering (…) speculative commitment to think how things could be different' (Puig de la Bellacasa 2017: 17) or not. In the following vignettes, I focus on how figurings of knowing relate to such speculative commitment and what I keep on learning from attending to the figure of the knower for the question of how to care for scholarly accounts.

This attempt at 'thinking from the field' does not mean I am trying to address an issue *de novo*. On the one hand, preoccupations about speculative commitment fit feminist contributions to ANT-oriented scholarship 'for which this mode of thought about the possible is about provoking political and ethical imagination in the present' (Puig de la Bellacasa, 2017: 7); such contributions range from provoking the 'interplay between Scientific Fact, Science Fiction, and Speculative Fabulation' (ibid.: 7) to contributions on a non-idealised notion of care (Martin, Myers and Viseu, 2015) and a commitment to the neglected, the invisible, marginalised and maintenance work (Star, 1991, 2015a; Zuiderent-Jerak, 2016; Denis, this volume). On the other hand, an increasing body of scholarship expands on feminist commitments, complicating notions such as normativity, difference and marginalisation through 'risky accounts' as fragile interventions (Latour, 2015). After analysing the vignettes, I touch upon such examples of ANT-inspired intellectual practices accentuating figurings of the knower in their work to expand the argument suggested here.

Why I do it, I don't know: prioritising speculative commitments to neglected things

During my research on prostate cancer screening practices, I thought that it might be interesting to work with the notion of care to complicate the split between the *healthcare* practices provided to men with possible prostate cancer and the *research* done on screening for prostate cancer study. In line with broadening up of what the notion of care might mean for other domains than healthcare (Eidenskog, 2015), with complicating an unproblematic (feel) goodness of the notion of care (d'Hoop, 2018; Driessen et al., 2017; Latimer, 2000; Lindén, 2016; Lotherington et al., 2017; Martin et al., 2015; Mol, 2008; Mol et al., 2010; Pols, 2004) and embracing such ambivalence of care (Pols, 2015; Puig de la Bellacasa, 2017), my focus on prostate cancer research began with an interest in affective and ethico-political dimensions in scientific work (Fox Keller, 2003). In my research I wanted to explore 'neglected things' (Puig de la Bellacasa, 2011) and 'invisible work' (Bowker and Star, 1999; Star and Strauss, 1999) in prostate cancer screening research practices and scrutinise how such practices are entangled with knowing cancer. This endeavour got complicated and affected me and my questions through the ethnographic explorations.

Let me take you along in how this happened by analysing two vignettes; and let me focus on what figurings of (not) knowing and (not) moving/being moved emerge in the first vignette describing observations of biopsies of men during magnetic resonance imaging (MRI)-guided targeted prostate biopsies.

Peter, a urologist, is fully immersed in the pictures on the screen. He tries to get a sense of what the prostate of the old man who participates in the study looks like. With the new visualising application of combining the ultrasonic pictures with previous MRI scans made, he can map a digital landscape of the prostate, while the old man is lying on the examination bench, turned with the back towards us. I am struck by the friendliness and thoroughness of Peter in his work, but also towards me. Peter is not only extremely friendly and attentive to his patients and me, but also to the cleaning lady that we have met just before on our way to the examination room; to the assisting nurse and the trainee; actually, to everyone and at all time – at least while I am around. He also stays sincerely friendly with a rather unfriendly and almost aggressive man who comes with his partner to get a biopsy done.

This man interrupts Peter's explanations of how the biopsy will be done with 'use-less stuff.' 'What do you mean? Let me explain you how it works.' Peter stands up and walks around the examination table on the side of where the man and his partner sit on two black-leather chairs. Peter has to squeeze himself behind the couple, along the window-wall, towards the left side wall with one of the various posters in the small examination room. Pointing at the poster, Peter explains the physiology of the enlarged prostate and what such a follow-up examination like the current one does. 'Useless stuff these biopsies,' the man repeats and turns away from Peter not paying attention to Peter's still friendly explanations.

I decide to leave the examination room. It is obvious that the man feels awkward and literally put in a corner, considering that there are five persons attending his biopsy in this tiny room. I hope that me leaving will take off some awkwardness from this situation.

I re-enter the room when the door opens, and the man and his partner leave. 'This patient was ... let's say not so friendly,' Peter says and sits down behind the computer on which he can access the hospital patient records and MRI files. Peter continues typing while talking. 'Although the biopsy was much less complicated in comparison to the previous ones [he has done today]; which means that the suspected area was at a more convenient location to be targeted.' The assisting nurse Martha adds:He was very unpleasant and did not understand the importance and value of the examination. It is a very awkward and unpleasant examination; for everyone. But we do our best to ease the awkwardness. He does not understand that this very understandably awkward and uncomfortable examination is only for his good; and behaved as if we are against him." Martha looks puzzled while Peter continues to focus on one of the four screens to update the hospital information system with his notes. I later come to understand that this man had already had three biopsies before where nothing had been found, although his PSA-value keeps on being high and his prostate enlarged; findings that make urologists advise to do a biopsy analysing the prostate and the tissue around it.

This all seems pretty straight forward; only the frustrated man looks like 'just not getting it'; and as I will explain, I don't either – although in a different way than the frustrated man. Peter seems to know what he does when providing care and why he does so: An invasive examination of the (surrounding) tissue of the enlarged prostate given the higher PSA-value in the blood guided by a digitally visualising technique. There may be scientific controversies around the relation between an increased PSA-value and the risk of cancer (McNaughton-Collins and Barry, 2011), but such uncertainty does not feature in the care he provides. He is focussed on doing the biopsy. I am impressed by how much thoughtful attention the screens get from Peter. Based on the images, he develops a sense of where to position the needle while minutely manipulating the images. It is important that this delicate manoeuvring on the screen gets his full attention. However, when the man questions the use of biopsies, Peter does not waver and turns all his attention toward the questioning man and explains the benefits of such biopsies. But how come that I feel rather troubled about someone being so caring?

Before exploring this question further, let me first contrast these observations of *healthcare practices* around biopsies for screening for prostate cancer with some observations on *research practices* done on screening for prostate cancer.

We stand in the office of Patricia, the principle investigator of a large prostate cancer screening study. Carol, one of the two project members responsible for entering the study data, browses the folders standing in metal office cupboards, searching for the records of the men she needs to follow up.

Carol explains to me: 'C means the control men [as opposed to the screener men].[3] So, these are all the control men that have gotten prostate cancer. Now I have all the men on this list that I want to take with me [to the hospitals where she needs to complete their research data with follow-up details]. I have them ready. After that I run them always all still through the computer for myself. [...] I go through them. To see if they are still alive. [...] Why I do it, I don't know. But I find it always important to do so. Maybe just my curiosity. [...] Look," she says, pointing towards a small photo displayed in the upper left corner on her computer screen next to other textual entries, 'he has been here at the cardiology department in October last year still. [...] This running all the men through the computer, I really do not know why I do this.'

Petra who, together with Carol, is responsible to feed the database looks up from her work from her desk opposite Carol's: 'What are you doing?'

Carol: 'I run all my participants through [the digital patient record system of the hospital]. To see if they are still alive. [...] Yes, and this man [...],' she continues pointing towards a record with a photo on her screen, 'I am curious... And now I can have a look. Just last year he was still undergoing radiation treatment with the oncologist. He has a really annoying metastasis. [...] I hope he is doing well.'

Sonja: 'How old is he?'

Carol: 'He is 74 [...].' She points to a link on the screen, 'He must have immense pain in his back, I think,' she states while she clicks from one link to another in the digital record reading the scanned letters. When screening the next man Carols says, 'This is also a whole story. He has gone through an awful lot. He has diabetes, had a cerebral haemorrhage, is 81 already and had an operation of the stomach and an anomic adrenal gland. [She sighs] And his aortic valves don't work anymore. God! Oh God! How much pain! It is great that you get so old these days, but if you see how people decay. Not sure if it is so much of a progress that you get 100 years these days. Don't you think so? Gosh, also when you see what a battery full of medicine these people swallow.

Why Carol does what she does, she does not know, but she always considers it important and insists on doing so. This 'not knowing', but still 'insisting on doing so', catches my ethnographic attention. Let's peruse what is happening here and contrast it with the first vignette: Carol does something, although she does not know why. Running the men through the hospital information system and searching for whatever bits of their stories she can trace is not necessary to do the work she *needs* to do for the screening study. She herself wonders why she does it. Wondering herself does, however, neither stop her from doing it, nor even sharing with and showing to me *how* she does it. Does it make any sense at all? Other than just doing it out of curiosity? Carol switches from the routine database entering tasks actively to searching for more details about these men – photos and bits of stories of the men – that are not necessary for the screening study; she is moved by what she finds out about the lives of the men and prioritises all that over knowing why she does it; only later this makes me realise that this could have caused ethical concerns about sharing such information with me. Emphasising mere curiosity or potential issues of privacy breaching would neglect the moving/being moved by an 'other' although not knowing why, and always still doing it. What I rather want to accentuate is the practice of prioritising motile not-knowing versus solid knowing, that is, prioritising still doing so versus knowing why *and* facing the normative ambiguity – is this empathy or mere curiosity?[4]

Carol deviates actively from what she is supposed to do. She makes time for it before continuing with what her usual tasks are. She takes the time and space to being moved by an 'other,' changes her practice and the men at the same time by details that hint at what the men go through – at least as far as she is able to trace that. Why she does so, she does not know. She also does not gain anything else than a sense for a few painful moments of the lives of these men. Her being moved is tightly interwoven with and would not be possible without the databases, the devices, codes, lists, metal cupboards, handwritten lists, digital records, etc., and all the 'invisible work' (Star and Strauss, 1999) for and with such databases. Carol's being moved, I understand later, also extends to literally driving with a suitcase full of blue folders to hospitals and running codes through a database. She moves/is moved by an 'other' *through* these very materialities and doing something that she is not asked to do, which is to research the men in her way. This moving/being moved by an 'other' changes all and everything involved irreversibly: Research practice, Carol in her doing, the men as they are known, the devices, etc., by going '"along with things to see where they lead"' (Garfinkel, 1967 – quoted in Munro, 2012: 71).

In the vignette with Peter and Martha instead, the care practices I point out seem to order the situation with the frustrated man by shaping 'clear or fixed' entities of what the issue is (high PSA values and enlarged prostate), whom or what it concerns (the frustrated man) and in which way and how to deal with it (more biopsies); imposing such care practices on the anger or rather separating them from it as someone or something that 'just does not understand.'

When frictions emerge, for example, explicit frustration, or questions, or doubts towards such care practices, Peter and Martha seem to face such pushbacks with an almost stoic and calm friendliness grounded in firmly knowing how to care. The friendliness seems to make it harder for the frustrated man of the first vignette or even for Peter and Martha to attend to neglected things that move and change all and everything involved without knowing quite yet why that would matter – but *still* doing so. Such solid knowing excludes 'useless biopsies' protected by knowing how to care. How come this made me feel so uncomfortable despite the friendly care displayed? Did I really just leave the examination room to take off the pressure for the man in this tiny room, or was there something else that moved me?

Why I left the room I don't know, but it felt important to do so. Perhaps I left because I am just less able to keep the balance Peter displayed in the face of anger. But attending to figurings of knowing and (not) moving/being moved in the aforementioned vignettes, it seems I might have left because the figure of the knower that emerged through Peter's healthcare practice made the 'how to care' question unaskable. So, although I don't know why I left the room, attending to such easily neglected unease in the analysis of scholarly accounts might avoid 'the loss which comes upon us when we overlook other possible worlds hidden or silenced in marginalized spaces   including the fun of them! (…) spaces (…) not only about pain, violence, and survival (…) [but] also about possibility' (Puig de la Bellacasa, 2015: 49). Bringing such uncomfortable moments into the analysis while navigating the delicate balance of not sacrificing 'the love for stringent analysis' for a 'touchy-feely' kind of scholarship might also help to avoid 'divorc[ing] me from my life experience, feelings, and feminist commitments' (Star, 2015b. 122). Just to clarify: I am neither interested in essentialising Peter or Carol's practice of care nor am I interested in polarising solid (not) knowing and motile (not) knowing. I rather suggest that attending to figurings of knowing and (not) moving/being moved helps to understand the difference speculative commitment to neglected things might make for the question of how to care for scholarly accounts.

The figure of the knower and the intellectual project of ANT

Attending to the figure of the knower in scholarly accounts resonates with other work within the intellectual project of ANT in different ways. This can be done through reflexively including the analyst, for example by drawing on established literary figures (i.e. dialogue), the first, third, etc., voice and presence of the narrator in a scientific contribution. A well-known example is Bruno Latour's dialogue between a professor and a PhD student in *Reassembling the Social: An Introduction to Actor-Network Theory* (2005), where, in a fictional and humorous conversation, Latour creates space for counterarguments to ANT, critique and frustration, through a PhD student who leaves the conversation wondering why he ever visited this ANT professor. Another strategy in the same volume is Bruno Latour's plea for 'writing down risky accounts' by bringing the very practice of writing scholarly accounts into the foreground (2005: 121–140). In both examples, the figure of the knower instantiated through the (writing of risky) accounts differs from Carol's practice in that it excludes moving/being moved by an 'other' through, for example, unease, disconcertment, joy or (speculative) commitments; in Carol's example, this means doing something because one considers it important to do so without knowing why.[5]

The Zimbabwe Bush Pump: Mechanics of a Fluid Technology attends in another way to the figure of the knower. Annemarie Mol and Marianne de Laet show that the success and distribution of the Bush Pump they study do not require an engineer that 'masters the situation and subtly subdues everyone and everything involved' (2000: 227). The figure of the knower that gets foregrounded is a 'serviceable inventor' moved by surprises, one that prioritises serving and listening to people before standing out as a 'solid statute' (ibid.) of knowing. In a symmetrical move, the figure of the knower instantiated through the authors themselves allows them 'to be *moved by*' (ibid. 253, italics in original) this Bush Pump. Moreover, in doing so, their article is 'an attempt to move you, reader, too' (ibid.). This seems much closer to Carol's practice of motile not-knowing.

Another example problematising the priority given to knowing versus not knowing is the article *Gino's Lesson on Humanity: Genetics, Mutual Entanglements and the Sociologist's Role* by Michel Callon and Vololona Rabeharisoa (2004). They pause at the unconventional refusal by Gino, a man suffering of limb-girdle muscular dystrophy, to comprehend the disease through their interview questions on genetics, thereby also rejecting a version of agency and subjectivity 'in which the individual is considered as an autonomous subject forced to choose between a number of pre-established options and responsible for the consequences of his choices' (ibid.: 1). While this conventionally understood 'failed' interview could have easily ended up on the graveyard of research mishaps or at best be turned into an undertheorised anecdote (Michael, 2012), Callon and Rabeharisoa pause at this 'failure' to speculatively explore the possible commitments Gino holds. Inspired by Gino's unconventional refusal to know, Callon and Rabeharisoa proceed to problematise a conventional, scholarly device to know – the interview – and argue that 'the interview situation constructs a situation that reduces the patient to silence' (2004: 1). By foregrounding the refusal to know as an insightful part of the figure of the knower, Callon and Rabehoarisoa point towards the possibility of different forms of humanity and morality.

Attending to what Helen Verran calls 'ontological disconcertment' (1999, 2001) and the practice of 'canny deferral' (2018) seems to offer both and resists disentangling (not) knowing from (not) moving/being moved by an 'other' including disconcertment into the analysis. Disconcerting moments refer to fleetingly subtle moments of immediate uncomfortableness, often physically experienced, that manifest during fieldwork. It occurs at moments when you stumble across empirical moments that do not fit your analysis because of a different ontology being invoked in that very moment.

These references show that pausing at the figure of the knower is by no means a novel idea within the intellectual project of ANT. The approaches, however, differ in how (not) knowing and (not) moving/being moved by an 'other' are entangled and whether the (not) moving/being moved extends the figure of the knower also through something like disconcertment or wonder that are fleetingly subtle and easily neglected. Expanding on such an intellectual practice, let me conclude by specifying what difference speculative commitment to neglected things can make for keeping on asking how to care for scholarly accounts.

Cultivating 'how to care' for scholarly accounts as an open question

The aforementioned vignettes offer perhaps a generative explosion of multiple ways to care for scholarly accounts. Where I was disconcerted about Peter and about the disconcertment I experienced when he was so attentive and friendly, disconcertment was far from what I experienced when Carol showed me what she did and how she did not know why she did it, while still doing it. Perhaps I was intrigued. Or rather moved by her apparent commitment to moving/being moved by an 'other' without knowing why. Perhaps that was because, while Carol was not talking about her 'moral intention' or what she 'cared about,' she was all the more able to do the 'maintenance work' (Puig de la Bellacasa, 2017: 5 drawing on Tronto, 1993) that turns her commitment into caring *for* the details of the men. Accentuating

Carol's commitment to move/being moved by the details of the men precludes a 'bifurcated consciousness' (Smith, 1987) in the analysis. In the same line, caring for unease or wonder in the analytical work prevents 'the splitting of affective involvements from the researcher's experience' (Puig de la Bellacasa, 2017: 62) in scholarly accounts and might help keeping the question of *how* to care open. This might not be possible without challenging what Susan Leigh Star calls 'The Wall of Transcendental Shame' one runs into when 'we are silently shamed – either within academia or within the swamps of convention' (Star, 2007: 229) as one exposes what one cares for in her everyday technological (research) lives (Star, 2007: 228) and what gets otherwise easily neglected in scholarly accounts.

Although there is no guarantee that this specific attentiveness to the neglected will be reciprocated (Puig de la Bellacasa, 2017: 120), this is by no means 'useless.' Such speculative commitment seems to be rather entangled with a 'thinking [of] reciprocity through a collective web of obligations' cultivating a 'multilateral circulation of agencies of care' (Puig de la Bellacasa, 2017: 120) and going along with 'neglected things' and seeing where such going along leads. It therefore rather instantiates a practice of what Puig de la Bellacasa calls 'caring obligations' (Puig de la Bellacasa, 2017): Commitment to 'motile not-knowing an "other"' in scholarly accounts seems to instantiate an 'obligation to reciprocate attentiveness to others, but one that is quite different from that of a moral contract or the enactment of norms' (Puig de la Bellacasa, 2017: 120); it also 'happens asymmetrical (…). Because people who care, caregivers, cannot give with the expectation for it to be symmetrically reciprocated' (Puig de la Bellacasa, 2017: 121). Such speculative commitment to motile not-knowing in scholarly accounts may not offer an answer in the form of solid knowing to the opening question of this chapter of how to care for our scholarly accounts. But that does not imply that speculative commitment is about 'abandoning oneself' (Gomart and Hennion, 1999) and being left to the agency given by the dispositif either. It rather accentuates the difference a commitment to the work of motile not-knowing an 'other' can make in the making of the world, embracing 'the gift of agency from the dispositif' through attending to the involvement in one's scholarly accounts.

Acknowledgements

This contribution has benefitted from the generosity and availability of the healthcare organisations and residents, healthcare researchers and practitioners who allowed me to interview them and observe them in their everyday doing. I thank all the editors of this companion, particularly Ignacio Farías, for his precious readings, comments and encouragements with writing this chapter. I would also like to thank the participants of the Berlin ANT companion workshop, particularly David Pontille and Jerôme Denis for their helpful comments to the first draft. Others who have contributed in a variety of ways are Jeannette Pols and Teun Zuiderent-Jerak. Last but not least, I would like to thank the prostate group and P6 at Linköping University for being committed to affectionate research – particularly Ericka Johnson who has been inspiring and supportive all along. The research for this paper was funded by the Swedish Research Council and part of *A Constant Torment. Discursive Contours of the Aging Prostate* (Principal Investigator: Ericka Johnson).

Notes

1 Quoted in (Puig de la Bellacasa 2017: 16).
2 Inspired by the contrast of solid and fluid (De Laet and Mol, 2000), I contrast solid and 'motile' (Latimer, 2013; Munro, 1996; 2012).

3 Carol and Petra refer to the men participating in a randomised study of screening for prostate cancer as the 'screen man' as opposed to the 'control men,' that is, the men that are either part of the
screening or control-arm of the study.
4 Interestingly, curiosity is etymological related to care; in the 14th century, it seems to mean 'careful
attention to detail' – that sense is now obsolete. See www.etymonline.com/word/curiosity (accessed 4 July 2018). Many thanks to Ignacio Farías for pointing out the possibility of such a link.
5 For a thrilling intervention drawing on literary technologies foregrounding (not) moving/being
moved, see Mann and Watts 2016.

References

Bowker, Geoffrey C and Susan Leigh Star (1999), *Sorting Things Out: Classification and Its Consequences*.
Cambridge, MA: MIT Press.
Callon, Michel and Vololona Rabeharisoa (2004), 'Gino's Lesson on Humanity: Genetics, Mutual
Entanglements and the Sociologist's Role', *Economy and Society* 33(1): 1–27.
De Laet, Marianne and Annemarie Mol (2000), 'The Zimbabwe Bush Pump: Mechanics of Fluid
Technology', *Social Studies of Science*, 30(2): 225–263.
D'Hoop, Ariane (2018), *Modest Attachments: An Inquiry into the Potentialities of Material Spaces in a
Psychiatric Day Care Centre*. PhD, University of Bruxelles/ University of Amsterdam, Bruxelles/
Amsterdam.
Driessen, Annelieke, Ilse van der Klift and Kristine Krause (2017), 'Freedom in Dementia Care? On
Becoming Better Bound to the Nursing Home', *etnofoor*, 29(1):29–41.
Eidenskog, Maria (2015), *Caring for Corporate Sustainability*. PhD, Linköping University, Linköping,
644.
Fox Keller, Evelyn (2003), *A Feeling for the Organism: The Life and Work of Barbara Mcclintock*. New York:
Henry Holt and Company.
Garfinkel, Harold (1967). *Studies in Ethnomethodology*. Engelwood Cliffs, NJ: Prentice Hall.
Gomart, Emilie and Antoine Hennion (1999), 'A Sociology of Attachment: Music Amateurs, Drug
Users', in John Law and John Hassard (eds.), *Actor Network Theory and After*, 220–247, Oxford:
Blackwell Publishers.
Latimer, Joanna (2000), *The Conduct of Care: Understanding Nursing Practices*. Oxford: Blackwell Science.
Latimer, Joanna (2013), *The Gene, the Clinic and the Family: Diagnosing Dysmorphology, Reviving Medical
Dominance*. London, UK: Routledge.
Latour, Bruno (2005), *Reassembling the Social: An Introduction to Actor-Network-Theory*. Oxford: Oxford
University Press.
Lindén, Lisa (2016). *Communicating Care: The Contradictions of HPV Vaccination Campaigns*. PhD,
Linköping University, Linköping 682.
Lotherington, Ann Therese, Aud Obstfelder and Susan Halford (2017), 'No Place for Old Women: A
Critical Inquiry into Age in Later Working Life', *Ageing and Society* 37(6):1156–78.
Mann, Anna and Laura Watts (2016), 'Letters from Wanna Wonder and the Electric Nemesis', *EASST
Review* 35(4): 67–74.
Martin, Aryn, Natasha Myers and Ana Viseu, eds. (2015), 'The Politics of Care in Technoscience',
Social Studies of Science, 459(5): 625–641.
McNaughton-Collins, Mary F. and Michael J. Barry (2011), 'One Man at a Time: Resolving the PSA
Controversy', *The New England Journal of Medicine* 365(21): 1951–1953.
Michael, Mike (2012). 'Anecdote', in Celia Lury and Nina Wakeford (eds.), *Inventive Methods: The
Happening of the Social*, 25–35, London: Routledge.
Mol, Annemarie (2008). *The Logic of Care: Health and the Problem of Patient Choice*. New York: Routledge.
Mol, Annemarie, Ingunn Moser and Jeannette Pols (2010), *Care in Practice: On Tinkering in Clinics,
Homes and Farms*, Bielefeld: transcript.
Munro, Rolland (2012), 'Agency and "Worlds" of Accounts: Erasing the Trace or Rephrasing Action?'
in Jan-Hendrik Passoth, Birgit Peuker and Michael Schillmeier (eds.), *Agency without Actors? New
Approaches to Collective Action*, 67–86, New York: Routledge.
Pols, Jeannette (2004), *Good Care: Enacting a Complex Ideal in Long-Term Psychiatry*. Amsterdam: Uni-
versity of Amsterdam.
Pols, Jeannette (2015), 'Towards an Empirical Ethics in Care: Relations with Technologies in Health
Care', *Medicine, Health Care and Philosophy*, 18(1): 81–90.

Puig de la Bellacasa, María (2011), 'Matters of Care in Technoscience: Assembling Neglected Things', *Social Studies of Science*, 41(1): 85–106.

Puig de la Bellacasa, María (2015), 'Ecological Thinking, Material Spirituality, and the Poetics of Infrastructure', in Geoffrey C. Bowker, Stefan Timmermans, Adele E. Clarke and Ellen Balka (eds.), *Boundary Objects and Beyond: Working with Leigh Star*, 47–68, Cambridge, MA: MIT Press.

Puig de la Bellacasa, María (2017), *Matters of Care: Speculative Ethics in More Than Human Worlds*. Minneapolis and London: University of Minnesota Press.

Smith, Dorothy E (1987), *The Everyday World as Problematic: A Feminist Sociology*. Boston: Northeastern University Press.

Star, Susan Leigh (1991), 'Power, Technology and the Phenomenology of Conventions: On Being Allergic to Onions', in John Law (ed.), *A Sociology of Monsters: Essays on Power, Technology and Domination*, 26–56, London: Routledge.

Star, Susan Leigh (2007), 'Interview', in Jan-Kyrre Berg Olsen and Evan Selinger (eds.), *Philosophy of Technology: 5 Questions*, 223–231, New York: Automatic Press.

Star, Susan Leigh (2015a), 'Revisiting Ecologies of Knowledge: Work and Politics in Science and Technology', in Geoffrey C. Bowker, Stefan Timmermans, Adele E. Clarke and Ellen Balka (eds.), *Boundary Objects and Beyond*, 13–46, Cambridge, MA: MIT Press.

Star, Susan Leigh (2015b), 'Living Grounded Theory: Cognitive and Emotional Forms of Pragmatism,' in Geoffrey C. Bowker, Stefan Timmermans, Adele E. Clarke and Ellen Balka (eds.), *Boundary Objects and Beyond*, 121–41, Cambridge, MA: MIT Press.

Star, Susan Leigh and Anselm Strauss (1999), 'Layers of Silence, Arenas of Voice: The Ecology of Visible and Invisible Work', *Computer Supported Cooperative Work*, 8: 9–30.

Suchman, Lucy (2007), 'Feminist STS and the Sciences of the Artificial', in *New Handbook of Science and Technology Studies*, 139–163, Cambridge, MA: MIT Press.

Verran, Helen (1999), 'Staying True to the Laughter in Nigerian Classrooms', in John Law and John Hassard (eds.), *Actor Network Theory and After*, 136–155, Oxford: Blackwell Publishers.

Verran, Helen (2001), *Science and an African Logic*. Chicago, IL: University of Chicago Press.

Verran, Helen (2018), 'Dancing with Strangers: Imagining an Originary Moment for Australian STS', Dyason Lecture, *The Australasian Association for the History, Philosophy, and Social Studies of Science*, Sydney, Australia: State Library of New South Wales.

Zuiderent-Jerak, Teun (2016), 'If Intervention Is Method, What Are We Learning?', *Engaging Science, Technology, and Society*, 2: 73–10.

19

Is ANT an artistic practice?[1]

Francis Halsall

ANT might have actually been called ART, if its original name 'Acteur-Reseau Théorie,' where Reseau means network, had remained untranslated.[2] The acronym would have been quite fortunate, for ANT resembles more an artistic practice rather than a specific theory. More specifically, I argue that this equivalence has four dimensions. Firstly, both practices are a consequence of the conditions of contemporaneity. Secondly, as a result of this consequence, both describe and exemplify those conditions which are characterised by the logic of information as it is distributed over systems of communication and control. Thirdly, these conditions are underwritten by a particular understanding of what the human subject is, that is, as one distributed and dispersed over those social systems. So, finally, ANT is as much an aesthetic as it is an epistemic practice.

In this chapter, I explore ANT as a practice, claiming that this practice both describes and manifests the conditions of contemporaneity which, for the purposes of this discussion, are identified as the conditions of late capitalism named by Castells as *The Network Society* (2000). These contemporary conditions are evident in forms of subjectivity that are distributed and *dispersed* (to use a term from the artist Seth Price (2002)) across systems of communication and control. I argue that models of art are underwritten by a particular understanding of what the human subject is. Contrary to the Humanist or Expressive model of art and humanity, the subject of the Network Society is expressed through cultural forms that are distributed and dispersed across mediums, platforms and networks. My discussion concludes with some specific case studies from contemporary art and John Law's version of ANT in *Aircraft Stories*. In all cases, a model of practice is presented in which a 'fractionally coherent subject' (Law 2002: 3) is in play and *dispersed* across the systems of communication constituting the contemporary world.

Like the broad and eclectic field of contemporary art, ANT is fluid and diverse, reflecting the different positions, sensibilities, commitments, subjects and objects produced by its practices. This equivalence relies on a model of the dispersed and distributed human subject of contemporaneity. In both fields, there is a blurring of the historical and discursive distinctions between artistic and philosophical methods. That is, both ANT and contemporary art are predicated on practices of eclecticism in choice of means, methods and mediums; promiscuity in collaborative partnerships and a decentring of existing subject and object relationships.

In other words, the strategy of both much ANT and contemporary art is to use aesthetic practices to complicate the world and reconsider its structures, relations and subjects.

The practices of ANT and art

Despite their myriad disagreements, the practitioners associated with ANT might agree that there are no fixed theoretical methods or coordinates shared by them all. ANT is a sensibility or way of looking at the world. It is a practice where no particular outcomes are guaranteed, and that its practitioners seem to acknowledge is fluid and *dispersed*.

Compare, for example, Mol:

> ANT is not a theory: there is no coherence to it. No overall scheme, no stable grid, that becomes more and more solid as it gets more and more refined. The art is rather to move – to generate, to transform, to translate. To enrich. And to betray.
>
> *(Mol 2010: 257)*

Or Law:

> The actor–network approach is not a theory. Theories usually try to explain why something happens, but actor–network theory is descriptive rather than foundational in explanatory terms […] it is a sensibility to the messy practices of relationality and materiality of the world. Along with this sensibility comes a wariness of the large-scale claims common in social theory: these usually seem too simple.
>
> *(Law 2008: 141)*

This methodological sensibility is comparable to the spirit of invention of contemporary artistic enterprise, which gives licence to work without the constraints of discipline or medium. The dispersion of ANT is very similar to the condition of contemporary art which is also characterised by an eclecticism of methods, mediums and outcomes. From the 1950s onwards, in the historical move from modernism, there was a shift in artistic sensibility from medium specificity to medium agnosticism. As a result, contemporary art can take anything as its medium and subject matter and the means of production available to practitioners are similarly multiple and open-ended. For philosopher Arthur Danto, there is no 'particular way that contemporary art has to be.' It is the 'paradigm of no paradigm' (Foster 2010: 128).

This equivalence between ANT and art prompts the question of how to judge the validity of their practices. If ANT is like contemporary art, then must the validity of its ontological and epistemological claims be evaluated in the same way? This requires, pace Habermas, recognising aesthetics as a sphere of validity that is distinct from the two other spheres of validity in modernity, namely scientific truths underwritten by empirical observation and the normative values of judicial and ethical claims. This raises a further question about the claims arising from ANT practices which will be posed but ultimately not answered here. If these claims are the result of an aesthetic practice, then they will not be verifiable in terms of other modes of philosophical, sociological, epistemological or empirical discourses. This does not mean to negate the claims that ANT might make. It does, however, require accepting that the way in which their meaning is achieved is not through conceptual or empirical verification or even by conforming to normative principles. Instead, it will be via aesthetic criteria in which an objective truth is not guaranteed and, like art, ANTs claims to validity will be regulated by judgements of taste rather than through appeal to specific and agreed concepts.

ANT and contemporaneity

The conditions of contemporaneity have been called: 'The Network Society' by Manuel Castells (2000); the 'age of the world system' by Frederic Jameson (1992) and 'The Postmodern Condition' by many others, including JF Lyotard (1984). These phrases refer to the economic, social and technological circumstances of late capitalism and the forms of knowledge and subjectivity that they produce. These circumstances, it is argued, arise from social conditions developing in the second half of the 20th century coupled with the cultural and historical influences of electronic technologies such as computing and telecommunication systems. These lead to both the subsequent dominance of information as a metaphor for communication and organisation and particular configurations of those social systems that were established in modernity including: Economy; law; education and forms of modern liberal democracy.

The post-Fordist and post-industrial practices that emerged in the aftermath of the Second World War are inseparable from a global system of the transfer of materials, services and data. Consider, for instance, the phenomena of containerisation following the invention of the modern container (1956). This standardised shipping according to a module that could be easily transferred between ships, trains and trucks. This offered greater flexibility as to where manufacturing took place, allowing site of production to be located close to cheaper resources. Containers rendered everything transferable in a global system – raw materials, products and people – and made capital migratory on a global scale. This radically changed the nature of trade in a global network that transcended nation states. Dramatic changes in labour and employment were brought about by new technologies including mechanised production and computing. In the early 1970s, the United States abandoned the gold standard which, along with the collapse of the Bretton Woods System of international economic management, presaged the subsequent market fundamentalism of neo-liberalism and its logic of deregulation and privatisation in a system of economic trading supported by the logic of speculation. It is no coincidence that this occurred simultaneously to the military–industrial–cultural complex's development of telecommunication and computing networks including the Internet (and its predecessor Arpanet) and the World Wide Web (in the early 1990s), which facilitated the rapid and massive exchange of information across global systems of communication and surveillance.

These effects of production and power create new experiences of subjectivity which, like information, is distributed and dispersed across different communicative networks and also produced and mediated by them. Hence, in the Network Society, human identity does not exist a priori to the processes of its production but rather emerges from the ecology of social, historical and material conditions within which it is positioned (such as economic transactions, communication systems and social media). In other words, subjectivity does not pre-exist processes of power and production but is instead constituted by them. It is *dispersed*.

Practices of the subject

So far, following a broadly Marxian model, my argument is that the historically contingent conditions of subjectivity in the Network Society can be mapped in cultural forms including creative practices such as art *and* ANT. That is, both practices provide ways of finding equivalences between models of practice and models of subjectivity.

This equivalence is most clearly evident in art practices. For example in the humanist model of art, exemplified in the Western European Renaissance, the artwork is understood as an expression of transhistorical values predicated on the universality of human reason. For

Erwin Panofsky, for example, the history of art was a 'Humanistic Discipline' (1940) enacted though the method of Iconology (the meticulous decoding of the original meaning of symbols in art). Thus, Panofsky's Iconology is illustrative of his own humanism, a humanism he saw as being paralleled in Renaissance values and worldview. On the one hand, this view understands works of art as being meaningful historical documents that can be meaningfully interpreted precisely because they are the product of a process of human rationality. And on the other hand, it understands human subjects as also participating in a process of human rationality that is not culturally or historically specific, hence their ability to understand the meanings of objects that are not culturally or historically specific to their own milieu.

Conversely in the Expressionist model, art is understood not as the product of shared human values but the index of a unique and individual expressive act. The primacy of self-expression is a dominant trend within artistic modernism, exemplified by German Expressionism such as the art of Die Brücke of the early 20th century. In contrast to the humanist subject who participates in some collective project, the modern subject here is the autonomous and expressive agent who is also offered the articulation of their identity in the systems of modernity including politics, law, education and the economy.[3]

The art and the human subject that exemplifies the Network Society is neither the expression of a set of universal human values, nor an autonomous instance of individual expression. Instead, both subjects and objects are radically distributed and dispersed across networks of communication and control. Furthermore, just as ANT challenges the primacy of humanity as actors in social networks, so too contemporary art is characterised by a lack of confidence in the autonomy of art as a special focus of human attention. The condition of contemporary art is the one framed by the legacy of Duchamp and the Readymade. Art is now in a condition of ontological instability that is performed through the move to eclecticism in uses of medium and techniques. Just as anything can now be art and as a result, its status as art is entirely dependent upon it being recognised as being positioned and observed within a network of relations. In ANT, I suggest, one also finds the same understanding of the human subject as a contingent one. That is, in ANT, one also finds a model of the human subject that is dispersed.

The subject of art

Right now, anything can be (and probably is) art. In the following examples, the limits of the art object are demonstrated to be porous and ambiguous. Where the specific art object is located is unclear and so too, as a result, is the specific labour of the artist or their role as subject of the work.

Consider the ongoing E-Flux project created by Anton Vidokle as a:

> publishing platform and archive, artist project, curatorial platform, and enterprise which was founded in 1998. Its news digest, events, exhibitions, schools, journal, books, and the art projects produced and/or disseminated by e-flux describe strains of critical discourse surrounding contemporary art, culture, and theory internationally. Its monthly publication e-flux journal has produced essays commissioned since 2008 about cultural, political, and structural paradigms that inform contemporary artistic production (E-flux).

E-Flux began as a means of using a relatively new means of communication – email – to disseminate information about forthcoming art events. It grew from this into a system of information exchange involving over 90,000 subscribers who receive the information about activities in the artworld (exhibitions, events, launches and so on) which is subsequently archived and made available for reference. Receiving the information is free and the agencies, such as art institutions, wanting to make an announcement pay a fee. This network has now grown to include a variety of different stake holders including public institutions as well as commercial galleries (through Art-Agenda) and educational institutions (through Art and Education, which is run in collaboration with Artforum International).

Vidokle insists that E-Flux is an art practice stating:

> Everything can be a work of art, including a company: in its totality, e-flux is a work of art that uses circulation both as form and content. I would not claim sole authorship. ... It has always been a collaboration, a series of collaborations with a very large group of people... So it's not some object you can display in a museum alongside a wall label with my name and date – it's a very different model of an artwork.
>
> *(Vidokle 2009)*

Such a practice relies on the systems of distribution throughout which it is dispersed. Likewise, Vidokle, as an artist, becomes a subject who is similarly dispersed through the same systems.

dOCUMENTA (13) (2012) was directed by Carolyn Christov-Bakargiev, as the 13th iteration of a large-scale international exhibition of contemporary art held in Kassel, Germany, every five years since 1955. Ryan Gander's work *I need some meaning I can Memorise (The Invisible Pull)* (2012) was exhibited in the ground floor of the central space, the Kunsthalle Fridericianum. It was nothing more than a soft breeze which wafted through the space. It was barely perceptible, not so strong as to be immediately obvious and artificially created, yet not weak enough to be ignored. Within this same space was a horizontal vitrine containing several pages of a handwritten letter and the title label reading: 'A letter to Carolyn Christov-Bakargiev by Kai Althoff, May 24, 2011. Exhibited on the initiative of Carolyn Christov-Bakargiev and with the permission of the artist.' This gesture was a statement of aesthetic withdrawal in which the painter explains how, after a process of deliberation, they were not able to participate in the exhibition giving, amongst the reasons, the explanation that "'life' was more important."'

Outside of this central exhibition space, in Karlsaue Park was Pierre Huyghe's *Untilled*. Huyghe claimed that the medium for his work was 'live things and inanimate things, made and not made.' It included a compost heap and the replanting of many plants, including marijuana and poisonous fruits providing an environment for: A lifelike stone sculpture of a nude woman with a live beehive for a head; an emaciated, white Podenco Canario dog with a pink leg that seemed to live at the site and all the curious human visitors to the installation.

Christov-Bakargiev explained how the whole exhibition was:

> driven by a holistic and non-logocentric vision that is shared with, and recognises, the knowledges of animate and inanimate makers of the world. The attempt is to not put human thought hierarchically above the ability of other species and things to think or produce knowledge... What these participants do, and what they exhibit, in dOCU-MENTA (13), may or may not be art... The boundary between what is art and what is not becomes less important.
>
> *(Christov-Bakargiev 2012: 31)*

My final case study is the practice of Seth Price, which takes as its primary concern the interrogation of the various mediums, platforms, networks and systems through which the work becomes distributed and dispersed. *Dispersion* (2002 – ongoing) began as an academic essay drawing on art historical and theoretical sources to discuss conditions of dispersion against the horizon of the emerging predominance of the Internet. It was initially available in printed form but has since become widely available in digital form as a PDF and via online sources. It has also formed the basis for sculptural work such as *Essay with Knots* (2008) in which the essay is represented across nine wall-mounted panels of vacuum-formed plastic incorporating knotted cord. In each iteration, *Dispersion* performs its distributed nature as it declares its distribution over the numerous platforms by which it is presented. In the essay, Price poses:

> Suppose an artist were to release the work directly into a system that depends on re-production and distribution for its sustenance, a model that encourages contamination, borrowing, stealing, and horizontal blur?
>
> *(Price 2002)*

Redistribution (2007–present) is an audiovisual recording of a typical artist's talk. How-ever, any singular meaning of the talk is undermined by the use of jarring graphics, background music, voice-overs and repetition which sometimes make it hard to follow. Furthermore, the piece remains in a process of adaptation as Price edits, adds and removes parts for different manifestations (public talks, video recordings, etc.). In *Redistribution*, the form and content of the artwork become indistinguishable from one another, that is, the means by which the work is recorded and subsequently distributed is also its primary subject matter.

In *Hostage Video Still with Time Stamp* (2005–08), Price uses images taken from the Inter-net of the head of Nicholas Berg, the American Jewish freelance radio-tower repairman who was captured and then beheaded by Islamic militants in 2004. This was a notorious example of how images are circulated through online sources, in this case via a five and a half minute video which showed Berg first addressing the camera before being decapitated. Price used a low-resolution image of Berg's head which was then printed on Mylar (a clear polyester film). David Joselit identifies three modes in which the piece is dispersed:

> First, a computer file – the germ of an artwork, as in many of Price's pieces – is rendered nearly illegible, the result of several generations of reproduction, as Price digitalizes, compresses, downloads, blows-up, and then screen-prints original footage. Second, while bolts of the printed Mylar are sometimes unrolled flush to the wall, at some point in its installation the material is twisted or tied into crumpled configurations that serve as a spatial metaphor for the ostensibly 'immaterial' traffic of images online – as though successive screen views on a monitor had piled up continuously lie a disorderly comic strip rather than being constantly 'refreshed.' Finally, third, the grisly and horrible phys-ical violation of Berg is an explicitly biological form of 'dispersion,' in which a head is parted from its torso.
>
> *(Joselit, 2011: 84–85)*

Such contemporary practices are undergirded by a particular model of the human subject as dispersed. Price has acknowledged this in discussing his work *http://organic.software* (2015), which is an online database that he had originally produced anonymously. It contained a data set of over 4,000 high-profile art collectors including images of faces, addresses, political donations, net worth and other information available online. When asked about the act of using data as art, and whether this constituted a form of Institutional Critique of the systems related to art practices, Price replied:

> The hypocrisy would come from someone who thought I am condemning a system, or individuals, while benefiting from it, and I recognize that's a risk in making something like this. But I don't think of myself as a critical voice, in doing this. *This is more like a self portrait.*
>
> *(Price 2018, emphasis added)*

The subject of ANT (*Aircraft Stories*)

One of ANT's foundational tenets is that there is no observable reality or phenomena external to the relations of elements. It thus considers social, cultural, historical and natural entities, or actors, as constituted through networks of interrelations. Law describes his own version of ANT as assuming:

> that nothing has reality or form outside the enactment of those relations… [it] thus describes the enactment of materially and discursively heterogeneous relations that produce and reshuffle all kinds of actors including objects, subjects, human beings, machines, animals, 'nature', ideas, organisations, inequalities, scale and sizes, and geographical arrangements.
>
> *(Law 2008: 141)*

In *Aircraft Stories*, Law gives an account of the British military aircraft TSR2 to think 'about modernism and its child, postmodernism – and about how we might think past the limits that these set to our ways of thinking' (Law 2002: 1).

The TSR2 (or *Tactical Strike and Reconnaissance, Mach 2*) was a military jet-powered aeroplane developed by the British Aircraft Corporation (BAC) for the British Royal Air Force (RAF) from the late 1950s to the early 1960s. It was intended for both attack and reconnaissance duties in the context of the Cold War and also, potentially, available for sale to allied and 'friendly' countries (the RAF having committed to buying). It was to be the product of the most advanced developments in aviation technology of the time and be the very best aircraft of its type in the world. In particular, it was intended to be able to fly especially low and fast in order to evade hostile radar and, this, mitigate potential enemy counter-attacks.

Yet, despite the military and technology aspirations for the plane, the TSR2 was a failure and was scrapped in 1965 when it was, as Law says, subject to crucial 'Decisions' (Chapter 7): 'behind schedule and much more expensive than anticipated, cancellation, long debated, became a real possibility with the election of a Labour government in late 1964 – a government that replaced thirteen years of Conservative administration' (Law 2002: 143).

In Law's account, the jet plane becomes both subject to the conditions of globalisation in the Age of the World System and an object through which to trace the nature of some of the conditions of that system. The TSR2 is a complex object that is situated within a variety of different networks that are particular to late capitalism. For example, it has identities as a product of the post-war military and industrial complex in the United Kingdom and all of the various systems implicated in that network including: Weapon systems (being capable of carrying a nuclear warhead); communications systems; fuel delivery systems and so on. But it was also an agent and actor in other networks including those of strategic deployment in the post-war system of global power relations and political relations, both globally in the context of the Cold War and domestic politics in the UK (such as Harold MacMillan's Tory government). It was a commodity within market systems and hence occupied a very particular position within networks of industrial and economic production. For example, as Law describes, the aircraft would be constructed through a collaboration between the companies Vickers and English Electric with Vickers building the front half, EE the rear (Law 2002: 80).

Law's account moves beyond a direct description of military technology to a more comprehensive description of the networks comprising the social system of the 'Euro-American World' after the Second World War. For Law, the TSR2 is not merely a plane but rather an object best understood as an actor positioned within, *and constituted by*, a complex set of network relationships (involving technology, culture, economics, weapons, telecommunications and so on). It is 'a fractionally coherent subject or object [...] that balances between plurality and singularity. It is more than one, but less than many' (2002: 3).

This balance 'between plurality and singularity' is nicely reflected in the chapter headings which demonstrate how the narrative is constructed using a decentred set of coordinates by which the TSR2 becomes the 'fractionally coherent subject' Law claims it to be. These chapters are:

2   Objects
3   Subjects
4   Cultures
5   Heterogeneities
6   Aesthetics
7   Decisions
8   Arborescences
9   Pinboards (2002)

But what unfolds during the text is that there is not one but actually three things being described in Law's project: The TSR2; the social, historical and economic networks from which it emerged; and, perhaps most importantly, the aesthetic conditions of Law's own method. The book, he says, is:

> written in a way that has much to so with the logic of the pinboard. Though individual chapters are not primarily visual in form, in the narrations I have often tried remake a version of this visual logic by working in a more or less broken or juxtapositionary mode. Or in some cases in a more linear manner, by describing the effects of juxtaposition as something other than a loss. And the book as a whole is no different. A set of partially connected narrations, it is also in some measure a pastiche.
>
> *(Law 2002: 192)*

For Law, the subject of his inquiry is dispersed. On the one hand, this dispersion refers to the TSR2, which is presented as a: 'fractionally coherent subject or object [...] that balances between plurality and singularity. It is more than one, but less than many' (2002: 3).

But on the other hand, this dispersion mimics the decentred human subject of the Network Society which is, just like the TSR2, distributed across a variety of systems of communication and control. As Law says of his project, in terms which leave it tellingly ambiguous as to which subject he is explicitly referring:

> So the idea of the decentered subject is scarcely new [...] But if the idea of the decentred subject is not new, then what of the decentred object? What of the object that does not hang together? Or holds together only partially?
>
> *(Law 2002: 32)*

Dispersed subjects

Both art and ANT are contemporary practices that emerge from the particular network effects of late capitalism. They not only describe but also exemplify the effects those conditions have on the subjects they create. So, the subjects constituted through the Network Society, contemporary art and ANT are distributed throughout various systems of communication and control. They are dispersed subjects.

One problem this raises has already been briefly mentioned although not tackled directly. If ANT is a practice regulated by judgements of taste, then what claim to validity does it have beyond appealing to aesthetic choices? What, for example, if it's just not very good or to our taste? The same problem stands for contemporary art, of course, and my hunch there is that a great deal of it is not very good at all either as spectacle or critique. But perhaps this is the necessary consequence of the position of art in the aesthetic milieu of the age of the world system.

More significant is the question of the role of practice in acts of critique. If, through acts of dispersion, cultural practices exemplify contemporary conditions of subjectivity, then in what sense can this exemplification be seen as anything more than mere passive reflection in lieu of critique? Or, in what ways can the practitioners of both art and ANT respond to accusation that in celebrating dispersion they are complicit in the logic of neo-liberal market fundamentalism? Art can, all too often, seem to be performing as the R&D department for late capitalism. Without lapsing into nostalgia for past forms of aesthetic identification and expression, contemporary practices face the challenge of how they can offer meaningful and effective critical engagement. Without this ability to facilitate critique, they risk appearing as mere compensatory or deflationary gestures in the face of the dispersion of subjectivity in contemporary society.

Notes

1 This paper explores themes first outlined in Halsall (2016).
2 Mol explains: 'In the early eighties, in an article in French, he [Michel Callon] was the first to speak of acteur-reseau. A short while later this term was translated and transformed to become actor-network in English' (Mol 2010: 253).
3 These two different understandings of art and the human subject may not be mutually exclusive. The act of individual expression, for example, might be related to a general human impulse shared by all, just as the participation in a universal human rationality can be understood as a means of moving from the individual to the general which is a mainstay of Kantian aesthetics. However, they do represent how different ways of understanding what a human is become manifest in different aesthetic practices that are specific to the historical and cultural contexts within which they are situated.

References

Castells, Manuel (2000) *The Rise of the Network Society*. 2nd ed. Oxford: Blackwell Publishing, 2000.

Christov-Bakargiev, Carolyn (2012) *dOCUMENTA 13*. Ostfildern: Hatje cantz Verlag.

E-flux, www.e-flux.com/about [accessed May 2018].

Foster, Hal, (2010) *Design and Crime: And Other Diatribes*. London: Verso.

Halsall, Francis, (2016) 'Actor-Network Aesthetics: The Conceptual Rhymes of Bruno Latour and Contemporary Art', *New Literary History*, Vol. 47 pp. 439–461.

Jameson, Frederic (1992) *The Geopolitical Aesthetic Cinema and Space in the World System*. Bloomington: Indiana University Press.

Joselit, David (2011) 'What to do with Pictures' *October*, no. 138, pp. 81–94.

Law J (2002) *Aircraft Stories*. Durham and London: Duke University Press

Law J (2008) 'Actor Network Theory and Material Semiotics in Turner, B. (ed.), *The New Blackwell Companion to Social Theory*. Hoboken, NJ: Wiley-Blackwell, pp. 141–158.

Lyotard, Jean-François (1984) *The Postmodern Condition: A Report on Knowledge*, trans. Bennington, G. and Massumi B. Manchester: Manchester University Press.

Mol, Anne-Marie. (2010). 'Actor Network Theory: Sensitive Terms and Enduring Tensions' *Kölner Zeitschrift für Soziologie und Sozialpsychologie*. Sonderheft vol. 50, pp. 253–269.

Panofsky, Erwin. (1940), 'The History of Art as a Humanistic Discipline' in Greene. T. (ed.), *The Meaning of the Humanities*. Princeton, NJ: Princeton University Press, pp. 89–118.

Price, Seth, (2002) 'Dispersion' in Lauren Cornell and Ed Halter (eds.) *Mass Effect*. Cambridge MA: MIT Press, (2015) pp. 51–69.

Price, Seth, (2018) 'Organic Software: An Interview with Seth Price' *Data Matters*, https://data-matters.nyc/?p=18790 [accessed May 14th 2018].

Vidokle, Anton (2009) 'Interview: Anton Vidokle of e-flux' originally published at: www.dossier-journal.com/read/interviews/interview-anton-vidokle-of-e-flux/ archived at: http://archive.is/rREiH#selection-335.0-335.34 [accessed May 14th, 2018].

20

How to stage a convergence between ANT and Southern sociologies?

Marcelo C. Rosa

The narrow scope of subjects and ontologies considered legitimate topics in the field of sociology is a trait that is not limited to national borders and aims to secure a certain uniformity, as noted by former presidents of the International Sociological Association (Archer, 1991; Sztompka, 2009, 2011). Intentionally or otherwise, sociologies from the South, though they may be pluralistic, are bound by social limitations in their quest to narrate collective lives (Latour, 2005). While hegemonic intellectual practices have acknowledged the effects of colonialism and colonisation in a perfunctory way, these remain deeply ingrained in their readings, writings and presentations of theoretical concerns, preventing reflections on the possible disconcertments provoked. Therefore, what is here referred to as 'Southern effects' do not therefore necessarily differ from the limitations of sociological knowledge and practice faced elsewhere as suggested by Law (2004) and Latour (2005). Nevertheless, as I will show, these Southern obliterations lead to specific ontological outcomes.

It is precisely when criticising the effects of this restrictive homogeneity that Actor-Network Theories (ANT) converge with several Southern social theories to challenge current trends in the discipline, as suggested by Yehia (2007). In order to strengthen these convergences, I would like to suggest understanding sociology as a form of 'internal colonialism' (Casanova, 1965) that has imposed a defined set of subjects as the only true realms of the social in the South. A central issue that both ANT and Southern social theories face is the ontological constraints of the hegemonic modern sociological approaches in assembling non-Euro-American collectives.

This chapter outlines the connections between the goals of the two set of theories, focusing on work identified with ANT by Law (2004, 2015), Law and Benschop (1997), Mol (1999) and Verran (2001, 2002) and those collected by Connell (2007) and Santos and Meneses (2009) in their anthologies of Southern theories. Following this discussion, I turn to the works of two authors not usually included in the literature on Southern theories, Viveiros de Castro (2004) and De La Cadena (2015), as examples of an effective ontoformative convergence between ANT and Southern theories. The term 'ontoformative' is borrowed from the works of Connell (2011; 2012) and describes the necessary sociological construction of expanded ontologies after what she characterises as the colonial encounter. Finally, I propose two possible outcomes for this convergence: (a) the building of Southern sociologies as an ontoformative theory-methodology within the discipline; and (b) the recognition of geo-ontological-politics in shaping contemporary sociology.

Southern sociologies

As noted in Rosa (2014, 2016), the term 'South' has no easy definition. Its use as a category to analyse social theory – and especially sociology – is, in many senses, problematic. In principle, the 'South' lacks both clear borders and a single encompassing intellectual agenda. Nonetheless, a group of works has recently invoked the 'South' to refer to a limited and strategic (and thus self-consciously 'temporary') project within debates on the prospects of the discipline for achieving a more symmetrical academic labour division (Connell, 2007; Santos and Meneses, 2009; Comaroff and Comaroff, 2011; Rosa, 2014).

Within sociology, 'the South' has not been used to describe a self-contained platform but instead a complex positionality. According to this perspective, Southern sociologists have made important contributions to a global mode of knowledge despite the daunting material conditions for research they sometimes face. In terms of theory and methods, however, the research produced and assembled in the South has been unable to expand its reach in the face of Northern scholarship. In my view, these efforts have not resulted in the reconsideration of any major sociological theory in Euro-America.

No one can deny that sociology today is understood as a mode of knowledge consecrated by the Euro-Americans and their institutions. If we accept Latour's description of this process (2005), it is a knowledge built on a hegemonic and linear account of modernity as the facilitator of our theories, research and subjects.[1] A purified modernity became the universal nomothetic narrative, while others were relegated to provincial and ideographic peripheries (Hountondji, 1997; Mafeje, 2000). This is one of the possible explanations of the non-theoretical effects of Southern research in hegemonic narratives. Since modernity and its 'derivatives' (late modernity and postmodernity) are the only metanarrative tolerated in academic debates, Southern realities work by adding components to an existing exemplary model, as in, for example, the work of Domingues (2011) on the phases of modernity in Latin America, South Asia and China.

In recognising their peripheral and non-hegemonic positions in the academic world, certain authors have addressed challenging questions. Maia (2011) has analysed how some authors use their peripheral position to advance and challenge the discipline of sociology. Research on modernity-and-coloniality (Quijano, 2002), post-coloniality (Go, 2016), subaltern studies (Chakrabarty, 1995) and indigenous sociologies (Adesina, 2002, Akiwowo, 1999) are examples of these intellectual agendas. These works have all incorporated agencies, actors, processes and histories that are conventionally classified as idiosyncrasies and backwardness in modernist sociology (Latour, 2005). In this perspective, the Southern intellectual project shares the theoretical exploration of objects often considered local eccentricities in order to challenge the simplistic employment and narrow reach of hegemonic categories. In this regard, a proactive South draws attention to geopolitical differences that cannot be absorbed by the hegemonic disciplinary agenda.

But what would be distinctively Southern in such theories? History is a central issue as noted by subaltern studies and the modernity/coloniality project in India, Latin America and Africa. The crucial turning point in most of these works is the inclusion of colonialism as a central and problematic fact of modernity as a theory (Connell, 2007). This chapter posits that connections with ANT can be found in Southern theorists' efforts to expand the agencies considered relevant to our analyses (Latour, 2005). When we look at challenges posed by the Yoruba indigenous sociologies (Akiwowo, 1999) or by Muslim influences in Asian social sciences (Alatas, 2010), we encounter narratives that introduce new ontologies into sociological debates by demonstrating their effects in the everyday lives of various 'Southerners.' Along the same vein, in a shared quest to increase the understanding and the universe of accountable agencies, a group of STS researchers (Latour, 2005; Law, 2004; Mol, 1999; Verran, 2001) have begun to think about the ontological dimension of the problem. In both sets of work, the colonial dimension of analytical scope is at the centre.

By placing geopolitical differences and inequalities in knowledge production at the forefront of their analysis, some of these sociological texts are often interpreted through the classical dualism of local (difference) versus universal (commonalities) as in the works of Keim (2008), Patel (2010), Reed (2013) and Albrow and King (1990). Even in works that aim to increase the circulation and distribution of Southern perspectives, we can observe certain scepticism with regard to the possibility of generalising and universalising Southern initiatives like the Yoruba indigenous sociologies (Akiwowo, 1986, 1999). Connell (2007) is focused on the possibility of directly applying the concepts introduced by Akiwowo (1999) to Australia and other countries, while other authors such as Adesina (2002) have instead opted to stress the instability of the ontological categories present in the Yoruba project.[2] Beyond the limits of any general concept, the methodological procedure for treating ontologies as unstable and performative should lead such local approaches to eventually challenge a major trend in the discipline, allowing the multiplicity of existences to emerge.

Division of academic labour, theoretical colonisation and internal ontological colonisation

The challenges presented in recent literature invoking the notion of the South suggest potential commonalities with at least certain aspects of ANT. As mentioned earlier, one crucial issue that both projects have addressed is the ontological constraint of the hegemonic modern sociological approaches in assembling non-Euro-American collectives. The Southern literature cited earlier has described this hegemony and challenged it geopolitically in several ways. Here, I particularly focus on two approaches: The notions of internal 'colonialism' and the 'potted plants in greenhouses' metaphor of positionality.

A vociferous critique of the limited ontological scope of merely applying sociological theories in many countries in the South can be read from the perspective of 'internal colonialism.' Casanova (1965) used this term to describe how the political and economic control of the colonies by European metropoles later morphed into internal practices of exploitation and expansionism by a group of settlers at the forefront of the independent states. The transmutation and transplantation of the modern state's hegemonic notion of governing from a distance have been replicated through administrative principles of racialising the nation and dismissing native or indigenous roles and existential perspectives. Limiting its application to the context of this chapter, we cannot overlook the fact that sociology was – and still is – a kind of national project in many countries (Alatas, 1977; Peirano, 1981).[3] A central motive for these schemes is the nation state's inability to deal with the irreconcilable internal differences that could dismantle the project of sociology itself.

In the colonised regions where sociology was established as a legitimate form of public and academic narratives, the discipline replicates the domination of certain 'cultural heterogeneous groups,' to use the 1960s term (Casanova, 1965:167). National academic discourse is therefore shaped by a racial–cultural perspective (Rodney, 1981). The limited set of agencies, subjects and relations deployed in academic analysis (such as the use of the farm and slavery/

rural waged labour as the only legitimate forms of existence) has also been fundamental to pigeonholing national contexts in the Euro-American hegemonic scopes of interpretation. This is about more than a lack of alternatives or bad faith; most of the researchers are certain that intellectually, they belong to the world of these white-Western moderns' categories.

In the African context, Nyamnjoh (2012) has referred to local academics who view Euro-American theories and methods as their only possible toolboxes as 'potted plants in greenhouses.' The greenhouse is an interesting allegory to describe the artificial modern scenario of global sociology where the production and reproduction of a globalised agenda oblige academics to ignore empirical, political and theoretical processes occurring at the local level. Somewhat ironically, for Nyamnjoh (2012), 'potted' academics would benefit from local research funds to develop their countries into the dominant and ideal Euro-American types of (supposed) backwardness and commiseration.

From my perspective, most of the sociologists of white-European descent working in the South – but others as well – have replicated the aims and desires of local liberal academic elites. They have played the role of intellectual colonisers of local existences while adopting and adapting categories that resemble those available in the theoretical agendas established within the sociological knowledge controlled by North Atlantic institutions such as religion and politics. Alatas (2000) has described this condition as the 'captive mind.'

Even critical studies aimed at understanding how Southern lives have been politically emancipated tend to seek out processes that replicate specific forms of social transformation already trialled in the academic centres (i.e. social movements, industrial sociology and sociology of labour). Important sociological works in the South are thus often dedicated to development, revolution, rationalisation, secularisation and modernisation, as if these were not merely transversal but unquestionably the most important themes in any experience worldwide.

It is not surprising, then, that many exemplary intellectual tools taken from the South emulate and rely on dualisms such as centre–periphery or modern–traditional practices. In short, peering at local research problems through 'civilized' methodological and theoretical glasses is akin to considering reality a disputed field between what is (still) traditional (processes predominately lived by non-whites and indigenous groups) and its desired and desirable modern dimension (reserved for certain former colonial elites and urban groups).

Convergence with ANT: the quest for ontological autonomy and an ontoformative agenda

The effects of hegemonic theories and methods in sociology lead me to consider the disconcertment that serious empirical research often provokes. Scholars typically connect their research findings with some established theoretical perspective. When such connections are not obvious, such processing tends to focus on the empirical material and not on the theories, which are assumed to be stable or permanent. According to Law, this is the 'general way of moving effortlessly from place to place without attending to specificities' (2004: 155), required by the Euro-American social-science tradition. In the South, it has led to a mainstream methodological trend of only researching objects that already fall within the scope of available Euro-American theories.

This trend of standardising subject matters can also be read as an ontological constraint connected to the politics of knowledge production in sociology. Under an arrangement where precision and stability are regarded as values, the inability of established categories and concepts (with their logical requirements) to adequately describe Southern and non-Western ontologies is interpreted as both empirical and (local) theoretical failures from the South. An

213

important example is the reluctance on the part of Sztompka (2009) and Connell (2007) to accept the indigenous sociologies' initiative developed by Akisola Akiwowo for the International Sociological Association.

Moving away from the North Atlantic in a more symmetrical approach would mean making an effort to establish *ontological politics* that ponder the conceptual 'incompleteness' of the South as a legitimate and qualified subject (Law, 2004; Mol, 1999). This implies, first, that Southern existences do not meet the requirements of the purified ideal Northern theoretical types; and second, that they are not stable or permanent either in space or time. Both may have the potential to challenge the work of sociologists in the North and South, although in most Southern contexts, they can be considered a specific circumstantial advantage. Emergent practices and agencies that Northern theories dismiss or avoid due to their theoretical imprecision could become a prolific field for Southern theories. My own research with the *abahlali basemapulazini* (black families living in white farmers' land) in South Africa, for example, describes a contemporary indigenous existence that cannot be captured by the dualistic modern notions of rural worker or resident in the traditional reserves created by the apartheid (Rosa, 2012). The *abahalali* are outside the theoretical borders of the hegemonic Euro-American social sciences and any research on and with them should be sociologically *ontoformative,* including, for example, ancestors and dreams.

A specifically Southern ontological politics here should be enacted throughout the confrontation between stable and unstable, completeness and incompleteness, and, certainly, centre and periphery. As observed elsewhere (Rosa, 2016), this confrontation may also create an option to give privilege to non-exemplary subjects already engaged in subverting theory-making. The *abahlali basemapulazini* do not do politics only in modern unions or social movements: Their ontopolitical gatherings may include such Euro-American entities but never only them.

The recognition and importance of the 'epistemic disconcertment' caused by emergent ontologies both comprise an interesting movement within ANT and also characterise certain of the Southern sociological claims. Law and Lin call on postcolonial STS and ANT scholars to cultivate and dwell on disconcertment as a way to narrate 'and enact differences or seemingly improbable, impossible, or unreal worlds' (2010:148). When studying land ownership disputes in Australia, Verran (2002) describes the Western and Aboriginal knowledge systems as 'disparate.' These authors conceive of either disparity or disconcertment as opportunities to continuously enlarge the disciplinary scope of sociology in a situation in which knowledge is in dispute. More specifically, such opportunities allow for the consideration of different and incommensurable existences, located in either different (geopolitical) regions or in the same nation state, thus casting welcomed doubt on the nation state as a perennial and integrative entity.

Within recent sociological debates, the question has been crafted in discussions on our subject as parts of one world (Archer, 1991), whole world (Connell, 2011) or many plural worlds (Law, 2015). Smart (1994) praises the debate as a political and moral undertaking that avoids treating the internationalisation of sociology as a legislative process or, I would add, as a means for the theoretical domestication of existences. Law (2015) takes on this debate to reflect a contemporary conjuncture where we should think of difference as a matter that exceeds the conceptions of a liberal, Euro-American mindset. According to Law, we have 'to wrestle with the implications that worlds in the plural are enacted in different and power-saturated practices' (2015:128). An important contribution from these ANT debates is that no single ontology, when commonly treated as a disciplinary prerogative in sociology, can regulate others without being performed as a theoretical power struggle.

The consideration of the production of a less colonial and imperial world or an internationalised environment for sociology inspires the ANT/Southern Sociology encounter to enact a knowledge politics that is not ontologically bounded. In my view, such an encounter brings the 'geo' of geopolitics to the already known ANT debate on ontological politics. Enacting ontological geopolitics is here understood in the way described by Latour (2016:16) in dialogue with Chakrabarty (2000) as not taking for granted the grid of the 'the older Globe of the first imperial history.' In this approach, departing from the notion of South would necessarily lead us to overcome the classic ways in which the global order was assembled and ultimately challenge its current meanings.

Based on this debate, a tentative answer to the question, 'How to stage a convergence between ANT and Southern Sociologies?' could be that a platform is created for the consideration and introduction of the notion of *geo-ontological politics*. The notion, of course, would only come into play when authors establish the South as a performative oppositional platform (Sandoval, 2000) and as an ontoformative effect on any research outside the imagined Euro-American territory.

Ontological innovations from the South

It is important to note that an ontological factor is not typically central to the Southern theories presented in the collections cited earlier. In the works of both Connell and Santos, ontological quests are only presented in the cases of African intellectual movements (Yoruba indigenous sociologies included). However, researchers from other regions dealing with local intellectual traditions have moved the ontological dimensions of their objects to the central stage of their knowledge production.

The convergence between Southern ontologies and theoretical challenges for the discipline has been found, among others, in the works of South American researchers like Viveiros de Castro (2004), who deals with the Amazonian cosmologies of South America, and De la Cadena (2015), who researches those of the Andes. Verran (2001, 2002) has done something similar in her research in West Africa and Australia. The work of these authors has translated what Connell defines as the social knowledge of colonised groups (most of them oral) into a theoretical–methodological approach. A shift of this kind partially arises from an intensive dialogue with some of the key principles of ANT, including the symmetry between academic and non-academic knowledge.

According to Viveiros de Castro, 'the question is how to configure the [indigenous] people as theoretical agents rather than as passive "subject"' (2004:4). This configuration does not occur as a simple translation of the 'native' perspectives into established 'Euro-Christian' academic notions; instead, it should subvert and reshape – or in my opinion, expand – the conceptual toolbox available in the disciplines. This is what the author calls 'controlled equivocation,' a perspective also used by De La Cadena (2015) to discuss the encounters between scholars, land struggles, shamanism and politics in the Peruvian Andes. It is an approach that privileges (ontological) difference rather than commonalities in the production of research.

In her studies of Andean indigenous land politics and nature, De La Cadena (2015) uses the term 'the comedy of equivocations' to describe the incommensurability of the mutual interpretations of Quechua activists and North American/Peruvian scholars. The activists are at the centre of the author's narrative as she describes their conscious approach to differences when dealing with the richer and more powerful scholars. The Quechua were familiar with the stereotypes the academics hold and were also aware that these stereotypes served as a scaffold for continued relations with outsiders. In keeping with Viveiros de Castro (2004),

De La Cadena is not interested in presenting the 'correct' or precise translation of the Quechua agencies of Peru, but she does take them seriously enough to challenge her own academic interpretation.

Yet, why aren't Viveiros de Castro and De La Cadena regarded as Southern theorists? Even when taking 'the [indigenous] people' as theoretical agents, neither author appears to emphasise the classic geopolitics of knowledge. The bulk of the theoretical–methodological debate they each present is influenced by established Euro-American debates (ANT among them) and both have opted to avoid any discussion on inter-academic imperialist relations in Latin America. Again, nation states may not function as self-sufficient containers, but their part in assembling these associations is hardly negligible.

In the definition used at the beginning, I argued that Southern theories should continuously challenge the disdain for Southern ontologies arising from the imperialistic application of Northern theories to non-Euro-American existences. When pointing out the differences between the worlds of scholars and indigenous worlds, Viveiros de Castro and De La Cadena do not necessarily address the sociological equivocations present in critical literature of the South. From my own perspective, most Southern scholars simply decide to apply Euro-American theories and methods as unequivocally and universally inspirational. Since we are using hegemonic conceptual tools, we tend not to problematise the equivocations. Southern sociologies, inspired by some studies in anthropology, should perhaps consider their local and empirical theoretical corpus as such when describing and reclaiming specific agencies and ontologies, including their own.

Conclusions

In my view, the ontological quest described here would be fruitful for sociologists in the South. However, it would not have the same effect on the social knowledge described by Connell (2007) as Southern theory. It would be meaningless to affirm that ANT or any other academic proposal would directly affect indigenous thinking across the world. Instead, the proposed ideas here are meant to address what is considered an ambiguous and highly colonised mode of thinking known as academic sociology.

Casanova's notion of 'internal colonialism' (1965) could be used to refer to how mainstream Southern sociologists were obliged or goaded into ignoring non-modern effects in southern political lives. As 'potted plants in greenhouses' (Nyamnjoh, 2012), some of us have been raised under the artificial but practical atmosphere of Euro-American modernity and its ontological pillars, which colonise existences across the globe. Accordingly, the resulting standard methodological procedures involve adapting and expanding consecrated ontologies to fit Southern subjects in work which has undoubtedly secured a mode of internationalism for the discipline.

Under this form of international consolidation, the ambivalence of certain Southern arguments becomes very explicit. How can we claim to be part of a discipline without having the same subjects? How can we speak of Southern theories if our theoretical scopes are the same as those in the North? Inspired by ANT and Southern literature, I have attempted to show here that our common disciplinary ground is not (modern) ontology itself, but the continuous and variable process of assembling them to create unpredictable effects.

However, it remains challenging to research and think sociologically about certain relatively autonomous subjects performed by indigenous existences. Indigenous people are present in the North, but only in some parts of the South do they represent the majority. It is no coincidence that most of the Southern works cited here aim to sociologically enact the life of indigenous subjects with the consequent controlled equivocations of an asymmetrical solidarity.

If we can find a substantial critique to this limitation within the discipline (Casanova, Connell, Quijano and Santos, among others), it is mostly in the works situated outside traditional sociological borders where native ontologies are enacted as subjects (De la Cadena, Viveiros de Castro and Verran among others). In citing such non-conventional sociologists, I aim to show just how far sociologists have to travel – all sociologists, but especially those of the South – to embrace ontoformativity as a theoretical–methodological distinction. This is, in my perspective, one possible way to claim autonomy while being part of a global discipline, and could also serve to deter our continuous production of internal colonialism.

As Verran (2002) and later Law and Lin (2017) suggest, when ANT meet the postcolonial collective and Southern theories, the path opens for a necessary methodological experiment. In Southern theories, this would lead to a shift towards ontoformativity as a method and identity. For ANT scholars, the encounter with Southern perspectives would genuinely bring the 'geo' to the table of ontological politics in order to rethink the hegemonic grid and our system of coordinates (Latour, 2016). As the aforementioned works have suggested, there are more political collectives and agencies in the South than the classic ANT academics ever imagined, provoking welcomed disconcertments in academic colonial mentalities.

Notes

1 Works like Alatas and Sinha (2017) have added new gendered, historical and geographical outlooks to sociological theory, though most of the courses, books and journals in the South continue to insist on masculinist 19th-*century* and Euro-American narratives.
2 Adesina (2002) would certainly not agree with Connell (1997: xii) that 'it is only written texts that allow the communication of complex social knowledge across planetary distances.' In Adesina's perspective, the orature of Yoruba poetry is precisely the kind of source that can break with Western tradition and the limitations on what is supposed to be known.
3 Quijano (2007), an important figure in the sociology of modernity/coloniality movement in Latin America, has highlighted the limits of the internal colonialism notion because it is limited to the Eurocentric notion of national state. While considering his admonition, it is important to note – as does Alatas (1977) – that national-state projects, despite their European roots, are still central pillars for peripheral critiques of mainstream sociological theories. I find it difficult to dissociate the current dominant sociological understanding of the nation state *from* the classic Eurocentric theoretical exemplars.

References

Adesina, J.O. (2002). Sociology and Yorùbá Studies: Epistemic Intervention or Doing Sociology in the 'Vernacular'? *African Sociological Review* 6 (1): 91–114.
Akiwowo, A. (1999). Indigenous Sociologies Extending the Scope of the Argument. *International Sociology 14*(2): 115–138.
Akiwowo, A. A. (1986). Contributions to the sociology of knowledge from an African oral poetry. *International Sociology, 1*(4), 343–358.
Alatas, S.F. (2010). Religion and Reform: Two Exemplars for Autonomous Sociology in the Non-Western Context. In Patel, S. *The ISA Handbook of Diverse Sociological Traditions*. Los Angeles: Sage, 29–39.
Alatas, S.H. (2000). 'Intellectual Imperialism. Definitions, Threats and Problems'. *Southeast Asian Journal of Social Sciences 28*(1): 23–45.
Alatas, S.H. (1977). *Intellectuals in Developing Societies*. London: Frank Cass..
Alatas, S.F. and Sinha, V. (2017). *Sociological Theory Beyond the Canon*. London: Springer.
Albrow, M., & King, E. (Eds.). (1990). *Globalization, knowledge and society: readings from international sociology*. London: Sage.

Archer, M.S. (1991). Sociology for One World: Unity and Diversity. *International Sociology* 6(2): 131–147.

Casanova, P.G. (1965). Internal Colonialism and National Development. *Studies in Comparative International Development (SCID)* 1(4): 27–37.

Chakrabarty, D. (2000). *Provincializing Europe: Post-Colonial Thought and Historical Difference*. Princeton, NJ: Princeton University Press.

Chakrabarty, D. (1995). Radical Histories and Question of Enlightenment Rationalism: Some Recent Critiques of 'Subaltern Studies'. *Economic and Political Weekly* 30(14): 751–759.

Comaroff, J. and Comaroff, J. (2011). *Theory from the South. Or, How Euro-America is Evolving Toward Africa*. London: Paradigm Publishers.

Connell, R. (2012). A iminente revolução na teoria social. *Revista Brasileira de Ciências Sociais*, 27(80): 9–20.

Connell, R. (2011). 'Sociology for the Whole World'. *International Sociology* 26(3): 288–291.

Connell, R. (2007). *Southern Theory: The Global Dynamics of Knowledge in Social Science*. Cambridge: Polity.

De la Cadena, M. (2015). *Earth beings: Ecologies of practice across Andean worlds*. Durham, Duke University Press.

Domingues, J.M. (2011). Beyond the Centre: The Third Phase of Modernity in a Globally Compared Perspective. *European Journal of Social Theory* 14(4), 517–535.

Go, J. (2013) For a Postcolonial Sociology. *Theory and Society* 42(1): 25–55.

Hountondji, P.J. (1997) *Endogenous Knowledge: Research Trails*. Dakar: Codesria.

Keim, W. (2008). Social sciences internationally: The problem of marginalisation and its consequences for the discipline of sociology. *African Sociological Review/Revue Africaine de Sociologie*, 12(2): 22–48.

Latour, B. (2016). Onus Orbis Terrarum: About a possible shift in the Definition of Sovereignty. *Millennium* 44(3): 305–320.

Latour, B. (2013) Biography of an Investigation: On a Book about Modes of Existence. *Social Studies of Science* 43(2): 287–301.

Latour, B. (2005) *Reassembling the Social: An Introduction to Actor-Network-Theory*. Oxford: Oxford University Press.

Latour, B. (1993). *We Have Never Been Modern*. Cambridge, MA: Harvard University Press.

Law, J. (2015). What's Wrong with a One-World World? *Distinktion: Scandinavian Journal of Social Theory* 16(1): 126–139.

Law, J. (2004). *After Method: Mess in Social Science Research*. New York: Routledge.

Law, J. and Benschop, R. (1997). 'Resisting Pictures: Representation, Distribution and Ontological Politics'. In Hetherington, K. and Munro, R. (eds.), *Ideas of Difference: Social Spaces and the Labour of Division, Sociological Review Monograph*, Oxford: Blackwell, 158–182.

Law, J. and Lin, W. (2010). Cultivating Disconcertment. *The Sociological Review* 58: 135–153.

Mafeje, A. (2000). Africanity: A Combative Ontology. *CODESRIA Bulletin* 1: 66–71.

Maia, J.M. (2011). 'Space, Social Theory and Peripheral Imagination: Brazilian Intellectual History and De-Colonial Debates'. *International Sociology* 26(3): 392–407.

Mol, A. (1999). 'Ontological Politics: A Word and Some Questions'. In Law, J. and Hassard, J. (eds.) *Actor Network Theory and After*. Oxford: Blackwell, 74–89.

Nyamnjoh, F. (2012). 'Potted Plants in Greenhouses': A Critical Reflection on the Resilience pf Colonial Education in Africa. *Journal of Asian and African Studies* 47(2): 129–154.

Patel, S. (2010). The Imperative and the Challenge of Diversity: Reconstructing Sociological Traditions in an Unequal World. In Burawoy, M., Chang, M.K. and Hsieh, M.F. (eds.) *Facing An Unequal World: Challenges for a Global Sociology*. Tapei, Taiwan: Institute of Sociology, Academia Sinica and International Sociological Association, 48–60.

Peirano, M.G. (1981). The Anthropology of Anthropology: The Brazilian Case. PhD Dissertation. Department of Anthropology, Harvard University.

Quijano, A. (2007). Coloniality and modernity/rationality. *Cultural studies*, 21(2-3), 168–178.

Quijano, A. (2000). 'Coloniality of Power and Eurocentrism in Latin America'. *International Sociology* 15(2): 215–232.

Reed, I. A. (2013). 'Theoretical Labors Necessary for Global Sociology: Critique of Raewyn Connell's Southern Theory'. *Political Power and Social Theory* 25: 157–71.

Rodney, W. (1981). Plantation Society in Guyana. Review (Fernand Braudel Center), 4(4), 643–666.

Rosa, M.C. (2016). Sociologies of the South and the Actor-Network Theory: Possible Convergences for an Ontoformative Sociology. *European Journal of Social Theory* 19(4), 485–502.

Rosa, M.C. (2014). Theories of the South: Limits and Perspectives of an Emergent Movement in Social Sciences. *Current Sociology 62*(6): 851–867.

Rosa, M.C. (2012). A terra e seus vários sentidos: por uma sociologia e etnologia dos moradores de fazenda na África do Sul contemporânca. *Sociedade e Estado 27*(2): 361–385.

Santos, B. and Meneses M.P. (2009). *Epistemologias do Sul*. Coimbra: Almedina/CES.

Sandoval, C. (2000). *Methodology of the Oppressed* (Vol. 18). Minneapolis: University of Minnesota Press.

Smart, B. (1994). Sociology, Globalisation and Postmodernity: Comments on the Sociology for One World Thesis. *International Sociology 9*(2), 149–159.

Sztompka, P. (2011). Another Sociological Utopia. *Contemporary Sociology 4*(4): 388–396.

Sztompka, P. (2009). 'One Sociology or Many?' In Patel, S. (ed.) *The ISA Handbook of Diverse Sociological Traditions*. Los Angeles: Sage, 21–29.

Verran, H. (2002). A Post-Colonial Moment in Science Studies: Alternative Firing Regimes of Environmental Scientists and Aboriginal Landowners. *Social Studies of Science 32*(5–6): 729–762.

Verran, H. (2001). *Science and an African Logic*. Chicago, IL: University of Chicago Press.

Viveiros de Castro, E. (2004). Perspectival Anthropology and the Method of Controlled Equivocation. *Tipití: Journal of the Society for the Anthropology of Lowland South America 2*(1), 1, 2–20.

Watson, M.C. (2011). Cosmopolitics and the Subaltern: Problematizing Latour's Idea of the Commons. *Theory, Culture and Society 28*(3): 55–79.

Yehia, E. (2007). Descolonización del conocimiento y la práctica: un encuentro dialógico entre el programa de investigación sobre modernidad /colonialidad / decolonialidad latinoamericanas y la teoría actor-red. *Tabula Rasa 6*: 85–114.

What might ANT learn from Chinese medicine about difference?

Wen-Yuan Lin

ANT-immobile?

Actor–network theory (ANT) has been used in many different ways. From technoscientific entrepreneurship through environmental conflicts and organisation studies to healthcare practices, it has mobilised empirically oriented methods to explore heterogeneous ordering, fluidities, ambivalences and multiplicities. Moving from its earlier rather monolithic terminology (Callon 1986; Latour 1988; Law 1986), it has decentred how it thinks about objects; it has shifted and itself become multiple (Mol 2002).

At the same time, however, most of the cases important to its conceptual development have continued to come from Euro-America. The consequence of this is that the Western naturalism (Descola 2013) of those cases has remained embedded in its intellectual apparatus (think, for instance, of concepts such as 'heterogeneity,' 'hybridization' and 'relationality' that aim at destabilising, while still being infused with, the naturalistic version of the empirical).

This tendency is particularly obvious in ANT-inflected work on Chinese medicine (CM). Thus, though ANT has changed as it has explored CM, it has also continued to draw on a relatively stable infra-language. In what follows, I use the term '*ANT-immobile*' as a shorthand to characterise this stability. I explore the power of this 'ANT-immobile,' and ask how far ANT has shifted and is capable of shifting what it takes to be 'the empirical,' as well as a possible non-Western alternative, by asking what ANT might learn from how CM handles difference.

World-mattering

There is a large literature on the translations, hybridisations and purifications of CM. For instance, Lei (1999) has described how the socio-material networks of *Changshan* (常山), a Chinese herb, were re-networked by laboratory science into the biomedical relations of a new antimalarial drug. Karchmer (2010) has considered how techniques of 'pattern differentiation and therapy determination' (辯證論治) have been used to translate, hybridise and purify CM in ways that simultaneously integrate it with, and distinguish it from, biomedicine.

In such studies, ANT has sometimes been used to understand CMs (I put these in the plural, for reasons discussed below), but this leaves us with an outstanding problem. This is because in such studies, CMs have been doubly domesticated. First rendered ungenerative by modern technosciences as these have translated CM into their own language, they have subsequently been further domesticated in the theoretical redescriptions created by ANT-immutable. Thus, as Karchmer (2010, 230) puts it:

> We can't understand the hybridity of contemporary Chinese medicine—the "integration of Chinese medicine and Western medicine"—without also understanding its opposing purifications—the constructions of "Chinese medicine" and "Western medicine" that doctors attempt to combine…. The dual processes of hybridization and purification are a central dynamic in the modern development of Chinese medicine….

But CMs and their relations with biomedicine do not have to be understood in this standard ANT language of hybridity and purification. Instead, ANT-immutable may be put at risk in its encounter with concepts and ontologies that are radically other. Zhan (2009) has traced the material and symbolic circulation of CM between China and the USA since Maoism. She has shown how CMs have multiplied in the processes of knowledge production, and has explored the worlding, temporalising and locating of CMs in this circulation. As a part of this, she has argued that we should examine how (medical) knowledges are produced in uneven, interactive and translocal processes that constantly push and disrupt a set of 'great divides' (including those of 'West' and 'other' inadvertently upheld by ANT-immobile).

Somewhat relatedly, Mol has explored how differences in biomedical practices are handled in the different practices of a Dutch hospital. Indeed, at first glance, the multiplicity of CMs is not very different from those of Mol's post-ANT (Ward 2012). But despite the similarities, it is the disparities in which empirical differences are *specified* and *handled* that are most important for my argument. Mol (2002, 33) writes that:

> [a]n ethnographer/praxiographer out to investigate disease never isolates these from the practices in which they are, what one may call, *enacted*. She stubbornly takes notice of the techniques that make things visible, audible, tangible, knowable.

Thus, 'the "disease" that ethnographers talk about is never alone' (Mol 2002, 31). For instance, in examining thickened intima in the pathology department, the ethnographer is not following the intima-in-itself, but rather the intima under a microscope or some equivalent of the microscope.

However, whilst it is possible to explore CMs in the same biomedical style by following observable empirical traces (Kim 2006), Mol is also suggesting that we think through the enactments of the world in the languages of the practices themselves. Studies drawing on CM resources of world-mattering such as *duixiang* (對象, 'targeted appearance'), *shi* (勢, 'propensity') or *zheng* (證, 'pattern') tell us that the clinical expressions of the body in CM practices are specified and handled in ways quite unlike those of biomedicine (Farquhar 2012; Lin 2017; Lin & Law 2018; Scheid 2014). These are studies that have used CM's own conceptual characteristics including 'correlativity,' 'the empirical as conceptual' and Daoist worlding logics to explore biomedicine–CM intersections. They have shown that CMs differ methodologically, politically and ontologically from ANT-immobile as much as they do from biomedicine. They have also reversed the direction of conceptual transfer and started to undo the naturalism of Euro-American infra-language by finding CM-inflected ways of world-mattering (Farquhar 2015; Kim 2006; Lin & Law 2014; Zhan 2011, 2014). This chapter works in the same way to ask how ANT might travel well and benefit from provincialising itself as it intersects with CM (Chakrabarty 2000).

Lost in translation

How to think about and approach this topic? One possibility comes from anthropologist Eduardo Viveiros de Castro (2012). He starts from the assumption, not far removed from ANT, that realities in different locations are different. If translation is betrayal (*traduttore, traditore*), then any translation between realities is also and necessarily a mistranslation. De Castro highlights words in common, homonyms, that might appear to bridge these realities, but instead conceal differences and betray the reality of the Other world. But what should be done about this? Answering this question, he recommends the method of *controlled equivocation* (2004). We should, he says, scrutinise homonyms, recognise that mistranslations are inevitable and decide which world or worlds we are going to betray.

This is a suggestion that has been picked up as ANT thinks about its mis/translations across empirical and ontological difference (Jensen 2014). But de Castro's work also suggests that we might ask how to hold onto the *empirical* realities of CM instead of those of ANT-immutable. How might we mistranslate the latter rather than the former? What might count as a better '*balance of betrayal*'? Latour (2005, 49) offers one solution. He calls for an empirical metaphysics that bases itself on 'infra-language':

> [a]nalysts are allowed to possess only some infra-language whose role is simply to help them become attentive to the actors' own fully developed meta-language, a reflexive account of what they are saying.

However, as Haraway warns:

> It matters what thoughts think thoughts. It matters what knowledges know knowledges. It matters what relations relate relations. It matters what worlds world worlds. It matters what stories tell stories.
>
> *(Haraway 2016, 35)*

The implication is that ANT-immutable may have flattened most of the metalanguage of Western actors and theorists but still holds on to a particular 'empirical' in its infra-language. Mol has made a similar argument, observing that the use of English-language intellectual resources betrays Dutch realities (Mol 2014), a concern that resonates with postcolonial critiques of the supposed universalism of Western conceptual and methodological apparatuses as these render the irreducibility of the Other (Cheah 2001).

The question, then drawing on Descola's (2013) specifications of various ontologies, is whether ANT infra-language is not itself a version of an abstract of naturalistic Euro-American universality. Is it not based on and being confused with 'the concrete world hegemony derived from Europe's position as center' (Escobar 2008, 167–168)? And if this is so, then instead of flattening CMs with such an infra-language, it might be better to ask how CMs' intellectual resources could be used to inflect the infra-language of ANT – or inflect it further.

In what follows, I explore this first by searching for empirical, conceptual and ontological differences hidden in ANT's translations of CM, before considering how CM attends to those differences and how this might inform the ways in which these are handled. I then consider the hidden empirical differences brought out by CM's attention to correlativity, propensity and patterning. Finally, I explore the implications of this CM-inflected ANT.[1]

Correlativity

Dr. Hsu is a popular CM doctor in a major CM hospital, holds a PhD in epidemiology and is one of the most productive CM professors in a research university in Taiwan. In one of his studies, he uses questionnaires to differentiate CM patterns, using patterns of symptoms to identify types of problems among patients suffering from various diseases. Like most contemporary CM researchers, he publishes his findings in SCI (science citation index) journals. In clinical practice, Dr. Hsu specialises in cancer treatment, and he has designed a herbal decoction, Kuan Sin Yin (KSY), to help patients survive the rigors of chemo and radiation therapy.

There are straightforward ANT reasons for exploring this combination of treatments. For instance, a material semiotic reading of his practices tells us that these are heterogeneous and exhibit specific patterns of relations. At the same time, if we look at their specificities, then we discover that this translation is also a betrayal. Thus, KSY is a herbal complex formula (複方) based on the famous Four Gentlemen Decoction (四君子湯, hereafter FGD), designed for 'supporting right [qi] (扶正[氣])' which is composed of four herbs: *ginseng* (人參), *baizhu* (白朮), *fuling* (茯苓) and *gancao* (甘草). Dr. Hsu:

> Because spleen and stomach, [which are] the earth (土) of the five phases, are the major generators of *qi* in the body after birth (脾胃屬土,主後天之氣) .... FGD ... uses the four herbs that supplement *qi* and fortify earth....

Obviously, this is not a composition derived from a biochemical analysis of the properties of herbs. Instead, says Dr. Hsu, it works by matching herbs so that they can play the roles of sovereign, minister, assistant and courier in a team created to correct biased disease propensities. So in FGD, *ginseng* is the sovereign for supplementing *qi*, while *baizhu, fuling* are, respectively, minister and assistant for fortifying the spleen, and *gancao* is the courier that harmonises the whole composition.

To consider how this differs from ANT-immutable, we need to immerse ourselves in the world of CM. We need to understand that in this world, the 'ten thousand things' (萬物) (meaning everything) is at work, including diseases, people, decoctions and the Buddha, and the *qi* of *yin-yang* circulates through the ten thousand things in the dynamics of the five phases. We also need to understand that in the circulation of *qi*, the ten thousand things are always in specific, local metamorphoses (Nappi 2009). CM medications such as herbs are classified by property and taste (性味). They are 'correlated' in a huge canon of metaphorical links with their entry into specific meridians (歸經) (Unschuld 1986).[2] And it is these links that make up *correlativity.*

Crucial to this is the complex movement of *qi* through this correlative web. Here, lines of correspondence and practical resonances can be drawn between meridians, directions, colours, climates, musical notes, emotions, tastes, sense organs and parts of the body. Importantly, given its composition with different herbs, the roles that each herb plays may be different. So, *fuling* may have the inclination to go, in different circumstances, to the kidney (water), heart (fire) and spleen and stomach (earth) meridians. When matched with the three herbs in the FGD composition, the correlative formation of *qi* will facilitate *fuling* to balance the biased inclination of spleen and stomach. To make CM work for each patient is therefore to tinker between correlative patterns of *qi* while attending to each patient's specific condition. And this is why Dr. Hsu not only modifies FGD into KSY but also adapts the treatment to the specific conditions of particular cancer patients, a point to which I return below.

So how do the descriptions of CM and ANT compare here (Lin & Law 2014)? One answer is that in insisting on the heterogeneity of actors, ANT-immutable points to something important (Kim 2006, 2969). Even so, this does not really catch the significance of the *specific* ways in which ten thousand things transform in CM. The argument is that ANT infra-language has inherited a certain naturalism in its cosmology which allows it to imagine difference with the notion of heterogeneity. However, the idea of difference in kind implied in this concept is foreign to CM. So, for instance, the earth, the spleen and stomach and *fuling* are not qualitatively different in terms of their *qi*. That idea makes no sense in CM. But neither are they homogeneous, for they are also different in their specific embodiments. As in the case of *fuling*, things are defined in their *correlativity,* such as the movement and composition of *qi*. This means that heterogeneity and homogeneity are simply not relevant as registers or realities – and that we are far from the infra-language of immutable ANT. The issue, then, is how to take empirical differences seriously whilst travelling across ontological difference with an ANT sensibility (Law & Lin 2014). To think about this, consider the following.

First, social science writ large shares the empirical legacy of the great divide between matters of fact (in scientific and experimental forms) and matters of opinion first institutionalised in the creation of experimental science in the Royal Society of London by Robert Boyle in 17th-century post-Restoration England. To put it simply, this divide later spread worldwide in the colonialisation of the following centuries.

Second, this empirical legacy goes along with assumptions about the relations between how we think on the one hand, and reality on the other. This works most straightforwardly in natural science empiricism which seeks good ways of discovering that reality. It is more complex for social science where data include what people think, i.e. meanings and cultures. Here, the 'cultural,' together with ontological and empirical difference, may become data to be explained, while 'reality' is turned into whatever affords explanatory purchase.

Third, despite the different ways of handling the relations between culture and reality, ANT-immutable holds on to particular kinds of 'empirical reality' in its infra-language, implicitly prescribing what is 'really going on' in its own terms of art, and so redescribes the metalanguage of the actors. Working in this way, ANT-immutable brings parts of the reality embedded in its terms of art into the world of the actors. The example I just gave for this operating procedure is the very notion of heterogeneity.

But what if the world we are trying to describe is not a single reality, and there are different realities in different places? Of course, ANT knows that different practices enact different realities, but what if its infra-language is affiliated to one particular naturalistic version of reality (Law 2011)? And what if we want to do without that particular reality? In short, the issue is probably less about whether to elide *correlativity* with *heterogeneity* or CMs with ANT in general, but more about *the grounds* on which ANT might work well with CMs. To put it in de Castro's terms, what we are witnessing in ANT-immutable is an *uncontrolled* equivocation that conceals 'empirical' differences that are – or at least may be – important ontologically, theoretically and methodically. This suggests the need for the CMs to speak for their correlative version of the 'empirical' and so to rearrange the tools of description.

Propensity

So how do CMs know the correlative transformability of *qi* between the ten thousand things? This leads us to a second difference between CM and ANT-immobile. This has to do with '*propensity*,' an awkward translation for the Chinese word *shi*. *Shi* specifies the disposition, inclination, movement and momentum of *qi* of the body in a particular situation.

> Dr. Hsu: Cancer patients undergoing chemo or radiation therapy suffer from deficiency in both *qi* and *yin* (陰虛). … [But modified FGD] *ginseng* was far too strong for cancer patients and usually caused dryness–heat (燥熱)… [so] it is replaced with *dangshen* at a lower dose, and I added *huanqi* as the sovereign herb for enhancing qi…. [I] added *maimendong* and *nuzhenzi* to nourish *yin* (滋陰) in the first version. But … radiation therapy damaged yin more severely and caused more heat symptoms than chemotherapy. …Nourishing yin too fast might retain the evils, … and *huoxiang* was good enough for waking up the spleen and stomach. So I designed the second version [of KSY] for patients undergoing chemotherapy.

Here, the two versions of KSY have different propensities for different types of patients, and they work to 'support … the patients' right [qi]' by initiating the propensity of enhancing *qi* and nourishing their *yin* without directly supplementing *yang* with *ginseng* which might dry out *yin*. Thus, 'the same' herbs work in different ways in different formulae, while different prescriptions transform 'the same' disease propensities. This *shi*-as-reasoning also applies to the *shi* of radiation therapy and chemotherapy: The former damages *yin* more and causes more heat than the latter.

But this describes only a small part of the propensities at work. Dr. Hsu thinks of cancer treatment as a balance between attacking evil and supporting right *qi*. This works in various ways. First, he acknowledges that biomedicine attacks evil and that most patients also use biomedicine. He thus encourages patients to use biomedicine and has developed KSY to support patients' right *qi*. Second, he thinks of cancer treatment as being like climbing a mountain. Patients need good guides and good company. So, he and colleagues organise groups to lead patients away from the evil *qi* trapping their spirits and bodies. The groups are named after KSY, since 'Kuan Sin' means to 'broaden the heart.' He also teaches patients *Lotus Sutra*, especially *The Universal Gateway of the Bodhisattva*, and Buddhist meditation, and asks them to pray to Bodhisattva. And to add to this list, he publishes Science Citation Indexed (SCI) papers to prove that CM patterning, KSY and meditation are effective. Overall, then, he works by tackling a wide range of propensities including those to do with patients' bodies, the domination of biomedicine and science, and the social and spiritual conditions of his patients.

How does this intersect with the infra-language of ANT? One possible answer is that the working of CM herbs can be explored with ANT's insights on networked *relationality* (Lei 1999), but this is a poor translation because it conceals crucial differences between CM and ANT-immutable. This is because *shi* is not a 'network of relations.' Instead, it is an expression of (local) dispositions, directions of movement, tendencies or inclinations. Dr. Hsu's interventions all depend on knowing the disposition of things, and how to exert influence over unbalanced propensities. The body of the patient, biomedical and scientific domination, and patients' social and spiritual conditions are involved, and need to be rebalanced. KSY, patients' groups, SCI papers and Buddhism – each is important.

Although ANT also attends to things that are not atomistic or individual, a sensibility to *shi* reveals that things do not arise from specific patterns of relations. Rather, they lie within the metamorphosis of the myriad things. They are part of *shi*. Thus, though the ANT notion of 'relationality' indexes part of the web of connections, it hides the crucial differences implied by propensities, differences that lie within constantly changing, correlative configurations.

Patterning

So, the issue is not only ontological difference. Rather, it is how to make productive use of the *differences hidden* by ANT's CM equivocations. Here is Joseph Needham on Chinese thought:

> Conceptions are not subsumed under one another but placed side by side in a pattern, and things influence one another not by acts of mechanical causation [so] things behave in particular ways not necessarily because of prior actions or impulsions of other things, but because their position in the ever-moving cyclical universe was such that they were endowed with intrinsic nature which made that behavior inevitable for them.
>
> *(Needham 2005, 280–281)*

The key here is pattern. To play with Mandarin, the pronunciation of 'knowledge' (zhī shì, 知識) is identical to both '知勢' (which literarily means 'know propensity') and '之勢' (which literarily means the 'propensity of' something). Indeed, Taiwanese people use *shi* to describe everything about the tendencies, situations and inevitable driving forces of daily life. This coincidence can be used, less coincidently, to say that 'to know' in CM is to know propensity or, more exactly, the propensity of something. And pattern differentiation is one of the arts CM uses to comprehend the endless differences that manifest themselves in the appearances of *shi*. For instance, patterning in CM diagnosis:

> [O]pens a range of possibilities that are variously deployed according to the conditions of the moment…[and] include the habits and the training of the doctors as well as the manifestation of illness with which he is dealing.
>
> *(Farquhar 1994, 134)*

How does this work? Dr. Hsu says that cancer patients undergoing biomedical treatment usually exhibit the pattern of '*qi* deficiency,' and he further differentiates between those receiving radiotherapy and chemotherapy. This seems to fit well into ANT's description of the multiplicity in CM practices. But as I have hinted, the nuances of CM multiple are in danger of being lost in ANT infra-language. Despite the similarity in practices, different doctors trained with different schools and working in different situations have their distinctive techniques of pattern differentiation, concerns and particular ways of combining herbs and resources as they help patients. And all of these vary according to the specific propensities at work. Without the help of the CM intellectual apparatus that differentiates these patterns, the 'empirical' specificities and differences become incomprehensible.

Thus, in clinical practices, the propensities of a specific disease caught in the art of pattern differentiation are not 'out there' disease objects.[3] But neither are they subjective interpretations by the doctors, nor simply disease representations. In the diagnosis of specific conditions, patterns lie *between* doctors and diseases. The specificities of describing patterning have to grasp the existence of *specific shi as an object of practice* by engaging in *shi*-generating practices (Farquhar 2012). By patterning manifested *shi* as objects, CM multiplies the arts of patterning by accumulating medical cases from different places and times. But this also means that CM practitioners working with *shi* do not bother with distinctions between theory and practice, or the empirical and the theoretical (Scheid 2008). The cases and their patterns are *experiential conceptualisations* that add to CM knowledge, and they do so without abstraction. Zhan (2014, 241) observes that

> [r]ather than subscribing to the naturalized scale and order that underlies deductive or inductive thinking, metaphors and analogies in traditional Chinese medicine work sideways and in the specific, requiring practitioners to think relationally, critically, and creatively while confronted with particular clinical situations.

Thus, described in ANT's terms, CM 'hybridises' without 'purifying' (Zhan 2009). Working with the empirical as conceptual, it is not only that patterns might vary from condition to condition, but that the art of patterning might also differ from school to school. There are various ways of diagnosing and a whole range of strategies of treatment. In short, and taken together, the insights of Mol and de Castro therefore suggest that CM doctors are not tackling *'the same' disease,* but that the different arts of patterning are *making 'empirical' equivocations* contingent on *specific flows of shi as 'objects' of specific diagnoses.*

What does this mean for ANT? Despite the fact that both ANT and CM explore contingent contexts of equivocation, ANT-immutable frames its cases using its own toolbox, and the logic of its intellectual apparatus remains relatively stable. Its material semiotics is not completely unfamiliar with the kinds of variability that I have just described, and it has the merit of extending the multiplicity and infra-language of empirical metaphysics. But if we attend to the alternative 'empirical' facts such as *shi* in the arts of CM patterning, we are no longer entirely in the world of material semiotics. We have moved to worlds of *shi* that are specific to the particular moments of particular diagnostic practices, such that the 'same' things and words index different ways of patterning propensities in different contexts.[4] The lesson lies in the similarities, contrasts and convergences between how ANT and CM 'apply' their methods to differentiate the endless differences in their specific intersections, i.e. *shi*.

Where is ANT?

How can ANT travel well across the differences between itself and CM? If that is the initial question, then if we follow the lead of post-ANTs and read ANT-immutable through a CM-inflected equivocation, the question begins to change. It becomes: How will ANT learn from the hidden differences between CM and itself as they encounter one another? As we have learned from CM, the answer will not be stable. Indeed, only one thing is certain. Such an ANT will not rely on a limited and presumptively transportable intellectual apparatus. Following Mol and de Castro, ANT practitioners will need to be cautious about their conceptual toolkit, and avoid drawing their models from a particular version of reality while delegating the practices of others to the periphery.

And if the balance of betrayal were to shift in this way, what might this mean for a CM-inflected ANT? One obvious suggestion is that this would need to explore how CM handles empirical differences in the manifestation of *shi* in practices. If we focus on the intellectual apparatus for detecting empirical and ontological differences (Mol 2014), then we can explore how ANT might trace world-mattering in different places where *shi* is in play.

As we have seen, Viveiros de Castro (2004, 5) advises us to allow alien concepts to deform and subvert the translator's conceptual toolbox. Similarly, I have tried to displace ANT's intellectual apparatus by working with/on the specific *shi* of contemporary Taiwanese medical practices (Lin 2013, 2017; 林文源 2014). So doing ethnography of Chinese medical practices in Taiwan while working with material semiotics and writing in English in an attempt to be faithful to ANT is also to betray ANT-immutable – in the service of another ANT, and perhaps one that is 'CM-inflected.' This is because it attends to the lost empirical and ontological specificities of *shi* (Law & Lin 2017). This resonates with Mol's (2002, 54) emphasis on the importance of situated specificity:

> The praxiographic "is" is not universal, it is local. It requires a spatial specification. In this ontological genre, a sentence that tells what atherosclerosis is, is to be supplemented with another one that reveals *where* this is the case.

Following the traces of CMs by moving from particular to particular and working within the specific practices in different locations (Zhan 2014: 240), ANT would have to open itself to the empirical mistranslations and displace itself by following local *shi* case by case and location by location. Applying the arts of patterning also suggests that ANT might be transformed, provincialised and flourished with the *shi of its own interventions* in practices.

So what does this mean? What might it mean to think and do research if ANT's conceptual toolboxes were deformed and subverted in this way? Again, the logic of *shi* suggests that any answer will also be situated and specific. CM has transformed and reformed itself time after time by differentiating the specificities into various schools of practices (Scheid 2002). But this can also be said about the differentiations and differences between ANTs' displacements. And this implies, I am afraid and as Mol (2014) has warned, that no universal infra-language can travel across all empirical specificities. The 'where' that specifies where the praxiographic conceptual toolbox is located always needs to be supplemented with a 'what' and a 'how.' More generally, this means that doing ANT research is about *how* to make situated empirical and conceptual equivocations contingent on *which* specific flows of practices become 'objects' of material semiotic diagnoses, and on what ultimate 'ground.'

And then, neither is the intellectual apparatus of *shi*-as-reasoning universal. Indeed, there are different schools of thought that find *shi* useful and deliver different patterns of 'empirical' accounts. For instance, mobilising Sun Tzu's strategic understanding of *shi* might lead to a familiar, conventionally representational research, while a Lao Tzu-informed commitment to *shi* would probably end up with normative epigrams that highlight the contexted and immanent flows and counterflows of things in the world (Law & Lin 2018).

Conclusion

Drawing on CM, this chapter has reflected on ANT-immutable's propensity to flatten empirical and ontological differences with its infra-language. Warning against the politics of assimilation, I have argued neither for segregation nor essentialisation. Instead, I have sought to make differences in how CM handles difference explicit. Describing CM resources of world-mattering such as correlativity, *shi* and patterning in practice, I have sketched how CM allows itself to be mistranslated and absorbed into the empirical realities that it encounters. But these are just a few of the implications that come from a specific ANT encounter with Taiwanese CM practices. What are the patterns that might appear along the journey of a CM-inflected ANT? What might count as 'theory' in such an alternative CM-inflected ANT? In what sense would this be more productive than other versions of ANT? And how?

Perhaps Mol's (2008, 55–56) words on doctoring are useful in this context:

> Mis/translation is not a matter of implanting ANT intellectual apparatus, but of experimenting with them....
>
> Let us experiment, experience and tinker together-practically... Shared mis/translation requires that everyone concerned should take each other's contributions seriously and at the same time attune to what bodies, machines, conceptual tools and other relevant entities are doing.

All of this is yet to be explored, for here I have simply shown how CM correlative reality is mis/translated in the implicit commitment of 'ANT- immutable' to a naturalistic version of the empirical. But these interferences between ANT and CM in these Taiwanese encounters are just one set of possibilities. There will be many more challenges on the many journeys of equivocation that lie ahead.

Notes

1 See Lin (2017) for details of the case and part of the argument.
2 Correlativity has nothing to do here with statistical correlation.
3 See Lin and Law (2018) for the implications of the art of patterning for STS.
4 Saying that *shi* is manifested in specific moment of practice does not mean that *shi* is isolated from moment to moment as it would be in 'position first' Western theoretical preconceptions (Massumi 2002).

References

Callon, Michel (1986) 'Some Element s of a Sociology of Translation: The Domestication of the Scallops and the Fishermen of St. Brieuc Bay', in J. Law (ed.), *Power, Action, and Belief: A New Sociology of Knowledge?* (London: Routledge & Kegan Paul): 196–233.

Chakrabarty, Dipesh (2000) *Provincializing Europe: Postcolonial Thought and Historical Difference* (Princeton, NJ and Oxford: Princeton University Press).

Cheah, Pheng (2001) 'Universal Areas: Asian Studies in a World in Motion', in N. Sakai & Y. Hanawa (eds.), *Traces 1: Specters of the West and the Politics of Translation*: Hong Kong: University of Hong Kong Press): 37–70.

De Castro, Eduardo V. (2004) 'Perspectival Anthropology and the Method of Controlled Equivocation (translation of Introdução ao método do perspectivismo)', *Tipiti* 2(1): 3–22.

De Castro, Eduardo V. (2012) *Cosmologies: Perspectivism in Amazonia and Elsewhere* (Manchester: HAU Network of Ethnographic Theory).

Descola, Philippe (2013). *Beyond Nature and Culture* (Chicago, IL and London: The University of Chicago Press).

Escobar, Arturo (2008) *Territories of Difference: Place, Movements, Life, Redes* (Durham, NC and London: Duke University Press).

Farquhar, Judith (1994) *Knowing Practice: The Clinical Encounter of Chinese Medicine* (Boulder, CO: Westview Press).

Farquhar, Judith (2012) 'Knowledge in Translation: Global Science, Local Things', in L. Green & S. Levine (eds.), *Medicine and the Politics of Knowledge* (Cape Town: Human Sciences Research Council Press): 153–170.

Farquhar, Judith (2015) 'Metaphysics at the Bedside', in H. Chiang (ed.), *Historical Epistemology and Making of Chinese Medicine* (Manchester: Manchester University Press): 219–236.

Haraway, Donna (2016) *Staying with the Trouble: Making Kin in the Chthulecene* (Durham, NC: Duke University Press).

Jensen, Casper Bruun (2014) 'Continuous Variations: The Conceptual and the Empirical in STS', *Science, Technology & Human Values* 39(2): 192–213.

Karchmer, Eric I. (2010) 'Chinese Medicine in Action: On the Postcoloniality of Medical Practice in China', *Medical Anthropology* 29(3): 226–252.

Kim, Jongyoung (2006) 'Beyond Paradigm: Making Transcultural Connections in a Scientific Translation of Acupuncture', *Social Science and Medicine* 62(12):2960–2972.

Latour, Bruno (1988) *Pasteurization of France* (Cambridge, MA: Harvard University Press).

Latour, Bruno (2005). *Reassembling the Social: An Introduction to Actor-Network-Theory* (New York: Oxford University Press).

Law, John (1986) 'On the Method of Long Distance Control: Vessels, Navigation and the Portuguese Route to India', in J. Law (ed.), *Power, Action and Belief: A New Sociology of Knowledge?* (London: Routledge & Kegan Paul): 234–264.

Law, John (2011) 'What's Wrong with a One-World World', *Distinktion: Journal of Social Theory* 16(1): 126–139.

Law, John & Wen-yuan Lin (2014) 'Chakrabarty's Problem', in, *Critical Sociology: Current Issues and Future Challenges: An international Conference June 23–25* (Goethe-University Frankfurt am Main).

Law, John & Wen-yuan Lin (2017) 'Provincialising STS: Postcoloniality, Symmetry and Method', *East Asian Science, Technology and Society: an International Journal* 11(2): 211–227.

Law, John & Wen-yuan Lin (2018) 'Tidescapes: Notes on a Shi(勢)-Inflected Social Science', *Journal of World Philosophies* 3(1): 1–16.

Lei, Hsiang-lin (1999) 'From Changshan to a New Anti-Malarial Drug: Re-Networking Chinese Drugs and Excluding Chinese Doctors', *Social Studies of Science* 29(3): 323–358.

Lin, Wen-yuan (2013) 'Displacement of Agency', *Science, Technology & Human Values* 38(3): 241–443.

Lin, Wen-yuan (2017) 'Shi (勢), STS and Theory: Or What Can We Learn from Chinese Medicine?', *Science, Technology & Human Values* 42(3): 405–428.

Lin, Wen-yuan & John Law (2014) 'A Correlative STS? Lessons from a Chinese Medical Practice', *Social Studies of Science* 44(6): 801–824.

Lin, Wen-yuan & John Law (2018) 'Working with Binaries: Patterning in a Postcolonial World', in http://heterogeneities.net/publications/LinLaw2018WorkingWithBinaries.pdf.

Massumi, Brian (2002) *Parables for the Virtual: Movement, Affect, Sensation* (Durham, NC: Duke University Press).

Mol, Annemarie (2002) *The Body Multiple: Ontology in Medical Practice* (Durham, NC and London: Duke University Press).

Mol, Annemarie (2008) *The Logic of Care: Health and the Problem of Patient Choice* (London: Routledge).

Mol, Annemarie (2014) 'Language Trails: 'Lekker' and Its Pleasures', *Theory, Culture & Society* 31(2–3): 93–119.

Nappi, C. S. (2009). *The Monkey and the Inkpot : Natural History and Its Transformations in Early Modern China* (Cambridge, MA: Harvard University Press).

Needham, Joseph (2005) *Science and Civilization in China, vol.2: History of Scientific Thought* (Cambridge: Cambridge University Press).

Scheid, Volker (2002) *Chinese Medicine in Contemporary China : Plurality and Synthesis* (Durham, NC: Duke University Press).

Scheid, Volker (2008) *The Mangle of Practice and the Practice of Chinese Medicine: A Case Study from Nineteenth-century China* (Durham, NC: Duke University Press): 110–128.

Scheid, Volker (2014) 'Convergent Lines of Descent: Symptoms, Patterns, Constellations, and the Emergent Interface of Systems Biology and Chinese Medicine', *East Asian Science, Technology and Society* 8(1): 107–139.

Unschuld, Paul U. (1986) *Medicine in China: A History of Pharmaceutics* (Berkeley: University of California Press).

Viveiros de Castro, Eduardo (2004) 'Perspectival Anthropology and the Method of Controlled Equivocation (translation of Introdução ao método do perspectivismo)', *Tipiti* 2(1): 3–22.

Viveiros de Castro, Eduardo (2012) *Cosmologies: Perspectivism in Amazonia and Elsewhere* (Manchester: HAU Network of Ethnographic Theory).

Ward, Trina (2012) 'Multiple Enactments of Chinese Medicine', in V. Scheid & H. MacPherson (eds.), *Integrating East Asian Medicine into Contemporary Healthcare* (Edinburgh and New York: Churchill Livingstone Elsevier): 55–74.

Zhan, Mei (2009) *Other-Worldly: Making Chinese Medicine through Transnational Frames* (Durham, NC: Duke University Press).

Zhan, Mei (2011) 'Worlding Oneness: Daoism, Heidegger, and Possibilities for Treating the Human', *Social Text* 29(4): 107–128.

Zhan, Mei (2014) 'The Empirical as Conceptual: Transdisciplinary Engagements with an "Experiential Medicine"', *Science, Technology & Human Values* 39(2): 236–263.

林文源 (2014) 看不見的行動能力: 從行動者網絡到位移理論 (台北: 中央研究院社會學研究所出版社).

Section 4

Translating ANT beyond science and technology

Celia Roberts, Anders Blok and Ignacio Farías

'Traditionally,' if such a word can be used to speak about something so young, actor–network theorists have focused on science and technology. After all, the core discipline in which ANT was born – which Bruno Latour emphatically declared 'the best discipline' at the 2004 European Association for the Study of Science and Technology (EASST) Conference in Paris – is called 'Science and Technology Studies.' However, not only are the limits of what gets to count as 'science and technology' eminently stretchable but also, ANT can arguably be used to study and analyse almost any sociomaterial arrangements. For this section, then, we invited authors whose work takes ANT into new and possibly unexpected encounters with concepts, materialities, practices and spaces that are not conventionally classified as 'science and technology.' We asked them to describe how they translate ANT in such settings, and to explore and explain what happens both to ANT and to other academic disciplines and ideas when such encounters occur.

Some of the chapters articulate scenes and practices that are closer to science and technology than others. Amade M'Charek and Irene Van Oorschot's chapter on race is probably the closest: Although their direct empirical focus is on criminology and public appeals for assistance in solving crime, the chapter explores the close links between such discourses and the science of genetics. Our question to them was simple: 'What about race?' Their answer, although clear, is far from simple. Race, they argue, is material and relational; any particular claims about racial differences – whether made in science or in criminological texts and images – inevitably fold multiple historical textual and other practices. While ANT has proved immensely useful for theorising racialising discourses, scientific and otherwise, considering race also has profound – often under-considered – significance for all STS scholarship.

Uli Beisel's chapter on public health in the developing world has a similarly close relation to science and technology – to biomedicine in particular – but her focus here is on how ANT helps us to understand the complexities and failures of antimalarial campaigns. Referring to her extensive ethnographic work in Ghana, Beisel shows how an ANT sensibility illuminates the ways in which pharmaceutical companies, philanthropists and health activists configure relations between mosquitoes, drugs and bodies that may not work in the best interests of human health. Taking ANT into the multiple and highly differentiated sites of 'global health,' she argues, facilitates new ways of seeing and acting against seemingly entrenched, life-threatening inequalities.

Both Liliana Doganova and David J. Denis, writing from the Parisian heartlands of ANT, brilliantly demonstrate how ANT concepts and methods can illuminate practices that are technical but do not relate directly to science: Namely, economic valuation, and maintenance and repair, respectively. In both cases, the authors provide clear and concise summaries of their fields, explaining precisely how ANT has contributed to new theorisations of endogenous key concepts. Doganova argues that while ANT has already made strong contributions to valuation studies, it has further to go in exploring economic devices beyond those typically associated with markets. Denis similarly suggests that ANT has much to offer to those interested in repair and maintenance of objects and systems. While ANT has a rich history of studying resilience, solidification and strength, Denis proposes, ANT-inspired studies of repair and maintenance highlight the necessity of exploring investigate fragility, breakage and diversity. How do things and systems – on scales ranging from national economies, through city transport systems to handheld devices – hold together in difficult circumstances?

Alexa Färber also takes us to the city in her exposition of the extensive and energetic debates around the use of ANT to study urban contexts. Building on the work of Bender, Blok and Farías, she introduces the notion of 'promissory assemblage,' engaging with a range of disciplines, including anthropology and literary studies to argue that ANT can provide rich insight into many key issues of urban studies, including modernisation and infrastructuring. The concept of promissory assemblage, she shows, may help scholars to theorise the promise-oriented elements of cities: Her case studies are not-yet-built, endlessly-promised, urban public transport systems. How do people live with the promise of better transport that never seems to arrive? What kinds of subjectivities of endurance are produced? And how might ANT help to theorise cities as places of both hope and failure?

Finally, Arthuro Arruda Leal Ferreira takes us to psychology and psychotherapy – perhaps two of the least comfortable spaces for many advocates of ANT. In an intriguing account of an ANT-inspired ethnographic and interview-based study of patients' and therapists' experiences of a particular therapy programme in Brazil, Ferreira explores ANT's complex, somewhat obscured but yet promising, relationship to subjectivity. This chapter tells a nuanced and many-pleated story about how ANT might – despite its desire to eschew psychological interiority – facilitate our capacity to understand what is to be a human self. Exploring a set of lesser-known ANT texts and arguments, Ferreira gives those who think that ANT has nothing to do with minds delicious reason to pause.

Together, the chapters in this section showcase just how far ANT has come since its early days. They demonstrate both the endurance and the flexibility of its key ideas and the ingenuity of its various proponents in taking these further than their originators may have thought possible. Albeit a small and eclectic set of possible examples, the chapters will, we hope, both leave readers wondering where ANT might go next and give them tools for taking it wherever they want to go.

22

What about race?

Amade M'charek and Irene van Oorschot

Something ghostly

Let us start right where we always already are: in the middle of things.

Let us start, then, with the snapshot (see Figure 22.1) of an unknown suspect, released on the 9th of January, 2015, by the Columbia SC Police Department. It shows the face of an individual whose DNA was found at a crime scene. On the basis of this DNA, Parabon Nano-Labs, a commercial company, produced a DNA-photofit: An image of what the perpetrator would look like. Striking is the suspect's face, which is presented here in different shades (dark and light) and from different angles in a mug-shot-like fashion. Below the three faces, the image details information drawn from the DNA at the crime scene, such as geographical ancestry, skin, hair and eye colour as well as the absence of freckles. Categories are explicated by bars indicating colours or a map of the world marking places of origin.

There is something *arresting* about this image: Its portrait-like qualities ask us to consider the suspect's face as a singular, identifying characteristic. However, as a collage, it details and disaggregates that face into measures of likelihood on several dimensions, showing how the 'face' is assembled out of a variety of measures. It also takes us from face to place, when a link is made between the suspect's looks and his likely ancestry, and from individual to population, when it details the suspect's likely ancestry in the categories of West African and North West European. It connects what is within the suspect – his DNA – with the surface of his body, his face.

A lot is going on here: Both on the surface of the image and *somewhere else*. Something seems to be haunting the image in the jump from individual to population, from probability to typology (and back). What is made absent here? What places and times does this image implicate – fold within itself – only to produce this immediately legible surface? *What ghost is here at work?*

We propose to think of this ghost as the spectre of *race*, and this contribution is a way to untangle the many ways race flickers between absence and presence, in and out of our historical moment. With Callon and Law, we understand that which is present to be irrevocably entangled with productive and generative absences (2004); we also understand that which is 'of the now' to be produced by intricate foldings of multiple temporalities which are never entirely

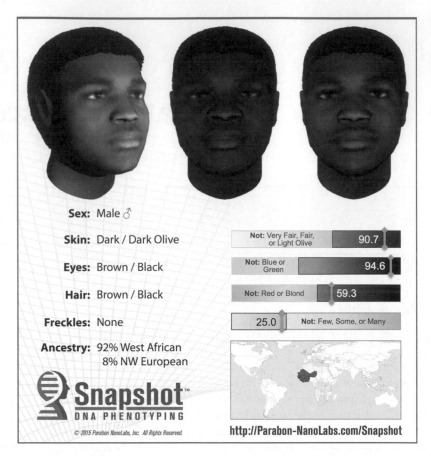

Figure 22.1   Nanolab snapshot

lost (Derrida 1993; Bhabha 1994; Latour and Serres 1995; M'charek 2014). Thinking with this image, we contend, means thinking with the continuing and troubling realities of *race*.

This is not immediately obvious, perhaps. Since the 1951 UNESCO statement on race declaring that there is no scientific basis for classifying people racially, many have taken race to be a pseudoscientific concept. While the UNESCO statement has rendered some ways of understanding race quite unpopular or obsolete, we seem, however, to be quite unable to leave it behind (Lipphardt 2012). Indeed, as human biological difference continued (and continues) to be an object of scientific concern in practices of genetics and forensics, biological differences have become subject to novel forms of articulation and configuration (Abu El-Haj 2007; Duster 2005; Fullwiley 2007; Koening et al. 2008; M'charek 2005, 2010; Nelson 2008; Whitmarsh, et al 2010; Schramm et al. 2011). In a way, race 'haunts' these practices and makes unexpected appearances.

In this snapshot, for instance, race rears its head when we consider that the face demanding our attention is much more precise about the suspect's phenotype than the genetic data seem to allow. For instance, there is no genetic test for the suspect's hair texture, nor is there a measure of the relative thickness of his nose or lips. His interpolated West African ancestry does the rest of the work, it seems: It is that titbit of information that gives shape to his lips, nose and hair.

Race is not only a continuing (if sometimes elusive!) reality, it is also a troubling reality in the sense that it troubles ready-made, binary distinctions and conceptualisations with which we apprehend, and act in, the world. On the one hand, the snapshot evokes a conception of human difference as rooted squarely in our bodies, in our DNA. It is also a conception of race that has led researchers in a variety of fields to investigate to what extent distributions in genetic material influence intelligence, aggression, risk-taking behaviour or the relative risk of certain diseases (Duster 2005). On the other hand, these conceptions have histori-cally been countered by emphasising the fictional, 'constructed' character of race. In such accounts, race has predominantly been viewed not as a thing in and of our bodies but rather as a social construction, taken more as a matter of ideology than of 'real' science. However, neither of the two conceptualisations of 'race' is adequate to grasp race in the political sense (M'charek 2013; Nelson 2008).

The emphasis on the fictional, socially constructed character of race leaves us empty-handed to apprehend and intervene in the burgeoning study of human difference, most crucially as it takes place in the practices of population genetics, biomedical research or forensic science. After all, while geneticists themselves may be quick to point out that the human genome displays more similarity than difference, precisely these genetic differences assume political salience in research and discussions regarding medical practices and healthcare (see Braun 2002) or social order and crime (Raine 2008).

Against this background, we ask: What are we to make of this picture? It is too simple to write race off as politics by scientific means – especially if that means we end up lacking a language to talk about novel configurations of race. Meanwhile, treating human difference rather as an unquestionable fact rooted in nature runs the grave risk of recuperating or rein-venting racial hierarchies – a move that overlooks how the body itself is an effect of practices of genetic 'corporealization' (Haraway 1997: 141). Contending with the political, ethical, cultural and economic impacts of the field of genetics requires a sustained engagement with the question of human difference: How such differences are made, where they become salient and consequential, and how they may be unmade.

In the following pages, we demonstrate how certain actor-network theoretically inspired moves are able to generate important diffractions in the study of 'race.' Central in this section are the distinctions between fact and fiction and nature and culture, distinctions that 'race' continually evokes as well as disavows. Here, ANT and especially feminist studies of techno-science serve as 'tools for thinking' (Stengers 2005: 185), that is, as tools to think with, and through, the spectre of race. At the same time, we discuss how 'race' requires us to question and adapt our tools, as with the emphasis on the material-semiotic 'network' often comes a largely unattended kind of *presentism,* prioritising both that which is present spatially (in the network) and temporally (as a present). Race asks us, indeed, to attend not only to presence but that which is (made) *absent,* not only to the here and now, but to the forgotten or erased. As ghosts are not simply 'of the past' nor something 'of the present' (Buse and Scott 1999), race asks us to consider *temporalities* of various sorts.

Diffracting race

If the preceding responses to race have concentrated on the question, what *is* race (fact or fiction? Nature or culture?), the primary contribution of ANT to the study of race has been to let go of such definitional exercises and instead ask: *Where and how is race done?* It calls us to attend to images like these not as a representation, but to ask instead: How does this image *do* race (M'charek 2013)?

With this performative rather than representational approach comes the possibility that we are not dealing with a singular, underlying reality, but instead with multiple realities. Mol's (2002) conception of the object multiple has helped to attend to the specificity and situatedness of objects as they are made and remade in specific practices. In other words, at stake in knowledge-making practices may not be different perspectives on essentially the same objects (e.g. disease or biological difference), but rather different ways of enacting an object. These different enactments may 'hang together' – for instance, when coordination or hierarchisation takes place – yet, we cannot assume, therefore, that these objects are essentially the same thing, or related to a singular referent.

The snapshot is a case in point. It shows 'race' to be simultaneously something legible on the surface of the body as well as a matter of genes. It shows race to be tied to place, yet at the same time treats race as a probabilistic value. We are dealing, then, with a multiplicity. These different enactments of race 'hang together' in this case because they are brought together in one image, offering an individual up for inspection.

But how was this image made? It is one thing to talk about the realities it enacts, but it is another to carefully scrutinise how it came about. This is another crucial contribution of ANT to the study of race: It cautions us against understanding scientific facts as simple representational statements. It prevents us from forgetting that facts were made in the first place.

Claes and colleagues (2014) are a prime example of such fact-making – which, as we will see, is also partially a practice of race-making. In their paper, which reports on the scientific work for the Parabon Snapshot-tool, they are interested in facial composite construction on the basis of 24 single nucleotide polymorphisms (SNPs). Focusing on these SNPs, they argue, can help to generate an individualised image of a potential suspect. Their work is based on a study of the facial features and the genes of 592 individuals, 18–40 years old, clustered in three populations: US-American, Brazilian and Cape Verdean. They report that the three populations were genetically clustered according to European and African ancestry facial forms. In this research based on the study of genetic markers and facial landmarks facilitated by 3D images, genetic ancestry is *the* pivotal operator in their face-making technology:

> We first use genomic ancestry and sex to create a base-face, which is simply an average sex and ancestry matched face. Subsequently, the effects of 24 individual SNPs1 that have been shown to have significant effects on facial variation are overlaid on the base-face forming the predicted-face in a process akin to a photomontage.
>
> *(Claes et al. 2014: 208)*

Analyses of the 3D images of 592 individuals, the authors argue, have suggested 44 principal components (landmarks) on the basis of which all facial variation can be explained. Recombining these data, the technology promises to uncover patterns that are invisible to the bare eye.

Moving from the surface to the molecular, population genetics is called upon to produce a '*base-face*.' In theory, this would be the average face compiled on the basis of 591 individuals. But the 24 SNPs aimed at individualising face are statistically accurate only when analysed within homogeneous populations. The individual can't quite do without a population (M'charek 2000)! Hence the African and European ancestry populations. This means that the 'base-face' is the base-face of the population to which the specific subject is said to belong (African or European). In accordance with the genetic clustering of the 592 individuals based on ancestry markers, their faces are allocated to one of these populations. The 44 landmarks can then be used to correlate and determine the base-face of one of these populations.

Of course, this is an operation not dissimilar to the making of racial types. The base-face thus becomes a racial type.

Specific ways of relating DNA and face are here at work. These ways of relating DNA and face can only occur in practices mobilising statistics, advanced genetic testing procedures, digital visualising software – and, we note, practices mobilising a conception of internally homogeneous yet distinct populations. Thus, race becomes a device that mediates between DNA and face. In that capacity, it is also a relational object: Not only made *through* relations, but active in making relations *anew*, too. Indeed, the mobilisation of population genetics in the production of an individualised face means that there might be *other* ways to make individual faces; ways, perhaps, that instead of mobilising population types involve scalar measures (a possibility the authors themselves mention).

Materiality, relationality and multiplicity are key concepts in ANT-inspired approaches and mobilised in this context which teach us important lessons with regard to race. Characterised by a 'pragmatic respecification' (van Oorschot 2018) of the definitional question, 'what is race?' this approach is more interested in the question: 'how and where is race done?' This gesture is indebted in many ways to feminist studies of technoscience, and in particular the non-dualist thinking of Donna Haraway, whose thought, especially her figure of the cyborg (1991), continually resists the logic of either/or that structures our habits of thought (see Johnson, this volume). As we will see, with this pragmatic sensibility comes a different way to attend to the reality of race, and the ways it is 'natural' or 'cultural.'

Instead of defining ANT, the previous discussion has instead aimed to demonstrate its uses in the case of race. More than a theory in any substantive sense, it is first and foremost a pragmatic sensibility (see also Alcadipani and Hassard 2010; Law 1991). ANT is better apprehended as a lever to set things in motion, a tool to be 'passed from hand to hand,' than as a theory or fixed collections of instruments (Mol 2010). However, as we will see, race itself sets ANT in motion, too, and particularly the emphasis, in ANT, on *present doings* and an attendant relative neglect of *absent* or *past* doings. Indeed, paying close attention to race means running up against certain tendencies in the now canonical ANT studies of laboratories, clinics and other sites of knowledge-making. These tendencies are *presentist* in that the emphasis on material-semiotic network has often led analysts to concentrate on that which is present rather than absent, and secondly, presentist in the sense that they concentrate on that which is going on 'in the present' at the cost of developing an understanding of networks or objects as they make, mobilise or fold multiple histories and temporalities. In a way, ANT has not itself been impervious to a metaphysics of presence (Heidegger 1962; Law 2006), prioritising the temporal present and what is present over the forgotten, illusive, erased or suppressed (Law and Singleton 2004). Both forms of presentism, we will demonstrate, handicap our thinking about and thinking with the spectre of race.

Going underground and surfacing again: race as an absent presence

Scientific practices are a matter of making relations between people, things, ideas and the recalcitrances of objects of study. The network, the oft-evoked metaphor to understand these imbrications, has been a crucial concept to attend to precisely this relationality, and has been crucial to uncover this sometimes fleeting and messy relational realities from narratives that would 'black-box' the network.

Reading our image in relation, as it were, to relationality, we would be able to attend to the practices taking place in between a crime scene and the Snapshot. We would be able to ask questions about the gathering of evidence and the way legal standards of evidence collection, the distribution of tasks over officials in the chain of custody, the practices of forensic geneticists, the design of the genetic testing and the design of the software used in generating the snapshot. We could trace this image further, trying to follow it around as it mobilises interest and is taken up in practices of identification, investigation and legal truth–telling.

Indeed, a second snapshot (see Figure 22.2) based on the same genetic material, this time produced in 2017, is more suggestive of its implication in forensics and, by extension, in the legal system. The individual, we learn here, has now become part of a 'case.' The snapshot

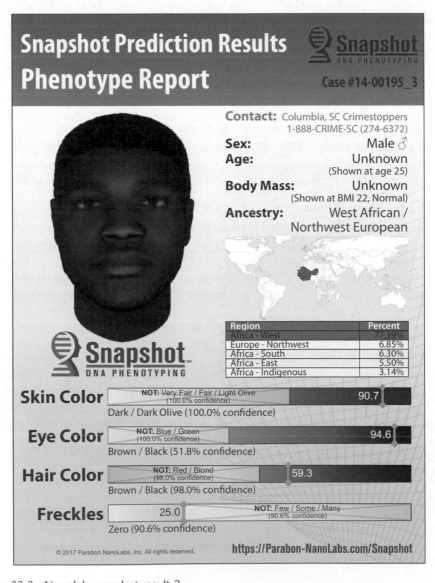

Figure 22.2 Nanolab snapshot result 2

displays a case-number and a phone number to call with the police. This image, more so than the first, folds the individual into a circuitry of both scientific and forensic forms of truth-making; perhaps, it exists at their nexus, at the place the two circuits cross each other.

The image's implication in forensics draws our attention to a specific way of asking questions about our research materials that can teach us not only about race, but also about the presentist limitations of ANT. In forensics, the task is to use available materials as traces and clues to that which for now tends to escape our observations (see also van Oorschot 2018); it is particularly interested in following trajectories and uncovering paths. But forensics is also a practice of making relations, of drawing lines and connections even where these seem to be cut off. Following these traces and clues may take us to surprising places and actors that may not be immediately present, but do play a role in the production of this image. Actants shimmer in and out of existence, are themselves 'pattern[s] of presences and absences' (Law and Singleton 2004: 12) or go underground as soon as we try to grasp them, only to appear just in the periphery of our vision (see also M'charek et al. 2014).

In our image, 'race' is shimmering just like this. In tying face to place, the image is on the one hand suggestive of race as *type*. Locating this suspect's face in 'West Africa,' evoked is a world in which different populations occupy different, circumscribed spaces in this world. On the other hand, this interpretation is belied by the probabilistic measures detailing the suspects' ancestry, which falls apart into '92% West African' and '8% North West European,' suggesting that at stake is not typological 'race' but a probabilistic categorisation based on in *ancestry* populations. The tension is evident: 'race' flickers in and out of existence in the oscillation between these different ways to 'read' the image: as type and as ancestry. In this sense, 'race' acquires characteristics of an absent-presence: It becomes present at the expense of things that are made absent. Of course, the problem is not that things are rendered absent, but rather that this proceeds without further reflection (Law 2004) on what is being made absent and to what effect. Think, as an example, of something that is rendered unimportant and absent, for instance, of our own modes of seeing and witnessing. Would we have readily 'seen' race if the unknown suspect had been white? Or is our way of seeing 'race' part and parcel of sedimented infrastructures of knowledge? The suspect's blackness alerts us to race in ways that suspend questions about the ways our very modes of seeing and glancing are racialised. However, the suspect's race does not exist in and of itself. It is relational and its presence depends on many things that are made absent. Race, always in flight, forces us to ask precisely the question: What about race?

Folds of time

Studies within ANT have developed various ways to attend to the question of time, sensitising us to how scientific practices both take time and make time. The notion of 'projectness' (Law and Singleton 2000), for instance, alerts us to the way the relations between humans and nonhumans can be given a definite shape and a pre-established duration; how, in other words, the question of chronology and phasing is crucial to the way scientific projects are narrated and imagined (see also Law 2002). We also encounter temporalities as a salient ingredient of what we could call scientific self-descriptions: Scientific practices that like to distance themselves from traditions and are quite comfortable siding with 'modernity' in a narrative of historical rupture; or else, we might encounter a progressive teleology in which scientific advances are made standing on the 'shoulders of giants.' The fact that this is only one way to conceive of the way scientific practices relate to time is asserted, of course, in Latour's *We Have Never Been Modern* (1991), in which he makes the case that modernity, and the narrative of scientific progress, is a poor way to attend to the actual practices of us, Moderns.

The study of human difference is similarly a field in which narratives of scientific progress can be encountered, while being haunted by histories of eugenics and colonialism. A narrative of historical discontinuity is one way to make race absent: Insisting that at stake in genetics is not 'race' but population, for instance, is a way to enact a break between the scientific now and the pseudoscientific, racist past. These narratives resonate with appeals to a post-racial present, in which we have moved beyond race… However, these pasts continue to *haunt* (Derrida 1993) the present. Colonial and imperialist histories are evoked, for instance, as we are dealing with a Snapshot from a suspect in the US, whose geographic, genetic origins likely lie in Western Africa – a fact that obliquely addresses the era of transatlantic slavery and its consequences. More insidiously, yet another history is evoked in the 'interpretative jump' from measures of hair, eye and skin colour to the phenotype presented in the image: A history of phenotypical classifications not wholly unrelated to 19th-century scientific practices and imperialism. A *topological* (rather than solely linear and chronological) understanding of time allows us to understand the absent-presentness of 'race' as an effect of temporal cuts and folds (M'charek 2014).

A particularly striking demonstration of the relationship between scientific practices and racial histories is the so-called Anderson sequence (M'charek, 2005, 2014). The Anderson sequence was up until the late 1990s *the* sequence of mitochondrial DNA, used widely in genetic research as a standard reference against which other strands of mitochondrial DNA could be compared. These comparisons made it possible to measure differences between populations (in their divergence from the Anderson sequence), or to correct technical errors in the sequencing process. Importantly, it was also used to measure genetic variations in mitochondrial DNA so as to trace genetic lineage (M'charek 2005). For quite a time, the Anderson sequence was largely an unproblematised and unmarked, ready-to-hand tool in these genetic research practices. However, a closer look at its history suggests that racial histories are folded into this object in intricate yet revealing ways. First of all, the Anderson sequence was and is not, if we may use the problematic phrase, 'naturally occurring.' Instead, it is a collage of mtDNA drawn from three differences sources: Bovine DNA, placental tissue and the HeLa cell line. And this cell line, used especially in cancer research, was named after the woman whose cancerous cells were taken, in 1952, to be used in medical research: Henrietta Lacks (or, as she was known by a pseudonym in the medical community: Helen Lane).

Henrietta Lacks, Landecker (2000) shows, was a black woman in the early 20th-century US. Her life as well as her encounters with a white medical establishment were significantly shaped by her blackness: This was a time in which doctors would routinely neglect to inform patients, especially black patients, of the exact nature of their disease as well as their proposed treatment. It was also a time in which medical tissues were taken for research without informed consent. For decades after her death in 1952, it was that tissue that would significantly advance cancer research and acquire a life of its own: The HeLa cell line proved to agree with laboratory conditions and would replicate itself quickly, and offered researchers the opportunity to experiment almost limitlessly with cancerous cells. Again, however, race reared its head once when it turned out that the HeLa cell line had been contaminating a wide variety of labs worldwide, a realisation that led to an uneasy discourse linking these cervical cells with racially (and sexually) charged notions of promiscuity and pollution (Landecker 2000; M'charek 2005). The HeLa cell line and the Anderson sequence with it, then, is a *folded object*: While its historicity may not be written on its surface, it is nevertheless folded into it. 'History can be recalled in objects. History is never left behind' (M'charek 2014: 31).

Objects, then, can be conceived of as folds. Take another example of the relation between race and temporalities. Contrast our first snapshot with the second one presented earlier.

While the first snapshot was made in 2015, the second one was produced more recently, in 2017. Aside from the way the image has been now more visibly implicated into a forensic circuitry, we see a few additional changes. While the first image spoke of a suspect with 92% West African and 8% North West European 'roots,' the second image disaggregates the suspect's DNA in more detail. The second image tells us that the suspect still shares most DNA with people from West Africa, but he has also ancestry in other parts of Africa. It now reads: West Africa, 77.39; Europe, 6.85; South Africa, 6.30; East Africa 5.5; Indigenous Africa, 3.14. To be sure, the DNA of the suspect did not change over time. What did change was the number of entries in the database to which the DNA of the suspect was compared a few years later. The larger the collection of DNA samples, the larger the diversity in ancestry produced – which reminds us of the fact that genetic difference is not *in* the DNA but *of* the DNA.

Curiously, we see that the suspect's face, which is now based upon more precise and disaggregated measures of ancestry, has not changed one bit. Indeed, by contrasting the more individualising specificities with the collectivising-type features, the face gains in reliability. Individuality moves front-stage, while race fades away, to become a matter of fact of diversity. Contrasting the first and the second images allows us to appreciate the image as a folded object, in which time is an operator rather than a parameter or a chronological line on which to allocate events (Rheinberger 1997). The second image works precisely as a folded object as it superimposes two moments in time, two knowledge practices that are as far apart as two years. Indeed, denying the passage of a certain kind of time is also, of course, a way of doing and evoking time.

The notion of the folded object offers us a glimpse of what it could mean to treat practices as involving the partial silencing and partial mobilisation of histories, within which not only human beings, but objects too are complex folds of time. Presents, it turns out, are always made by intricate foldings of different pasts (Serres and Latour 1995). As such, the incorporation of such a sensitivity to temporality and history also provides a way into the question of politics, itself the articulation of possible futures. In particular in the case of politically sensitive topics – race, sex, belonging and the nation – an emphasis on the way histories are implicated in their making offers us a way out of presentisms or teleologies that are characteristic of scientific practices and the erasure of racial histories.

Ghostly ANTics

If race is productively refracted using the combined ANT sensibilities of relationality, multiplicity and relationality, race itself productively addresses a largely implicit kind of presentism within actor-network theory. Crucially, race demonstrates that objects or networks are not simply spatial entities or accomplishments, but may flicker back and forth between presence and absence, and may be themselves temporally folded. Thinking with the object of race asks us to (re)consider other sites, networks and objects within these non-presentist terms. Can we have an eye for different and multiple temporalities as these are folded within ostensibly black-boxed, 'ready-to-hand' objects? Can we allow ourselves to trace not simply what is made present in networks, but those objects that contribute to the making of networks in more ambiguous ways?

And what would happen to our understanding of other concepts of the critical tradition using the emphasis on absences and time? Is it possible to conceive not only of race, but also of sexuality, gender or even class in such terms? Thinking along with these suggestions, we can not only consider, for instance, for the way sexuality is a matter of both cultural and scientific modes of ordering and comparing bodies and objects (chromosomes, brain structures, hormones, etc.), but also for the way these modes of enacting sexuality evoke and disavow multiple temporalities, such as the progressive temporality of child–adolescent–adult as much as a transhistorical temporality of sexual essentialism, i.e. the eternal feminine… Or think of the way class is not only a matter of bodies plugged more or less securely in certain circuitries of value (money, esteem, etc.), but also something that is selectively made invisible or suppressed in networks ostensibly relying on performance. Thus, it may become possible, too, to refract the matter of intersectionality, for can we not understand intersectional 'identities' as congealing out of highly situated patternings of absences and presences (M'charek 2010)? Thinking with the spectre of race alerts us to the ways scientific practices are crucial to the doing of both objects and subjects, and the ways these congeal and dissolve in complex patternings of times and spaces. The spectre of race thus alerts us to the ways folds matter to the world as it is – and what it may become.

Acknowledgements

We thank the editors of and contributors to this ANT-companion for feedback and helpful suggestions during one of the workshops dedicated to this volume held in Copenhagen. We are also grateful to the European Research Council for supporting our research through an ERC Consolidator Grant (FP7-617451-RaceFaceID-Race Matter: On the Absent Presence of Race in Forensic Identification); see also https://race-face-id.eu.

References

Abu El-Haj, N. 2007. 'The genetic reinscription of race.' *Annual Review of Anthropology* 36: 283–300.

Alcadipani, R. and Hassard, J. 2010. 'Actor-network theory, organizations and critique: towards a politics of organizing.' *Organization* 17(4): 419–436.

Bhabha, H. K. 1994. *The Location of Culture*. New York: Routledge.

Braun, L. 2002. 'Race, ethnicity, and health: can genetics explain disparities?' *Perspectives in Biology and Medicine* 45(2): 159–174.

Buse, P. and Scott, A. eds. 1999. *Ghosts: Deconstruction, Psychoanalysis, History*. London: Macmillan.

Callon, M. and Law, J. 2004. 'Introduction: absence – presence, circulation, and encountering in complex space.' *Environment and Planning D: Society and Space* 22(1): 3–11.

Claes, P., Hill, H. and Shriver, M. D. 2014. 'Toward DNA-based facial composites: preliminary results and validation.' *Forensic Science International: Genetics* 13: 208–216.

Derrida, J. 1993. *Spectres of Marx*. New York: Routledge.

Duster, T. 2005 'Race and reification in science.' *Science* 307(5712): 1050–1051.

Fullwiley, D. 2007. 'The molecularization of race: institutionalizing human difference in pharmacogenetics practice.' *Science as Culture* 16(1): 1–30.

Haraway, D. 1991. 'A cyborg manifesto: science, technology, and socialist-feminism in the late twentieth century.' In Haraway, D. (ed.), *Simians, Cyborgs, and Women: The Reinvention of Nature*. London: Free Association Books, pp. 149–183.

Haraway, D. 1997. *Modest Witness@SecondMillennium* [etc.] New York: Routledge.

Heidegger, M. 1962. *Being and Time*. New York: University of New York Press.

James, W. 1995[1907]. *Pragmatism*. Cambridge, MA: Harvard University Press.

Koening, B.A., Lee, S.-J. and Richardson, S.S. (eds.) 2008. *Revisiting Race in a Genomic Age*. New Brunswick, NJ: Rutgers University Press.

Landecker, H. 2000. 'Immortality, in vitro: A history of the HeLa cell line.' In Brodwin, P. (eds.) *Biotechnology and Culture: Bodies, Anxieties, Ethics*. Bloomington: Indiana University Press, pp. 53–72.

Latour, B. 1991. *We've Never Been Modern*. Cambridge, MA: Harvard University Press.

Latour, B. and Serres, M. 1990. *Conversations on Science, Culture, and Time*. Ann Arbor: University of Michigan Press.

Law, J. 1991. 'Introduction: monsters, machines and sociotechnical relations.' In Law, J. (ed.) *A Sociology of Monsters: Essays on Power, Technology and Domination*. London: Routledge, pp. 1–23.

Law, J. 2004. *After Method: Mess in Social Science Research*. London: Routledge.

Law, J. and Singleton, V. 2000. 'Performing technology's stories.' *Technology and Culture* 41: 765–775.

Law, J. and Singleton, V. 2004. 'Object lessons.' *Organization* 12(3): 331–355.

Lipphardt, V. 2012. 'Isolates and crosses in human population genetics; or, a contextualization of German race science.' *Current Anthropology* 53(5): 69–82.

M'charek, A. 2000. 'Technologies of population: forensic DNA testing practices and the making of differences and similarities.' *Configurations* 8 (1): 121–158.

M'charek, A. 2005. *The Human Genome Diversity Project: An Ethnography of Scientific Practice*. Cambridge: Cambridge University Press.

M'charek, A. 2010. 'Fragile differences, relational effects: stories about the materiality of race and sex.' *European Journal of Women's Studies* 17(4): 307–322.

M'charek, A. 2013. 'Beyond fact or fiction: on the materiality of race in practice' *Cultural Anthropology* 28(3): 420–442.

M'charek, A. 2014 'Race, time, and folded objects: The HeLa error.' *Theory, Culture and Society* 31(6): 29–56.

M'charek, A., Schramm, K. and Skinner, D. 2014. 'Technologies of belonging: the absent presence of race in Europe.' *Science, Technology & Human Values* 39(4): 459–467.

Mol, A. 2002. *The Body Multiple: Ontology in Medical Practice*. Durham, NC: Duke University Press.

Mol, A. 2010. 'Actor–network theory: sensitive terms and enduring tensions.' *Kölner Zeitschrift für Soziologie und Sozialpsychologie* 50(1): 253–269.

Nelson, A. 2008. 'Bio science: genetic genealogy testing and the pursuit of African ancestry.' *Social Studies of Science* 38(5): 759–783.

Raine, A. 2008. 'From genes to brain to antisocial behaviour.' *Current Directions in Psychological Science* 17: 323–328.

Rheinberger, H. J. 1997. *Toward a History of Epistemic Things. Synthesizing Proteins in the Test Tube*. Redwood, CA: Stanford University Press.

Schramm, K., Skinner, D. and Rottenburg, R. eds. 2011. *Identity Politics after DNA: Re/Creating Categories of Difference and Belonging*. Oxford: Berghahn Books.

Serres M. and Latour B. 1995. *Conversations on Science, Culture and Time*. Ann Arbor: University of Michigan Press.

Stengers, I. 2005. 'Introductory notes on an ecology of practices.' *Cultural Studies Review* 11(1): 183–196.

Van Oorschot, I. 2018. *Ways of Case-Making*. Dissertation, Erasmus University Rotterdam.

Whitmarsh, I. and Jones, D. S. 2010. *What's the Use of Race? Modern Governance and the Biology of Difference*. Cambridge, MA: MIT Press.

23

What might we learn from ANT for studying healthcare issues in the majority world, and what might ANT learn in turn?

Uli Beisel

When reading the title of this chapter, you might have stumbled upon the term 'majority world.' So let me address this first, also because it might teach us something about the issues at the heart of this chapter. Majority world is a term introduced by the photographer and activist Shahidul Alam as an alternative to 'third world' or 'developing countries' (Alam, 2007). I like the term, as it immediately makes clear that in terms of numbers, 'the third world' is actually in the majority: Many more people live in countries that are economically worse off than in countries usually connoted as 'first world' or 'developed.' It also brings the privileges of the few and the continuing inequalities in terms of quality of life and health, as well as healthcare provision to the fore.

This chapter is concerned with how ANT can help us understand some of these issues better than, or at least differently to, established disciplinary perspectives on health in the majority world. I find one of Latour's examples in his classic book *We Have Never Been Modern* a useful starting point:

> The smallest AIDS virus takes you from sex to the unconscious, then to Africa, tissue cultures, DNA and San Francisco, but the analysts, thinkers, journalists and decision-makers will slice the delicate network traced by the virus for you into tidy compartments where you will find only science, only economy, only social phenomena, only local news, only sentiment, only sex.
>
> *(Latour, 1993, p. 2)*

It is exactly in this way that ANT has and continues to inspire social studies of global public health and medicine. The academic task, following Latour, is to draw things together, to map heterogeneous associations that together constitute our bodily, physical, political and social environments. This means that we need to adopt an ethos of being 'in the midst of things' rather than assuming distinct domains (Bingham, 2006).

For my own writing, working in an after-ANT 'set of sensitivities' (Mol, 2010, p. 13) has meant trying not to divide the realms that make up the complex knots of realities. In this sense, I understand ANT, and importantly the developments that became known as 'ANT and after' (Law and Hassard, 1999), with Mol as a set of 'sensitive terms and enduring tensions' (Mol, 2010). In what follows, I will use a case from my own empirical work to explore what it means to follow an object through societies. I use my case study to focus our attention on, and weave together, key aspects of ANT-inspired scholarship on health in the majority world. I first introduce my case and focus on how one might best understand the bioeconomy of healthcare and its effects on local health and global science infrastructures. Secondly, I concentrate on discussing the role of evidence and the social construction of numbers, as well as human–non-human dimensions and ecologies of biosciences for health interventions in the majority world. At the end of this chapter, I make a case for how this might constitute thinking in the midst of things, and what benefits it might yield if we do not reduce phenomena but follow their heterogeneous actor-networks.

The case that I will discuss in this contribution is the development and clinical trials of a vaccine against malaria. Until today, no vaccine against malaria is licenced and the development has been notoriously difficult. Research into vaccine possibilities and the development of vaccine candidates' dates back to the early 20th century, but progress has been very slow for various reasons. The involvement of the Bill and Melinda Gates Foundation (BMGF) in the early 2000s accelerated the vaccine development of one particular candidate called MOS-QUIRIX considerably. The Gates Foundation saw it as a worthwhile investment because the development of a malaria vaccine has the potential to have a 'huge impact' on saving lives and so fits to the foundation's overall mission of solving the grand challenges humanity is facing. But what exactly does such a framing of 'huge impact' do on the ground? How might it shift priorities in global health funding? Which effects might it have on local health systems and global science infrastructures? And how sustainable might the solutions these 'new tools' offer be? It is these questions I suggest we need to ask if we are interested in an ANT-inspired analysis of the politics of big solutions, of philanthropy's role in global health and in the regulation of global science and the pharma industry.

The promise of such grand solutions chimes with a phenomenon that Adriana Petryna, in her work on globalised clinical trials, calls 'experimentality' (Petryna, 2009). Petryna shows that in searching for profits and clinical subjects willing to take part in clinical trials, the pharmaceutical industry increasingly outsources and offshores trials to the economically worse-off majority world. She argues that patient protection in trials is tied to the market's competitiveness and therefore 'risks are unevenly distributed and its costs, unjust' (ibid, p.188). Kaushik Sunder Rajan calls this 'pharmocracy,' a global hegemony of the multinational, Euro-American pharmaceutical industry (Rajan, 2017). Both scholars have been important in understanding the effects that the meeting of global science and bio-capitalism has in an unequal world, and press us to inquire into what exactly globalised clinical trials mean for patients and health infrastructures in specific contexts. As Charlotte Brives demonstrates in her analysis of a mother-to-child HIV transmission drug trial in Burkina Faso, trial participation not only involves unequal conditions globally, but also ideas and norms travel with and are embedded in the trial design and practices. Following Mol, Brives discusses these as 'onto-norms' (Mol as quoted in Brives, 2013) and shows how the ideas and practices measured and encouraged in the trial shape participants' ideas of what constitutes 'a good (HIV-positive) mother' (Brives, 2013). Clinical trials can form intimate subjectivities of participants, and the broader conditions of access to care can also alter wider social relations, as Nguyen argues in his analysis of access to HIV drugs in times of scarcity in the 1990s in Burkina Faso and the Ivory Coast (Nguyen, 2010). Nguyen shows that in times when HIV drugs were not available for everyone, the ability to talk about oneself could determine if one got access to drugs or not. In the following sections,

I will trace some related threads connected to the development of the malaria vaccine MOS-QUIRIX and inquire into its effects on global and local health infrastructures, as well as the bodies of humans and parasites alike.

MOSQUIRIX™ and the search for the Holy Grail: vaccine trials, global infrastructures and subjectivities

> For decades, a malaria vaccine has been one of the holy grails of modern medicine. Dr Pedro Alonso, Director of the WHO Global Malaria Programme (http://www.who.int/malaria/news/2016/gavi-malaria-vaccine/en/

Research on a vaccine against malaria started as early as the beginning of the 20th century. But to date, there is no vaccine against any parasitic disease in humans. This is not for lack of trying but is largely owed to the high biological complexity of parasites (in comparison to bacterial or viral organisms). In recent years, new scientific prospects have arisen from the possibility of mapping the genomes of humans and parasites. Due to recent advances in scientific knowledge, about 30 potential malaria vaccine candidates are currently being developed. Here, I will use the most advanced of these vaccines, MOSQUIRIX™, to illustrate what an ANT-inspired analysis can offer understanding global health and vice versa. I draw on ten years of ethnographic research on malaria control interventions, inter alia the malaria vaccine trials in Ghana and their attendant national and global science and health policy processes.

MOSQUIRIX™ was initially developed in GlaxoSmithKline's laboratories in Belgium in the 1980s. In the 1990s, the vaccine went through initial test phases in Belgium and at the Walter Reed Army Institute in the USA, as well as a first proof-of-concept study test in Gambia and later Mozambique, where it was tested on 2,022 children aged 1–4 years in a double-blind, randomised, controlled trial. Vaccine development and trials started to accelerate substantively from 2001, when the public–private partnership PATH Malaria Vaccine Initiative (MVI) started to finance and coordinate the development and trials. MVI's involvement accelerated the development and trial of the vaccine considerably. By 2018, the vaccine had successfully finished phase II and III clinical trials and was in implementation trials. MVI is initiated and funded by the BMGF. For the Gates Foundation, a malaria vaccine has been attractive as it makes the promise of solving the notorious malaria puzzle for good:

> While the best job he could have in the world is the one he has, says Gates, the second best would be discovering new medicines. 'It is a field that is changing. You get new tools all the time. You can have a huge impact. The kind of work and thinking that goes on is very like software', he says.
>
> (Sarah Boseley, Guardian, 2006)

It is exactly this thinking in superlatives that underlies the Gates Foundation's work in malaria. In 2007, the Foundation announced as its aim nothing less than the eradication of the disease. This is what Gates called 'creative capitalism' in his 2008 speech at the World Economic Forum in Davos: Namely, the idea that where there are big problems, there lie great opportunities for impact, and therefore good financial investment. It is clear that this notion of 'creative capitalism' is – at least for Gates himself – connected to 'doing good.' However, as Linsey McGoey (2015: 235) has suggested, philanthropy as creative capitalism might be a 'selfish gift' in the sense that it keeps the capitalist power balance intact while addressing inequalities through business innovations. Read in this way, Gates and the BMGF promise that creative capitalism can unite markets and morals (ibid, p. 16) and so solve the big problems of the world more efficiently than can state actors or private companies.

But how does this big ambition translate on the ground, in places where the malaria vaccine was tested? What have the phase II and III clinical trials meant for patients and health facilities in one of its trial locations, namely in Ghana? Between 2007 and 2014, phase II and III trials were conducted in seven African countries. Ghana had two vaccine trial sites: Agogo and Kintampo. The malaria vaccine trials were among the first large-scale vaccine trials conducted on the African continent in history. As I also discuss elsewhere (Beisel, 2017), in order to adhere to international standards of trial regulation, a so-called 'capacity building component' has been important to the international consortium conducting the trials: Funding from BMGF enabled the establishment of a Malaria Clinical Trial Alliance (MCTA), which is run under the umbrella of the INDEPTH Network, a Ghana-based organisation that is also funded by BMGF. In practice, MCTA and MVI conducted 'Good Clinical Practice' training for health workers involved in all stages of the trials – from doctors, nurses, fieldworkers and malaria microscopists to trial management staff. This training was highly valued by participants and offered, for instance, local laboratory workers a rare opportunity to participate in international training and networking (interviews at malaria vaccine trial sites in Ghana, 2008). Trial locations have also benefitted from new physical infrastructure, such as better equipment for laboratories. Such capacity building is an advantage for the participating hospitals, and contributes to biomedical capacity building of the trial sites in the seven participating countries (interviews at malaria vaccine trials' sites in Ghana, 2008).

However, such investments in selected hospitals are tied closely to research-trial institutions. While they might offer better services to local populations, they at the same time reinforce unequal geographies of healthcare provision within countries. Such investments create what has been called 'archipelagos of science' on the African continent (Geissler, 2013; Rottenburg, 2009). In Ghana, extensive medical research has in the past been conducted in four key sites. It is these four regions where people have extensive experiences with clinical trials, and arguably the attendant health centres offer superior quality medical care. This medical privilege, however, in the case of the research hotspot in and around Agogo, dates back to colonial times and is tied to missionary efforts: The Agogo Presbyterian Hospital hosting the studies (inter alia phase II and III clinical trials of MOSQUIRIX™) was established by the Basel Mission in 1931. By 2007/8, when I spent time in Agogo, the MOSQUIRIX™ trials had meant that a German paediatrician was based in Agogo fulltime, that the laboratory was upgraded significantly and that staff involved in the trial had undergone additional training. By 2018, Agogo Hospital had 13 medical doctors. In contrast, the close-by governmental hospital in Konongo, a town with around double the inhabitants, had only one medical doctor. This is just one example of the patchy and fragmented infrastructures and their colonial legacies that are built upon and sustained by large-scale clinical trials and other project-based global health investments.

This observation suggests that an ANT-inspired analysis asks: How do biomedical projects relate to general health infrastructures present; how might inequalities (inadvertently or not) be reinforced and what kinds of social effects might trial (non)participation have? It is also crucial to investigate what kinds of practices and projects are *not* funded under the logics of large biomedical investments. For instance, Ann H. Kelly and I analysed a mosquito larviciding project using the microbial insecticide Bti in Tanzania that initially attracted support by

BMGF but was discontinued when the project moved from trialling to sustaining the project (Kelly and Beisel, 2011). We suggested that the case study

> raises the critical question of how global attention shapes its object. The strategic global map leaves out the malaria that begins where the pavement ends, in blocked drains and discarded plastic cups. Commitment to controlling this malaria might seem like a bad business decision, but we argue it is a good long-term investment.
>
> *(ibid., pp. 84–5)*

In line with Petryna, Sunder Rajan and others, then, I suggest that ANT-inspired analyses should plot the patterns of global health inequalities, as health research in the Global North often receives more funding than African governments (can) spend on their basic public health infrastructure. This means that the political economy and social values embedded in trials and other global biomedical interventions need to receive careful attention in our analysis. However, in an ANT-inspired sensitivity, one needs to not only attend to the 'bigger' picture of health infrastructures, global R&D and their intended and unintended effects, but also weave this together with the seemingly 'small' changes and shifts in onto-norms and practices of these projects and interventions. Reading these seemingly 'small' or 'unintended' effects and politics of these 'big' and 'simple' technological fixes together helps us to understand the effects of the politics of global health and to contribute to 'diffracting' these by shining light in unexpected places (Haraway, 1994).

But, you might ask, is this really ANT, or an ANT-inspired sensibility? Where are the non-human actants? Where are the heterogeneous assemblages? So far, this sounds pretty much like an established STS analysis of science as culture (and economy). Where is a hybrid analysis of following the virus (or in my case, the parasite) as Latour suggested it? Indeed, the story have I told so far is crucial in understanding global health, but it is not enough. It keeps the compartments of science, policy and economy intact and does not pay enough attention to complex entanglements. The next section, then, offers another take on the malaria vaccine trials. In the conclusion, I will attempt to draw these analyses together and then as a result to provide an answer to the question this chapter poses.

Vaccine or 'malaria reducer'? The prize of the Holy Grail

In response to the earlier section, Gates would probably concede that every biomedical intervention has drawbacks and 'unintended consequences' (in other words, things that fall through the cracks and are beyond control), but he would surely argue (and he is not alone in this) that big projects like the malaria vaccine also potentially offer significant gains. After all, a successful vaccine could be a game changer in disease control and elimination. As Pedro Alonso, the current head of the WHO Malaria Programme, put it, 'a malaria vaccine has been one of the holy grails of modern medicine.' So, the question then becomes: Has MOSQUIRIX™ been able to lead to the holy grail?

This question invites us to include questions of efficacy, (un)certainty as well as socioecological dimensions of the vaccine into our analysis (for an extended discussion of these issues see Beisel, 2017). When in 2014 the MOSQUIRIX™ malaria vaccine completed phase III of clinical trials, it became clear that it will not have the protective effect for which many were hoping. The aim of the vaccine was never to provide full protection from malaria, but to offer partial immunity of around 60%, simply put meaning that the immunity levels of immunologically naïve individuals will be elevated to the level of having successfully lived through a malaria infection. Sixty per cent is considered significant in public health logic, as it would avert six out of ten malaria infections (Interview, Vaccine Scientist, 2012). Considering that the large majority of malaria deaths occur in immunologically naïve children under five, it is assumed that this would have a significant impact on malaria mortality. The results of the trials, however, showed that the vaccine efficacy against clinical malaria in infants was 27%, 'with no significant protection against severe malaria, malaria hospitalization, or all-cause hospitalization' (RTS,S Clinical Trials Partnership, 2015). Furthermore, initial protection rates declined quickly, making a four-dose vaccination regime necessary.

A 2012 editorial in the *New England Journal of Medicine* stated that the vaccine's efficacy results were 'disappointing' and that it remained doubtful that the vaccine will be licenced at the end of the trial (Daily, 2012). Other commentators were more optimistic and expected a positive licencing decision by the World Health Organisation (WHO). In 2015, WHO decided to insert an additional step before licensure, namely a pilot implementation study. The reasoning of the WHO's 'Strategic Advisory Group of Experts on immunization' (SAGE) was as follows:

> One primary outstanding question with regard to RTS,S/AS01 use in 5–17 month old children is the extent to which the protection demonstrated in the Phase 3 trial can be replicated in the context of the routine health system because of the challenge of implementing a four-dose schedule that requires new immunization contacts.
>
> *(WHO/SAGE, 2015)*

WHO worried that a four-part routine immunisation is neither practical nor affordable in many contexts in sub-Saharan Africa.

The MVI had already anticipated this dilemma in a marketing document from 2001:

> It might be necessary to get governments and individuals to change their thinking about what constitutes a 'vaccine.' At present many people regard vaccines as 100% effective; anything less is considered a failure, and thus of little or no interest. Conduct small market studies to determine if it is better to promote this new product as a 'malaria inhibitor' or 'malaria reducer' — in order to focus purchasers' minds on the actual benefits.
>
> *(MVI, 2001: 20)*

But how useful is such a 'malaria reducer' for local populations, when, for instance, a study in Ghana showed that parasites are routinely present in approximately 58% of the population (Owusu-Agyei et al., 2009)? The people taking part in this study did not show any symptoms of disease; they were randomly chosen and then tested for parasite prevalence. In other words, in areas with high malaria transmission, people cohabit with parasites most of the year without necessarily developing symptoms of malaria. This is due to acquired specific immunity, which humans develop through continued exposure to malaria parasites and lived-through malaria infections. Specific immunity is partial, meaning that it does not prevent all malaria infections, but makes malaria episodes less frequent and severe. In this regard, naturally acquired specific immunity is quite similar to the immunity the MOSQUIRIX™ vaccine provides. This, or so epidemiologists argue, could be a life-saving intervention for children under five years whose immune systems are still developing and who have not yet had an opportunity to develop specific immunity. However, even for these groups, infection with parasites does not necessarily lead to disease. As malariologist Kevin Marsh points out, 'even during the period of maximum susceptibility to severe disease, children spend the majority of their time *parasitized but healthy*' (Marsh, 2002, p. 256, emphasis added).

Despite much scientific research on the topic, the concrete factors that render an individual vulnerable to developing malaria are still unclear. Disease interactions and adaptations are manifold but what remains uncertain is which factors make a human body vulnerable to malaria or protect against it. This uncertainty is also expressed in people's reasoning and disease prevention practices that are based on their observed experiences with fever, as one of my research assistants reflected on our interviews in Ghana:

> One man among the opinion leaders present for the workshop said 'everyone is bitten by mosquitoes day in day out but those, who are more likely to get malaria, are those who eat a lot of oily food and stay in the sun for longer hours'. Among the pregnant women, there was a wide range of causes for malaria which included mosquitoes, unkempt environment, staying in the sun for long hours, drinking dirty water (from stagnant source), eating a lot of oily food such as groundnut, eating excess sugar, houseflies and working around fire (…)
>
> *(Interview, Martin Agyemang, 2008)*

Such reasoning could, of course, be dismissed as ignorance and evidence of lack of formal education as is regularly done in the biomedical literature and in health education approaches, but this does not do justice to what is going on. An unhealthy diet (eating too much sugar or fatty oils), or a weakened immune system (standing too long in the sun, being close to breeding places) could well be factors contributing to someone's vulnerability to malaria. These are triggers of malaria that people observe in their bodies, their lives and their practices. While malaria gets transmitted by mosquitoes, it is not only caused by the presence of parasites in the blood. In high transmission areas then, malaria regularly lingers in human bodies and the body's falling ill or not is determined by a complex chain of factors. Read in this context, a 30%–40% effective malaria vaccine, or 'malaria reducer' as MVI put it, would offer one additional tool in the complex game of malaria triggers, prevention and treatment.

In addition, there is another dynamic at play that destabilises the rendering of the malaria vaccine as 'the holy grail': this is related to biology. As the editorial in the *New England Journal of Medicine* quoted earlier puts it: 'the success of this pathogen is related to prolonged coevolution with humans, allowing the development of cunning biologic strategies to resist immune-system clearance' (Daily, 2012, p. 1). In other words, Plasmodium parasites are biologically highly adaptable and mutable due to their long and successful history of cohabiting with humans, which makes them rather tricky to 'defeat.' As studies in mice models have shown, a partially effective vaccine might at best 'buy us time.' Indeed, a malaria vaccine would considerably increase evolutionary pressure in Plasmodia, which has in mice models resulted in the emergence of more virulent parasites (Mackinnon and Read, 2004). A partially effective vaccine might be even less effective once 'cunning' parasites have developed tolerance to it. Considering this, and the high investments vaccine development requires, one might wonder if the hope for such a grand solution is the best way to limit the deadly circulations between humans, mosquitoes and parasites.

Conclusion

My case study has, I hope, helped to showcase an important locus of analysis for global health technologies. Namely, the task to critically inquire into what 'efficacy' means, or to put it more generally, how global health science constructs its objects, as well as indicators and measures of success or failure. In this context, a whole body of work in science studies has destabilised and differently contextualised calculative practices in global health, such as indicators, standards and evidence-based medicine in global health and the majority world (Adams, 2016; Brives, Le Marcis and Sanabria, 2016). To give one concrete example, Emily Yates-Doerr analyses how global health policy understands and measures 'hunger.' Juxtaposing this with understandings of hunger by Guatemalans' and local physicians, she shows that the category is 'not solid but is made and unmade variously' (Yates-Doerr, 2015, p. 229). This means that while global health policy aims to measure hunger with mathematical certainty, 'situated uncertainty' at all three locations persists (ibid.). Yates-Doerr argues that the excavation of 'persistent uncertainty' offered by ANT-inspired analysis of global health policy and practice is crucial in offering up alternative responses to hunger.

The malaria vaccine case study also resonates with a second field of scholarly engagement in global health, namely work on multispecies entanglements (Haraway, 2008). In the spirit of the Latourian quote with which I opened the chapter, this strand of scholarship has followed viruses or parasites through landscapes and labscapes, economies and ecologies, to map the interwovenness of species and the effects this might have on human and animal health in the majority world (Kelly and Lezaun, 2014; Nading, 2014; Porter, 2016). In my own work, I have aimed to show how mosquito net markets and mutations in the genome of Anopheles mosquitoes are linked through humanitarian health commodities. Global health in mosquito net distribution, I argue, only values the nets' immediate health benefits, and ignores the wider ecological dynamics and so inadvertently contributes to the development of insecticide resistance (Beisel, 2015).

The entanglements of humans and non-humans, and of science, economy and ecology, are evident in malaria vaccine politics. Trying to understand the potential and pitfalls of a malaria vaccine for the communities who would need them the most, I have sought to suggest that we need to indeed find ways 'not to slice the delicate network' (Latour, 1993, p. 2) but rather to weave together science policy, philanthropy-economics, the sociopolitical effects on local health infrastructures with a critical reading of vaccine efficacy and the health ecology of the development of (potential) resistances. To paraphrase Latour, understanding malaria vaccines requires us to follow the parasite to the Gates Foundation's offices in Seattle, GlaxoSmithKline's development labs in Belgium and the vaccine trial sites in Ghana, as well as neighbouring towns that do not benefit from the trial infrastructure, and, finally, to closely attend to the ecological dynamics in which parasite bodies are implicated. It is in tracing these complex and heterogeneous assemblages of science, economy and ecology that the contribution of ANT-inspired thinking and writing for understanding healthcare issues in the majority world lies. In weaving unlikely stories together, in transcending the boundaries between established disciplines and genres of thinking, a different kind of thinking about health science and human/non-human well-being can emerge. What makes the context of health in the majority world particular is surely that a special focus on health inequalities, the continuing problems of access to basic healthcare and a dominance of pharmaceutical R&D from the Global North is paramount not only to analysing, but also to ethically engaging with healthcare issues as scholars. In this vein, I have suggested that for ANT-inspired analyses, it is crucial that we also ask what kinds of projects and health infrastructures do *not* receive funding and attention due to grand approaches like malaria vaccine development? What kinds of practices of caring for bodies, patients and environments fall off the agendas of philanthropists, pharmaceutical industry and global health policies? It is exactly here, where the study of health in the majority world can inspire STS, namely in taking more seriously the effects that situations of (often prolonged) economic, environmental and social vulnerabilities and insecurities have on bodies (human and parasite alike), minds, infrastructures, and our living environments and landscapes more broadly. In times of rapid environmental change that will unarguably have significant effects on health, it is urgent that we find ways to weave together asymmetrical and (seemingly) unconnected realities to deepen our understanding of what is at stake in global health.

References

Adams, V. (ed.) 2016. *Metrics: What Counts in Global Health*. Durham: Duke University Press Books.

Alam, S. 2007. Reframing the Majority World. Available via: http://auroraforum.stanford.edu/files/transcripts/Aurora_Forum_Essay_Shahidal%20Alam_Reframing%20the%20Majority%20World.10.01.07.pdf (last accessed 01/03/2018)

Beisel, U. 2015. Markets and Mutations: Mosquito Nets and the Politics of Disentanglement in Global Health. *Geoforum* 66: 146–155.

Beisel, U. 2017. Resistant Bodies: malaria and the question of immunity/resistance. In Herrick, C. and Reubi. D. (eds.) *Global Health Geographies*. London: Routledge, pp. 114–134.

Bingham, N. 2006. Bees, Butterflies, and Bacteria: Biotechnology and the Politics of Nonhuman Friendship. *Environment and Planning A* 38: 483–498.

Boseley, S. 2006. Wealth and Experience. www.theguardian.com/society/2006/jul/20/internationalaidanddevelopment.ethicalliving

Brives, C. 2013. Identifying Ontologies in a Clinical Trial. *Social Studies of Science* 43(3): 397–416.

Brives, C., Le Marcis, F. and Sanabria, E. (eds.) 2016. The Politics and Practices of Evidence in Global Health. *Medical Anthropology* 35(5): 369–451.

Daily, J.P. 2012. Malaria Vaccine Trials—Beyond Efficacy End Points. *The New England Journal of Medicine* 367(24): 2349–2351.

Geissler, P.W. 2013. The Archipelago of Public Health. Comments on the Landscape of Medical Research in 21st Century Africa. In Prince, R.J. and Marsland, R. (eds.) *Making and Unmaking Public Health in Africa: Ethnographic and Historical Perspectives*. Athens: Ohio University Press, pp. 231–256.

Haraway, D.J. 1994. A Game of a Cat's Cradle: Science Studies, Feminist Theory, Cultural Studies. *Configurations* 2(1): 59–71.

Haraway, D.J. 2008. *When Species Meet*. Minneapolis: University of Minnesota Press.

Kelly, A.H., and Beisel, U. 2011. Neglected Malarias: The Frontlines and Back Alleys of Global Health. *BioSocieties* 6(1): 71–78.

Kelly, A.H. and Lezaun, J. 2014. Urban Mosquitoes, Situational Publics, and the Pursuit of Interspecies Separation in Dar Es Salaam. *American Ethnologist* 41(2): 368–383.

Latour, B. 1993. *We Have Never Been Modern*. Cambridge, MA: Harvard University Press.

Law, J., and Hassard, J. 1999. *Actor Network Theory and After*. London: Wiley-Blackwell.

McGoey, L. 2015. *No Such Thing as a Free Gift: The Gates Foundation and the Price of Philanthropy*. London: Verso.

Mackinnon M.J. and Read A.F. 2004. Immunity Promotes Virulence Evolution in a Malaria Model. *PLoS Biology* 2(9): e230.

Marsh, K. 2002. Immunology of Malaria. In: Warrell D.A. and Gilles H.M. (eds.) *Essential Malariology*. London: Arnold, pp. 252–267.

Mol, Annemarie. 2010. Actor-Network Theory: Sensitive Terms and Enduring Tensions. *Kölner Zeitschrift Für Soziologie Und Sozialpsychologie*. Sonderheft 50: 253–269.

MVI (2001). Malaria Vaccine Market Consultation. Downloadable via: www.malariavaccine.org/down-publications.htm (last visit: 24/03/2007)

Nading, Alex M. 2014. *Mosquito Trails: Ecology, Health, and the Politics of Entanglement*. Berkeley: University of California Press.

Nguyen, Vinh-Kim. 2010. *The Republic of Therapy: Triage and Sovereignty in West Africa's Time of AIDS*. Durham, NC: Duke University Press.

Owusu-Agyei, S. et al. 2009. Epidemiology of Malaria in the Forest-Savanna Transitional Zone of Ghana. *Malaria Journal* 8(220): 1–10.

Petryna, A. 2009. *When Experiments Travel: Clinical Trials and the Global Search for Human Subjects*. Princeton, NJ: Princeton University Press.

Porter, N.H. 2016. Ferreting Things Out: Biosecurity, Pandemic Flu and the Transformation of Experimental Systems. *BioSocieties* 11(1): 22–45.

Rajan, K.S. 2017. *Pharmocracy: Value, Politics, and Knowledge in Global Biomedicine*. Durham, NC: Duke University Press.

Rottenburg, R. 2009. Social and Public Experiments and New Figurations of Science and Politics in Postcolonial Africa. *Postcolonial Studies* 12(4): 423–440.

RTS,S Clinical Trials Partnership. 2015. Efficacy and Safety of RTS, S/AS01 Malaria Vaccine with or without a Booster Dose in Infants and Children in Africa: Final Results of a Phase 3, Individually Randomised, Controlled Trial. *The Lancet*, 386(9988): 31–45.

World Health Organisation/SAGE. 2015. *Summary of the October 2015 Meeting of the Strategic Advisory Group of Experts on Immunization (SAGE)*, Geneva: World Health Organisation, accessible via: www.who.int/immunization/sage/meetings/2015/october/sage_report_oct_2015.pdf?ua=1

Yates-Doerr, Emily. 2015. Intervals of Confidence: Uncertain Accounts of Global Hunger. *BioSocieties* 10(2): 229–246.

24

What is the value of ANT research into economic valuation devices?

Liliana Doganova

Introduction: ANT and the problem of value

"What is the value of…" is a question relentlessly asked today, and increasingly addressed in the language of quantification and economic calculation. Consider a few examples. What is the value of nature? The economic value of the "ecosystem services" that it provides to society. What is the value of pollution? The price of CO_2 on emissions trading markets. What is the value of environmental regulation? The value of the human lives that it will help saving expressed in monetary units, and balanced against the costs that it will impose on "economic actors." What is the value of scientific research? The number of publications and citations that it has generated, but also the amount of licencing fees that universities have been able to collect from the patents protecting research, or the amount of the investments that scientists–entrepreneurs have been able to attract from venture capitalists or industrial companies.

How can ANT research make sense of such questions and the peculiar kind of answers that they propel? Although attempts to define ANT tend to fail, most would agree with the statement that ANT is wary of nouns such as "society" or "value." Echoing ANT's broader move from a "sociology of the social" to a "sociology of associations" (Latour 2007), ANT scholars who have delved into the question of value have done so not by attempting to identify, describe or measure value(s), but by looking at processes and practices of valuation. The shift from value to valuation pertains to a pragmatist stance (Hennion 2016) that has been widely shared in the emerging literature on valuation, which spans STS, sociology and organisation studies (Berthoin Antal et al. 2015, Helgesson and Muniesa 2013, Kornberger 2017, Muniesa 2011, Stark 2009). Studies of valuation inspired by ANT embrace a conception of agency as sociotechnical and distributed across heterogeneous agencements. In line with ANT's prior ventures into the fields of science and technology (Latour 1987), and later, markets (Callon 1998), these studies have paid particular attention to devices. They have broadened the focus on "market devices" proposed by Callon et al. (2007), and brought to light the manifold *valuation devices* that are mobilised in queries about the value of things and attempts at making things valuable.

What is the value of ANT research into economic valuation devices? I will use this question as a starting point to sketch the perspectives opened by ANT's incursions onto a problem that has long been considered as pertaining to the domains of disciplines like economics and management. I will start with a particular instance of the problem of value – the economic value of scientific research – in order to position ANT research in comparison with other research traditions that have addressed the same kind of problem. I will then discuss what ANT research into valuation devices can bring to the vision of the economy in ANT-inspired accounts – a vision which, I will argue, tends to narrow the locus of economic activities to markets. I will conclude by exploring whether and how ANT research into valuation devices can be of any help in contemporary political debates, such as those on sustainable development and climate change, which deeply engage the question of what is valuable.

On the economic value of scientific research

How has ANT addressed the problem of the economic value of scientific research? Science is the empirical field in which ANT was born and from the study of which it forged its most powerful and challenging claims, namely the agency of non-human entities and the dissolution of the technology/society and nature/society dichotomies. Latour and Woolgar's (1979) study of an endocrinology laboratory in California and Callon's (1984) study of marine biologists introducing scallop harvesting in Britany are certainly the most famous examples of ANT's take on science and technology. The main problem they were concerned with was the production of knowledge, and related questions of truth, objectivity and power. They described researchers as building credit and alliances, making investments and acting as entrepreneurs, but such economic terms were used rather metaphorically. The metaphor that dominated their understanding of the economic was that of the market (see Muniesa, 2020). As they appeared in these accounts, researchers were certainly concerned with securing funding, but hardly concerned with demonstrating the economic value of their research in terms of the returns that it could bring to investors. What made research economically valuable was not the key issue at stake.

One might wonder whether the place given to the problem of the economic value of scientific research was the result of the theoretical inclination of early ANT accounts, which were engaged in vivid debates with the sociology of scientific knowledge, or a mere reflection of the absence of this problem in the practices they were observing. Whatever the answer, one can hardly observe the work of scientists today without constantly bumping into the problem of demonstrating value, and economic value in particular. The valuation of scientific work is certainly not a novel phenomenon; what is novel, however, is the extent of formalisation of the scientific value of research, through measures such as the number of citations, h-indexes, impact factors, journal rankings, etc. (Pontille & Torny 2015). Novel, also, is the institutional embedding of the quest for the economic value of research. The scientist–entrepreneur is no longer a metaphor, nor the portrait of a few heroes of the past, but what every researcher is urged to become by demonstrating the economic value of her research and providing her contribution to innovation and growth. One example, among many, is the recurrent complaint that France "isn't the hotbed of innovation it would like to be" because its scientists, mainly public servants, "rarely start a company to turn their discoveries into new products or services" (Pain 2017). Similar complaints, urging scientists to become entrepreneurs to help their countries, can be heard all around the world.

One of the canonical examples of academic entrepreneurship is the birth, in 1976, of Genentech, often described as the first biotechnology start-up company. Interestingly, the events that led to the creation of this company, and more broadly to the "science business" of biotechnology (Pisano 2006), occurred not too far from the Salk Institute and precisely in the period (1975–1977) when Bruno Latour was doing fieldwork there for his famous ethnography of scientific work (Latour & Woolgar 1979). That was a troubled period for the Californian biomedical community, which was grappling with a groundbreaking but troublesome invention: Recombinant DNA. In the early 1970s, geneticist Stanley Cohen from Stanford and biochemist Herbert Boyer from the University of California developed a method for transplanting genes. While the controversy over the promises and threats of genetic engineering was raging in the scientific and the public arenas, Stanford University and the University of California applied for a patent on the recombinant DNA process, and biochemist Herbert Boyer partnered with venture capitalist Robert Swanson to found Genentech. On October 14, 1980, a few months after the Supreme Court ruled that living organisms engineered by man are potentially patentable, in the famous Diamond v. Chakrabarty case, Genentech went public, raising $38.5 million in a few hours and giving Boyer and Swanson a paper profit of $60 million (Hughes 2001). On December 2, the US Patent Office granted a patent to the Cohen–Boyer application, and on December 12, the Bayh–Dole Act was passed, giving the right and incentive to universities to hold patents on innovations arising from federally funded research.

The success of Genentech did not go unnoticed. It was seen as yet another sign that Europe "lags behind" the US: As one analysis explained, while "European scientists" played their part in the discoveries that led to the development of biotechnology, "American entrepreneurs were much quicker to exploit the new techniques" (Owen 2016). Reacting, the European Union set about building a regulatory environment that could favour the development of biotechnology while meeting European citizens' fears and doubts (Jasanoff 2007). Many European countries implemented policies aimed at fostering academic entrepreneurship and the creation of start-ups that would "valorize" public research and "transfer" its results from university to industry (Mustar et al. 2008). This gave rise to profuse research on the evaluation of these public policies and the success factors explaining the transformation of science into money. The results were disappointing: Europe was still lagging behind the US; it was unable to produce success stories able to compete with companies like Genentech, and while scientists had decidedly engaged in the creation of start-ups, the companies they built remained small, created few jobs and provided little financial returns for their "parent" research organisations. This, in turn, triggered even more research on how scientific knowledge can be turned into economic value.

Beyond the diversity of backgrounds and vantage points, research on the "valorization" of research has shared a common assumption, envisaging economic value as something that science inherently had and which only needed the right conditions to be unleashed. When the economic value of science was problematised, it was problematised "from the outside," often from the standpoint of the policymaker who is concerned with measuring the contribution of science to economic growth in order to fine-tune the right level of R&D spending and government support. ANT-inspired analyses have opened novel perspectives in this discussion in so far as they have problematised the economic value of science "from the inside," that is, by espousing the viewpoint and concerns of scientists, who attempted to demonstrate the value of their research, and of investors, to whom these demonstrations were addressed. Instead of asking "what is the value of research?" (and, as a corollary, how can this value be augmented?), ANT scholars have been more likely to ask "how is the value of science demonstrated, negotiated and contested?" or rather "how is the value of science constituted trough practices involving a variety of (human and nonhuman) actors?"

I will take one example from my own work (Doganova & Eyquem-Renault 2009, Doganova 2012). What do scientists do when they "valorize" their research and "transfer" it to other territories typically referred to as "industry" or "the market?" They craft business models, write business plans and pitch their ideas to investors. They demonstrate the technologies they have developed at various events, conferences, meetings, networking breakfasts, etc. They calculate, by means of more or less sophisticated formulas, how much the future products that might result from their research are worth today, what revenues they are likely to generate and at what price they can be sold at those willing to buy them. These models and plans, demonstrations and formulae, can be analysed as "valuation devices." How these devices operate, what kind of work they do and what effects they produce are the kinds of questions that an ANT perspective leads us to ask in order to understand how science is made valuable and why this matters.

From market devices to valuation devices

The notion of "valuation devices" has been modelled after that of "market devices" developed by Callon et al. (2007) to refer to "the material and discursive assemblages that intervene in the construction of markets" (p. 2). The notion of "market device" was itself inspired from ANT's notion of "socio-technical device." In line with ANT's perspective on action as a distributed process, the focus on devices was not a focus on objects, be they objects with agency, but a focus on "agencements." As Muniesa et al. (2007) put it in the introduction to the edited volume *Market Devices*, 'instead of considering distributed agency as the encounter of (already "agenced") persons and devices, it is always possible to consider it as the very result of these compound agencements' (p. 2). Rather than a unit of analysis, devices can thus be seen as an entry point for the sociologist who seeks to understand how markets are constructed and operate.

The variety of objects and situations embraced under the label of market devices (in that edited volume, these include securities analysts' reports, the financial chart, a purchasing centre, merchandising techniques, supermarket settings, focus groups, consumer tests, fishing quotas, index-based contracts and capital-guarantee products in financial markets, pharmaceuticals classification schemes, cotton prices and consumer credit scoring) raised an interesting question: What makes a device a *market* device? The answer sketched by the authors was that devices can be qualified as "market devices" when they contribute to the construction of markets or reconfigure market situations such as shopping or trading. For example, a shopping cart is a material device that is also "enacted" as a market device "because it reconfigures what shopping is" (p. 3). This definition of market devices pertains to a broader definition of economic agencements as agencements that render things economic, or perform processes of "economisation," which was developed further by Çalışkan and Callon (2009, 2010) and Callon (2017). Marketisation is one modality of economisation, and markets and market devices are one form of economic agencements.

The notion of valuation devices can serve as an entry point for the analysis of different forms of economisation, beyond marketisation. Economic valuation is indeed involved in a variety of settings some of which can hardly be described as markets. Courts are one of these settings. In courts, economic valuation can serve to calculate the monetary compensations that victims should receive. Focusing on oil spills and their damage on nature, Fourcade (2011) has highlighted the differences between the valuation methods used in different cases (the Amoco Cadiz and Erika ships in France, and the Exxon Valdez in Alaska), and the concomitant differences in the conceptualisation of the entities that are being compensated ("victims" and "nature"). That courts are not markets is a rather straightforward claim. More interestingly, the methods used to perform economic valuation in courts differ in whether they embed a market logic. Fourcade notes that while French experts used the estimation of prices and costs, thereby creating fictitious market situations in which the biomass destroyed was priced, or the restoration of the environment destroyed was contracted out, US experts used a method known as "contingent valuation" which relies on surveys to estimate the worth of the environment to individuals. Contingent valuation can be analysed as a valuation device that conceptualises the value of nature as a subjective experience, rather than as something that results from or is revealed by market transactions. It is a device that does not contribute to the construction of markets, but to solving a political problem: The distribution of blame and monetary resources across actors.

Another interesting setting in this respect is public policy, where economic techniques such as cost–benefit analysis have been introduced in the quest for objectivity and rational decision-making (Porter 1995). Cost–benefit analysis can be described as a valuation device that translates a political problem (for example, should environmental regulation be implemented?) into a decision-making exercise in which benefits and costs are compared (for example, the benefits that environmental regulation will produce for the population, and the costs that such regulation will incur for firms). It involves complex equivalences which transform lives saved in the future into amounts of money today. For example, the discussions on regulation over asbestos that took place in the 1980s in the USA involved controversial claims such as this one: The value of a human life saved by banning asbestos is 1 million dollars, but if the cancer provoked by asbestos remains latent and causes death 40 years later, this value should be "discounted" and brought down to 22,000 dollars, because an event that occurs in 40 years is worth less than an event that occurs today (this example is discussed in Doganova, forthcoming).

Here again, the analysis should move beyond noting that making public policy is not constructing markets, to examine the modalities of economisation performed by different valuation devices. Cost–benefit analysis can thus be put in perspective with other devices that entangle environmental and economic valuation, such as certification schemes for "sustainable" products, lists of "clean" technologies or carbon markets where CO_2 emissions are exchanged (Doganova & Laurent 2016, Lohmann 2009). Such comparisons reveal that valuation devices may differ in the way in which they articulate the economic and the political – as separate domains or enmeshed processes – and lead to problematising the very qualification of certain settings as "markets."

Using valuation devices as an entry point, the analysis does not need to distinguish a priori between market and non-market settings. Following the foundational insights of ANT should rather lead us to look into the devices themselves and examine their "script" (Akrich 1992), that is, the assumptions that they embed on what is valuable, who is entitled to value and whom is to be valued. Some valuation devices do indeed act as market devices, in so far as they configure valuation situations as situations in which a good is bought and sold, a price is set and property rights are exchanged. Other valuation devices, however, act in a different way: The valuation situations that they configure are not situations of market exchange, but situations of investment. One interesting example in this respect is Neyland's (2018) analysis of social impact bonds – a new public policy instrument which combines the logics of social investment and payment by results – as an "anti-market" device: In other words, a device that configures a valuation situation that does not follow the laws of the markets, but that of investment.

To take again the example of scientific research and biotechnology discussed earlier, patenting the process of replicating DNA and selling licences to those wishing to employ this process enacts an economic agencement that can be described as "marketisation." Crafting a business model in which recombinant DNA is transformed into an asset able to generate future revenues for the investors who take a stake in it enacts a different economic agencement that can more aptly be described as "capitalisation" (Muniesa et al. 2017). Business models can be analysed as a peculiar kind of valuation devices, which value things by turning them into objects of investment, rather than objects to be bought and sold through market transactions (Doganova and Muniesa 2015). ANT research into valuation devices thus opens a novel path into the study of capitalisation (see also Birch 2016, Leyshon & Thrift 2007), understood as a specific modality of economisation which is distinct from other modalities typically referred to as marketisation, commercialisation or commodification.

Conclusion: valuing valuation devices

The observation that there are many kinds of valuation devices, each of which carrying specific assumptions about what is valuable and performing different modalities of economisation, suggests that valuation devices themselves may be the object of valuation. A recurrent question addressed to valuation devices deals with their capacity to accurately measure the value of things. The questions that ANT research addresses to valuation devices take us in a different direction. They invite us to look into valuation devices and their "scripts," and to explore the work that valuation devices do, the relationships that they build and the effects that they produce.

How should we value valuation devices? Callon's (2009) recent work on "civilizing markets" opens a promising path. Markets, Callon suggests, are neither good nor bad intrinsically; a good market is a market that is "designed well," treated as an ongoing experiment and subjected to collective scrutiny that allows for concerns to be expressed and taken into account. Could we imagine, in a similar fashion, economic valuation devices designed in such a way that they embrace a variety of concerns and can be engaged with by a variety of actors?

Ongoing efforts to reform economic valuation devices in response to concerns related to sustainable development and climate change shed light on two caveats that should be kept in mind when embarking on the path suggested by Callon. First, there are a number of actors, among which policymakers, economists, consultants and NGOs, who are involved in the production of alternative valuation devices, and a number of competing devices which all claim to perform better valuations. For example, a genuine market has developed around metrics for evaluating the environmental performance of companies and implementing sustainable investment strategies (Chiapello 2015). Second, discussions on the reform of valuation devices are often expressed in the language of economics and require a degree of equipment that may hinder the collective scrutiny open to many different voices that Callon envisages. For example, current debates on how the future is to be taken into account in the present, the extent to which future generations should be cared for and the costs that we are ready to incur today to avoid future climate damage are still expressed in the language of discounting – a language inherited from economics and finance which few will find appealing or even intelligible, and which carries with it underlying assumptions about the definition and sources of value (Doganova, forthcoming).

So what is the value of ANT research into economic valuation devices? It probably lies in its ability to intervene in contemporary debates on what is valuable by "de-scribing" (Akrich 1992) the valuation devices involved in these debates. Such descriptions should be valuable if they fulfil a twofold role: Help concerned groups to engage with economic expertise, and render visible and debatable the assumptions embedded in different valuation devices, the effects that they induce, and the modalities of economisation that they enact.

References

Akrich, M. 1992. The de-scription of technical objects. In W. Bijker & J. Law (eds.), *Shaping Technology, Building Society: Studies in Sociotechnical Change* (pp. 205–224). Cambridge, MA: MIT Press.

Berthoin Antal, A., Hutter, M. & Stark, D. 2015. *Moments of Valuation: Exploring Sites of Dissonance.* Oxford: Oxford University Press.

Birch, K. 2016. Rethinking value in the bio-economy: Finance, assetization, and the management of value. *Science, Technology, & Human Values* 42(3): 460–490.

Çalışkan, K. & Callon, M. 2009. Economization, part 1: shifting attention from the economy towards processes of economization. *Economy and Society* 38(3): 369–398.

Çalışkan, K. & Callon, M. 2010. Economization, part 2: a research programme for the study of markets. *Economy and Society* 39(1): 1–32.

Callon, M. 1984. Some elements of a sociology of translation: domestication of the scallops and the fishermen of St. Brieuc Bay. *The Sociological Review* 32(1_suppl): 196–233.

Callon, M. (ed.). 1998. *The Laws of the Markets.* Oxford: Blackwell.

Callon M. 2009. Civilizing markets: Carbon trading between in vitro and in vivo experiments. *Accounting, Organizations and Society* 34: 535–548.

Callon, M. 2017. *L'emprise des marchés: Comprendre leur fonctionnement pour pouvoir les changer.* Paris: La Découverte.

Callon, M., Millo, Y. & Muniesa, F. (eds.) 2007. *Market Devices.* Oxford: Blackwell.

Chiapello, E. 2015. Financialisation of valuation. *Human Studies* 38(1): 13–35.

Doganova, L. 2012. *Valoriser la science. Les partenariats des start-up technologiques.* Paris: Presses des Mines.

Doganova, L. & Eyquem-Renault M. 2009. What do business models do? Innovation devices in technology entrepreneurship. *Research Policy* 38(10): 1559–1570.

Doganova, L. & Muniesa, F. 2015. Capitalization devices: business models and the renewal of markets. In M. Kornberger, L. Justesen, J. Mouritsen, & A. Koed Madsen (eds.), *Making Things Valuable* (pp. 109–125). Oxford: Oxford University Press.

Doganova, L. & Laurent, B. 2016. Keeping things different: coexistence within European markets for cleantech and biofuels. *Journal of Cultural Economy* 9(2): 141–156.

Doganova, L. Forthcoming. Discounting the future: a political technology. In J. Andersson & S. Kemp (eds.), *Futures.* Oxford: Oxford University Press.

Fourcade, M. 2011. Cents and sensibility: economic valuation and the nature of "nature". *American Journal of Sociology* 116(6): 1721–77.

Helgesson, C-F. & Muniesa, F. 2013. For what it's worth: an introduction to valuation studies. *Valuation Studies* 1(1): 1–10.

Hennion, A. 2016. From ANT to pragmatism: a journey with Bruno Latour at the CSI. *New Literary History* 47(2): 289–308.

Hughes, S. 2001. Making dollars out of DNA: the first major patent in biotechnology and the commercialization of molecular biology, 1974–1980. *Isis* 92(3): 541–575.

Jasanoff, S. 2007. *Designs on Nature: Science and Democracy in Europe and the United States.* Princeton, NJ: Princeton University Press.

Kornberger, M. 2017. The values of strategy: valuation practices, rivalry and strategic agency. *Organization Studies* 38(12): 1753–1773.

Latour, B. 1987. *Science in Action: How to Follow Scientists and Engineers through Society.* Cambridge, MA: Harvard University Press.

Latour, B. 2007. *Reassembling the Social: An Introduction to Actor-Network-Theory.* Oxford: Oxford University Press.

Latour, B. & Woolgar, S. 1979. *Laboratory Life: The Social Construction of Scientific Facts.* Beverly Hills, CA: Sage.

Leyshon, A. & Thrift, N. 2007. The capitalization of almost everything: the future of finance and capitalism. *Theory, Culture & Society* 24(7–8): 97–115.

Lohmann, L. 2009. Toward a different debate in environmental accounting: the cases of carbon and cost–benefit. *Accounting, Organizations and Society* 34(3–4): 499–534.

Muniesa, F. 2011. A flank movement in the understanding of valuation. *The Sociological Review* 59(2): 24–38.

Muniesa, F. (2020). Is ANT a critique of capital? In A. Blok, I. Farías, & C. Roberts (eds.), *The Routledge Companion to Actor-Network Theory* (pp. 56–63). New York: Routledge.

Muniesa, F., Millo, Y. & Callon, M. 2007. An introduction to market devices. *The Sociological Review* 55: 1–12.

Muniesa, F., Doganova, L., Ortiz, H., Pina-Stranger, A., Paterson, F., Bourgoin, A., Ehrenstein, V., Juven, P-A., Pontille, D., Saraç-Lesavre, B. & Yon, G. 2017. *Capitalization: A Cultural Guide.* Paris: Presses des Mines.

Mustar, P., Wright, M. & Clarysse B. 2008. University spin-off firms: lessons from ten years of experience in Europe. *Science and Public Policy* 35(2): 67–80.

Neyland, D. 2018. On the transformation of children at-risk into an investment proposition: a study of social impact bonds as an anti-market device. *The Sociological Review* 66(3): 492–510.

Owen, G. 2016. Biotechnology: why does Europe lag behind the US? *LSE Business Review*, Nov. 17. http://blogs.lse.ac.uk/businessreview/2016/11/17/biotechnology-why-does-europe-lag-behind-the-us/

Pain, E. 2017. France should do more to help scientists become entrepreneurs, report says. *Science*, Feb. 16. www.sciencemag.org/news/2017/02/france-should-do-more-help-scientists-become-entrepreneurs-report-says

Pisano, G. P. 2006. *Science Business: The Promise, the Reality, and the Future of Biotech.* Cambridge, MA: Harvard Business Press.

Pontille, D. & Torny, D. 2015. From manuscript evaluation to article valuation: the changing technologies of journal peer review. *Human Studies* 38(1): 57–79.

Porter, T. 1995. *Trust in Numbers: The Pursuit of Objectivity in Science and Public Life.* Princeton, NJ: Princeton University Press.

Stark, D. 2009. *The Sense of Dissonance: Accounts of Worth in Economic Life.* Princeton, NJ: Princeton University Press.

25

How does ANT help us to rethink the city and its promises?

Alexa Färber

Why is ANT good for studying the city? In this chapter, I will argue that although the incorporation of ANT into the field of urban studies has been highly contentious, it has been variously productive and offers useful conceptual perspectives on the city and the urban (see Farías & Bender 2010, Blok & Farías 2016). The originality of an ANT approach to urban infrastructuring or urban planning has, for example, demonstrated how engineering interacts with governance and both contribute to urbanisation and the (post-)modern city as an object of diverse and competing knowledge practices (Doucet 2015; Hommels 2005). Working with ANT shows how agency is distributed within the socio-material situations of creating a city and highlights the contingency and multiplicity of these socio-material situations. Viewed in this way, infrastructures contribute to the governance of how people and goods, production, services, consumption and aspirations are centralised as a city. At the same time, unfolding infrastructures as processes of infrastructuring also reveals that they can hardly be 'governed' as such because they are constantly in the making: Contested, expanded and maintained (see Denis this volume).

'Cities,' according to Bender, 'invite such expansion of notions of agency and contingency' (Bender 2012, 137). By foregrounding that the city is materialised/objectified in, for instance, the ideal of technocratic functionality *and* simultaneously remains irreducible to it, ANT's redescription of the very premises of modernity (Latour 1993) meets and advances urban studies' basic assumptions about the modern city's complexity, relying on the interplay of density and heterogeneity (Simmel 1903; Wirth 1938; for a critique, see Roskam 2011). Reflecting the affinities between urban history and ANT, Bender and Farías (2010) suggest conceptualising the city as urban assemblages.

Conceiving the city as urban assemblages does not merely rely on its 'capacity to incorporate more urban actors, many of whom we conventionally and routinely exclude from our analysis' (Bender 2012, p. 129). The concept of assemblage provides to my mind, as an urban anthropologist, an awareness of the potentialities of the city. It is more interested in the *emergence* of difference than in difference itself, in how these differences are virtual and therefore how the socio-material composition of the city contains a 'present absence' (see Farías 2010, p. 15). This version of ANT clearly refers to Gilles Deleuze' and Félix Guattari's (1980) use of assemblage/agencement and bears close affinities to various disciplines in the humanities (philosophy, history, literary studies, cultural/social anthropology and cultural studies).

These kinds of elective affinities may not be relevant to other fields of ANT research. Blok and Farías (2016, p. 3) remind us that 'ANT does not provide a general theory of the social that could be applied to different cases and fields.' Nevertheless, referencing Annemarie Mol, they describe ANT-informed research as an 'experimental concept-machine finely attuned to the specificity of the objects and cases' (Blok & Farías 2016, p. 3). In their volume, these cases and concepts range from the articulation of political protest in street assemblies, and the publicness of urban space in face of 'mega events,' to my chapter on economic practices as they spell out in low-budget mobility and sewage infrastructures. The notion of 'urban cosmopolitics' aims to 'redistribut[e] the political across a broader spectrum of urban settings, material devices, and forms of embodied experience in the city' (Blok & Farías 2016, p. 240), pointing out the irreducibly incomplete 'common world.' Urban coexistence becomes the focal point that articulates – and confirms – the political dimension of 'urban cosmopolitics' as ANT-informed urban studies (see Blok & Farías 2016, pp. 2–3). Furthermore, profoundly embracing philosophical scholarship (mainly Isabelle Stengers 2005 for conceiving cosmopolitics, and Peter Sloterdijk 2016 (2004) for spherology) and encouraging a multidisciplinary intellectual practice, urban cosmopolitics stems from and reaffirms ANT's position in the social sciences and urban studies.

Following this trajectory, I conceptualise the city as promissory assemblage, highlighting the temporality and virtuality of urban assemblages. This concept grasps the urban as simultaneously attuned articulations in and of the city: The promise of the city that centralises resources and opportunities; the promised city as an object of desire in 'urban imagineering' (Färber 2011b, 2014a) and urban promises that articulate these and other expectations of the city. In this chapter, I will, firstly, revise different conceptual takes on the promise – from speech act and social theory, to literary and anthropological approaches – to unfold the analytical potential of the city as promissory assemblage around notions such as elusiveness and endurance. Secondly, I will suggest examples of how these articulations of promise in/of the city implicate the project of Western modernity and therefore invite analysis with an ANT-orientation. In order to do so, I discuss studies of infrastructuring and show how its promissory quality adds to the centralising dimension of cities.

Arguing for a theoretical perspective and claiming the relevance of such a perspective, everyday intellectual practices should never be left unquestioned – even less so when we are committed to an analytical perspective such as ANT that is founded on curiosity about how scientific truth is fabricated (Latour & Woolgar 1979). Thus, questioning ANT as an intellectual practice in urban studies in the final section reveals that in environmental, architecture and design faculties, ANT-orientated urban research has been received quite positively (see Guggenheim in this volume). This trajectory is quite opposite to the controversy surrounding the introduction of ANT to urban studies in 2011. I therefore conclude the chapter with a discussion of the potential – or promises – of multidisciplinary collaborations with ANT and the institutional settings that enable an integrative intellectual practice where the notion of promissory assemblage might seize – and thus challenge – the role of endurance as a condition for the city's promise as a centralising power.

Unfolding the city as a promissory assemblage

Cities are neither simply objects nor are they subjects, although we often talk about them as if they were single actors doing something wilfully. Nevertheless, they may be considered objects of desire/s and enigmatic with regard to one's attachments to them when we conceive of them as 'clusters of promises' (Berlant 2011, p. 24), or promissory assemblages. This approach can be a means of studying the improbable centralising power or attractiveness of big cities and yet the evident ecological damage they engender. Interrogating this cruel 'triumphalism' of the city (see Roskamm 2017, p. 9) with the notion of the promise allows us to understand how cities are articulated and enacted as the place for (better) living, despite the fact that, for example, ecological concerns point to the harm to life that occurs there. Lauren Berlant's literary analysis of the ways in which such contradictions are lived in the mode of endurance (in a mode of 'despite') demonstrates that it is a driving force for the continuity (rather than transformation) of social difference and inequality in (late) capitalism (Berlant 2011). The promise appears to be at the heart of endurance as an activity.

My research on low-budget urbanity (Färber 2011a, Färber & Otto 2016) and urban imagineering (Färber 2011b, 2014a) explores the possible connections between the often unfulfilled promises *in* the city, such as a variety of public goods facilitating low-budget housing, consumption and mobility, and the enduring promises *of* the city, enclosed in the imagined city as a place of equal opportunities and rights;[1] between 'the city as a promise' of modernisation and the 'promised city' as a desired object and destination for migration. Here, the coexistence of different articulations of promise seems to be a general condition of experiencing and thinking the modern city. I draw on urban assemblage analysis and humanities' concepts of the promise to problematise the unbroken centralising force of, and conditions for, cities.

How does the promise as a concept in speech act theory and literary studies as well as philosophy of law relate to an ANT notion of urban assemblages? Conceived of as illocutionary speech acts, promises are 'organizing acts,' as Bruno Latour writes, referring to John Austin (1962), that create a temporal difference 'between an ensemble of antecedent and posterior conditions' (www.modesofexistence.org, entry 'Austin'). Furthermore, promises create (only) *rather* stable, *quasi*-contractual relations between different actors (in Schneider 2005a): The relation between the promise as articulation, the promisor, the promisees and possibly different forms of what can count as 'promissory notes' (Wittrock 2000) such as a rendering in urban planning, or an agreed budget is later evaluated according to the un/fulfilment of the promise. Promises therefore become attached to values such as responsibility, reliability and trustworthiness (Lahno 2005; Schneider 2005b). The quality of the promise also organises temporal difference. These organised relations are monitored based on moral, rather than juridical judgements.

Interested in constellations of the social and how, for instance, newness emerges, urban assemblage scholars question the power relations and distribution of agency. Their interest in contingency also considers virtuality an integral part of urban assemblages. I suggest that studying the city as urban promissory assemblage highlights the everyday experience of contingency and virtuality. It points to how the promise organises rather loose social connections, while enabling affective modes of endurance that articulate the city as an object of desire.

Let's take the projection of urban public transport as an example of an urban promissory assemblage. Public transport articulates promises to city dwellers about their possible connectedness to and within the city but often fails to meet their expectations (Ureta 2015). It materialises promises about the metropolitan character of a city as well as subjectivities attached to the urban imaginary of equal opportunities and thereby encourages endurance regarding everyday experiences of inequality (see Farías & Höhne 2015; Harvey & Knox 2015; Höhne 2012). Public transport often appears as a promised object of urban politics and as a hard-to-realise project of urban development.

Sometimes, these promises of future urban transport are realised unexpectedly decades later – like the Second Avenue line in the NYC Subway that was partially finished in 2017 after almost 100 years of projection. How do people live (in) the city as a terrain of not-yet-realised promises? How do they relate to a place that has become incomplete once the promise of something to come has been uttered – a not-yet-realised-place through the promise? How do we interpret the time passing while the promise is not fulfilled? And what does waiting entail in such a situation? Kemmer and Simone (forthcoming) address how residents of Rio de Janeiro and Jakarta manage to detach from the transitivity of urban promises and, in this way, to transform a passive waiting for a better future into an active mode of endurance. They describe how the city is assembled under the impact of the promise, by highlighting how inhabitants of Rio and Jakarta's urban cores maintain the everyday forms of cooperation and activism that generated promises of access to the city in the first place, and how this keeps them politically relevant to those who claim the authority to make promises.

This approach is not focused on un/fulfilment of promises; it instead opens up a specific time-space that social anthropologists Simone Abram and Gisa Weszkalnys (2013) describe as a gap when studying the 'elusive promises' of planning:

> The 'gaps' between ideal, ideology and practice fill themselves with things unplanned, unexpected and inexplicable, and with things that get overlooked and forgotten. … the ethnographer's task is to chart how people deal with these gaps and mismatches, and to understand how they are significant to, and are occasionally elided by, the work of planning.
>
> *(Abram & Weszkalnys 2013, p. 23)*

These gaps are time-spaces attuned by the promise and therefore qualified by a certain elusiveness. To return to the Second Avenue line in the NYC public transport project: People may forget about the promise even though the promise may have made them move to the area. They may never have heard of the promise, or have other reasons for staying. Perhaps they cannot move away or do not want to miss the moment when it is finally realised and meets their initial expectations and imagined lives. Situated in a 'not yet,' things are related to each other loosely 'as if,' and sometimes they are (re)assembled despite better knowledge, that is, 'nonetheless.' These multiple loose modes also allow us to rethink the important common understanding of the promise as something binding, as something that coincides with a relatively stable set of (moral) values: We might view it as an organising agencement that is only somewhat monitored. This almost uncontrollable presence in absence (or virtuality) of the promise attunes the time-space in a perhaps reliably loose way – and demands endurance to keep it staying.

Infrastructuring the modern city: the centralising agency of promises in/of the city

As projects of modernity, urban infrastructures have articulated the promises of different forms of society: From liberalism and capitalism, to communism and fascism. Based on the expectations of engineering expertise, the determinism of technological progress over nature and economic wealth (=growth), these promises have often failed while the urban situations that they have articulated have shown some 'obduracy' (Hommels 2005). This inevitably makes infrastructuring an emblematic field to study the above-mentioned potentialities, the expanded notions of agency and contingency central to urban assemblage research. From this perspective, Ash Amin and Nigel Thrift suggest conceiving the city in a Deleuzian vein through its relational modality or mode of connectivity, as a machine, mainly stressing the multiplicity of assembling sociotechnical systems with a 'combinatorial and rhizomatic' agency (2017, p. 160; also see Jensen in this volume).

ANT-oriented urban studies focus on how agency is distributed and the contingent quality of infrastructures and infrastructuring alongside urban, social and environmental histories. William Cronon's (1991) work on the connective character of Chicago's urbanisation in the 19th century (retrospectively deemed an ANT-study of the city *avant la lettre* by Bruno Latour 2005, p. 11), for example, demonstrates how the centralising effect of infrastructuring (mainly via railroads) is based on an effort to stabilise the connection between future commodities from the 'Great West' and the growing city of Chicago, and how 'nature' is objectified in this infrastructured metropolis. For Cronon, the urban promise of growth (profit and prosperity) is connected to 'nature' as its main resource (be it grain, lumber or meat), and yet claims a major difference/frontier at the same time: 'If a city's growth was assured by nature or – better yet – ordained by God, then only a fool could doubt its future promise' (Cronon 1991, p. 35).

Similarly concerned with modern urbanisation processes and related promises in/of the city is Patrick Joyce's (2003) work on how liberalism materialises (in) the modern city. An affinity with ANT-oriented urban studies or assemblage theory is evident, for example, in an expanded notion of agency, when he writes that infrastructures

> set up the conditions of possibility in which freedom might be exercised. This freedom is where I take 'liberalism' to reside. And the action of these pipes, sewers, roads and lights comprised what I have called the agency of material things. 'Conditions for possibility' is a relevant term here, for these things did not determine outcomes (... the clear, unobstructed road had constantly to be dug up for maintenance, for example), so that these things carried a certain capacity for action, and built into them were certain kinds of agency.
>
> *(Joyce 2003, pp. 11–12)*

Joyce's interest in the potentiality provided by these promising infrastructures coincides with an insistence on the distributedness of agency and most importantly on the contingency involved (the emergence of unforseeable possibilities, the breakdown and non-functioning of infrastructures) and its handling in the mode of maintenance. He thereby preconceives the subject of maintenance and repair studies that have become important to ANT-oriented urban research (see Denis this volume). Joyce shows that infrastructures of built and maintained roads or sewage pipes do not act as simple mediators of modernity – here in the shape of liberal ideas. Rather, they are 'constantly creating problems for themselves in the very acts of operation,' and therefore 'political reason, in order to govern, forever had to create 'problems' and 'questions' around which governance might be thought and prosecuted' (Joyce 2003, pp. 71–72). Governing by infrastructuring meant materialising liberalism in ways that promised the modern city. Maintaining infrastructures and therefore trying to govern potentialities account not only for how engineering and technocratic expertise co-produce the multiplication of dis-/functionality. These modes of complicating the world as an arena of modern divides add to the complexity of the city as a promise of modernity.

Today, these studies in ecological and social histories resonate with the ways in which infrastructures have been conceived and studied in the vein of STS and in urban assemblage research. These have turned our attention from infrastructure as a stable object of engineering, design, governance and management to infrastructuring as historical instances (Bestor 2004, de Boeck & Plissart 2004, Höhne 2012, 2017, Simone 2015). These instances unfold anything but smoothly. Concepts such as 'splintering urbanism' emphasise the extent to which commons are the result of former struggles and therefore under threat in the neoliberal era of 'networked infrastructures' (Graham & Marvin 2001). The privatisation of formerly common good infrastructures (streets, water, etc.) has led to spatial fragmentation of the city and enhanced social marginalisation in the city. In my view, splintering urbanism offers a possible refinement of the relation of promises *in* the city (infrastructures as a common good) and *of* the city (the place of equal opportunities, or, here, access).

Concerning the multiplicity of goods subjected to these 'vast socio-technical meshworks' (Höhne 2012, p. 163) (food, different energy resources, housing, education, etc.), the study of water and sewage infrastructuring in cities in the global south seems to count among the most urgent and revealing cases. The work of Colin McFarlane (2011c) in urban India and South Africa, for example, contributes to the problematisation of publicness/commonness. He studies conflicts around water infrastructures that question the integration (inclusion or exclusion) from the infrastructures and collective action around infrastructures (Graham et al. 2015). These conflicts rearticulate the centralising effect of unequal infrastructures and point out the elusiveness of the infrastructural promise of the city. However, these collective actions also challenge the everyday endurance of living with this elusiveness.

Back to the future: promises and threats of ANT engagements in urban studies

The question of how to describe 'the world while keeping it open' (Law & Mol 2002, p. 16; Latour 2005; see Pontille and Ossandón this volume) is, of course, not exclusive to ANT-informed research on the city, but the use of ANT in design-oriented institutions does promote new forms of representational work. The appointment of ANT-scholars in departments of architecture and planning – in Hamburg, Maastricht, Manchester, Munich or Stockholm – gives indirect credence to certain affinities and expectations attached to their intellectual practice (see Yaneva and Mommersteeg this volume). Moreover, the professional knowledge of architects and planners, for instance, regarding visual or three-dimensional representations might be attractive for collaborating on questions of complexity and complication. If ANT is 'a theory heavy in gerunds' (Bender 2012, p. 136), ANT-oriented research of the city might be(come) an intellectual practice 'heavy in visuals' that keep the world open. Thus, one benefit – or promise – of researching the city and the urban with ANT lies in its accessibility to those disciplines confronting complexity in a visual and three-dimensional way. It is worth noting that such an overlap inevitably bears theoretical implications emanating from both sides that do not translate 'well' (see Guggenheim, this volume).

Latour also presents various useful experiments in this regard (for his collaborations in exhibitions, see Färber 2016, Petersen 2016, Thorsen & Blok 2016). Some of these experiments are listed on his website, others on the platform of the Programme d'Experimentation en Arts Politiques (SPEAP). On his website, we find Latour's city-related research under the section publication and 'media mix': 'Paris, ville invisible.' This photo-text essay by Latour (text), Emilie Hermant (photos) and Susan Shannon (layout) was published as a book in 1998 and made into a website by Patricia Reed in 2004. Although nowadays the website (indexed as 'Sociological web opera' by Latour) is a beautiful reminder of how more than a decade ago web design was able to mobilise text and image (Zitouni 2004), the book retains its surprising originality.

The experimental and multifarious design of *'Paris, ville invisible'* was ahead of its time in a very contemporary way, most prominently in the case of Rem Koolhaas' publication 'S, M, L, XL' (Koolhaas & Mau 1997). In this particularly chaotic way of squeezing together images and texts, the city is articulated as a 'haphazard material organization' (Harper 2000, pp. 156). Today, graduates and faculty from the academic contexts mentioned earlier are motivated to find new ways of tracing urban assemblages. They echo, for instance, Law and Mol's (2002, pp. 16–17) claim for listing, walking and telling stories when they work with notions such as 'unfolding' (Danusiri 2008), 'randonner' (Michaelis 2015), 'itinérer' (researchingacity.com) and 'mapping' (Yaneva 2012).

Reflecting on ANT-writing as an institutionally embedded intellectual practice not only points to existing and future spaces of ANT engagement in urban studies. It also sheds light on possible reasons for conflict in the broader field of urban studies. In this chapter, I have argued from the perspective of an urban anthropologist, situated more in the humanities than the social sciences. As such, I can embrace urban assemblage research and ANT-oriented studies of the city from a position where notions of agency, contingency, emergence and virtuality, as put forward by Farías and Bender (2010), do not pose theoretical threats. This is important to note because in 2011, the attempt to conceptualise the city as 'urban assemblages' (Farías & Bender 2010, McFarlane 2011) was received in quite an astonishing way. Within the broader field of urban studies, the introduction of ANT as assemblage research of the city sparked a polarising debate. Urban studies with theoretical roots in political economy (and a strong Anglo-North-American-German institutional background in social sciences) initiated a quite powerful response to ANT-inspired urban research that then had divergent, even partly contradictory, philosophical grounds (as well as institutional backgrounds in different European countries within cultural and social sciences). In the core forum for this debate – a set of 13 articles spread over five issues of the Journal *'City'* (2011/15/2–15/6) – the differences were mainly debated along the notions of power and structure vs. agency and composition, and theoretical vs. empirical capacities – with assemblage in urban theory (McFarlane 2011a, 2011b) being minimised from the critical urban studies' counterpart as a methodological approach with a certain descriptive capacity and value but no further analytical reach towards questions of power and society (Brenner et al. 2011a, 2011b).

This somewhat cacophonous backlash was perhaps partly fuelled by 'the narcissism of se-mantic difference – the enjoyment of a skillful argument between people who basically agree with each other,' as sociologist Fran Tonkiss stated (Tonkiss 2011, p. 587). Certainly, it ensured that the very beginnings of an explicit ANT-orientation for theorising the city were polarised. ANT-orientated scholars working on the city are now well aware of its potential for creating conflict within urban studies. However, as an urban anthropologist who did not take part in this debate but commented on it (Färber 2014b, 2014c), it was astonishing to observe how this boundary work and its antagonism became an almost inevitable reference point in urban stud-ies. For a while, other positions within the field of urban studies garnered much less attention, which is why I have emphasised humanities' and cultural studies' perspectives here.

Asking what ANT analysis can offer urban studies, I have suggested the notion of the city as 'promissory assemblage.' This concept acknowledges the multiple ontologies of urban as-semblages as a promise of the city, as the promised city and urban promises. To study the city as promissory assemblage gives analytical access to: The promises attached to modernity that are articulated in urban infrastructuring; the elusiveness of promises that we confront through-out modernisation and urbanisation; and to the connection between urban opportunity (and misfortune), subjectivation and endurance as a mode of activity that does not facilitate change. Herein lies the political relevance of researching the city with an ANT lens: Allowing us to detect 'immanence and change' (Acuto & Curtis 2014, p. 10) *in* and *of* the city, it also helps us to find enduring means to collectively address the false promises in and of the city.

Acknowledgement

Many thanks to the editors of the companion for their helpful feedback during the workshop in Berlin and helping me to write-up this chapter. I also thank Michèle Perry for the trans-lation and editing.

Note

1 Distinguishing promises in and of the city echoes the paradigmatic claim of urban anthropology in and of the city (Hannerz 1980).

References

Abram, S. & Weszkalnys, G. (eds.) (2013) *Elusive Promises. Planning in the Contemporary World*. New York, Berghahn Books.

Acuto, M. & Curtis, S. (2014) *Reassembling International Theory: Assemblage Thinking and International Relations*. Basingstoke, Palgrave Macmillan.

Amin, A. & Thrift, N. (2017) *Seeing like a City*. Cambridge, Polity Press.

Austin, J. L. (1962) *How to Do Things with Words*. Oxford, Clarendon Press.

Bender, T. (2012) History, theory and the metropolis. In: Brantz, D., Disko, S. & Wagner-Kyora (eds.) *Thick Space. Approaches to Metropolitanism*. Bielefeld, transcript-Verlag, pp. 125–140.

Berlant, L. G. (2011) *Cruel Optimism*. Durham, NC, Duke University Press.

Bestor, T. (2004) *Tsukiji: The Fish Market at the Center of the World*. Berkeley, University of California Press.

Blok, A. & Farías, I. (2016) *Urban Cosmopolitics. Agencements, Assemblies, Atmospheres*. Basingstoke, Taylor & Francis Ltd.

Boeck, F. d. & Plissart, M.-F. (2004) *Tales of the Invisible City*. Antwerpen, Leuven University Press.

Brenner, N., Madden, D. J. & Wachsmuth, D. (2011a): Assemblage urbanism and the challenges of critical urban theory. *City*, 15(2), 225–240.

Brenner, N., Madden, D. J. & Wachsmuth, D. (2011b): Between abstraction and complexity. Met-theoretical observations on the assemblage debate. *City*, 15(6), 740–750.

Cronon, W. (1991) *Nature's Metropolis: Chicago and the Great West*. New York, Norton.

Danusiri, A. (2008) *The Fold*. Documentary Educational Resources. 12 min.

Deleuze, G. & Guattari, F. (1980) *Mille plateaux*. Paris, Les Éditions de Minuit.

Doucet, I. (2015) *The Practice Turn in Architecture: Brussels After 1968*. Farnham, Ashgate.

Färber, A. (2011a) Competing desires for mobility: Low-cost airlines and the transformation of European cities. *Korean Journal of Urban History*, 6, 147–170.

Färber, A. (2011b) Constructing successful images of failure: Urban imagineering in Berlin after 1989. In: Cerda, A. et al. (eds.) *Metropolis Desbordadas: Poder, Culturas y Memoria en el Espacio Urbano*. México, Universidad Autonóma de la Ciudad de México, pp. 303–341.

Färber, A. (2014a) Low-budget Berlin: towards an understanding of low-budget urbanity as assemblage. *Cambridge Journal of Regions, Economy and Society*, 7, 119–136.

Färber, A. (2014b) Potenziale freisetzen. Akteur-Netzwerk-Theorie und Assemblageforschung in der interdisziplinären kritischen Stadtforschung. *sub\urban. Zeitschrift für kritische Stadtforschung*, 2(1), 95–103.

Färber, A. (2014c) Replik. *sub\urban. Zeitschrift für kritische Stadtforschung*, 2(1), 135–140.

Färber, A. & Otto, B. (2016) Saving (in) a common world: Studying urban assemblages through a low-budget urbanities perspective. In: Blok, A. & Farías, I. (eds.) *Urban Cosmopolitics. Agencements, Assemblies, Atmospheres*. London, Routledge, pp. 25–43.

Färber, A. (2016) Gedankensprüngeausstellung. *EASST Review*, [Online] 35(2). Available from: https://easst.net/article/gedankensprungeausstellung/ [Accessed June 20, 2018].

Farías, I. (2010) Introduction: decentring the object of urban studies. In: Farías, I. & Bender, T. (eds.) *Urban Assemblages: How Actor-Network Theory Changes Urban Studies*. London, Routledge, pp. 1–24.

Farías, I. & Bender, T. (eds.) (2010) *Urban Assemblages: How Actor-Network Theory Changes Urban Studies*. London, Routledge.

Farías, I. & Höhne, S. (2015) Humans as vectors and intensities: Becoming urban in Berlin and New York City. In: Frichot, H., Gabrielsson, C. & Metzger, J. (eds.) *Deleuze and the City*. Edinburgh, Edinburgh University Press, pp. 22–44.

Graham, S., Desai, R., & McFarlane, C. (2015) The politics of open defecation: Informality, body and infrastructure. *Antipode*, 47(1), 98–120.

Graham, S. & Marvin, S. (2001) *Splintering Urbanism: Networked Infrastructures, Technological Mobilities and the Urban Condition*. London, Routledge.

Hannerz, U. (1980) *Exploring the City: Inquiries toward an Urban Anthropology*. New York, Columbia University Press.

Harper, D. (2000): The image in sociology: Histories and issues. *Journal des anthropologues*, 80–81, 143–160.

Harvey, P. & Knox, H. (2015) *Roads: An Anthropology of Infrastructure and Expertise.* Ithaca, NY, Cornell University Press.

Höhne, S. (2017) *New York City Subway. Die Erfinung des urbanen Passagiers.* Köln, Böhlau Verlag.

Höhne, S. (2012) An endless flow of machines to serve the city – infrastructural assemblages and the quest for the machinic metropolis. In: Brantz, D., Disko, S. & Wagner-Kyora, G. (eds.) *Thick Space: Approaches to Metropolitanism.* Bielefeld, transcript-Verlag, pp. 141–164.

Hommels, A. (2005) *Unbuilding Cities: Obduracy in Urban Socio-Technical Change.* Cambridge, MA, MIT Press.

Joyce, P. (2003) *The Rule of Freedom: Liberalism and the Modern City.* New York, Verso Books.

Kemmer, L. & Simone, A.M. (forthcoming) Standing by the promise. Arts of Endurance and Exposure in Rio and Jakarta. Unpublished Manuscript, last modified March 30, 2019. Microsoft Word File.

Koolhaas, R. & Mau, B. (1997) *S, M, L, XL.* New York, The Monacelli Press.

Lahno, B. (2005) Treue als künstliche Tugend: Humes Theorie der Institution des Versprechens. In: Schneider, M. (ed.) *Die Ordnung des Versprechens.* München, Wilhelm Finkg Verlag, pp. 201–218.

Latour, B. (2005) *Reassembling the Social – An Introduction to Actor-Network-Theory.* Oxford, Oxford University Press.

Latour, B. (1996) *Aramis, or the Love of Technology.* Cambridge, MA, Harvard University Press.

Latour, B. (1993) *We Have Never Been Modern.* New York, Harvester Wheatsheaf.

Latour, B., Hermant, E. & Shannon, S. (1998) *Paris. Ville Invisible.* Paris, La Découverte.

Latour, B. & Woolgar, S. (1979) *Laboratory Life: The Construction of Scientific Facts.* Beverly Hills, Sage Publications.

Law, J. & Mol, A. (2002) *Complexities: Social Studies of Knowledge Practices.* Durham, NC, Duke University Press.

McFarlane, C. (2011a) Assemblage and critical urbanism. *City* 15/2011, 204–224.

McFarlane, C. (2011b) Encountering, describing and transforming urbanism. Concluding reflections on assemblage and urban criticality. *City* 15/2011, 731–739.

McFarlane, C. (2011c) *Learning the City: Knowledge and Translocal Assemblage.* Chichester, Wiley-Blackwell.

Michaelis, T. (2015) *Showtime Wilhelmsburg – A Randonnée of Possibilities.* Leipzig, Spectorbooks.

Petersen, A. (2016) Procedures to deal with modernity without irony. *EASST Review*, [Online] 35(2). Available from: https://easst.net/article/gedankensprungeausstellung/ [Accessed June 20, 2018].

Roskamm, N. (2017) *Die unbesetzte Stadt: Postfundamentalistisches Denken und das urbanistische Feld.* Gütersloh, Birkhäuser.

Roskamm, N. (2011) *Dichte. Eine transdisziplinäre Dekonstruktion. Diskurse zu Stadt und Raum.* Bielefeld, transcript-Verlag.

Schneider, M. (2005a) Dem Versprechen ent-sprechen. Kontraktuelle Sprachmanöver. In: Schneider, M. (ed.) *Die Ordnung des Versprechens.* München, Wilhelm Fink Verlag, pp. 395–419.

Schneider, M. (2005b) (ed.) *Die Ordnung des Versprechens.* München, Wilhelm Finkg Verlag.

Simmel, G. (1903) Die Großstädte und das Geistesleben. In: Petermann, T. (ed.) *Die Großstadt. Vorträge und Aufsätze zur Städteausstellung*, Vol. 9. Jahrbuch der Gene-Stiftung, Dresden, pp. 185–206.

Simone, A. (2015) Sociability and endurance in Jakarta. In: Frichot, H., Gabrielsson, C. & Metzger, J. (eds.) *Deleuze and the City.* Edinburgh, Edinburgh University Press, pp. 224–240.

Sloterdijk, P. (2016 (2004)) *Foams: Spheres Volume III: Plural Spherology.* Cambridge, MA, MIT Press.

Stengers, I. (2005) The cosmopolitical proposal. In: Latour, B. & Weibel, P. (eds.) *Making Things Public: Atmospheres of Democracy. [Exhibition at ZKM, Center for Art and Media Karlsruhe 20.03. – 03.10.2005].* Cambridge, MA, MIT Press, pp. 994–1003.

Thorsen, L. M. & Blok, A. (2016) Reset Latour! *EASST Review*, [Online] 35 (2). Available from: https://easst.net/article/gedankensprungeausstellung/ [Accessed June 20, 2018].

Tonkiss, F. (2011) Template urbanism. Four points about assemblage. *City*, 15(2), 584–588.

Ureta, S. (2015) *Assembling Policy. Transantiago, Human Devices, and the Dream of a World-Class Society.* Cambridge, MA, MIT Press.

Wirth, L. (1938) Urbanism as a way of life. *American Journal of Sociology*, 44(1), 1–24.

Wittrock, B. (2000) Modernity: one, none, or many? European origins and modernity as a global condition. *Daedalus*, 129(1), 31–60.

Yaneva, A. (2012) *Mapping Controversies in Architecture.* Farnham, Ashgate.

Zitouni, B. (2004): Paris ville invisible – un diorama sociologique, Compte-rendus de sites internets. Available from: www.ethnographiques.org/2004/Cr-Zitouni [Accessed June 21, 2018].

26

How to study the construction of subjectivity with ANT?

Arthur Arruda Leal Ferreira

The question about the presence of the issue of subjectivity (or the modes of its production) in Actor-Network Theory (ANT) won't receive as generous an answer as it would if it concerned Gilles Deleuze's and Felix Guattari's schizoanalysis or Michel Foucault's archaeologies and genealogies. Amongst the writings of the best-known, earlier proponents of ANT, the clearest references to subjectivity are to be found in the work of Latour (1987, 1991, 2004, 2005, 2011), even if relevant methodological discussions can be found in both Law (2004) and Mol (2002). Clearer still are the references one finds in the work of Isabelle Stengers (1989, 1992a, 1992b, 1999) and Vinciane Despret (1999, 2004, 2009, 2011a) and in the field of Political Epistemology (PE). Here, the contributions of Stengers and Despret will be considered in their connections and their interface with the work of Latour, especially due to their shared interest in concepts such as risk, generalisation and recalcitrance.

Because Latour is the key author (amongst the classics of ANT) when it comes to exploring the issue of subjectivity, I take his writings as the basis for the present chapter. There are four relevant threads running through his work:

1   Criticism of the modern division between a human/social/political/subjective domain and a natural/universal/scientific/objective domain, where the function of psychological knowledge would be to work as a stabiliser for the first group (especially devoted to subjectivity) and the function of epistemologies would be to work as a stabiliser for the second group (Latour 1987, 1991, 1996, 1999). According to him, nature, society, objectivity and subjectivity are all effects of processes of purification, and they are the last to arrive, like Elizabeth II of England when the Parliament makes a decision (Latour, 1987).

2   An amodern alternative – represented by the ethnopsychiatric work of Tobie Nathan (see Latour, 1996, 1998b, 2011) – to the accounts of subjectivity that result from the 'modern divide' and purification work.

3   An approach to subjectivity as an effect produced by processes of various kinds within a realist–constructivist perspective (Latour 2003, 2004, 2005).

4   A description of concrete and specific situations whereby subjectivities are produced by sociotechnical devices (Plan 49 in Latour & Hermant, 2009).

A Berlin wall in modernity

Let us start with the first aspect. Even if the writings of Latour (1987, 1991, 1996, 2002, 2011)[1] under consideration have different purposes and directions, they also have a common basis: They extend David Bloor's (1976) principle of symmetry, which calls into question the opposition between modes of analysis for 'successful' ('correct') and 'failed' ('incorrect') forms of science. Thus, further dualities, or further alleged essential oppositions, are symmetrised and called into question by ANT: Culture X Nature; Subjectivity X Objectivity; Modernity X Pre-modernity; Civilisation X Primitivity. Those oppositions do not intersect but they do touch, especially the first two. Resistance to the division between subjectivity and objectivity – of most interest here – is the central subject of *Pétite Réflexion sur le Culte Moderne des Dieux Faitiches* (Latour, 1996). Indirectly and by approximation, we can also rely on the criticism of the division between culture and nature as it appears centrally in many texts (Latour, 1989, 1991 1999, 2011). The ontological thesis is that all of these divisions are peculiar to specific, localised historical settings, such as the 16th-century debate between Robert Boyle and Thomas Hobbes, as in Latour's (1991) reading of the work of Shapin and Schaffer (1985).[2] Before the modernist purification of culture and nature, fetishes and facts, all what exists are various configurations of hybrids, or *factishes*.[3]

It is important to link this debate to Latour's (2003) realistic constructivism, according to which the more complicated the practices and arrangements constructing a thing, the more real it gets, thus eschewing the traditional division between realism (according to which reality is objective and not constructed) and constructivism (which points to processes of artificial production). In this line of reasoning, subjectivity and objectivity, nature and society, are not starting points but consequences of processes of purification (Latour, 1991) or drainage (Latour, 1996), regulated as they are by the *Modern Constitution* (Latour, 1991), the main effect of which is to multiply the hybrids (Latour, 1991) and *factishes* (Latour, 1991, 1996).[4]

In *Pétite Réflexion sur le Culte Moderne des Dieux Faitiches*, psychology is called into question as the great heir, together with epistemology, of the modern separation of subjectivity and objectivity. This division acts as a suction pump, attracting entities to each of these poles of purification (subjective and personal versus objective and real). On the epistemological side, we find the debate about the production of facts; on the psychological side, we find the debate about the ways in which subjectivity – understood in terms of beliefs, that is, in terms of our errors relative to scientific reason – is produced. As Canguilhem (1966) puts it, psychology emerges in the modern project, as the mind's apology to scientific reason.

The fabrication of the selves

The alternative lies, according to Latour (1991), in fostering an amodern attitude (not anti-modern, nor premodern, nor yet postmodern). Such an attitude is characterised by a positive view of the hybrids and of the processes of fabrication of entities that occur in various networks, such as our modes of subjectivity. The alternative to the purification that is associated by Latour with most psychological traditions lies, according to him, in the ethnopsychiatric work of Tobie Nathan (Latour, 1996, 1998b, 2011). This leads us to the second aspect of Latour's approach to subjectivity.

For Latour, Nathan swims against the ethnopsychiatric current formed by the work of other experts, such as Georges Deveraux, who have always attempted to understand the experience of non-Western groups in terms of internal, unconscious entities, more in line with republican, modern French ways (Latour, 1998b). The way in which Nathan handles issues related to subjectivity avoids purification; he does not convert variegated experiences into internal entities. According to him, non-Western human groups share with Westerners the ontological belief in the existence of invisible beings (or beings of metamorphosis; see Latour, 2011). However, for non-Westerners, these invisible beings are not internal, personal or governed by universal psychological laws. They aren't bound by the subjective limits of a world of beliefs and fantasies, but cohabit a common vital world. Nathan's work is achieved by acknowledging and welcoming, in French territory, the non-modern experiences reported especially by his immigrant, diverse clientele, without translating them into modern terms or reducing them to some ultimate, true view of our subjectivity. Even better, he multiplies versions of the fabrication of our subjectivity, and of our selves (Latour, 1998b), a process in which intervention and suggestion are key aspects.[5]

Accordingly, it is possible to think of subjectivity as a hybrid production, in much the same way in which Latour thinks of the production of reality by technical–scientific devices (Latour, 1987; Latour and Woolgar 1979). Here, we find the third and fourth meanings that open up new leads for studies in subjectivity production, especially as it is achieved by psycho-techno-social, and therefore less modern, devices (a theme that has developed in the Iberian-Latin American literature)[6]. In these third and fourth meanings, we find references to rather peripheral parts of Latour's books and papers (Latour, 2004, 2005, 2009).

Constructing the subject

The best leads as to how Latour conceives of this manifold production of subjectivity in a positive way (the third sense)[7] can be found in two texts: *How to talk about the body?* (2004) and the thesis about subjectivity plug-ins developed in *Reassembling the Social* (2005), where he discusses a version of the social in terms of face-to-face meetings. The first question that comes to mind is: What could a paper about the body teach us about the production of subjectivity? By eschewing every modern division (such as that between subjectivity and body), we can, even if this isn't easy, understand those two modes of the same process of constructing hybrids or *factishes* as constantly co-articulating and co-affecting entities.[8] Here, the body cannot be defined as 'an essence, a substance (...), but rather, (...) an *interface* that becomes more and more describable as it learns to be affected by more and more elements' (Latour, 2004, p. 205). As an example of this power of being affected, Latour considers the learning process of specialists in the perfume industry (also called 'noses') with a kit of odours, during which they are trained to make entirely new distinctions of smell. For the author, this learning process is not the simple acquisition of an ability by a natural body, but its reciprocal construction with the world: 'Acquiring a body is thus a progressive enterprise that produces at once a sensory medium and a sensitive world' (Latour, 2004, p. 207). This articulate body allows one to experience and obtain knowledge about the world (and of other bodies) by being co-produced together with them, registering differences in new modes and expanding contrasts between phenomena which were previously considered homogeneous: 'Articulation thus does not mean ability to talk with authority (...) but being affected by differences' (Latour, 2004, p. 210). Without a division between subjectivity and the body (or other materialities), it is not difficult to understand subjectivity as produced by multiple modes of articulations.

In *Reassembling the Social* (2005), Latour even more directly reflects on the composition of subjectivity in a comparison with plug-ins, discussing the possibility of thinking the social in an elementary form as the face-to-face meeting of people. However, for Latour, even the psychological abilities that are supposedly more pure than others (be they affective or cognitive) are not natural entities in people, but rather are incorporated in the same way that a plug-in is installed in a computer: 'They could be called subjectifiers, personalizers, or individual-isers, but I prefer the more neutral term of plug-ins, borrowing this marvellous metaphor from our new life on the Web' (Latour, 2005, p. 207). All the components that constitute these abilities flow and are distributed in a series of materialities and practices that compose the psycho-morphs, which are the creators of our subjectivity: 'floods, rains, swarms of what could be called psycho-morphs, because they literally lend you the shape of a psyche' (Latour, 2005, p. 212). Thus, in the same way that novels and soap operas can constitute our affections, calculation practices and linguistic distinctions can generate our cognitive forms: 'Of course I am, but only as long as I have been individualized, spiritualized, interiorized' (Latour, 2005, p. 212).

Maybe the model of subjectivation through plug-ins is too simplified: In the same way that for Latour objective reality is produced through extensive controversies in action (Latour, 1987), the production of subjective reality is certainly a process that is as much or even more unstable and conflicting. What makes ANT interesting in this case, more than its strict doc-trine, is the possibility of presenting a strong constructivist and controversial approach to the production of subjectivity (Latour, 1999). Research in subjectivity production must take this plural and frictional composition into account. However, before we develop this reflection, we must explore the fourth meaning proposed by Latour.

Inside psycho-techno-social devices

The forth possible approach to the issue of subjectivity is even more peripheral in Latour's work. We can briefly mention *Plan 49* in *Paris: Ville Invisible*, written in collaboration with Emilie Hermant (Latour and Hermant, 2009). Here, Latour clearly refers to the way in which Parisians subscribe to 'the psy sciences' and the humanities. But the two fast pages of that plan function more like a tasting menu or a free sample describing these practices.[9] How do those practices of subscription happen? How do processes of subjectivity production take place here? How can we carry out such studies with what Latour (2005) calls the light equipment of ANT, following the actors in their modes of composition without forcing on them the weight of the great explanatory categories, shaped by the modern dualities? Here, we move away from Latour's writings and into the work of other ANT scholars (Law, 2004; Mol, 2002) and allies (Stengers 1989, 1992; Despret, 1999, 2004, 2011a) to explore processes of subjectivation engendered by sociotechnical devices (and especially the devices of the human sciences).

The production of worlds and modes of research

Starting from Latour's approach, which is at the same time constructivist and realist and attuned to controversy, how can we think about the ways in which we can study the pro-duction of subjectivity? Before answering this question, we need to better understand the performativity of scientific knowledge. For this, I will rely on ANT-related approaches that focus on various political modes of scientific knowledge production, such as the work of Mol and Law, as well as Stengers' and Despret's PE.

For ANT and PE, scientific knowledge is constituted of the traces of a plural performance, in the articulation and coaffectation between various actors and in the unexpected produc-tion of effects, permeated by controversies (Latour, 2004). If scientific knowledge can no longer be distinguished in terms of good and bad representations, assuming an external term of evaluation (as traditional epistemologies assume), how can it be evaluated? The answer lies in the specific modes of connection between entities, in a microphysical analysis of the modes of articulation. What characterises bad or good articulations? In the first case, we have a situation where the articulation between various entities is extorted or conditioned to an expected response, without any risk. In the second case, we have an articulation where the testimony goes beyond a mere response, opening up to the risk of invalidating the research-er's own questions and propositions and letting the research raise new questions: In other words, this constitutes a recalcitrant relationship. Scientific knowledge thus does not lead to a thin and uniform reality but to the unfolding of diverse possible worlds and subjectivities, including researchers and researched entities (Despret, 2004; Latour 2004).

From this stems a singular characteristic of ANT and PE regarding the multiplicity of scientific knowledge. Most epistemological approaches identify multiplicity as a sign of a pre-scientific state, while the presence of a unified project and its rationality (Canguilhem, 1966) or a paradigm (Kuhn, 1962) is a sign of scientific activity. In ANT and PE, in contrast, multiplicity is considered more positively. Referring especially to PE, Latour (2004, p. 220) points out this positive approach: 'Generalization should be a vehicle for travelling through as many differences as possible – thus maximizing articulations – and not a way of decreasing the number of alternative versions of the same phenomena.' This completely inverts certain critiques of the scientific status of human sciences, such as that established by Canguilhem (1966), referring to psychology.

Similarly, Mol (2002) and Law (2004) take multiplicity as a positive aspect of many sci-entific and technical devices in their concept of ontological politics. For these authors, more than being a representation of a pre-existing reality in different perspectives, scientific prac-tices produce distinct (multiple) worlds, without any ultimate unity (singularity), but are also not entirely inarticulate (plurality). Here, the term 'multiplicity' is specified: It is not an anomaly in a unique and singular world, as is the perspective of Euro-American metaphysics[10] (Law, 2004, p. 25), and neither does it point to a plurality of events without connections: 'We are in a world where bodies, or organisations, or machines are more than one and less than many. In a world that is more than one and less than many. Somewhere in between' (Law, 2004, p. 62).

An example of this performed multiplicity can be found in Mol's study (2002) of arte-riosclerosis, which is not seen as a pathological state that is inherent to the body that is rep-resented in different perspectives (in the laboratory or in clinical exams). Instead, each one of these scientific practices enacts a mode of arteriosclerosis, a pathological reality that does not necessarily converge, but that is also not entirely disjoined. Mol uses the term political ontologies, since each scientific practice constructs a particular reality among other possible ones. This, in an interplay with other scientific practices, constitutes a multiverse: More than one and less than many.

Clinical devices and modes of production of subjectivities

How can we conduct research on the modes of subjectivity production present in diverse technoscientific networks? First, we need to consider the remarkable diversity of political ontologies present in the many scientific devices, and expressed in their multiple capacities of fabricating subjects or 'artificial selves' (Latour, 1998b). As we have seen before, this produc-tive aspect of subjectivities is not considered by ANT or PE as a parasitic remnant, but as an actual aspect of the production of knowledge. The problem of certain scientific devices (such as those used in most psychological research) is that they very frequently present a mode of knowing based on the extortion of their witnesses (Stengers, 1989), not only in the way the tasks are demanded, but especially how these observers are positioned, rarely introducing problems or questionings. Here, Latour, Stengers and Despret all highlight the relative obedience and docility of human beings towards scientific authority:

> Contrary to non-humans, humans have a great tendency, when faced with scientific authority, to abandon any recalcitrance and to behave like obedient objects, offering the investigators only redundant statements, thus comforting those same investigators in the belief that they have produced robust 'scientific' facts and imitated the great solidity of the natural sciences!
>
> *(Latour, 2004, p. 217)*

For Latour (1997, p. 301), the human sciences (psychology included) would only really become sciences if they did not imitate the objectivity of the natural sciences, but the possibility of recalcitrance. In other words, we can say that their major problem is not their capacity to influence and produce subjects, but to extort and create docility, inhibiting the possibilities of recalcitrance. In this way, while seeking the 'objectivity of the natural sciences,' the human sciences frequently extort their research subjects, thus producing standardised subjectivities.

Despret (2004) provides a very interesting discussion about this, focusing on the design of psychological research. She points out that much of the history of psychological methods can be told as the change from a design in which the research subjects should be trained experts (as it was done in the first psychological laboratories at the end of the 19th century based on practices of introspection), to one in which the participants should be deprived of any information about the objectives, questions, resources and hypotheses of the research. The point of depriving subjects of information was to avoid the researchers' influence in the production of experimental results. Thus, the research subjects are shifted from being experts, with an importance frequently exceeding that of the experimenters, to being 'anyone.'[11] Importantly, Despret proposes a reversal of the way most manuals in the history of psychology read this shift: Moving from the trained subjects of the first psychological laboratories to the naïve subjects present in most of our present laboratories is not a step *forward* in the direction of objectivity and control, but a step *backward* from the possibility of recalcitrance, creating docile, asymmetric and limiting articulations in relation to its witnesses: 'Subjects without the excellence of expertise do not pose a risk of taking up a position in investigations' (Despret, 2004, p. 97).

Lost in a labyrinth: walking through a division of applied psychology

Rather than continuing this historical analysis of the changes in research design in the psy-disciplines into the present, in the remainder of this chapter, I'd like to draw attention to the micropolitical effects of research practices: Producing worlds or subjectivities involves great political risks. More specifically, I'll dwell on a research programme that our group has been developing at the Division of Applied Psychology (DAP) of the Institute of Psychology of the Federal University of Rio de Janeiro (UFRJ), in order to follow the diverse ways in which certain psychological guidelines entail distinct types of production of subjectivity (Ferreira et al., 2013, 2014). The DAP offers some psychotherapeutic services of different orientations (psychoanalysis, cognitive-behavioural therapy (CBT), gestalt therapy, institutional analysis and existential analysis) to the community as a low-cost service, using as therapists student interns, who are supervised by a professor or a psychologist of the Institute of Psychology.

Our research is being done through interviews and field observations of the participants' actions (patients, interns and supervisors) when they are put in contact with different psychological services, following the many ways in which these subjects are formed through different kinds of articulations. Our research makes use of some parameters of the ANT and PE: The participants are taken as experts on the topic, without any division between common and scientific knowledge; and, since they are considered experts, we ask them to describe their practices and experiences, as well as the clinical processes. In other words, we follow the actors and their descriptions of their actions (Latour, 2005, Chapter 7, part 1).

We have been observing supervision sessions (where the clinical cases were discussed by supervisors and interns)[12] and interviewing patients, interns and professors–supervisors across clinical services of different orientations. We also attend to further actors in this scenario such as the architecture of the building, the laws that regulate internship, the rooms where the sessions are held and so on. A base questionnaire works as a general script, allowing an open space for the hints and indications of the participants. In order to prevent the questions from being seen as a test of knowledge, we often ask the participants to share their thoughts about what they consider to be the best research topic for a group studying the DPA to pursue (as proposed by Despret, 2011b). In this way, we expect that the participants can assume a position of expertise that can favour a more recalcitrant effect. One of the objectives we are pursuing with this research is to understand how the negotiation of meanings between such different lines of thought with different parameters can happen in this common space. Latour (1999), when speaking about the creation of scientific concepts, coined the concept of 'immutable mobiles,' which allow 'new translations and articulations, while keeping some forms of relationships intact.' Generally in psychological technical devices, we point out that, unlike other sciences and technics, there would be several 'mutable immobiles' (Ferreira et al., 2013); they are immobile because they are restricted to a certain orientation, and mutable thanks to their potency for producing different subjectivities in connection with their sociotechnical arrangement.[13] In other words, there is the unlikely possibility of articulation between different psychological orientations in the DAP, which leads us to suspect that this set of practices is much better characterised by plurality than by multiplicity in Law's (2002) terms.

We have searched for a point in which these practices connect and articulate concepts and experiences between interns and patients or even the circulation of the last between different therapies. What we observed is that there is no form of contact that can guarantee even the slightest articulation as multiplicity: The interns are not allowed to discuss the cases during triage and, even though the various teams are open to the circulation of patients from team to team (as they claim in the interviews), this rarely occurs (according to the same interviews). How does this state of plurality engender processes that produce subjectivity? We can investigate this both from the perspectives of the interns and of the patients.

On the intern-therapist's side, much of their education consists in the constant criticism of other orientations; it is almost as if part of the pedagogy of becoming a psychologist of a certain school was to accept its share of criticism against the other schools. This is in line with Foucault's (1994) claim that psychology grows by denouncing the myths of other orientations (concerning the duration and focus of therapies as well as the concepts of the orientations). An interesting effect of subjectivation (Ferreira, 2013) takes place when an intern inhabits more than one orientation: For instance, there was an intern working in a psychoanalysis team who reported having worked in a CBT research group. Even if she wasn't constrained in a more problematic way, she reported a series of daily acts of prejudice, involving stereotyped views of both orientations: Of CBT as a practice of self-help and of psychoanalysis as linked to the question of sexuality or to severe abnormalities. She even reported being questioned for having both of those experiences on her CV. This leads us to the conclusion that the obstacles to the circulation and composition of a common world across different psychological orientations can be radical to the point of it being impossible for them to inhabit the same professional career or the same body. A process like one of purging (of the former practices) and of conversion (to the new ones) would seem to be necessary.

This question of plurality does not affect the patients when it comes to defining an identity: The association with a group that manages a specific therapeutic practice isn't an issue for them. Likewise, their processes of subjectivation are not limited to transformations in the therapeutic practices: Patients link therapy to very diverse practices in very active ways.[14] We can define patients in various ways, but not as passive and patient creatures. When analysing the interviews, rather than classifying the practices of a given orientation as either extortive or as favouring recalcitrance, we found a series of clues in the uses that patients made of the therapies that pointed to specific techniques of the self. These manifested themselves in attitudes of problematisation of the self and instances of collective life – such as prejudice, stereotypes and subliminal messages – which led to rather peculiar exercises, such as the constitution of diaries and the appropriation of therapists' discourses.

Rather than simply classifying the clinical practices as producing docility or as open to recalcitrance, we tried to use those concepts to assess the openness that research itself can operate in relation to recalcitrant discourses, acting on modes of subjectivity production. In this sense, reflecting on our practices should be constant. We have here intended this practice of becoming unfamiliar with ourselves to involve both the act of researching itself and the revision of concepts (such as the boundary between docility and recalcitrance). In this way, we can ask ourselves how mutually exclusive, when it comes to the modes of articulation, recalcitrance and docility are.

A short answer

This chapter offers a way of exploring in an ANT fashion some devices of the Psy-sciences, considering the third and the fourth approaches of subjectivity proposed by Latour with some methodological reflections of other ANT (Mol and Law) and PE authors (Stengers and Despret). Promoted also by some researchers of Ibero-American countries (An interesting map in Ferreira & Carrasco, 2017), this approach can, from a southern perspective (Santos, 2014; see also Rosa, this volume), foster new perspectives for studying the production of subjectivities, similar to what ANT did for research into the modes of production of realities. Only the polyphony of the field, in my view, can expand our versions of subjectivity.

Notes

1 While *Science in Action* (Latour, 1989) is clearly a textbook, *We have never been modern* (Latour, 1991) contains deeper philosophical and historical debate. *Pétite Réflexion sur le Culte Moderne des Dieux Faitiches* (Latour, 1996) is halfway between being textbooks and being works in which he develops his views conceptually.
2 In the final chapters of *Pandora's Hope* (1999), Latour gestures at the idea that the division between science and politics began in the debate between Socrates and Callicles.
3 Wordplay: in French, *faitishes* results from joining together *fait* (fact) and *fétiche* (fetish).
4 Here, the argument is that the modern divide of nature and culture would not stop the proliferation of hybrids and factishes, but, quite on the contrary, the modern constitution leads to their

unaccountable multiplication and extension. We have never been modern (Latour, 1991) gives us a lot of examples such as the Green Parties or the Carbon debate.

5 For the differences between views and versions, see Despret (1999) and Latour (2004), especially concerning the principle of generalisation in Political Epistemology. For the question of influence and its conviction in Modernity for being a kind of cure that happens for bad reasons, see Stengers (1992a).

6 This matter has had a strong presence in the Iberian-Latin American literature (see Ferreira and Carrasco, 2017).

7 In that case, there is a descriptive activity, which consists in following the actors (2005) instead of focusing only on criticising, as Latour himself seems to do in most of his texts about subjectivity.

8 A further virtue of that paper is that in it, Latour makes his proximity to Political Epistemology clear by providing a list of its ten guiding principles (such as the generalisation, the risk and the recalcitrance principles).

9 Perhaps some references need to be done to some medical-psychiatric social versions of alcoholism in Law (2004, Chapter 5).

10 John Law (2004, pp. 24–26) considers Euro-American metaphysics as the common assumptions of the major modes of scientific knowledge, such as the precedence and independence of the reality to be represented, besides its definability, singularity and universality.

11 In this book, *Hans, le cheval qui savait compter*, Despret (2004) notes that this change in the experimental designs can be observed in a specific laboratory where a researcher (Pfungst) tries to analyse the real ability of a horse (Hans) to make complex mathematical calculations. Suspecting that most of the horse's correct responses were due its ability to react to human expressions (supposedly present when the horse was reaching the result by tapping his foot on the floor), Pfungst decided to investigate by introducing in the experiment a person that did not herself know the question proposed.

12 Supervision sessions are weekly meetings between supervisors and interns aiming to discuss the cases conducted by the latter. Their function is not only to improve the approach of the cases, but mainly to train the students in clinical practices.

13 This use is in some sense similar to Guggenheim's (2009) analyses of buildings.

14 This is similar to what Foucault (1995) calls techniques of the self.

References

Bloor, D. (1976) *Knowledge and Social Imagery*. Chicago, IL: University of Chicago Press.

Canguilhem, G. Qu'est-ce que la psychologie. In: *Cahiers pour l'Analyse*, n. 2, 1966, pp. 77–91.

Despret, V. *Ces émotions que nous fabriquent*. Etnopsychologie de l'authenticité. Le Plessis-Robinson: Synthélabo, 1999.

———. *Le cheval qui savait compter*. Paris: Les Empecheurs de Penser en Ronde, 2004.

———. Dôssie Despret. *Fractal*: Revista de Psicologia, Niterói, UFF, v. 23, n. 1, pp. 5–82, jan./abr. 2011a. Os dispositivos experimentais.

———. Vinciane Despret comenta as apresentações de Heliana Conde e Arthur Arruda Leal. In: *Práticas e Pesquisas Psicossociais*, v. 6, n. 2, 2011b.

Ferreira, A. et al. A produção de subjetividades em rede: Seguindo as pistas de uma divisão de psicologia aplicada. *Universitas Humanistica.*, v.76, pp. 371–392, 2013.

———. Técnicas de si e clínica psi: um campo de estudos etnográficos. *Polis e Psique*, v.4, pp. 80–99, 2014.

Ferreira, A. & Carrasco, J. *Campo CTS na América latina e domínios psi: pequena crônica de encontros e desencontros*. In: *Backchannels* (4S). Accessed in: www.4sonline.org/blog/post/campo_cts_na_america_latina_e_dominios_psi_pequena_cronica_de_encontros_e_d Published in 2017.

Foucault, M. La recherche scientifique et la psychologie. In: Defert, D. e Ewald, F. (Orgs.). *Dits et Ecrits*. Gallimard, Paris, 1994.

———. Genealogia da ética. In: Dreyfuss, H.; Rabinow, H. (Org.). *Michel Foucault*: uma trajetória filosófica. Rio de Janeiro: Forense Universitária, 1995.

Guggenheim, M. Mutable immobiles: building conversion as a problem of quasi-technologies. In I. Farías & T. Bender (Eds.), *Urban Assemblages. How Actor-Network theory Changes Urban Studies* (pp. 161–178). London: Routledge, 2009.

Kuhn, T. *The Structure of Scientific Revolutions*. Chicago, IL: University of Chicago Press, 1962.

Latour, B. *Science in Action: How to Follow Scientists and Engineers through Society*. Cambridge, MA: Harvard University Press, 1987.

———. *Nous n'avons jamais été modernes*. Essai d'anthropologie symétrique, Paris, La. Découverte, L'armillaire, 1991.

———. *Petite réflexion sur le culte moderne* des *dieux faitiches*. Éd. Synthélabo, coll. Les Empêcheurs de penser en rond, 1996.

———. Des sujets recalcitrants. *Recherche*, p. 301, set. 1997.

———. O exótico homem das cidades. *Folha de São Paulo*, São Paulo, 12 April. 1998a. Mais!, p. 3.

———. Universalidade em pedaços. *Folha de São Paulo*, São Paulo, 13 September. 1998b. Mais!, p. 3.

———. *Pandora's Hope*. Essays on the Reality of Science Studies. Cambridge, MA: Harvard University Press, 1999.

———. The promises of constructivism. In: Don Ihde (ed.) *Chasing Technology: Matrix of Materiality, Indiana Series for the Philosophy of Science*, Indiana University Press, pp. 27–46, 2003.

———. How to talk about the body. *Body & Society*, v. 10, n. 2–3, pp. 205–229, 2004.

———. *Reassembling the Social: An Introduction to Actor-Network Theory*. Oxford: Oxford University Press, 2005.

———. *Cogitamus: six lettres sur les humanités scientifiques*. Paris: La Découverte, 2011.

———. *An Inquiry of Modes of Existence*. Cambridge, MA: Harvard University Press, 2013.

Latour, B. & Hermant, E. Paris: Ville Invisible. Les empêcheurs de penser en rond, 2009.

Latour, B. & Woolgar, S, *Laboratory Life. The Social Construction of Scientific Facts*. Beverly Hills and London, Sage Publications, 1979.

Law, J. *After Method*. New York: Routledge, 2004.

Mol, A. *The Body Multiple: Ontology in Medical Practice*. Durham, NC: Duke University Press, 2002.

Nathan, T. Entrevista com Thobie Nathan. *Cadernos de Subjetividade*, n. 4, 1996.

Santos, B. de S. *Epistemologies of the South: Justice against Epistemicide*. Boulder, CO: Paradigm, 2014.

Shapin, S. & Schaffer, S. *Leviathan and the Air-Pump: Hobbes, Boyle and the Experimental Life*. Princeton, NJ: Princeton University Press, 1985.

Stengder, I. *Quem tem medo da ciência?* São Paulo: Siciliano, 1989.

———. *La volonté de faire Science*. Les empêcheurs de penser em rond, 1992a.

Stengers, I. & Chértok, L. *A Critique of Psychoanalytic Reason: Hypnosis as a Scientific Problem from Lavoisier to Lacan*. Les empêcheurs de penser en rond, 1992b.

Stengers, I. & Nathan, T. *Médicines et Sorciers*. Les empêcheurs de penser en rond, 1999.

27

Why do maintenance and repair matter?

David J. Denis[1]

The rise of maintenance and repair studies

In 'Power, technology and the phenomenology of conventions: on being allergic to onions,' Star (1991) recalls how science and technology studies helped reconsider the question of power in social science. One of the main outcomes of STS, she states, is that they demonstrate that both science and technology are what Latour (1987) called 'politics by other means.' Identifying the contributions of different streams of research and the more or less passionate disputes that arose between them, Star specifically highlights the weaknesses and the gaps of ANT-oriented accounts. Most of those, she claims, following Haraway (1991) and Fujimura (1991), are a matter of standpoints. While Callon, Law and Latour, among others, aim to open the black boxes of facts and artefacts in order to describe hitherto invisible processes, they stand – and remain – on the side of the 'winners': Those whose translations and *intéressements* are successful, and those who delegate and discipline. ANT scholars don't devote many words, if any at all, to those who are translated, disciplined and delegated to. This is highly problematic, Star argues: Not only are the experiences of the latter worth examination in order to study 'politics by other means,' but the place of these 'losers' and their own invisible work are crucial to the very stability of the networks in which they participate. To illustrate her point, Star returns to Latour's famous 'sociology of a door-closer' (Latour, 1988) and dwells on one of his sentences:

> As a technologist, I could claim that, provided you put aside maintenance and the few sectors of population that are discriminated against, the groom does it job well, closing the door behind you constantly, firmly and slowly.
>
> *(Latour, 1988, p. 302)*

To that hypothetical statement, Star retorts that:

> *There is no analytic reason to put aside maintenance* and the few sectors of population that are discriminated against, in fact, every reason not to.
>
> *(Star, 1991, p. 42, my highlights)*

During the two decades that followed the publication of Star's paper, numerous works in STS have enriched the critique of ANT, thanks notably to feminist scholars. ANT 'itself' has evolved, and its most well-known contributors have taken some of these recriminations into consideration (Latour, 2005; Law & Hassard, 1999). Yet, most of these critiques and adjustments have echoed only the second part of Star's statement. It's first half has long been ignored, with maintenance remaining an obscure, largely unexplored domain. Recently, however, a body of academic work on maintenance and repair has been rapidly growing, providing important contributions to science and technology studies by shifting the attention from the processes of 'form-giving' to those of 'form-keeping' in Ingold's terms (Ingold, 2013).

Even though Akrich (1993), Orr (1996) and de Laet and Mol (2000) have paved the way, it is undoubtedly the contributions of Henke (2000) and Graham and Thrift (2007) that set the agenda[2] of what we can call today 'maintenance and repair studies.' In these seminal papers, the three authors shape the problematisation of maintenance and repair, building on Goffman's work on 'backstage' activities and the vulnerability of social interaction (Goffman, 1959), and on the ethnomethodological analysis of 'repair' in conversation (Schegloff, Jefferson & Sacks, 1977). Their demonstration is straightforward: If social order is constantly maintained by generally unnoticed gestures during interaction, we should also investigate the operations that daily shape and preserve material order. Whether accomplished by dedicated workers or lay persons, made on huge technological systems or small objects, these operations, which are 'the main means by which the constant decay of the world is held off' (Graham & Thrift, 2007, p. 1), are indeed countless. Their close and careful examination should, therefore, considerably help refine and strengthen our understanding of the role of objects and technology in the very constitution and continuation of modern societies. This is what Edgerton, another pioneering scholar on the subject, please for in his book *The Shock of the Old*, a chapter of which is entirely devoted to maintenance. Maintenance and repair, he states, though particularly difficult to grasp, are essential dimensions of the global history of technology (Edgerton, 2006, p. 75–102).

Since these early appeals, maintenance and repair have been of growing interest in various areas of research, and more and more scholars have brought to light an incredible variety of hitherto neglected objects and practices. In the last years, studies have documented maintenance and repair activities in ICTs (Cállen & Sánchez Criado, 2015; Houston & Jackson, 2016; Jackson, Pompe & Krieshok, 2012; Rosner & Ames, 2014), Arts (Domínguez Rubio, 2016), large infrastructures (Barnes, 2017; Ureta, 2014), software and information systems (Cohn, 2019; Fidler & Russel, 2018), urban settings (Denis & Pontille, 2014, 2019; Strebel, 2011), legacy buildings and heritage sites (Edensor, 2011; Jones & Yarrow, 2013), domestic consumption (Gregson, Metcalfe & Crewe, 2009; Rosner, 2013) and even corpse preservation (Yurchak, 2015). In the following sections, I propose to give hints on these emerging maintenance and repair studies by articulating, beyond their variety, two of their main contributions. First, I'll show that maintenance and repair studies help reconsider an old legacy of ANT: The opposition between breakdown (crisis, controversy) and routine (taken-for-grantedness). Second, I will suggest that these studies renew the way matter (or 'materiality') is generally treated in ANT accounts, namely as 'that which resists.'

Before breakdown, beyond closure

Most researchers investigating maintenance and repair are wary of what Edgerton calls 'innovation-centric accounts' (Edgerton, 2006, p. xi). They aim instead to produce more balanced depictions through the (re)discovery of the people, workers and users who participate in the daily life of technologies, long after their invention. Maintenance and repair studies thus append a constantly expanding gallery of portraits to the figure of the lonely innovator. Withdrawing from the innovation-centric approach by exploring maintenance and repair activities also helps to widen the understanding of objects themselves. Things that are maintained and repaired are far from what Jackson calls the 'bright and shiny tools' (Jackson, 2014, p. 227) that innovation-centric accounts exclusively focus on, and that studies of technology generally stage. As Domìnguez Rubio (2016) reminds us, the brand-new, perfectly shaped and fully functional objects some scholars seem to have in mind when they insist on material agency are sparse in our daily lives. We rarely deal with pristine artefacts. Rather, we use and live surrounded by countless things that age more or less dramatically. Actually, technologies usually live a long life, even the most complex ones: Ships, airplanes, cars (Edgerton, 2006), urban infrastructures (Graham, 2010), software (Cohn, 2019)... Yet, long-lasting objects are neglected in innovation-centric accounts, and in most STS reflections. So are the manifold operations, from routine maintenance to exceptional repair, that take part in the durability of these objects. Maintenance and repair studies aim at bringing this unseen territory to the fore.

In doing so, these studies also encourage reconsideration of ANT's traditional approach to what appears to be the opposite side of brightness and sheen: breakdown. Since ANT's inception, breakdown has played a crucial role in what we can call its Heideggarian posture. This posture distinguishes two conditions for technologies. The first one is characterised by ordinariness and describes a technology 'ready at hand' in the sense of Heidegger (Verbeek, 2006), that is taken for granted. Used without being questioned, artefacts here function as 'black boxes': Stable objects whose composition is mostly unknown. It is this very mundaneness that gives power to technologies, most ANT accounts argue. Since nobody knows what these artefacts are made of, the decisions and the assumptions that are inscribed in them become naturalised. Automation, routine and mundane uses are the 'other means' of politics here. But such a peaceful state is problematic for scholars who, under such conditions, cannot grasp the political and moral dimensions of technologies. In order to reveal these dimensions, ANT and most STS scholars claim, one has to find the moments when technologies become 'present at hand,' that is debated, discussed and problematised anew. Accidents, controversies and breaks are typical of these moments: They are occasions of breakdown that shatter black boxes, interrupt any taken-for-grantedness and bring numerous hitherto invisible components to light. Breakdown, in contrast to routine, brings the sociotechnical depth of technologies into light.

To call this binary reading of technology into question is one of the main contributions of maintenance and repair studies. Indeed, studying maintenance and repair practices precisely consists in paying attention to all the overlooked situations that take place in the interstices of routine and breakdown, situations in which technologies are never completely functional and never completely broken:

> It is in this space between breakdown and restoration of the practical equilibrium— between the visible (that is, 'broken') tool and the concealed tool—that repair and maintenance make its bid for significance.
>
> *(Graham & Thrift, 2007, p. 3)*

This has important consequences, for instance, for the way in which infrastructures can be questioned and analysed beyond the rhetoric of great catastrophes and collapses. The master narratives that focus attention towards the threats of disasters actually obscure the way infrastructures work on a daily basis. Yet, their mundane mode of existence is less characterised by the risks of technological apocalypse than by 'a vast and hidden economy of repair and maintenance ... continually at work' (Graham, 2010, p. 10).

Investigating things in such intermediary states inherently leads us to reconsider another early ANT bedrock: the widespread metaphor of closure. Even at rest, artefacts are not as sealed and as stable as they may appear, maintenance and repair studies show. The ability for technologies to remain the same and to be taken for granted by most of their users requires a constant work that such terms as 'black box' or 'immutable mobiles' seriously understate.

Studying dwelling maintenance, for example, Strebel (2011) coins the notion of 'living building' to highlight the mundane life of edifices, daily punctuated with repair and maintenance operations:

> The living building is another kind of architectural drama, one that is performed in and through the successive scenes of interruptions, troubles and disturbances that concierges continually encounter and the ways they solve these problems in order to keep the building going.
>
> *(p. 259)*

The interruptions and troubles Strebel describes remind us that, for some people and in some situations, even the apparently most stable artefacts cannot be totally taken for granted. Maintenance work engages certain people in very specific relationships with objects that hinder complete closure and opacity.

Exploring maintenance and repair also leads us to discover forms of troubles and malfunctioning that are not always, and not by everybody, considered as breakdowns. This is particularly visible in situations where specialised maintenance workers take care of artefacts that users are supposed to enjoy as if they were always available and functioning. The taken-for-grantedness these users can rely on is the result of the ceaseless attention and intervention of maintenance workers. The mission of the latter is to detect breaches and flaws before any breakdown can be experienced by the former (Denis & Pontille, 2015). Mentioning the different relationships a railroad engineer and a passenger have with rails, Star explains that 'one person's infrastructure is another's topic, or difficulty' (Star, 1999, p. 380). In a similar manner, maintenance and repair studies show that, in many situations, one person's breakdown, or 'repairable' in ethnomethodogy's terms (Schegloff, Jefferson & Sacks, 1977), is another's mundane functioning.

More generally, these studies also show that breakdown is not necessarily a clearly identifiable event. Rosner and Ames' comparative research on the 'One Laptop Per Child' operation in Paraguay and the Fixit Community in California foregrounds, for instance, the relationality of breakage, which is not a universal, univocal thing that happens to objects. Rosner and Ames (2014) demonstrate that both in Paraguay and in California, 'definitions of breakdown lay on a continuum' (p. 328). Similarly, Cállen and Sánchez Criado (2015) insist on the repeated trials and tests that allow the various menders they observed to gauge the flaws and fragilities of objects that are not clearly broken:

> Despite the fact that vulnerability is in many practical situations easy to identify – such as when a clear breakage happens while using something, after an accident or as a result of a disaster – it usually emerges out as part of an ongoing process of sensing and practical manipulation, hardly ever recognised at first glance.
>
> *(p. 22)*

These accounts of neither fully functioning nor definitely broken technologies resonate with de Laet and Mol's description of the Zimbabwe Bush Pump, the breakdown of which, they show, is nothing but an 'intermediate stage' (de Laet & Mol, 2000, p. 240). In fact, in numerous situations, the endurance of objects is negotiated (Cohn, 2019; Rosner & Ames, 2014), and overly limited definitions of breakdown are contested.

What is at stake, therefore, is not only the reconsideration of breakdown, but also of stability and order itself, as relational phenomena that draw on and are inscribed in specific repair and maintenance activities.

Rethinking matter: fragility and material ecology

Another important contribution of maintenance and repair studies consists in providing a particularly distinctive look at objects and their role in modern societies. Following ANT, maintenance and repair scholars refuse to reduce artefacts to signs or receptacles of external social forces (Latour, 1996). Yet, their extensive forays into the material dimensions of the world lead to unprecedented depictions, and prompt a decentring of the investigation of matter and materiality beyond the sole figure of immutability. Inspired by Mol (2008) and Puig de la Bellacasa (2011), several works have notably highlighted the relationships between maintenance, repair and care (Denis & Pontille, 2015; Houston & Jackson, 2016; Jackson, 2014). As care, maintenance and repair practices indeed take decay and vulnerability as a starting point. Fragility is here the primary – 'normal' – property of matter. Everything has, in a way or another, to be taken care of.

Such a stance markedly contrasts with how objects' agency is problematised in most ANT-inspired works, which have typically focused on solidity, resistance and permanence. If artefacts' abilities to shape our world are to be questioned, maintenance and repair studies claim, they also have to be investigated in the most mundane situations, including those in which these abilities appear uncertain, even failing. In other terms, maintenance and repair bring to light an overlooked side of objects that scholars interested in understanding material agency should not ignore.

During my own investigation on subway signs design in Paris (Denis & Pontille, 2014; 2015), I experienced such a revelation. I started my research following the lead of standardisation, eager to understand the genesis of the Parisian wayfinding system. At first, in an early ANT fashion, I recognised an ordering process, in which a new set of omnipresent and standardised signs had been given a crucial role. Yet, once I entered the maintenance department and encountered the maintenance workers, I discovered a completely different aspect of the mode of existence of these seemingly immutable and immobile signs, and was forced to seriously reconsider the way I had analysed the whole ordering process. Shadowing several maintenance runs, I progressively understood that the standardisation, stability and permanence of the set of signs actually drew on a constant care, a daily attention to the numerous mutations the signboards were subject to (wear, moisture, rust, strokes, thefts…). The signs were thus materially fragile and their strength was enacted by a daily maintenance work.

In his study of conservation of the Saint Anne Church in Manchester, Edensor (2011) investigated the 'materiality' of the stone, a solid entity *par excellence* that one might consider as the most immutable thing to be. Edensor explains that in the eyes and hands of the workers who strive for historical authenticity, Saint Anne's stone presents, on the contrary, unstable properties that vary over time. Far from being inert and impervious to its environment, the stone is subject to numerous deteriorations (i.e. discoloration, crumbling, cracking…) that challenge not only the aesthetic characteristics of the legacy building, but its very permanence.

Besides putting fragility to the fore and thus widening the understanding of material agency, maintenance and repair studies also directly echo Ingold's call for exploring the material heterogeneity of objects, technologies and infrastructures. Ingold writes, it is as though our material involvement begins only when the stucco has already hardened on the house front or the ink already dried on the page. We see the building and not the plaster of its walls, the words and not the ink with which they were written. In reality, of course, the materials are still there and continue to mingle and react as they have always done, forever threatening the things they comprise with dissolution or even 'dematerialization'.

(Ingold, 2007, p. 9–10)

If meticulously observed, activities such as mending, restoring, up-keeping, fixing or preserving offer manifold occasions to fathom the material ecology that is hidden when objects are apprehended as crystallised artefacts. These activities precisely deal with the crumbling plaster, the ink and more generally the way materials interact long after the processes of 'form-giving' (Ingold, 2013). Repair and maintenance operations bring to light the material 'intra-actions,' a term that Barad (2003) uses to characterise the kinds of material transformations that do not imply the encounter of previously recognisable discrete components.

The workers in charge of preserving Saint Anne's Church (Edensor, 2011) do not consider the building as a unified and stabilised entity. On the contrary, they deal with it as an unsettled assemblage of partly unknown elements, constantly subject to external and internal disruptions. Materials have been added to the structure of the church over time (iron staples, mortar…), and their behaviours have important consequences on the building's life. The quality of air, which has been changing over the years, is also critical to the church assemblage, as are 'birds, bats, rodents, insects, bacteria, plants, fungus, lichen and moss' (Edensor, 2011, p. 242). Preserving a heritage building requires a considerable amount of work which largely consists in exploring this particularly complex and uncertain material ecology and tackling its challenges.

It is important to note, though, that maintenance and repair studies do not fully endorse Ingold's normative stance: There is no such essentialist expressions as 'in reality' in their accounts. What maintenance and repair studies investigate are the situated enactments of the material properties of objects, through specific instruments, theories, gestures, skills, etc. Far from being the result of an academic interpretative work, the recognition that 'the artefact is not a discrete entity but a material form bound into continual cycles of articulation and disarticulation' (DeSilvey, 2006, p. 335) is always staged as a grounded and partial outcome of repair and maintenance practices themselves.

Notably, this means that the components of the material ecology are anything but a starting point in repair and maintenance activities. There are no such things as an exhaustive list of relevant material properties or a stabilised body of knowledge that would provide a strict framework for handling the things that are to be taken care of. Repair and maintenance practices are exploratory and imply a great amount of improvisation (Henke, 2000; Orr, 1996). This is telling in the contrasted cases of informal maintenance practices studied by Callén and Sánchez Criado (2015). The migrant waste pickers and the 'hackers' in collective repair workshops both engage in exploration processes. They progressively identify material properties and artefact vulnerability 'through attentive and careful 'tests' on matter' (p. 22) ranging from situated sensing and manipulating to informal experiments.

These material enactments obviously involve workers' bodies. Making 'matter speak' (Sanne, 2010) and 'listening' to it are embodied processes. Dant (2008; 2010) has regularly emphasised the importance of perception in maintenance and repair. He showed that mechanics who repair cars engage in a continuous interaction with matter, as they try to appraise the composition and condition of objects, not only touching, hearing and smelling (and sometimes tasting) them, but also configuring a convenient 'manipulatory zone' through their displacements (Dant, 2005). As one may imagine, such embodied explorations are far from always being gratifying and safe, and maintenance and repair activities should not be over-romanticised. Besides dealing with dirt (Dant & Bowles, 2003; Shaw, 2014), disassembling objects and exploring matter sometimes put workers' health at risk, especially when their activities take place in countries with fewer safety regulations (Gregson, 2011).

It is worth adding though that the interest in perception and embodied practices in maintenance and repair studies does not aim to revive any body/matter dualism. Neither bodies nor knowledge are isolated from an alleged external environment in this work. In his pioneering paper, Henke, for instance, describes the body of repair workers as a 'networked body' to emphasise the material and cognitive interdependencies of repair work:

> …in fact, it may be more appropriate to say that a given worker's body is 'networked' through performances of skilled repair. This does not mean, of course, that the body does not learn, but rather that the interaction of body and setting is the crucial relationship to understand repair.
>
> *(Henke, 2000, p. 63–64)*

Hence, human bodies are fully part of the material ecology that maintenance and repair activities assemble *in situ*.

Several works in maintenance and repair studies also insist on the organisational and geographical distribution of activities (Edensor, 2011; Jackson, Pompe & Krieshok, 2012; Domínguez Rubio, 2014). Repairing mobile phones in the workshops of downtown Kampala, for instance, is not as local an activity as it may seem. It is framed by different authorisation regimes that shape specific dependencies with manufacturers and network providers and involve:

> a complex global and local flows that sustain repair work, ranging from forms of local collaboration (though also competition) that connect technicians (…) to the wider distributions of knowledge, expertise and material resources to be found in the tools and online resources that local repair workers regularly draw on in tackling the breakdowns they confront.
>
> *(Houston & Jackson, 2016, p. 8)*

In addition to the sometimes chaotic material ecology of workshops, thus, 'an ecology of tool developers, who sometimes hack each other's devices in the pursuit of unlocking algorithms' (Houston & Jackson, 2016, p. 9), is at play. The extended view Houston and Jackson provide on mobile phone repair greatly illustrates how maintenance and repair studies help explore the multiple 'entangled agencies' (Edensor, 2011) that characterise the life of objects, way beyond the divide between alleged 'human' and 'non-human' agencies and the obsession around sturdiness and immutability.

Conclusions

Let me gather the contributions of maintenance and repair studies I underlined here with a quick summary of a case Domínguez Rubio has been investigating during his research on contemporary art conservation (Domínguez Rubio, 2014). Certain 'unruly' artworks, Domínguez Rubio explains, can be particularly challenging for Museums in their ongoing effort to control the unrelenting process of physical degradation that threatens to undermine the specific relationship between material form and intention that defines artworks as meaningful and valuable objects.

(p. 620)

This is the case of Nam June Paik's 'Untitled,' in the Museum of Modern Art (MoMA) in New York. The original technologies this installation was based on (U–Matic desks, CRT monitors, analogic cameras…) progressively became obsolete and its conservation involved the replacement of several of its components. As long as he lived, the artist supervised and validated all of these delicate 'updates' but when he died, the MoMa faced uncertainties and dilemmas. How could the team of curators and conservators decide to replace a piece without endangering the authenticity of the artwork? How far should the maintenance of the oldest technologies go? What Domìngez Rubio shows is that the discussions and negotiations deployed to tackle these issues not only have consequences on the material and symbolic composition of the artwork itself, but also transform the organisation of the museum, modifying the hierarchies between curators and conservators and expanding this traditional landscape to encompass new actors (including computer scientists and audiovisual experts). From the standpoint of its continuous preservation, Nam June Paik's 'Untitled' thus appears as a *vulnerable* artefact *never completely broken*, but *never opaque* nor taken for granted for its various caretakers, who constantly explore the *uncertain material ecology* in which it takes part in order to enact, again and again, its stability and authenticity.

The body of work that I have assembled here under the probably-too-fashionable expression of 'maintenance and repair studies' is, of course, heterogeneous and does not offer a coherent set of empirical results and theoretical developments. Yet, it does account for a certain sensibility, an 'ethos of care' (Puig de la Bellacasa, 2011) that directly echoes Star's call for taking the neglected entities of sociotechnical assemblages into account (Star, 1991). Moreover, the issues these 'studies' raise go way further the questions I isolated in this chapter in order to underline their contributions to the so-called 'post-ANT' landscape. What should be kept in mind is that they raise the curtain on what appears to be a vast continent of people, objects, sites and practices that, although familiar and mostly accessible, remained unnoticed and unexplored for decades. Considering the variety of objects and situations these first works scrutinise and the richness of their analysis, there are a few doubts that we are just at the beginning of a long-lasting trend that will lead to more exciting discoveries.

Besides this empirical variety, it is also important to recognise the theoretical potential of maintenance and repair studies, each of their explorations offering an occasion to document original ways of enacting order, identity and authenticity, but also inviting us to denaturalise the relationship between maintenance and order. Whilst one may think that maintaining objects or infrastructures always amounts to bringing things back to order, we saw here that things are not so simple. First, studying maintenance and repair directs attention to the relationality of order (if an order is repaired or maintained, which one is it, from which standpoint?). Second, certain forms of maintenance have less to do with order than with precariousness and the life that emerges in the interstices (or the ruins) of innovation and even capitalism (Tsing, 2015).

Maintenance and repair studies also bring ways to unfold and discover power relations from the starting point of the various object ontologies that are enacted in distinct regimes of maintenance, from generalised care to organised non–repairability (Denis & Pontille, 2017). They thus pursue the ontological inquiry that ANT instigated in social sciences, by collecting and describing hitherto unexplored situations in which the question 'what is (and remain) an object?' is practically addressed.

Notes

1 The author-in-the-text is David J. Denis, whereas the author-in-the-flesh is twofold: Jérôme Denis and David Pontille. Since the end of the 1990s, we have used the attachment of our two civil names as a way 'to diffract' our own authorship, following Haraway (1996), and to disrupt the emphasis of single individuals performed by most of research assessment frameworks. As the editors of the ANT Companion were concerned with two different chapters bearing the same name(s), we created a fictional author as a way to pursue the diffraction process. Such a gesture is perfectly in line with the semiotic foundation of ANT (see the second footnote in Latour, 1988), which both insisted on the generative process of writing practices that bring new entities to existence and participate in their maintenance. See also Jérôme D. Pontille's chapter in this volume.

2 Brand's book *How building learn* (1994) should also be mentioned. Although not academic, it has inspired early research on maintenance.

References

Akrich, M. 1993. Essay of Technosociology: A Gasogene in Costa Rica. In Lemonier, P. (Ed.), *Technological Choices. Transformation in Material Cultures since the Neolithic*. London, Routledge, p. 289–337.

Barad, K. 2003. Posthumanist Performativity: Toward an Understanding of How Matter Comes to Matter. *Signs: Journal of Women in Culture and Society*, 28(3), p. 801–831.

Barnes, J.E. 2017. States of Maintenance: Power, Politics, and Egypts Irrigation Infrastructure. *Environment and Planning D: Society and Space*, 35(1), p. 146–164.

Brand, S. 1994. *How Buildings Learn: What Happens after They're Built*. New York, Viking Penguin.

Cállen, B. & Sánchez Criado, T. 2015. Vulnerability Tests. Matter of 'Care for Matte' in E-waste Practices. *Tecnoscienza*, 6(2), p. 17–40.

Cohn, M.L. 2019. Keeping Software Present: Software as a Timely Object for Digital STS. In Vertansi, J. & Ribes, D. (Eds.), *Digital STS: A Fieldguide and Handbook*. Princeton, NJ, Princeton University Press, p. 423–446

Dant, T. 2005. *Materiality and Society*. New York, Open University Press.

Dant, T. 2008. The 'Pragmatic' of Material Interaction. *Journal of Consumer Culture*, 8(1), p. 11–33.

Dant, T. 2010. The Work of Repair: Gesture, Emotion and Sensual Knowledge. *Sociological Research Online*, 15(3), p. 1–22.

Dant, T. & Bowles, D. 2003. Dealing with Dirt: Servicing and Repairing Cars. *Sociological Research Online*, 8(2), p. 1–17.

de Laet, M., & Mol, A. 2000. The Zimbabwe Bush Pump: Mechanics of a Fluid Technology. *Social Studies of Science*, 30(2), p. 225–263.

Denis, J. & Pontille, D., 2014. Maintenance Work and the Performativity of Urban Inscriptions: the Case of Paris Subway Signs. *Environment and Planning D: Society and Space*, 32(3), p. 404–416.

Denis, J. & Pontille, D., 2015. Material Ordering and the Care of Things. *Science, Technology, & Human Values*, 40(3), p. 338–367.

Denis, J. & Pontille, D. 2017. Beyond Breakdown: Exploring Regimes of Maintenance. *continent.*, 6(1), p. 13–17.

Denis, J. & Pontille, D. 2019. The Multiple Walls of Graffiti Removal. Maintenance and Urban Assemblage in Paris. In A. M. Brighenti & M. Kärrholm (Eds.), *Urban Walls. Political and Cultural Meanings of Vertical Structures and Surfaces*. London, Routledge, p. 215–235.

DeSilvey, C. 2006. Observed Decay: Telling Stories with Mutable Things. *Journal of Material Culture*, 11(3), p. 318–338.

Domínguez Rubio, F. 2014. Preserving the Unpreservable: Docile and Unruly Objects at MoMA. *Theory and Society*, 43(6), p. 617–645.

Domínguez Rubio, F. 2016. On the Discrepancy between Objects and Things. *Journal of Material Culture*, 21(1), p. 59–86.

Edensor, T. 2011. Entangled Agencies, Material Networks and Repair in a Building Assemblage: the Mutable Stone of St Ann's Church, Manchester. *Transactions of the Institute of British Geographers*, 36(2), p. 238–252.

Edgerton, D. 2006. *The Shock of the Old: Technology and Global History since 1900*. London, Profile books.

Fidler, B.R. & Russel, A. 2018. Financial and Administrative Infrastructure for the Early Internet: Network Maintenance at the Defense Information Systems Agency, *Technology and Culture*, 59(4), p. 899–924.

Fujimura, J.H. 1991. On Methods, Ontologies and Representation in the Social Sciences: Where Do We Stand? In David R. Maines (Ed.) *Social Organization and Social Process: Essays in Honor of Anselm Strauss*. Hawthorne, Aldine de Gruyter, p. 207–248.

Goffman, E. 1959. *The Presentation of Self in Everyday Life*. New York, Anchor.

Graham, S. 2010. When Infrastructures Fail. In S. Graham (Ed.) *Disrupted Cities*. New York, Routledge, p. 1–26.

Graham, S. & Thrift, N. 2007. Out of Order: Understanding Repair and Maintenance. *Theory, Culture & Society*, 24(3), p. 1–25.

Gregson, N. 2011. Performativity, Corporeality and the Politics of Ship Disposal. *Journal of Cultural Economy*, 4(2), p. 137–156.

Gregson, N., Metcalfe, A. & Crewe, L. 2009. Practices of Object Maintenance and Repair: How Consumers Attend to Consumer Objects within the Home. *Journal of Consumer Culture*, 9(2), p. 248–272.

Haraway, D.J. 1991. *Simians, Cyborgs, and Women: The Reinvention of Nature*. New York, Routledge.

Haraway, D.J. 1996. *Modest_Witness@Second_Millennium. FemaleMan©_Meets_OncoMouse™. Feminism and Technoscience*. New York, Routledge.

Henke, C.R., 2000. The Mechanics of Workplace Order: Toward a Sociology of Repair. *Berkeley Journal of Sociology*, 44, p. 55–81.

Houston, L. & Jackson, S.J., 2016. Caring for the Next Billion Mobile Handsets: Opening Proprietary Closures through the Work of Repair. *Proceedings of the Eighth International Conference on Information and Communication Technologies and Development*. p. 10.

Ingold, T. 2007. Materials Against Materiality. *Archaeological Dialogues*, 14(01), p. 1–16.

Ingold, T. 2013. *Making: Anthropology, Archaeology, Art and Architecture*. London, Routledge.

Jackson, S.J. 2014. Rethinking Repair. In Gillespie, T., Boczkowski, P.J. & Foot, K.A. (Eds.), *Media Technologies - Essays on Communication, Materiality, and Society*. Cambridge, MA MIT Press, p. 221–240.

Jackson, S.J., Pompe, A. & Krieshok, G., 2012. Repair Worlds: Maintenance, Repair, and ICT for Development in Rural Namibia. *CSCW'12*. Seattle.

Jones, S. & Yarrow, T. 2013. Crafting Authenticity: An Ethnography of Conservation Practice. *Journal of Material Culture*, 18(1), p. 3–26.

Latour, B. 1987. *Science in Action. How to Follow Scientists and Engineers Through Society*. Harvard, Harvard University Press.

Latour, B. 1988. Mixing Humans and Non-Humans Together: The Sociology of a Door-Closer. *Social Problems*, 35, p. 298–310.

Latour, B. 1996. On Interobjectivity. *Mind, Culture, and Activity*, 3(4), p. 246–269.

Latour, B. 2005. *Reassembling the Social: An Introduction to Actor-Network-Theory*. Oxford, Oxford University Press.

Law, J. & Hassard, J. 1999. *Actor Network Theory and After*. Oxford, Wiley-Blackwell.

Mol, A. 2008. *The Logic of Care: Health and the Problem of Patient Choice*. New York, Routledge.

Orr, J.E. 1996. *Talking About Machines: An Ethnography of a Modern Job*. New York, Cornell University Press.

Puig de la Bellacasa, M., 2011. Matters of Care in Technoscience: Assembling Neglected Things. *Social Studies of Science*, 41(1), p. 85–106.

Rosner, D.K., 2013. Making Citizens, Reassembling Devices: On Gender and the Development of Contemporary Public Sites of Repair in Northern California. *Public Culture*, 26(1 72), p. 51–77.

Rosner, D.K. & Ames, M., 2014. Designing for Repair?: Infrastructures and Materialities of Breakdown. *Proceedings of the 17th ACM conference on Computer Supported Cooperative Work & Social Computing*. p. 319–331.

Sanne, J.M. 2010. Making Matters Speak in Railway Maintenance. In M. Büscher, D. Goodwin & J. Mesman (Eds.) *Ethnographies of Diagnostic Work: Dimensions of Transformative Practice*. Houndmills, Palgrave Macmillan, p. 54–72.

Schegloff, E.A., Jefferson, G. & Sacks, H., 1977. The Preference for Self-Correction in the Organization of Repair in Conversation. *Language*, 53(2), p. 361–382.

Shaw, R. 2014. Cleaning up the Streets: Newcastle-upon-Tynes Night-Time Neighbourhood Services Team. In Graham, S. & McFarlane, C. (Eds.), *Infrastructural Lives: Urban Infrastructure in Contexte*. London, Routledge, p. 174–196.

Star, S.L. 1991. Power, Technology and the Phenomenology of Conventions: on Being Allergic to Onions. In Law J. (Ed.), *A Sociology of Monsters? Essays on Power, Technology and Domination*. London/ New York, Routledge, p. 26–56.

Star, S.L., 1999. The Ethnography of Infrastructure. *American Behavioral Scientist*, 43(3), p. 377–391.

Strebel, I., 2011. The Living Building: Towards a Geography of Maintenance Work. *Social & Cultural Geography*, 12(3), p. 243–262.

Tsing, A.L. 2015. *The Mushroom at the End of the World: on the Possibility of Life in Capitalist Ruins*. Princeton, NJ, Princeton University Press.

Ureta, S. 2014. Normalizing Transantiago: On the Challenges (and Limits) of Repairing Infrastructures. *Social Studies of Science*, 44(3), p. 368–392.

Verbeek, P. 2006. Materializing Morality: Design Ethics and Technological Mediation. *Science, Technology & Human Values*, 31(3), p. 361–380.

Yurchak, A. 2015. Bodies of Lenin: The Hidden Science of Communist Sovereignty. *Representations*, 129(1), p. 116–157.

Section 5
The sites and scales of ANT

Anders Blok, Ignacio Farías and Celia Roberts

If the situatedness of knowledge and practice has always been a key actor-network theory tenet, and if ANT refuses to accept scales of 'local' and 'global,' 'micro' and 'macro,' as pre-given and static, then these insights may also profitably be turned back on ANT itself. What specific sites and scales of exploration and inquiry is ANT itself committed to; and how, in turn, do such specific sites and scales make a difference to ANT's own predilections and propositions? If the previous section commenced a journey outwards and away (as it were) from the laboratories and (other) testing grounds of science and technology, then the present section is meant to continue and attenuate that journey. Here, we invited authors to attend specifically to the question of how moving through divergent sites and scales of practice and assemblage, and taking their spatio-temporal specificities seriously, may work to both challenge and change ANT, sometimes in unforeseen and perhaps at times even troublesome ways.

Given the unboundedness of these very notions (site, scale), it comes as no surprise to note that such an endeavour amounts to a double-edged challenge: To question and rethink ANT's own sites and scales are also, simultaneously, to question and rethink the centrality taken on by specific versions of these notions in ANT itself. Here, the terms may be said to bear family resemblance to the notion of 'context' so famously and vehemently cast aside by ANT ('context stinks!' as Latour once quoted architect Rem Koolhaas). However, just as this was hardly the last word on architecture's sites (viz. Yaneva and Mommersteeg's contribution to this section), neither 'context' nor 'site' or 'scale' shows any propensity to disappear as empirical, methodological and conceptual challenges anytime soon, in ANT or elsewhere. Better, then, to address them head-on and with a view to their (self-)implications for ANT.

Based in part on his own ANT-informed inquiries into the contemporary stakes of politics, Endre Dányi sets the section off by discussing a tangible and symbolically loaded building type of equal importance to (so-called) political modernity as that of the laboratory: the parliament. Are parliaments, he asks, still privileged sites for studying politics and liberal democracy, particularly once politics is taken as a form of practice and a set of problems in the ANT vein? Drawing on Brian Rotman, Dányi proposes to think of parliaments not only as sites but also and importantly as 'meta-sites,' that is, as a set of organising logics that remain crucial to a whole range of contemporary practices, including within ANT's own imagination of politics. From here, he proceeds to sketch certain limits of parliamentary democracy as a mode of doing politics, calling for increased analytical and practical attention to the many 'aboves' and 'belows' of parliaments.

In their chapter, Albena Yaneva and Brett Mommersteeg continue this kind of analytical attention to a specific locus of practice, yet with a twist: This time around, as noted, the inquiry concerns the very 'becoming' of sites as built-up and inhabited locales in and via the expert practice of architecture. Having questioned the way architectural theory (and practice) tends to parcel up 'the site' in recognisably dualistic ways, as either (too) 'objective' or (too) 'subjective,' the authors proceed to sketch an alternative reading of what they term the movement of site-ing in architectural practice. It is in and through this movement, they show, that not only a new building but also and simultaneously its very site or location, with its specific relations and dimensions of (urban) economy, politics, culture and materials, come into being and stabilise. Adding a further twist to this argument, Yaneva and Mommersteeg end by stressing the active role played also by the (ANT-informed) ethnographer in achieving such collective site-ing.

While he adopts a slightly different vocabulary, collective work of site-ing is also at play in Robert Oppenheim's subsequent chapter, in which he explores the question of how the South Korean city of Kyŏngju helps ANT think place and scale. This very question, of course, presumes a backdrop of ethnographic engagement (itself sited and scaled, we might add). As an anthropologist working amidst wider political reconfigurations in 1990s South Korea, Oppenheim began studying a technical controversy on high-speed rail construction and the threats it had widely taken to pose to the historical and cultural city of Kyŏngju, once the country's capital. For these purposes, the so-called post-ANT conceptions of multiplicity and topology proved helpful, as he details, in rethinking the very notions of place and scale. Adding further layers to this point, Oppenheim proposes in his chapter to draw on wider anthropological resources to study those 'circulating normativities' whereby scale-making efforts come to attain persuasive force.

Emerging from a kindred ethnographic commitment, only this time set amidst Paraguay's vast and sprawling soybean fields, Kregg Hetherington's chapter turns in part to the site of the courtroom to better understand how slow-moving environmental harms (sometimes) become eventful political disruptions. In recounting the human story and legal case of Silvino Talavero, a young boy killed from pesticide poisoning in the early 2000s, Hetherington discusses the fine line separating politics from 'the uneventful' in a situation of routinised poverty, ill health and silently accumulating environmental destruction. Here, he argues, the political theorising of Jacques Rancière, amongst others, supplements ANT's own take on eventfulness in ways that help make sense of the lasting legacies of Talavero's legal case. At times, Hetherington asserts, courts may amplify disruptions already underway, thus exercising their own scalar effects.

In the sections' two final chapters, the notions of scale and site, respectively, may be said to reach a certain limit, or vanishing point, in ways that serve to question their utility as much as their ANT incarnations. Here, in her chapter, Kristin Asdal deploys her own work on the politics of nature(s), often set in the context of Norwegian parliamentary politics, to ask whether ANT is equally good at dealing with local, national and global natures. Beyond detecting the 'trick-like' character of the question itself (after all, when did ANT ever accept such scalar designations as 'local' and 'global'?), Asdal places the discussion centrally on the territory of 'Nature' itself or, rather, the way ANT has come to translate this term as so many always local and specific imbroglios of non-humans with humans and technologies. In the face of Gaia and other 'nature-wholes' worked-out-there, she argues, we encounter the challenge of respecifying the very sites where politics of nature is done, including that welter of bureaucratic offices, councils and (yes) parliaments, all of which are somewhat local, national and global at once.

If 'Nature' indeed proves difficult to scale, for the ANT analyst as for everyone else, then much the same might be said for 'the digital,' as the site (or 'meta-site'?) picked up by Carolin Gerlitz and Esther Weltevrede in the sections' final chapter. Here, drawing on their collective experiences in inventive research settings of digital method development, the authors ask what happens to ANT, and its telltale emphasis on the socio-material grounding of the social, in digital sociology? Digital sociology, they argue, works both as attention to technological media settings and a challenge to invent new digitally aware methods, and in doing so, it challenges ANT to move from simply 'following the actor' and into a more intricate terrain of having to reconfigure its own empirical apparatus via 'interface methods.' Using the example of bots and other devices of automation on Twitter, Gerlitz and Weltevrede conclude that not all digital relations are equally visible or traceable from the outset. As such, their intervention serves as a valuable reminder, relevant for the entire section (and indeed the companion volume as such), that what ends up as the sites and scales of ANT-informed inquiry is partly a matter of the method apparatus deployed.

28

Are parliaments still privileged sites for studying politics and liberal democracy?

Endre Dányi

This chapter addresses the question in the title in three steps.[1] In the first part, I present the key terms of the title – politics, parliaments and liberal democracy – more or less as they appear in the Actor-Network Theory (ANT) literature. In the second part, in order to highlight a number of current developments that pose major challenges for both political theory and ANT, I discuss a set of complementary terms – people, problems and worlds – through snippets from my recent empirical work. Finally, I offer a tentative answer to the title's question, the aim of which – in the best of parliamentary fashions – is intended not to close down, but rather to generate further debate.

Parliaments as sites and meta-sites

Politics

ANT has always been about politics (for a general overview, see Lezaun 2017; Michael 2016; for an explicit discussion on ANT and politics, see Harman 2014). From the outset, it has focused on how politics is being done 'by other means' – mostly by technical and scientific means that make it difficult to disagree, to resist or to develop alternatives.

In such a short chapter as this one, it is not possible for me to provide a comprehensive overview, but here are a couple of vignettes that illustrate my opening statement. In his classic study of scallop farming, Michel Callon (1986) used an explicitly political vocabulary (problematisation, interessement, enrolment and mobilisation) to talk about technoscientific developments in France and Japan. In his historical analysis of the Portuguese colonisation of India, John Law (1986) examined how control-at-a-distance had been established through sailing vessels as miniature versions of a European model of governance. In his portrait of Louis Pasteur as a cunning stage manager, Bruno Latour (1988) described how laboratories participate in politics by connecting otherwise disparate places such as dairy farms and academies. And in a series of works on healthcare, Annemarie Mol (2002, 2008) has shown how multiple realities are being not only *known* but also *done* through material practices, opening up ontological politics as a new mode of engagement.

As these vignettes show, the central component of ANT's politics – not unlike the politics of governmentality studies (Barry et al. 1996; Lemke 2016; Rose 1999) and feminist techno-science (Haraway 1991; Suchman 2006) – has been to make visible the political aspects and consequences of seemingly apolitical practices (science, engineering, planning, but also market-making, to name a few). In doing so, ANT also makes such practices potentially alterable again via public forms of inquiry and contestation (Marres 2013).

Parliaments

Despite the political character of ANT, there has been relatively little work on explicitly political institutions and processes (although see Barry 2001). The term 'parliament' first entered the vocabulary of ANT as a metaphor: In *We Have Never Been Modern*, Bruno Latour (1993) talked about the 'Parliament of Things' as a scheme that might replace the Modern Constitution, that is, the cosmology associated with the West (mostly Europe and North America) based on the separation of Nature and Culture (see also Latour 2004a). The metaphor is very similar to Michel Serres' (1990) 'Natural Contract,' which is a poetic and philo-sophical reflection on the end of the Cold War and the changing role of the environment and other non-human entities in politics-to-come.

In the 1990s, the 'Parliament of Things' as a metaphor generated a lot of interest, but it took quite a while until parliaments became actual sites and objects of ANT inquiries. The breakthrough was undoubtedly Bruno Latour and Peter Weibel's (2005) *Making Things Public* exhibition and catalogue, which tried to do to legislatures what ANT scholars had done to laboratories: Open them up and examine them as sites *and* 'meta-sites.'

I use the term 'meta-site' the way Brian Rotman (1987) used the term 'meta-sign' in his analysis of the semiotic status of 'zero.' He argued that 'zero' (introduced to European math-ematics only in the 13th century) is not simply one sign among many, but also a sign that indicates the absence of all other signs, and as such stands both *inside* and *outside* the semiotic system it is associated with. In his wonderful book, Rotman went on to analyse the historical development of two similar meta-signs, the vanishing point in perspectival art in the 15th century and paper money in economic exchange in the 17th century, without which neither a European way of seeing nor a European way of doing business could have achieved its he-gemonic status. In a similar vein, in my view, it makes perfect sense to consider the Western parliament, developed in the 19th century, as a 'meta-sign' or a 'meta-site,' that is, a set of organising logics that are constantly extended, transposed and inscribed into other places in order to regulate what may count as 'good politics.'

Since Latour and Weibel's exhibition, a couple of ANT-inspired works have further specified this line of inquiry. Here are some more vignettes: In her detailed historical and ethnographic work on the Assemblée Nationale, Delphine Gardey (2015) has analysed the workings of the French legislature quite literally as a political body, with very gendered characteristics. Based on archival work, Kristin Asdal has traced how various non-human bodies (such as whales) could enter the Norwegian Parliament as contested objects (Asdal 2008; Asdal and Hobæk 2016). In my own work on the Hungarian National Assembly, I have examined how a 19th-century version of parliamentary democracy was reinvented and reap-propriated after the collapse of state socialism (Dányi 2012; 2013). And in his research on the German Bundestag, Thomas Scheffer (2014) has extended his ANT-compatible ethnometh-odological gaze to faction offices to show how bureaucrats and politicians are collectively caught up in the making of political positions.

Liberal democracy

In one way or another, the aforementioned vignettes elaborate what it means to call parliaments the meta-sites of politics. By closely regulating the ways in which various political sites and moments may relate to each other, legislatures insist on the hegemonic status of parliamentary democracy across national borders, and on different scales. This, however, does not imply that parliaments are the *only* places where politics happens. In fact, a strong argument in ANT has been that in order to understand the state of contemporary democracy, we need to move beyond standard political institutions. Michel Callon and his colleagues' (2009) suggestion, for instance, has been to concentrate on 'hybrid forums.' Bruno Latour (2007) and Noortje Marres (2007) have shifted focus to various publics that develop around concrete issues. John Law and Annemarie Mol (together with many colleagues, see Law & Mol 2008; Law & Singleton 2013; Law et al. 2014; Mol et al. 2010) have made important attempts to articulate 'good politics' in terms of care, rather than those of choice or conflict.

While these research directions have been clearly productive, in one way or another, they have all operated under the assumption that liberal democracy, unchallenged by other models of governance since the end of the Cold War, is here to stay (at least in Europe and North America – more on this below). This is not to say that ANT scholars have been uncritical of liberal democratic politics; it just means that their critiques have been practised mostly against the background of liberal democracy (Latour 2004b; see also Brown 2009; Ezrahi 2012). Recent developments, however, suggest that this background is less robust than it seems: The cracks made visible by the Brexit referendum, the election of US president Donald Trump and the sustained support of Viktor Orbán's vision of illiberal democracy, to name only the most obvious examples, have exposed the vulnerability not only of standard political institutions (including opinion polling companies and mainstream news media), but also of *democratic politics as such*.

The limits of parliamentary democracy

If the aforementioned statement is right, then similar to the opening scene of *An Inquiry into Modes of Existence* (Latour 2013), where readers are asked to trust the institutions of science, rather than the indisputable certainty associated with *science as such*, we as ANT-inspired scholars might have to find new ways of engaging with the institutions of democratic politics without the sense of indisputable certainty that stems from 1989 as 'the end of history' (Fukuyama 1992). But how to do this? In the past couple of years, I have focused on a number of political developments that, instead of moving beyond parliaments, seek to advance practices that operate between the insides and the outsides of parliamentary politics. These practices, I argue, pose important challenges for political theory and ANT alike. Let me show what I mean through three examples.

People

In the ANT literature, there has been plenty of discussion about the importance of extending politics to non-humans – in a way, this is what the 'Parliament of Things' has been about (Braun & Whatmore 2007; Harman 2014). However, the ways in which humans and non-humans collectively constitute a demos – a 'people' in a political sense – have received much less attention. What the European refugee crisis (among other events) has shown is that, in addition to the problematisation of the non/human distinction in politics, it is crucial to examine who or what may be *political*, and how (Isin 2001).

Antoine Hennion and his colleagues (2016) have engaged with a version of this question by carefully documenting the uses and the dismantling of the Calais refugee camp (also known as 'the Jungle'), while Amade M'charek (2018) has drawn attention to the political significance of refugees' bodies crossing the Mediterranean sea – both dead and alive. In a historical and ethnographic research I have conducted with Sebastian Abrahamson, the focus has been on a hunger strike that took place in Brussels in 2012 (Abrahamsson & Dányi 2019). More than three years before the outbreak of the European refugee crisis, that hunger strike was initiated by 23 *sans-papiers* from North Africa and the Middle East, who desperately wanted to get work permits, but who – without any documents – stood no chance to find their way in the labyrinth of Belgian bureaucracy.

The hunger strike lasted for more than a hundred days, and – despite wide coverage in the media and strong support by many Belgian citizens and international activist groups – ended without any tangible results. Still, in a political sense, it would be a mistake to call it a failure. From a parliamentary perspective, the hunger strikers were humans, but not political beings. And yet, there they were, in the symbolic capital of Europe, doing politics without uttering a word. As their silent bodies were getting weaker and weaker, the standard vocabulary associated with democratic politics, based on the distinction between *zoē* and *bios* (bare life and politically qualified life – see Agamben 1998), was becoming less and less useful. Such dichotomies as us/them, activity/passivity, strength/weakness pointed at the limits not only of political theory, but also of ANT's relatively narrow understanding of what it is for humans and non-humans to be political, epitomised by such terms as 'interressement' (Callon 1986) or 'trials of strength' (Latour 1988).

Problems

In a democratic context, ANT appears much better suited to talk about the objects than the subjects of politics. Sometimes, these objects have been referred to as issues (Marres 2007), other times as matters of concern (Latour 2008). Here, I call them problems, partly because it is more common to talk about the making of such objects as problematisation (Callon 1986; Foucault 1984), and partly because it is easier to recognise parts of democratic politics as sets of institutions and processes organised around specific problems.

In ANT – and in Science and Technology Studies (STS) in general – such sets of institutions and processes tend to be discussed in terms of policy (see Gill et al. 2017; Gomart & Hajer 2003). Indeed, it is possible to say that policies are materialised ways of knowing and handling problems that mobilise particular assemblages (Ureta 2015). A good example is drug use as a characteristically modern problem (see Courtwright 2001; Schivelbusch 1993). In the European Union, drugs have been repeatedly on the agenda of national parliaments, but since the early 1990s, there have also been considerable efforts to know and handle drug use as a *European* problem. Perhaps the most important outcome of those efforts was the establishment of the European Monitoring Centre for Drugs and Drug Addiction (EMCDDA) – an EU agency based in Lisbon.

As part of my research on problematisation processes, in 2015, I attended a workshop organised by the EMCDDA, where I had the chance to learn more about the complex ways in which 'Europe' has been performed into being (partly) through statistical data associated with drug use (Dányi 2018 – for a similar argument in a different domain, see Laurent 2017). With various ANT references in mind, I could also observe how in the EMCDDA's reports, drug use has been subtly framed as a criminal *and* a medical problem, disassociating Europe from the United States and its war on drugs ('the greatest policy failure of all times,' according to a

UN expert). What I could not see at the EMCDDA workshop was how drug use also has the capacity to challenge what it means to call something a problem in the first place.

To open up the politics of problemness, I had to leave the air-conditioned rooms of the EU agency and join a group of social workers running a harm-reduction programme in the outskirts of the city. The Portuguese drug policy is one of the most progressive ones in the EU, so – unlike their colleagues, for instance, in Budapest – the social workers in Lisbon were able to do their rounds in abandoned factories and empty parking lots, distributing sterile needles, wet wipes, citric acid and other items that make intravenous drug use safer, without fear of being arrested or harassed by the police. This, however, does not imply that they knew how to solve drug use as a problem. What they knew was that formulating solutions, through legislation or by other means, is only one mode of relating to problems. Another one is to keep looking for better ways of living with them (see also Jensen 2017).

Worlds

So far, everything I have written in this chapter is about the politics of the 'West,' mostly Europe and North America. This is a bias very much present in both political theory and ANT (see Lin and Rosa, this volume). One attempt to remedy this has been triggered by a remarkable exchange between Ulrich Beck (2004) and Bruno Latour (2004c). The former, critically revisiting Kant's notion of cosmopolitanism, speculated how – confronted with climate change – a Western understanding of democratic politics could be extended beyond nation states without replicating the horrors of colonialism and totalitarianism. In response, the latter borrowed Isabelle Stengers' (2005) concept of cosmopolitics in order to re-politicise 'the cosmos,' that is, the world we in the West believe we inhabit. This, indeed, was Latour's point: If cosmopolitan politics is about different *worldviews*, cosmopolitics is about different *worlds* and everything that populates them. Unless we find a way to articulate such differences, there is no chance for democracy on a global scale.

How to engage in cosmopolitics? Latour's suggestion, working with and away from Stengers (see Savransky, this volume), is to proceed slowly and carefully, and take into account everything – every *thing* – that may be required for the composition of a common world. However, others, for example Mario Blaser (2016), Marisol de la Cadena (2010) and Helen Verran (2007), suspect that a gradual composition of a common world might be neither feasible nor desirable (see also Law 2015). The best we can do – as scholar-translators – is to help different worlds become momentarily visible to each other, generating new possibilities within already existing political practices and institutions. This was the advice I tried to follow with Michaela Spencer, when in 2016 we did our first round of fieldwork in the Northern Territory in Australia. Not trained as anthropologists, our aim was not to describe Indigenous politics as such, but to document various instances in which different worlds – Western and aboriginal – become visible to each other (Dányi and Spencer 2019).

Early on, we learned about the Northern Territory National Emergency Response (the 'Intervention'), which was a bundle of laws passed by the Australian Parliament in 2007. It was expected to solve a range of problems in aboriginal communities, from drug and alcohol use to child abuse and domestic violence. According to our respondents, the Intervention might have been well-intended, but from a cosmopolitical point of view, its implementation – which relied heavily on the military – had a disastrous effect. The forced medical examination of aboriginal children, for example, did not only trigger bad memories of a violent colonial past; it also undermined many local initiatives aimed at bringing together Western and Indigenous authorities.

In order to formally object to the way in which the Australian government had designed and executed the Intervention, the Yolngu – an aboriginal people in north-east Arnhem Land in the Northern Territory – organised a parliamentary ceremony called the *Ngarra*. Such parliamentary ceremonies, which consist of a series of performances by various clan groups, are normally not accessible to an uninitiated Western audience. This particular ceremony, however, was recorded and a shortened video of it was uploaded to YouTube (Riyawarray 2008). As some Yolngu elders explained in the video (and later to us in person), this *Ngarra* was not a ceremony for the sake of ceremony. It was a deliberate cosmopolitical act during which a world otherwise invisible to the Australian Parliament was momentarily made visible. The participants of the ceremony did not ask for a symbolic recognition of sameness, as humans or Australians. They asked for a very practical recognition of difference. And they redeployed the parliament to effect this claim, however temporarily.

Tentative answer

So, are parliaments still privileged sites for studying politics and liberal democracy? What I have tried to show in this chapter is that parliaments as *sites* might no longer be privileged, but the most interesting developments in democratic politics are very closely related to parliaments as *meta-sites*.

The three snippets of empirical work have indicated, albeit differently, moments when parliamentary politics breaks down or reaches its limits. We have seen the difficulty of operating according to an us/them division in the first case; the difficulty of doing politics through problem-solving in the second case and the difficulty of dealing with cosmological difference in the third case. At the same time, these moments also outline modes of doing politics – hunger strikes, harm reduction programmes and Indigenous initiatives – that have the capacity to transform parliaments as meta-sites.

In light of this, the aim of the chapter has been to help us, ANT-inspired scholars, rethink politics as 'an endangered mode of existence' (Latour 2013) by sensitising ourselves to practices that connect the insides and the outsides – or the 'aboves' and the 'belows' – of parliaments in interesting and unexpected ways.

Note

1 Earlier versions of this chapter were discussed at the Institute for European Ethnology at the Humboldt University in Berlin, at the Centre for Anthropological Knowledge in Scientific and Technological Cultures (CAST) at the Ruhr University in Bochum, and at the Department of Communication at the University of California San Diego. I am grateful to the organisers and the participants of these events for their helpful suggestions and to Anders Blok for his detailed comments on previous drafts. The arguments outlined in this chapter have been developed over the course of many conversations. Special thanks to Sebastian Abrahamsson, Róbert Csák and Michaela Spencer for the collaborative research, and to Thomas Scheffer for all the provocation.

References

Abrahamsson, S., & Dányi, E. (2019). Becoming Stronger by Becoming Weaker: The Hunger Strike as a Mode of Doing Politics. *Journal of International Relations & Development*, forthcoming.

Agamben, G. (1998). *Homo Sacer*. Stanford, CA: Stanford University Press.

Asdal, K., and B. Hobæk. (2016). Assembling the Whale: Parliaments in the Politics of Nature. *Science as Culture*, 25(1): 96–116.

Asdal, K. (2008). Subjected to Parliament: The Laboratory of Experimental Medicine and the Animal Body. *Social Studies of Science*, *38*(6): 899–917.

Barry, A. (2001). *Political Machines*. London: Continuum.

Barry, A., Osborne, T., & Rose, N. (1996). *Foucault and Political Reason*. Chicago, IL: University of Chicago Press.

Beck, U. (2004). The Truth of Others: A Cosmopolitan Approach. *Common Knowledge*, 10(3): 430–449.

Blaser, M. (2016). Is Another Cosmopolitics Possible? *Cultural Anthropology*, *31*(4): 545–570.

Braun, B., & Whatmore, S. (2010). *Political Matter: Technoscience, Democracy, and Public Life*. Minneapolis: University of Minnesota Press.

Brown, M. (2009). *Science in Democracy: Expertise, Institutions, and Representation*. Cambridge, MA: The MIT Press.

Callon, M., Lascoumes, P., & Barthe, Y. (2009). *Acting in an Uncertain World: An Essay on Technical Democracy*. Cambridge, MA: The MIT Press.

Callon, M. (1986). Some Elements in the Sociology of Translation: Domestication of the Scallops and the Fishermen of St. Brieuc Bay. In J. Law (ed.) *Power, Action, and Belief*. London: Routledge Kegan & Paul, 196–233.

Courtwright, D. (2001). *Forces of Habit: Drugs and the Making of the Modern World*. Cambridge, MA: Harvard University Press.

Dányi, E. (2018). Good Treason: Following ANT to the Realm of Drug Policy. In T. Berger & A. Esguerra (eds.) *World Politics in Translation*. London: Routledge, 25–38.

Dányi, E. & Spencer, M. (2019). Un/common Grounds: Tracing Politics Across Worlds. *Social Studies of Science*, forthcoming

Dányi, E. (2013). Democracy in Ruins: The Case of the Hungarian Parliament. In D. Gafijczuk & D. Sayer (eds.) *The Inhabited Ruins of Central Europe*. London: Palgrave Macmillan, 55–78.

Dányi, E. (2012). *Parliament Politics: A Material-Semiotic Analysis of Liberal Democracy*. PhD thesis, Department of Sociology, Lancaster University, UK.

de la Cadena, M. (2010). Indigenous Cosmopolitics in the Andes: Conceptual Reflections Beyond 'Politics'. *Cultural Anthropology*, 25(2): 334–370.

Ezrahi, Y. (2012). *Imagined Democracies*. Cambridge: Cambridge University Press.

Foucault, M. (1984) Polemics, Politics and Problematizations. In P. Rabinow (ed.) *The Foucault Reader*. New York: Pantheon Books, 381–390.

Fukuyama, F. (1992). *The End of History and the Last Man*. New York: Free Press.

Gardey, D. (2015). *Le linge du Palais-Bourbon*. Lormont: Le Bord de l'Eau.

Gill, N., Singleton, V., & Waterton, C. (2017). *Care and Policy Practices*. London: The Sociological Review & Sage.

Gomart, E., & Hajer, M. (2003). Is *That* Politics? In: B. Joerges & H. Nowotny (eds.) *Social Studies of Science and Technology: Looking Back, Ahead*. Dordrecht: Kluwer, 33–61.

Haraway, D. (1991). *Simians, Cyborgs, and Women: The Reinvention of Nature*. London: Routledge.

Harman, G. (2014). *Bruno Latour: Reassembling the Political*. London: Pluto Press.

Hennion, A., & Thiéry, S. (2016). 'Réinventer Calais' https://reinventercalais.org/ Last accessed: 1 September 2017.

Isin, E. (2002). *Being Political: Genealogies of Citizenship*. Minneapolis: University of Minnesota Press.

Jensen, T. E. (2017). Materiality, Performance and the Making of Professional Identity. In S. A. Webb (ed.) *Professional Identity and Social Work*. New York: Routledge, 51–61.

Latour, B. (2013). *An Inquiry into Modes of Existence*. Cambridge, MA: Harvard University Press.

Latour, B. (2008). *What is the Style of Matters of Concern? Two Lectures in Empirical Philosophy*. Department of Philosophy of the University of Amsterdam, Amsterdam: Van Gorcum.

Latour, B. (2007). Turning Around Politics: A Note on Gerard de Vries' Paper. *Social Studies of Science*, *37*(5): 811–820.

Latour, B., & Weibel, P. (2005). *Making Things Public: Atmospheres of Democracy*. Cambridge, MA: The MIT Press.

Latour, B. (2004a). *Politics of Nature: How to Bring the Sciences into Democracy*. Cambridge, MA: Harvard University Press.

Latour, B. (2004b). Why Has Critique Run Out of Steam – From Matters of Fact to Matters of Concern. *Critical Inquiry*, 30(2): 225–248.

Latour, B. (2004c). Whose Cosmos, Which Cosmopolitics? Comments on the Peace Terms of Ulrich Beck. *Common Knowledge*, 10(3): 450–462.

Latour, B. (1993). *We Have Never Been Modern.* Cambridge, MA: Harvard University Press.

Latour, B. (1988). *The Pasteurization of France.* Cambridge, MA: Harvard University Press.

Laurent, B. (2017). *Democratic Experiments.* Cambridge, MA: The MIT Press.

Law, J. (2015). What's Wrong with a One-World World? *Distinktion: Scandinavian Journal of Social Theory*, 16(1): 126–139.

Law, J., Afdal, G, Asdal, K., Lin, W., Moser, I. & Singleton, V. (2014). Modes of Syncretism: Notes on Noncoherence. *Common Knowledge*, 20(1): 172–192.

Law, J., & Singleton, V. (2013). ANT and Politics: Working in and on the World. *Qualitative Sociology*, 36(4): 485–502.

Law, J., & Mol, A. (2008). Globalisation in Practice: On the Politics of Boiling Pigswill. *Geoforum*, 39(1): 133–143.

Law, J. (1986). On the Methods of Long-Distance Control: Vessels, Navigation and the Portuguese Route to India. In J. Law (ed.) *Power, Action and Belief: A New Sociology of Knowledge?* London: Routledge & Kegan Paul, 234–263.

Lemke, T. (2016). *Foucault, Governmentality, and Critique.* London: Routledge.

Lezaun, J. (2017). Actor-Network Theory. In C. Benzecry, M. Krause & I. A. Reed (eds.) *Social Theory Now.* Chicago, IL: University of Chicago Press, 305–340.

M'charek, A. (2018). 'Dead-bodies-at-the-border': Distributed Evidence and Emerging Forensic Infrastructure for Identification. In M. Maguire, U. Rao & N. Zurawski (eds.) *Bodies of Evidence: Anthropological Studies of Security, Knowledge and Power.* Durham, NC: Duke University Press, 89–109.

Marres, N. (2013). Why Political Ontology Must Be Experimentalized: On Eco-Show Homes as Devices of Participation. *Social Studies of Science*, 43(3): 417–443.

Marres, N. (2007). The Issues Deserve More Credit: Pragmatist Contributions to the Study of Public Involvement in Controversy. *Social Studies of Science*, 37(5): 759–780.

Michael, M. (2016). *Actor-Network Theory.* Los Angeles: Sage.

Mol, A., Moser, I., & Pols, J. (2010). *Care in Practice.* Bielefeld: Transcript Verlag.

Mol, A. (2008). *The Logic of Care: Health and the Problem of Patient Choice.* London: Routledge.

Mol, A. (2002). *The Body Multiple.* Durham, NC: Duke University Press.

Riyawarray: Common Ground. (2008). www.youtube.com/watch?v=ytfqf1LWdQI Last accessed: 31 August 2017.

Rose, N. (1999). *Powers of Freedom: Reframing Political Thought.* Cambridge: Cambridge University Press.

Rotman, B. (1987). *Signifying Nothing: The Semiotics of Zero.* Stanford, CA: Stanford University Press.

Scheffer, T. (2014). Die Arbeit an den Positionen. Zur Mikrofundierung von Politik in Abgeordnetenbüros des Deutschen Bundestages. In B. Heintze & H. Tyrell (eds.) Interaktion – Organisation – Gesellschaft. Special issue of the *Zeitschrift für Soziologie*, 369–389.

Schivelbusch, W. (1993). *Tastes of Paradise: A Social History of Spices, Stimulants, and Intoxicants.* New York: Vintage.

Serres, M. (1990). *The Natural Contract.* Ann Arbor: University of Michigan Press.

Stengers, I. (2005). The Cosmopolitical Proposal. In B. Latour & P. Weibel (eds.) *Making Things Public: Atmospheres of Democracy.* Cambridge, MA: The MIT Press, 994–1003.

Suchman, L. (2006). *Human-Machine Reconfigurations.* Cambridge: Cambridge University Press.

Ureta, S. (2015). *Assembling Policy: Transantiago, Human Devices, and the Dream of a World-Class Society.* Cambridge, MA: The MIT Press.

Verran, H. (2007). Postcolonial Moment in Science Studies. *Social Studies of Science*, 32(5–6): 729–762.

29

How does an ANT approach help us rethink the notion of site?

Albena Yaneva and Brett Mommersteeg

As early as the 1990s, Science and Technology Studies (STS) scholars have engaged with urban and architectural themes. These first studies looked at town planning as a technology and the city as an 'enormous artefact' (Aibar & Bijker, 1997), the invisible networks that shape big metropolises (Latour & Hermant, 1998), and urban obduracy and change (Hommels, 2005). Meanwhile, the sociotechnical analysis of innovations introduced by Madeleine Akrich and Bruno Latour in the late 1980s (Akrich, 1987, 1992), which had inspired work on design, focused on how technical objects generate specific modes of social, political and juridical organisation. This scholarship gradually led to further explorations in industrial (Dubuisson & Hennion, 1995, 1996) and engineering designs (Bucciarelli, 1994; Ferguson, 1992; Henderson, 1999; Law, 2002; Vinck, 2003).

Yet, architectural design remained a completely unmapped epistemological territory of STS analysis until 1996. In that year, Michel Callon released a seminal paper advocating the importance of Actor-Network Theory (ANT) as a methodological perspective for deepening our understanding of architecture. Arguing that 'the results of the anthropology of science and technology are *transportable* to the field of architectural studies,' he set an agenda for the decades to come (Callon, 1996, 1997). The first STS-inspired empirical study of architectural practice appeared on the pages of *Social Studies of Science* in 2005 (Yaneva, 2005), followed by studies devoted to the ethnography of architectural practices (Houdart & Minato, 2009; Latour & Yaneva, 2008; Loukisass, 2012; Yaneva, 2009a; Yaneva, 2009b). These works redescribed the socio-material dimensions of designing architects and set an ANT-inspired research agenda. The studies of cities through an STS-informed approach multiplied as well, as witnessed in a number of edited volumes (Blok & Farías, 2016; Farías & Bender, 2010; Guggenheim & Söderström, 2011, see also Färber, this volume) and special issues of Science Studies journals (Yaneva & Guy, 2008; Yaneva, 2011) and architectural journals (Yaneva, 2018).

When the ANT methodology started travelling outside its privileged domains of action – science, engineering, medicine and markets – the list of non-humans began to expand from the repertoire of microbes, scallops, bacilli, metro systems and other technical objects to include Portuguese ships, sick bodies, African elephants, drug addicts and amateurs of music. This list has since stretched to include an even more varied assembly of non-humans commonly found in architectural and urban worlds: Architectural renderings (Houdart, 2008), scale models (Yaneva, 2005), city plans (Zitouni, 2010) and maps (Nadaï & Labussière, 2013; Vertesi, 2008), for instance. However, as the ANT methodology has travelled to increasingly more domains, site itself has not yet been subject to the scrutiny of an ANT enquiry. In this short article, we will prepare the groundwork for one possible account of site.

'Site' as an object of (architectural) analysis

Despite its prominent status in design and in architecture, site has received little attention in architectural theory. More often than not, notwithstanding the technical and professional-oriented literature on 'site analysis' and 'site planning' (Lynch & Hack, 1984), site is conflated and grounded in other notions that have played a larger role in architectural discourse such as context, region, locality, setting, territory and land or landscape; or it is explored through historical examples of architectural projects without conceptual reflection. While in practice, sites are subject to a variety of analyses, observations, visits, tests, studies, plans and reports, bursting with different problematics like noise pollution, constraints, histories, traffic and logistics, trees and vegetation, layers of soil, waste, neighbours, visual impacts, among others, in theory, sites have been simplified and digested in mainly three ways: As *tabula rasa*, as natural facts and as social constructions.

Typically, sites are characterised as an external reality, a physical space with varying physical attributes *out there*, or *within* a series of representations, appearances and social relations superimposed over that external reality. In the modernist discourse of architecture, the site loses these characteristics as they are actively ignored. The physical and social aspects of the site are removed from the scenography in order to posit an abstract, idealised understanding of design that eschews the specificity and particular conditions of a site in search of the universal conditions for design (Redfield, 2005). Thus, the aim was to remove all pre-existing qualities and values of a site, to arrive at a zero point, a *tabula rasa*, from which to build.

By reducing site to a *tabula rasa* free of influences and attachments, and focusing solely on the products and objects of architecture, modernism's effects on thinking about site in architecture are substantial and far-reaching. In fact, all the subsequent discussions of site can be understood as reactions to the modernist discourse, circling around two foci: Do buildings *fit into* their sites, the context, constraints and parameters? Or do they *emerge* from the site, as a set of influences and determinants?

Responding to the modernist idea of a *tabula rasa* site, such follow-up discussions of site present it either in terms of primary or secondary qualities.[1] For the former strand of analysis, site is no longer a nuisance to suppress, but a resource to actively draw from as inspiration. As Ian McHarg (1971) argues, for instance, in the tradition of *genius loci,* design *ought* to extract, draw out the values that are hidden within the physical and natural aspects of a site; site, still a passive resource, is nevertheless an important aspect for design.

Here, the site, while present, acts as the architectural representative of Nature, a mute objectivity whose hidden values require human actions in order to be drawn out. The natural values of a site are captured by technical processes, in their purity, not yet spoilt by the social (Lynch & Hack, 1984; McHarg, 1971). This understanding of a site *qua* Nature is commonly found in landscape architecture (Meyer, 2005), drawing from 19th-century garden theory (Hunt, 1992). By upscaling site to a multi-scalar complexity of interconnected forces, ranging from the biological and physical to the economic and the political, recent landscape architecture theory highlights the continuity of a site across scales and domains. Still, akin to site planning, it remains by and large a process by which such interconnected forces can be captured, rationalised and turned into an objectified pattern (Gissen, 2010; McQuade &

Allen, 2011; Mostafavi & Doherty, 2010). Even as a totality of interacting forces, both 'natural' and 'human' site remains a set of facts *out there*: Either as a rough ('wild,' 'bucolic') set of values or as a smooth, rationalised pattern.

The third interpretation of site moves onto secondary qualities: We jump from the objective to the subjective. Within this body of literature, we are led from a phenomenological understanding of place (Beauregard, 2005; Hornstein, 2011) to the socially constructed nature of a site (Burns, 1996; Burns & Kahn, 2005; Kahn, 1998). Site is thus reduced to a perspectivism, where its physical nature is lost behind partial individual perspectives and social relations. Here, the discussion of site coincides with the increasing importance of place, region and context in architectural theory, and acts as an architectural avatar for these concepts, mimicking the theoretical discussions of space and place in the 'spatial turn' in geography. Rendered as different kinds of 'appearances,' site, as a concept in this body of literature, becomes interchangeable with its conceptual avatars, and loses its architectural specificity.

From this all-too-brief exploration of the concept of site in architectural theory, we can see that site has not been able to escape the effects of the Modernist influence. Either it is a blank slate whereupon, in a 'double-click,' buildings emerge without influence; or it acts as the architectural equivalent of Nature, as a stubborn set of facts (always set against the urban, the city, which are human artefacts); or, alternatively, as a social construct (always set against the impossibility of 'true' objectivity). In other words, site functions as a mediating interface between architecture, nature and society, caught between objective reality and subjective perception or social relations – bifurcated between primary and secondary qualities.

In architectural theory, the concept of site has thus followed a similar trajectory as notions of space and place, from an absolute empty container, to a physical and external 'outside,' to a product of social relations or subjective perceptions. Yet, if we follow the process of planning and architectural design, site appears as situated in the midst of reports, analyses, visits, technologies, measurements, photographs and tests; site is actively dissected and observed, folded and unfolded – more than a container, an object or a social construct. Thus, site emerges as a set of movements to follow.

Movements of *Site-ing*

If we follow design and construction processes, how often do we encounter a site simply as nature 'out there': Physical, static and passively awaiting the intervention of an adventurous designer? Conversely, how often is a site a pure social construction: A cultural product that is fabricated alongside another cultural construction, a building or infrastructure? Against the stubbornness of a site *qua* set of facts, and against the relativity of a site *qua* social construct, we are interested in moving beyond the two, not by adding them as two absolutes, but by focusing on what is frequently forgotten in the literature: How site *matters* in design, planning and construction processes, and the very work of accounting for site, working with it, according to it. We will call this specific work *site-ing*; this includes the work of placing and spacing simultaneously the built (a building or an urban infrastructure) and the site itself.

In what follows, we will zoom into specific situations where the work of *site-ing* becomes visible, and is made visible with ANT tools. We will trace the *site-ing* moves as ways of rearranging and repositioning the elasticity of the malleable urban networks of a city. *Site-ing* appears as a major work of questioning simultaneously the site and the built, nature and culture, building technologies and meaning. Unpacking these moves of *site-ing* will also illustrate the work that is needed by an ANT-informed ethnographer to unravel the complex agency of sites in design making, and to recalibrate, subsequently, the techniques of the ethnographic enquiry. We will reflect on the figure of the ethnographer, and will account for the recalibration of her specific trajectory, practices and interview techniques as they constitute the work of *site-ing* the ethnographer.

Site-ing a building

In 2014, the then Chancellor of the Exchequer of the UK government George Osborne announced in the Autumn Statement a new scheme called the 'Northern Powerhouse,' aimed to enhance growth in the North of England. At the 'heart' of the scheme is 'Factory,' an arts and cultural building in Manchester designed by the Office for Metropolitan Architecture (OMA)[2], which would act as the 'home' of the Manchester International Festival (MIF), a peripatetic arts festival, and a 'cultural anchor' for the regeneration of St. John's neighbourhood in Manchester city centre.[3] Factory's site is situated in a historically complex location, full of imposing constraints, from 'listed' historic buildings, neighbouring urban elements to the future developments of St. John's. Due to this, planning Factory has shown to be a laborious movement of *site-ing*.

We first encounter the complexity of the site in the rhetoric used by the designers in their presentations of the project to different audiences. During a recent presentation by the head of OMA's UK office, Carol Patterson presented the site as an active and surprising participant in design: When they had visited the site in Manchester, they realised that, instead of the simple box shape in a simple location that was stipulated in the competition brief, how *complicated* and enfolded that site was. It was only through the process of planning and development, which unfolds these features of the site, and that they confront its complications.

In her presentation, Carol explained that 'there were a lot of things that interweaved between our site: some of the cultural history, the music history and the future'; the site, they discovered, was full of historical artefacts and references that play an important role in the future of the site, and the building-to-be.[4] The site, as suggested, is 'interwoven' with features that need to be untangled and unfolded. Following the movement of *site-ing* requires the crafting of a new context, the constrained act of site expansion and redistribution. Through these movements, we witness the Factory site growing and being defined *at the same time* that the building is outlined in design.

Expected to be completed by 2021, Factory provides us a rich opportunity to see the movements of *site-ing* up close. We can still witness the movements, the expansion and contraction of the Factory site, and the continual back and forth process from one locale–of–practice to another (that is, planning offices, design and engineering firms, acoustics labs, public consultations, client's headquarters, among others), as it expands to other entities, accommodating their agency into the larger aggregate that Factory site will become. *Site-ing* is made in between many locales and aims to 'anchor' the complexities of the building site, by untangling and retying the threads, through negotiating and recalibrating the connections that it enters into.

One simple way to explore these movements is to slowly move through the planning documents, which do not have a simple representative function, but give us access to the specific knowledge practices and modes of organisation involved in *site-ing* work (Hull, 2012; Riles, 2006). Here, as particular 'inscription devices' (Latour & Woolgar, 1986), the various injunctions of the site are unfolded. We do not only have various viewpoints onto a virtual building, but the documents themselves (and the studies therein) unearth (literally sometimes) these various locales and connected entities within the site. Each document acts like 'oligoptica' (Latour, 2005; Latour & Hermant, 1998) that capture different aggregates, entities and locales, whether the composition of soil, flow of traffic or the various histories of the site, and never against some common substantial nature. They not only unfold the multilayered strata of the site that gathers actors as mediators (Asdal, 2015), but the documents transport the Factory site to other locales-of-practice, modifying it as they move.

In other words, not only are there many different aggregates within the one place, but this site travels to and gathers different locales-of-practice into it: From OMA's office in Rotterdam to LevelAcoustic's acoustic laboratory in Eindhoven to the city council rooms in Manchester, Planit's UXB (unexploded bombs) labs in York, the warehouses of the 1800s, to BuroHappold's offices in Bath. In all these movements, something is gained, something is added to the living aggregate that the Factory site is about to become.

However, this is not merely a process of unfolding, but – to reiterate – of re-crafting connections and aggregating them into a new context. *Site-ing* becomes a movement that reconfigures connections to achieve a single voice: Turning the many different aggregates and locales that need to be visited in order to recollect the Factory site for design. In other words, the site cannot remain a pure multiplicity, but requires closure as a practical necessity for planning permission and to be used in design (Figure 29.1).

Returning to Carol's presentation, we hear her develop a narrative through the confluence of histories specific to Factory's site: Important features that not only impinge on the design, but are also enrolled in the trajectory of the building-to-be in order to allow it to move. It is situated within the old headquarters of Granada TV, sitting atop the brick-arches of the Grade II-listed Colonnaded Railway Viaduct, whose now disused railway lines lead to a non-listed yet important industrial Bonded Warehouse that abuts the site to the East. Thus, rising above the Manchester and Salford Junction Canal, an industrial canal from 1839, the site is full of recalcitrant constraints that have become enlisted in its trajectory. It is important

Figure 29.1 Factory, OMA – building site in Manchester (Authors' Own)

to note, though, that this history does not rest in the background as a grand narrative to justify the site's importance. Instead, it is a crafting of context that actively unfolds in the construction and recollection of the Factory site. With each document and each urban feature,
another actor is invited into the fold of the site, expanding its reach, enrolling others; thus,
we witness the composition of site. As planning unfolds, its aim is to 'compose' a site, not in
contradistinction to the context, but simultaneously with it.

At the groundbreaking ceremony, which took place on the site of Factory during MIF
months later, Rem Koolhaas informs a crowd of Mancunians, in response to a question about
Manchester's heritage, that, yes, nostalgia serves a purpose, but only up to a point.[5] Site-ing is
neither a process of collecting a series of facts about the site, nor is it a way to extract the spirit
of a place that pre-exists. It entails work that gathers actors and locales of practice into a single
voice for two reasons. First, as a rhetorical device, the site-ing work composes a site, which
makes an argument for the building; it is a way of doing words with things. Secondly, it folds
in a specific way the contours of the site into an active 'origami' that also reacts to the other
givens and constraints in the design and planning of the project. While these enlisted actors
need to be negotiated into the project, they are nevertheless recalibrated along its trajectory.

In both these senses, we see how site-ing is not a way of finding breathing room in a dense
urban location, but it is rather about tracing a project trajectory to witness the right composition of the site for design. In other words, it is not a passive ground that a building fits into, or
a determining context, but a composition that shifts and modifies as new entities are enrolled
or subtracted. Moving through the planning documents, as they travel from one locale of
design and construction to another, following designers as they engage in grasping the complexity of the site and composing it as a full-blown participant in the project, we can trace
the manifold of the Factory site, the necessity to unfold, to multiply and then to aggregate
it again in order for the project to succeed. The planning documents and design movements
reveal site not as a resource that exists *in potential*, waiting to be realised, nor as an exercise in
cultural fiction, but as a living 'trajectory of actualisation,'[6] a process of crafting, redefining
and aggregating a new context.

Amidst the site as a composition that shifts, how is the ethnographer supposed to navigate
through this shifting field, to walk on this shaking ground of dispersed locales-of-practice?
As we will now unfold, there is more to this than the tracing of planning documents and
designers' moves, in that the ethnographer participates in the very work of *site-ing*.

Site-ing the ethnographer

Unlike other standard ethnographic studies, less time was spent in this field. However, a
few specific techniques helped intensify the presence of the ethnographer, maximising the
time spent. The work of *site-ing* made the ethnographer move from one locale-of-practice
(of design, planning and construction) to another, equipped with camera, notebook and tape
recorder. Meanwhile, the participants in the ethnographic study continued to work in these
various locales, surrounded by their objects, documents and tools, knowing that another
colleague was being interviewed at *that* moment and that a specific part of their practices was
being revealed, documented and ethnographically captured. Their movements sometimes
overlapped with the ethnographer's moves and this led to the synchronic visibility of all the
participants in the study, a form of co-presence[7] that maintained the transparency of the dispersed locales of the ethnographic enquiry.

It is well known from STS' laboratory ethnographies that knowledge production is embedded in local environments (Merz, 2006). As Knorr-Cetina (1995, p. 151) argues: 'If construction

is wrapped up in bounded locales, the ethnographer needs to "penetrate the spaces" and the stream of practices from which fact construction arises.' While in other ethnographies where we have spent years living immersed in an office (OMA, Rotterdam) or frequently travelling to visit a practice (FOA, London), this study included the work of *site-ing*. When the ethnographer gained access to the 'stream of practices' of the Factory-making, most of the ethnographic interviews[8] were conducted as locale-based interviews in a specific place of practice and included elements of immersive observation.

As an ethnographic interviewing technique, this contextual inquiry implied going to the participants' offices to gather qualitative data in order to capture the natural setting of specifically situated design, engineering and planning practices, and witness how the participants work, asking specific site-related, object-focused questions about what they were doing and why. Observing different actors and questioning them in their environments was paramount for bringing important details about their practice to light. Locale-based interviews in design studios, planning and engineering offices, city government offices and public presentations acted as *site-ing* devices recalibrating the ethnographer's presence and the presence of others.

Only rarely would an interview have a script and a fixed scenario, a sitting plan that would not change. Instead of engaging in a directed question-and-answer interviewing style, a static situation where everyone and everything is seated, 'in place,' the ethnographer spends time exploring the specific locale-of-practice, wandering, pointing to specific objects, touching documents, folders and screens that were currently in use, thus collecting a series of specific features about the building-making. This 'wandering around a locale-of-practice' type of interview prompted a number of questions emerging from the contextual specificity of the practice that unfolds and the socio-material circumstantiality of that locale-of-practice, streaming right from the world of objects, instruments and documents. That is, the ethnographer also *sites* herself according to the rhythm of the enquiry that will lead to the composition of the Factory site.

Site-ing, as an ethnographic strategy, is thus a way of 'establishing co-presence.'[9] *Site-ing* is not a rhetorical device that aims to convince. Instead, it is almost the opposite: Cracking 'open' the black box of studios, labs and planning offices, fissuring practices typical for this locale into many different associations. It implies that nothing is still; *site-ing* moves us all forward through new connections and additions. The ethnographic interview reminds us of the *site-ing* process outlined earlier in the example of the Factory site composition through planning documents and tracing design moves. Through this *site-ing* process, we do not capture the world of the locale-of-practice *in toto*, but only fragments of it.[10] It entails work that gathers people, documents, bodies and skills and *sites* them according to the ANT ethnographic enquiry: The interview itself becomes a way of producing words with things. By the strolling in the locale, the stream of new questions, the interviewees wandering around the ANT ethnographer and the dance of the objects activated 'on the go,' *site-ing* folds in a very specific experiential, but always fragmented way the contours of the locale-of-practice, hosting the interview, in an active relational spatial 'origami.'

Much as in design practice, it is the 'active' nature of the locale that gathers talking objects, instruments, scripts and actors together, to the extent that the place of ethnography is not a passive backdrop for an ethnographer (just as site is not a *tabula rasa*), but rather 'activates' the interview; the entire collective assemblage of enunciation speaks. Neither the actors nor the researcher speaks; rather, the building-making processes are spoken through them. The *site-ing* interviews are thus reminiscent to the work of composing a musical piece, where all participants contribute to finding the right tones, the good arrangements and activate new connections, which, when aggregated together in the rhythmic flow of a stream of practice, will generate a full-fledged set of tones that make sense. In these interviews, the locales-of-practice *site* us all through co-presence; the stream of the interview experience operates symmetrically through the researcher, and the actors convoked, as well as the devices and questions that activate it further.

Walking through different locales-of-practice where the *site-ing* is preformed, we stroll, visit, stop and discuss specific objects, the interview stages change as we move, and different objects appear and disappear from the misty scenes. Stopping by, *site-ing* around, and engaging in discussions that last longer than a question-and-answer sharp rhythm allows creating specific settings where the objects can talk, and the interviewees will talk *with* the objects *to* the researcher *via* the objects, and in many different locales-of-practice. Thus, the ethnography of the Factory-making is a walk on its own, a long and interrupted walk capturing 'occasions' or events. In this walking process, both interviewer and interviewees are situated, their boundaries blurred and the 'being-in' of a passive space turns into an active regime of 'being-with' many locales-of-practice that unfold. If *site-ing* interviews involve strolling *within* a locale-of-practice and *with* it, walking with the Factory site relies on exploring and walking with many locales-of-practice in a pluriverse ethnographic journey of enquiry. In each of them a different feature of Factory site is extracted and added to the composition. This also *sites* the ANT ethnographer within the unfolding dynamics of design and construction, and vibrant urban processes that are all interconnected. In addition, it makes the ANT ethnographer complicit to *site-ing*, that is, to the composition of the Factory site, lending to it some stability and coherence.

It is impossible to witness all the locales-of-practice: Only so many fragments from projects, rhythms and time spans are witnessed. Not being able to visit all of the locales-of-practice, to be present there at the same time and in the key moments of decision-making, we abandon the ambition of establishing connections or comparisons with other building projects, architectural firms and urban developments. Instead, bringing experiences from different locales of the Factory-making together, assembling information from *site-ing* interviews and walking as a way to recreate simultaneity and synchronicity, revealed the multi-actorial and multifaceted nature of design, and allowed the ethnography to capture a building world as it *opens* itself up to scrutiny, and as it takes us, as Donna Haraway (1991) would say 'in the belly of the monster.'

Multi-Site-ing

If the site is never 'out there' but always requires the work of *site-ing*, grasping the site in design and construction processes also demands the tedious work of an ANT ethnography that constantly *sites* the researcher. This is the double work of multi-*site-ing* and re-*site-ing*, of going back and forth between the different locales-of-practice, over and over again, from which a notion of an 'ethnographic site' is generated as something that remains the same throughout all these movements.

Just as the site is neither 'there' ready to accommodate a building, nor appears as a purely social or material construction, so is the work of the ANT ethnographer: None of her interviews happen 'on-the-spot.' The locales-of-practice are not ready at hand, waiting to be studied. They all require the active work of *site-ing* and re-*site-ing* the researcher; each of them provides partial access to the composition of the Factory site as an object of analysis. They have to congregate. It is only when assembled into an ethnographic account that the ethnographer's work begins to 'resemble' the designers' and planners' *site-ing* work. Following the pitfalls of *site-ing* the researcher within this shifting ground of composing the Factory site, nothing is stable or could simply begin from a *tabula rasa*. The ethnographer begins in the middle, as it were. It becomes even impossible to allude to the seemingly similar concept of multi-sited ethnography developed by George E. Marcus (1995). With ANT in hand, confronting objects and devices that act back, facing unknown sources of agency, and immersed within contextual mutations that shape as we move, we are rather part of an enquiry of 'multi-locality.' The scope of this ANT enquiry goes far beyond a notion of 'multi-sitedness' as every single 'site' (termed here locale-of-practice) is also produced as a subtle and 'multi-local' work of composition.

None of the participants in the design and planning process of the Factory can grasp its site totally on the spot. The ANT ethnographer is just as myopic. It is through the act of joining the *site-ing* of the Factory that the ethnographer becomes an integral part of folding its spatial 'origami' further, along with many designers, planners, engineers and other participants in the different locales-of-practice. The Factory site composition becomes the ethnographer's own locale. Each move adds new tones. All of them eventually achieve stabilisation (albeit fragile): A Factory site gradually emerges as we reach the end of this account about its making. Coherent and stable, the site is required to eventually host a building that stands heavy and unbearably durable. Composed and sound, the ethnographer eventually completes her account that is light and inevitably re-writeable by other researchers who can also take up an enquiry into the question of site.

Notes

1 This old philosophical divide is most clearly articulated by John Locke in *Essay Concerning Human Understanding* (1997 [1689]). Primary qualities refer to the brute matter extended in space, whereas secondary qualities refer to phenomenal, subjective attributes. But it is also the foundation for modern Occidental philosophy as the split between appearances and reality, the phenomenal and noumenal worlds, as well as the split between being and becoming. For the diagnosis of this divide, see Alfred North Whitehead's lectures in *The Concept of Nature* (2015 [1920]), what he has called the 'bifurcation of nature.'

2 OMA is a Dutch architectural firm, based in Rotterdam, founded in 1975 by Dutch architect Rem Koolhaas and Greek architect Elia Zenghelis, along with Madelon Vriesendorp and Zoe Zenghelis. It is one of the leading practices in the world today. See Yaneva's ANT-study of design in the making (2005 and 2009a) based on material from this practice.

3 OMA. (October, 2016). "Design and Access Statement: Factory". Manchester, UK: Manchester City Council.

4 Patterson, C. (June 2017). *Manchester Architect's Symposium: Factory.*

5 Koolhaas, Rem, Ellen Van Loon, and Hans Ulrich Obrist. (July, 2017). *The Factory Panel: Spaces for Culture.*

6 See Gilles Deleuze's work on Bergson (1991 [1966]) for a philosophic account of the 'trajectory of actualisation,' the production of the new, as it differs from the realisation of pre-existing possibilities.

7 Anne Beaulieu (2010) put forward the concept of co-presence as a way to shape fieldwork. If other authors underline the desirability of physical co-location as a requisite for ethnographic investigation, Beaulieu shifts the focus on achieving co-presence by emphasising that interaction is a potentially rewarding but precarious achievement (Goffman, 1957), and that physical presence is not equivalent to availability for interaction (Goffman, 1971). While physical co-location can be a resource for participants, it is not in itself a sufficient criterion for co-presence. Co-presence decentralises the notion of space without excluding it. Beaulieu argues also that the possibility of co-presence might be established through a variety of modes, physical co-location being one among others. Co-presence as a starting point enables a more symmetrical treatment of forms of interaction.

8 For a discussion on ethnographic interviews, see: James Spradley, *The Ethnographic Interview*, 1979, Wadsworth and John Van Maanen, *Tales of the Field: Writing Ethnography*, Chicago: Chicago University Press, 2011.

9 Per Beaulieu's understanding of co-presence as an interactive accomplishment by participants and ethnographers alike that does not share the unidirectional and ocular-centric connotations of witnessing, or interrogating.

10 For a discussion about the problematic of contexts and wholes in cultural anthropology and ANT, see Tsing (2010). In our account of the Factory 'site,' we see the congregation of many wholes that come together, but not as 'parts' that fit within a greater 'whole' that encompasses them. Nevertheless, at the end of the planning process, there needs to be a relatively stable 'site' for the building to be granted planning approval, and for the building to be built, and likewise, the ethnographer composes an account that is more or less coherent, without assuming to have captured its object 'wholly.'

References

Aibar, E. & Bijker W.E. (1997) Construct a City: The Cerdà Plan for the Extension of Barcelona. *Science, Technology & Human Values* 22(1), 3–30.

Akrich, M. (1987) Comment décrire les objets techniques? *Techniques et Culture* 9, 49–64.

——— (1992) The De-Scription of Technical Objects. In *Shaping Technology/Building Society.* (Bijker, W. & Law, J., eds.), MIT Press, Cambridge, pp. 205–224.

Asdal, K. (2015) What Is the Issue? The Transformative Capacity of Documents. *Distinktion: Scandinavian Journal of Social Theory* 16(1), 74–90.

Beaulieu, A. (2010) From Co-location to Co-presence: Shifts in the Use of Ethnography for the Study of Knowledge. *Social Studies of Science* 40(3), 453–470.

Beauregard, R. (2005) From Place to Site: Negotiating Narrative Complexity. In *Site Matters: Design Concepts, Histories and Strategies.* (Burns, C. & Kahn, A., eds.), Routledge, Great Britain, pp. 39–58.

Blok, A. & Farías, I. (eds.) (2016) *Urban Cosmopolitics: Agencements, Assemblies, Atmospheres.* Routledge, Great Britain.

Bucciarelli, L.L. (1994) *Designing Engineers.* MIT Press, Cambridge.

Burns, C. (1996) On Site: Architectural Preoccupations. In *Drawing/Building/Text.* (Kahn, A., ed.), Princeton Architectural Press, New York, pp. 146–167.

Burns, C. & Kahn, A. (2005) Why Site Matters. In *Site Matters: Design Concepts, Histories and Strategies.* (Burns, C. & Kahn., A., eds.) Routledge, Great Britain, pp. Vii–xxix.

Callon, M. (1996) Le Travail de la Conception en Architecture. *Situations, Les cahiers de la recherche architecturale* 37(1), 25–35.

Deleuze, G. (1991 [1966]) *Bergsonism.* (Tomlinson, H. & Habberjam, B., trans.), Zone Books, Brooklyn, NY.

Dubuisson, S & Hennion, A. (1995) Le design industriel, entre création, technique et marché. *Sociologie d'art* 8, 9–30.

———. (1996). *Le Design: L'objet dans l'usage.* Presses des Mines, Paris.

Farías, I. & Bender, T., (eds.) (2010) *Urban Assemblages: How Actor-Network Theory Changes Urban Studies.* Routledge, New York, NY.

Ferguson, E. (1992) *Engineering and the Mind's Eye.* MIT Press, Cambridge.

Gissen, D. (2010) *Territory: Architecture Beyond Environment.* John Wiley & Sons, London.

Goffman, E. (1957) Alienation from Interaction. *Human Relations* 10(1), 47–59.

———. (1971) *Relations in Public: Microstudies of the Public Order.* Basic Books, New York, NY.

Guggenheim, M. & Söderström, O., (eds.) (2010) *Re-shaping Cities: How Global Mobility Transforms Architecture and Urban Form.* Routledge, New York, NY.

Haraway, D. (1991) Situated Knowledges: The Science Question in Feminism and the Privilege of Partial Perspective. In *Simians, Cyborgs and Women: The Reinvention of Nature,* (Haraway, D., ed), Free Association Books, London, pp. 183–201.

Henderson, K. (1999) *On Line and on Paper: Visual Representations, Visual Culture, and Computer Graphics in Design Engineering.* MIT Press, Cambridge.

Hommels, A. (2005) Studying Obduracy in the City: Toward a Productive Fusion between Technology Studies and Urban Studies. *Science, Technology & Human Values* 30(3), 323–351.

Hornstein, S. (2011) *Losing Site: Architecture, Memory and Place.* Ashgate, England.

Houdart, S. (2008) Copying, Cutting and Pasting Social Spheres: Computer Designers' Participation in Architectural Projects. *Science Studies* 21(1), 47–63.

Houdart, S. & Minato C. (2009). *Kuma Kengo: An Unconventional Monograph.* Éditions Donner Lieu, Paris.

Hull, M. (2012) *Government of Paper: The Materiality of Bureaucracy in Urban Pakistan.* University of California Press, Berkeley.

Hunt, J. D. (1992) *Gardens and the Picturesque: Studies in the History of Landscape Architecture.* MIT Press, Cambridge.

Kahn, A. (1998) From the Ground Up: Programming the Urban Site. *The Harvard Architectural Review* 10, 54–71.

Knorr-Cetina, K. (1995) Laboratory Studies: The Cultural Approach to the Study of Science. In *Handbook of Science and Technology Studies.* (Jasanoff, S., Markle, G.E., Peterson, J.C., & Pinch, T., eds.) Sage Publications, Inc., Thousand Oaks, CA.

Latour, B. (2005) *Reassembling the Social: An Introduction to Actor-Network-Theory.* Oxford University Press, Oxford.

Latour, B. & Yaneva, A. (2008) Give Me a Gun and I Will Make Every Building Move: An ANT's View of Architecture. In *Explorations in Architecture: Teaching, Design, Research.* (Geiser, R., ed) Birkhäuser, Basel, pp. 80–89.

Latour, B. & Hermant, E. (1998) *Paris, ville invisible.* Empécheurs de penser rond, Paris.

Latour, B. & Woolgar, S. (1986) *Laboratory Life: The Construction of Scientific Facts.* 2nd edn. Princeton University Press, Princeton, NJ.

Law, J. (2002) *Aircraft Stories: Decentering the Object in Technoscience.* Duke University Press, Durham, NC.

Locke, J. (1997 [1689]) *An Essay Concerning Human Understanding.* (Woolhouse, R., ed.) Penguin Books, London.

Loukissas, Y. A. (2012) *Co-Designers: Cultures of Computer Simulation in Architecture.* Routledge, New York.

Lynch, K. & Hack, G. (1984) *Site Planning.* 3rd edn. MIT Press, Cambridge.

Marcus, G.E. (1995) Ethnography in/of the World System: The Emergence of Multi-Sited Ethnography. *Annual Review of Anthropology* 24, 95–117.

McHarg, I. (1971) *Design with Nature.* Doubleday & Company, New York.

McQuade, M. & Allen, S. (2011) *Landform Building: Architecture's New Terrain.* Lars Muller Publishers, Baden.

Merz, M. (2006) Différenciation interne des sciences: constructions discursives et pratiques épistémiques autour de la simulation. In *La fabrique des sciences. Des institutions aux practiques.* (Leresche, J.-P., Benninghoff, M., Crettaz von Roten, F., & Merz, M., eds.) Presses polytechniques et universitaires romandes (PPUR), Lausanne, pp. 165–182.

Meyer, E. (2005) Site Citations: The Grounds of Modern Landscape Architecture. In *Site Matters: Design Concepts, Histories and Strategies.* (Burns, C & Kahn, A., eds.) Routledge, Great Britain, pp. 92–130.

Mostafavi, M. & Doherty, G. (2010) *Ecological Urbanism.* Lars Muller Publishers, Baden.

Nadaï, A. & Labussière, O. (2013) Playing with the Line, Channelling Multiplicity: Wind Power Planning in the Narbonnaise (Aude, France). *Environment and Planning D: Society and Space* 31 (1), 116–139.

Redfield, W. (2005) The Suppressed Site: Revealing the Influence of Site on Two Purist Works. In *Site Matters: Design Concepts, Histories and Strategies.* (Burns, C. & Kahn, A., eds.) Routledge, Great Britain, pp. 185–222.

Riles, A. (Ed.). (2006) *Documents: Artifacts of Modern Knowledge.* University of Michigan Press, Ann Arbor.

Spradley, J. (1979) *The Ethnographic Interview.* Wadsworth, California.

Tsing, A. (2010) Worlding the Matsutake Diaspora, or Can Actor-Network Theory Experiment with Holism? In *Experiments in Holism: Theory and Practice in Contemporary Anthropology.* (Otto, T. & Bubandt, N., eds.) Wiley-Blackwell, Oxford, pp. 52–76.

Van Maanen, J. (2011) *Tales of the Field: Writing Ethnography.* Chicago University Press, Chicago.

Vertesi, J. (2008) Mind the Gap: The London Underground Map and Users' Representations of Urban Space. *Social Studies of Science* 38(1), 7–33.

Vinck, Dominique (Ed.). (2003) *Everyday Engineering: An Ethnography of Design and Innovation.* MIT Press, Cambridge.

Whitehead, A.N. (1920) *The Concept of Nature.* Cambridge University Press, Cambridge.

Yaneva, A. (2011) Traceable Cities. *City, Culture and Society* 3(2), 87–89.

———. (2009a) *The Making of a Building: A Pragmatist Approach to Architecture.* Verlag Peter Lang, Germany.

————. (2009b) Making the Social Hold: Towards an Actor-Network Theory of Design. *Design and Culture* 1(3), 273–288.

————. (2005) Scaling Up and Down: Extraction Trials in Architectural Design. *Social Studies of Science* 35(6), 867–894.

Yaneva, A. (Ed.). (2018). Bottega. [Special Issue] *Ardeth*. 2.

Yaneva, A. & Guy, S. (Eds.) (2008) Understanding Architecture, Accounting Society: A Dialogue of Architectural Studies and Science and Technology Studies. [Special Issue]. *Science Studies* 21(1).

Zitouni, B. (2010) *Agglomérer. Une anatomie de l'extension Bruxelloise (1828–1915).* Brussels University Press, Brussels.

30

How does the South Korean city of Kyŏngju help ANT think place and scale?

Robert Oppenheim

Somewhat more than a decade ago, I wrote a book on the making or assembly of place at the intersection of urban development, heritage politics and the practice of popular historiography in the South Korean historic city of Kyŏngju (Oppenheim 2008). My disciplinary background is in anthropology, and this project was meant to be in dialogue with the anthropological literature on place-making of the era, which mostly hinged on the representation and social construction of places. ANT did several things for me in this project. Most simply, it helped me connect traditionally different anthropological topics, everything from festivals to heritage tourism to social movement organisations to development planning, via their own mutual connection to objects of concern such as railway infrastructures, historic artefacts and organisational forms. It helped me ontologise the question of place, to treat place not just as meaningful but as a setting for doing things. Most specifically, it helped me offer a definition of place as a pragmatic 'intertopological' effect, the dimensionality that exists when hinterlands of settled relations are drawn into new configurations, spanning several senses of 'proximity' and 'distance' all at once (see also Oppenheim 2014).

This focus on the intertopological aspect of place is not the emphasis of the present chapter, but it is worth tarrying here a moment, in part because the direction I sought to take the anthropology of place has been convergent with ANT-inspired theorisation of urban assemblages in the field of urban studies (Amin and Thrift 2002; Blok and Farías 2016; Farías and Bender 2010). Calling upon more complex spatialities more clearly evident in later ANT work – including Annemarie Mol and John Law's exploration of topologies and Bruno Latour's understanding of 'plasma' (Latour 2005: 241; Mol and Law 1994; Oppenheim 2007) – I had envisioned place as an 'intertopological' effect among overlying topological arrangements in interference with or acting as 'hinterlands' with respect to one another (Oppenheim 2008: 14–15; see also Law 2004: 27–35). In a more formal subsequent definition, place was presented as 'a simultaneity of terrains of actuality and potentiality and of the intertopological movements that inextricably connect and collect them' (Oppenheim 2014: 396) – in other words, a spatiotemporality not only of the connected and unconnected, but also of the potentially, or not yet, connected.

In parallel, urban assemblage theory, itself put forth as an flattening alternative to approaches that view urban politics as determined by underlying political economic forces

(Farías 2010:6), has conceptualised 'the city (or site) multiple' (Farías and Blok 2016: 2; Lepawsky et al. 2015: 189–190; drawing upon Mol 2002) as composed of overlapping assemblages – versions, articulations or enactments of the city – that may interact, interfere and become coordinated through the course of urban politics (see also Ureta 2014). Assemblage urbanism theorists have likewise turned to Latour's 'plasma' to conceive of the way assemblages exist as potentiality with respect to one another, 'defin[ing] a virtual horizon of indeterminate possibilities and possible becomings' (Farías and Blok 2016:14). Place in my original argument and 'cityness' in urban assemblage theory have made similar room for the copresent temporality of the not yet (Farías and Blok 2016:10; Lepawsky et al. 2015:185; Simone 2010; see also Färber, this volume).

The emphasis of this chapter is instead on what I call 'circulating normativities' as key sites at which the coordination of conflicting assemblages and the recombination of place occurred in 1990s Kyŏngju urban politics. Such normativities include visualisations, material points of view, obligatory procedures and conceptual objects raised to iconic importance. Among them are the 'panoramas' and 'oligoptica' that Latour (2005:175–190) advances as a way to flatten processes of visualising and seeing away from methods that regard either the politics of representation or panoptic physics of power as existing on a separate plane from the concrete negotiations of assembly. More broadly, this approach is in alliance with anthropological work that takes such normativities or representations-in-the-plural as themselves 'artefacts entangled' in associative dynamics of place-making (Hull 2012:4). In what, then, lies the capacity of such circulating normativities to connect and mediate among assemblages? The argument here is that as actants, normativities often do this additively, by being multiplied or redoubled themselves. The coordination of the city multiple is thus commonly an effect of the multiplicity or partiality of objects of concern (Law 2004; Mol 2002).

This is a consideration of ontopolitical process, part of the mechanics or physics or pragmatics of place-making. Yet following Farías and Blok (2016) further, the action of circulating normativities can also be examined in terms of their cosmopolitical aspect, as nodes for the articulation of ethical mappings of a shared political setting. What gives some circulating normativities, and not others, the force to rewire place? In the present example, one answer is the way such normativities were able to index and access compulsory scales. Scale-making as an emergent aspect of assembly processes is an important ANT claim, but one that Blok (2016: 607) notes 'arguably remains under-theorized in assemblage urbanism.' Kyŏngju in the 1990s offered a situation in which scale-making took place not in a wholly open-ended fashion (an analysis that the simple ANT maxim that 'actors make scales' would seem to invite), but rather in relation to a set of scales – 'local,' 'national' and 'global' – that themselves, in turn, are not analytically a priori but rather were taken as 'pre-given' and desirable by actors with various positions on urban politics (Slater and Ariztia 2010: 91). 'Linking up' with these scales was a powerful act in urban disputes, to do so successfully brought the force of a host of associations. An urban development successfully rendered 'global' enlisted, as if by short circuit, a host of other desiderata linked to 'globalisation' in 1990s South Korea.

This chapter thus treats scaling in parallel to the assembly of place as a whole; scales are not natural containers for place assembly, but assembly may draw upon hinterlands of scalar routinisation, and with consequence. The coordinating capacity of circulating normativities was mediated by their ability also to activate routinised scalar potentialities with a more properly ethical, cosmopolitical claim on the desirable. This is how a conflict in which different participants accused opponents of paying insufficient attention to necessary scales of Kyŏngju's being became one in which Kyŏngju was refigured as an appropriately, appealingly global–local–national common world.

Scales of compulsory reference

As was the case in Asturias as discussed by Slater and Ariztia (2010), mid-1990s Kyŏngju presented an analytical situation in which much place-making activity by residents and non-residents alike was conducted in relation to scales that were understood as 'pre-given.' The city, like other South Korean regional cities, was simultaneously called upon to scale up and down, to 'be (more) global' and 'be (more) local,' even as Kyŏngju also, owing to its historic character, confronted ongoing, more particular imperatives to maintain its unique 'national' character in a set of ways. What such moves might concretely mean was contested by various actors, but the necessity of them and their scalar points of reference were broadly accepted. In 'following the actors,' one frequently followed them to and through these touchstones. This section begins with the substance of these scales' compulsory character, leading to two questions for a post-ANT framework that continues to take scaling as a result of process and not its precondition. How are such compulsory scales themselves reinforced, and how are they invoked in the making of place?[1]

The government of President Kim Young Sam (1993–98) chose *segyehwa*, 'globalisation,' as its policy cornerstone and rhetorical dominant. *Segyehwa* could mean many things, and was linked to a variety of developments in diplomacy, economics and social policy (Kim 2000). Culturally, *segyehwa* supposedly inaugurated a new openness, and slogans such as 'the most global thing is the most Korean thing' permeated official publications. There was a focus on building the 'world-class.' Among the most talismanic of infrastructural developments to which the label was applied was South Korea's high-speed rail network, later christened the KTX, which was in its design and early build stages in the 1990s. As had been the case with Japan's *shinkansen* (Hood 2006), the KTX became and has since been iconic of South Korean modernity itself, as well as a consistent companion to other South Korean entries onto the world stage. It was the high-speed railway that would be the central object of a significant debate over development in Kyŏngju.

The same era, however, also produced new economic demands on regional cities and counties. Changes in the funding relationship between the national government and local municipalities led to a situation in which such municipalities were newly compelled to compete for state and provincial investment. They also attempted increasingly to attract international interest and investment directly. Neologisms like 'glocalisation' (either in English or as *chibangsegyehwa*) gained some currency in the media. Despite its established base of historical tourism, Kyŏngju was not immune to these dynamics. In the early 1990s, city officials attempted to attract a new provincial administration building, for instance, and new efforts were made to monetise tourism by developing food products suitable for Japanese-style *omiyage* travel gifts.

Interlaced with these developments also came expectations placed upon the local as potentially a newly democratic and newly responsive sphere of political action. The end of military rule and the democratisation of national politics beginning in 1987 brought, within a few years, local self-governance (*chibang chach'i*), with direct elections for city councils and mayors. Social movement organisations also targeted the local as a newly significant setting (Oppenheim 2008:197–221).

In Kyŏngju's case, however, these new dynamics unfolded against the backdrop of an extant set of demands to speak for the Korean nation. The city owes its historical landscape of temples, pagodas, stone Buddhas and other monuments to its status as the location of the capital city of the first millennium kingdom of Silla. In the 1960–70s, Kyŏngju was developed according to a plan that emanated from the South Korean presidential residence itself as a showpiece historical preservation and reconstruction zone and a tourism destination. But in the post-Korean War period, Kyŏngju also had a unique role with a specifically *South* Korean national narrative. In conventional historiography, Silla was the state that, in the 7th century, defeated two other Korean kingdoms and thus achieved the first unification of the Korean peninsula. This account of an original unification at the hands of a southern state, by a kingdom ascribed with martial valour, appealed to the military rulers of South Korea seeking reunification with the North on their own terms within a politics of hostile national division.

To reiterate, these global, local and national scales at play in the 1990s are not intended as sovereign contexts, natural scalar levels that produced effects in Kyŏngju out of their own inherent dynamics (see Carr and Lempert 2016). Rather, the point here is akin to Callon's (1998) contention about the formatting effects of economics upon the economy: They were routinised scales of reference performatively reproduced in laterally reinforcing knowledge practices. While beyond the scope of this chapter, their routinisation, their reinforcement as compulsory, can be theorised in terms of various relations of interdependence or 'inter-scaling' (Philips 2016), as when, for instance, a mayoral candidate in one of the novel local elections sought to demonstrate his local(ist) *bona fides* by showing up at a Kyŏngju festival that presented itself as organised by and for local residents as opposed to tourists. Such scales could, in turn, be indexed or invoked by circulating normativities, operating as 'scaling devices' in Slater and Ariztia's (2010:92) terms, as mechanisms of 'reconfigur[ing] and position[ing]' Kyŏngju in relation to their desirability and cosmopolitical persuasiveness. It is to these that the argument now turns.

Kyŏngju's high-speed railway controversy

From 1995 to 1997, Kyŏngju was embroiled in a controversy that provided a condensed case of 'place in the making' and of 'urban politics in action.' The issue was the inclusion of a Kyŏngju station in the plans for South Korea's first high-speed railway line, as well as the location of the station and the path of the railway through the city. As of early 1995, the initial proposal called for a routing and station very close to Kyŏngju's urban and historical core. When the Ministry of Culture publicly raised objections to this plan, it created a political aperture in which a host of demands and proposals by different actors – ranging from calls for the cancellation of the Kyŏngju station entirely on historical preservationist grounds, to defence of the initial plan, to ideas for alternative Kyŏngju track and station locations – became items of popular and technocratic debate. Eventually, a revised routing proposal, which maintained a Kyŏngju station while locating station and track along a pathway farther from the city centre, emerged from a coalition of Kyŏngju expert and civic organisations. Over the course of months, this revised routing and its advocates managed to coordinate urban politics and Kyŏngju things such that it became a sufficiently agreed-upon new design – an 'obvious' solution that somehow had not been obvious previously.

The process of coordination by which many (if hardly all) initial pro- and anti-Kyŏngju station forces, and the versions of Kyŏngju for which they stood, arrived at a settlement was complicated and multi-agentive, and a fuller accounting is beyond the scope of this chapter (see Oppenheim 2008:139–194). Here, the focus is on a small group of selected normativities and on the coordinations that they articulated. In a classical ANT vein, such normativities were mediations that helped (partially) align and (partially) enrol some actants into support while (partially) deflecting some others. In a post-ANT mould, meanwhile, they were also scaling devices that specifically drew in scales taken as pregiven and desirable, which thus also 'folded' in actors' 'theoretical' prefiguration of the proper shape of a solution (Gad and Jensen

2010:73). With slight exaggeration of the clarity of the distinction, we might say that if the coordinating effect of normativities helped make the Kyŏngju railway solution make sense, their performance of compulsory scales helped invest this solution with something deeper and more affective: A 'sense of appropriateness' (Strathern 2004:10), an aesthetic forcefulness.

Global–national scaling from equivalence to connection

Both initial advocates and opponents of the routing of the high-speed railway through Kyŏngju invoked themes of the globalisation discourse of the era to imbed their arguments within claims on the proper global–national relationship. 'The most Korean thing is the most global thing' was heard, first and foremost, as a statement that South Korea was ready to embrace the world and deserved the world-class, not least in technology and infrastructure, and the high-speed railway became iconic of this new order (see, for instance, the cover image of Kongboch'ŏ 1995). For celebrants of the railway, it articulated regimes of equivalence: Japan, Germany, France, China and other advanced or powerful nations all had high-speed trains, and so too should South Korea. More subtly, however, the Korean–global pairing in this central globalisation trope was also taken as a call for self-recognition that deeply Korean things were also (already) world class. Even as the train system would rely crucially on imported technology, the railway project was invested with historicity such that its realisation was pre-ordained as a national achievement. Government officials and other boosters were wont to characterise the high-speed rail project as 'the largest national undertaking since Tan'gun' – since, that is, the mythological foundation of Korea itself legendarily five thousand years in the past. Beyond the economic benefits it might bring, for advocates of a Kyŏngju high-speed rail station it was both appropriate and an affirmation that South Korea's world-class historical city should be included within a world-class historical enterprise.

Opponents of large-scale development in Kyŏngju, including initially those of the high-speed railway station, questioned these supportive interrelations. At moments, they articulated a different version of global equivalence in which Kyŏngju's potential world-class status figured as its participation in global practices of historic and natural preservation. Yet more often, and more simply, development opponents portrayed Kyŏngju's national historical significance and its transformation according to a globalist agenda as inimical. One archaeologist turned the historicising formula of advocates on its head: Bringing the high-speed railway to Kyŏngju would be 'the greatest destruction of cultural objects since Tan'gun' (Yi 1996:282).

As planning and construction of the high-speed railway as a whole advanced in the late 1990s, however, another icon, or normativity, of the railway's significance began to circulate. This vision of the railway deepened its efficacies in relation to the global and the national as compulsory scales of reference. In brief, it was one that redoubled the role of the high-speed railway not only as a symbol of South Korea's global coming of age, but also as an actual future agent of global connection. Figuratively, this futurist manifesto took the form of portrayals of the railway as part of a 'new Silk Road,' or fantasies of future rail travel from Seoul to Paris (and somehow it was, very often, Paris). Visually, it took the form of maps that envisioned the South Korean rail line as part of a larger Eurasian rail network, via which one might eventually be able to travel from Pusan at South Korea's southern tip over the Trans-Siberian Railway ultimately to a western European destination of one's choosing. Because this potential of the high-speed railway presumed the presently (then and now) impossible act of traversing North Korea, it also redoubled the project's aspect as an agent of national mission; it rendered national reunification in the future anterior and thus figured it as

Figure 30.1   Trans Eurasian Railway Network map, Torasan station, 2007. Photo by author

an assumed future, as that which will have happened. For Kyŏngju in particular, this version of the high-speed railway made connection more urgent, since it turned the question from the city's membership in a global club of culture cities to the potential of their physical communion. It also drew upon the historical assemblage in which Silla chartered the once and future unification of Korea from the south, interpellating its successor, Kyŏngju, as proper coparticipant in this action. A decade later, a map bespeaking this future could be found on the wall of an equally significant site, the station at Torasan, the last station north of Seoul before the border with North Korea (Figure 30.1).

The scalar deepening of the high-speed railway's global–national puissance helped make the railway irresistible, but it did so more by obviating conflictual politics, imposing a more multidimensional globalism over the concern of railway opponents, than by mediating conflict. Other normativities, however, would more directly intervene and rewire place in the process.

Coordinating objects, situating voice

The coalition of civic and expert organisations that took the lead in bringing about a resolution to the 1995–1997 Kyŏngju high-speed rail dispute advocated for passage of the line through the city, but along a revised routing that would place the tracks and the Kyŏngju station farther from the central urban district. Crucial in the ultimate success of this proposal were several circulating normativities. In this subsection, two are considered: The *wanggyŏng* ('royal capital') as visual icon and *tapsa* ('field study') as verifying empirical ritual. These were nodes at which positions, and differing assemblages of Kyŏngju, were mediated, at which their objects crossed and intermingled and thus at which the active and potential relationships among assemblages constitutive of Kyŏngju place as such were recast. They also drew upon and drew in multiple normative scales, largely through the mechanism of their own multiplicity.

By the time of the dispute, the local had gained depth as a compulsory scale of reference in Kyŏngju. Among other developments, resident amateur historians and other cultural actors created new civic festivals that positioned themselves as being by residents, for residents – as opposed to tourists or state legitimation. These events celebrated the Kyŏngju taun, the 'Kyŏngju-appropriate' or even 'Kyŏngju-like,' which thus gained currency as an ideological lodestone alongside 'the global' and 'the national' as cited in the reversible couplets of 1990s globalisation discourse.

The revised routing solution to the Kyŏngju railway controversy mediated initial pro- and anti-Kyŏngju routing sentiment by spatially dividing the city into a zone from which railway development should be excluded on historical and archaeological preservation grounds and an exterior zone in which it would be permissible. The key conceptual–historical object invoked to effect this spatial sorting was the wanggyŏng or 'royal capital,' the precinct at the heart of the ancient Silla capital city and thus the site of many of its treasures. By proposing a station and routing outside the wanggyŏng, advocates proposed to arrest a trajectory of historical de- struction; at the same time, the wanggyŏng implicitly defined areas outside itself as of lesser historical significance and thus appropriate for developments of all kinds. Thus did the wang- gyŏng coordinate conflicting assemblage politics and reassemble Kyŏngju. At the same time, the wanggyŏng gained some of its force and logical plausibility in Kyŏngju by drawing upon the moral universe of the 'Kyŏngju-appropriate.' The wanggyŏng was itself multiple, and among its versions were a specific vista over the central city from a nearby mountain frequented es- pecially by local historians and cultural enthusiast groups, a panoramic icon in the form of a poster of a painting done by a Kyŏngju artist that portrayed the ancient Silla wanggyŏng in all its glory from roughly the same imagined perspective, and a rule-of-thumb rendition as the area within the ring of five mountains (oak) that surround central Kyŏngju (Kang 1996b:59). In order to grasp the wanggyŏng railway solution and to feel its claims, residents could variously look down while hiking, look at their walls and look up while downtown.

In the course of promoting this wanggyŏng mediation, the revised routing solution also co- ordinated procedures and participation in such a way as to insist upon the appropriateness of, and impart itself with the quality of, a greater degree of local Kyŏngju citizens' involvement in planning the railway through their own city. This coordination hinged in significant part on the polysemy of the practice of tapsa. Tapsa literally means 'field study' and includes the sort of site visits engineers might make, but also includes the historiographical act, enframed by a metaphysical ideology of encounter, of travelling to historical monuments or sites in order to experience the past more directly. In this latter mode, tapsa had long been one of the signature procedures of resident amateur historians hiking the cultural landscape of Kyŏngju and the basis of their claim to know and represent the 'Kyŏngju-appropriate.' In the Kyŏngju railway debate, defenders of the initial planned line at first responded to advocates of the revised routing by noting that the established plan had been subject to field inspection – a tapsa – by development planners, while the revised routing had not. Revised line advocates responded by questioning the validity of any tapsa that involved only central development planners and not resident historians or professors at Kyŏngju universities, whose local knowledge they thus placed on par with planners' technical expertise (Kang 1996a: 9). And when, some months later, a highly public tapsa of routing options was undertaken, it prominently featured a bal- ance of types of experts in which Kyŏngju local knowledge was well represented (Kyŏngju Sin- mun 1996). As the wanggyŏng had done with embodied Kyŏngju perspective, tapsa coordinated positions in part by pulling in a newly valorised world of local practice.

At the same time, however, other coexisting versions of the wanggyŏng and tapsa worked to enrol or neutralise still other divergent positions and versions of Kyŏngju and also, simul- taneously, to draw in cosmopolitical imperatives cast at a more national scale. Nationwide professional archaeological and historians' organisations had been among the strongest initial opponents of any routing of the high-speed railway; they tended to reiterate Kyŏngju's significance to national history and identity, which implicitly trumped resident agendas. Towards this audience, there was another version of the *wanggyŏng*, itself an artefact of archaeological and philological synthesis: A grid plan reconstruction of the Silla *wanggyŏng* first undertaken by a colonial-era Japanese historical scholar and an object of discussion within specialised academic writing since then. While this redoubled the *wanggyŏng* as sorting mechanism through yet another visualisation, it more subtly also reiterated the value of nationally scaled academic expertise as such by enfolding it within a specialised academic knowledge artefact. The case of *tapsa* was yet clearer. At the time of the railway controversy, the dominant depiction of the practice nationwide was a series of bestselling books by the art historian Yu Hongjun, *Na ŭi munhwa yusan tapsagi* (The Chronicle of My Field Study of Cultural Remains) (Yu 1993). This series posited personal visitation and investigation of Korea's historical sites as a new project of critical-democratic national citizenship, a way for individuals to work beyond the pabulum and manipulation of history under preceding authoritarian regimes. The prominent role of *tapsa* in the resolution of the Kyŏngju high-speed railway also carried the suggestion that it would be animated by this same democratic empiricism.

Conclusion

Visualisations of the *wanggyŏng* and the trans-Eurasian railway of the future, as well as obligatory procedures such as *tapsa*, were nodes at which different versions of Kyŏngju and the high-speed railway were crossed and activated in relation to one another, recasting place relations of actuality and potentiality. The habits and vantage points of Kyŏngju amateur history were drawn into the question of the line's routing through the city, while Silk Road histories and futures became part of the high-speed railway as a whole. Circulating normativities worked in effecting coordination by speaking to multiple audiences through their own multiplicity. Different renditions of the *wanggyŏng* made the sorting of zones for preservation and development make sense for everyone from specialised archaeologists to walkers in the city; the polyreferentiality of *tapsa* sutured authority across a similar range. The rendition of the South Korean high-speed railway as a segment within a future continent-spanning network shifted it from being only an agent of international technologist competition – a value that could be countered with the proposition that countries can also compete to preserve their history – to one that also promised connection, which had no easy riposte.

Yet, the hinterlands drawn together in the course of the Kyŏngju railway issue included some that bore the colour of scalar participation, which carried along projects of realising one or more compulsory scales. Resolving the dispute was not merely about resolving technical questions – where the line should run, and who should be empowered to verify the decision. It was also about crafting a vision of appropriate politics and an appropriately multiscalar Kyŏngju, which would be world class and locally responsive, an avatar of global communion and an icon of (South) Korean cultural achievement along with once-and-future (re)unification, in a suitable mix. This is not to reinstate analytically prior scales (global/national/local, macro/micro or whatever) as contexts, but it is to recognise that classical ANT advice to follow the actors as the actors engage in scaling implies an open-endedness to this process that is frequently foreign to actors themselves. To account for actors' reflexivity in scale-making is to account for their touchstones, pregiven scalar commitments to which the performativity of their versions of assemblage is tethered, as well as the cosmopolitical force that successful navigation of these touchstones can impart to the act of coordination. To treat such scalar touchstones as (pre-)packaged within and articulated through *circulating* entities, such as the 'circulating normativities' of this chapter, meanwhile, is to foreground the possibility of folding post-ANT concerns with aesthetics and theoretical recursion back into a more classical ANT processualism. The problem with 'following the actors' is the futurelessness that the trailing metaphor, left unexamined, implies – a futurelessness that is not a property of the actors' own action. Thus, one may indeed follow the actors, but only provided that one remembers to follow them also to and through their (ethical, aesthetic or cosmopolitical) anticipations and orientations.

Note

1 'Post-ANT,' here, invokes several implications of my approach developed by Gad and Jensen (2010:72–74). First, they identify post-ANT with Marilyn Stathern's (1991: xvi) 'postplurality,' one consequence of which is an attentiveness to the recursive participation of actors' 'theoretical' mappings in 'empirical' processes – and thus, a dissolution (in the spirit of ANT commitments to symmetry) of priorist theoretical/empirical or theoretical/descriptive distinctions themselves. But the entanglements of theory also imply the entanglements of scholarship: A second consequence is that one cannot 'follow the actors' while escaping responsibility for following them in some directions and not others, as I do here and did in my original Kyŏngju study in making place and scale the point of the exercise rather than, for instance, seeking to offer my own solution to the city's development woes.

References

Amin, Ash and Nigel Thrift. 2002. *Cities: Reimagining the Urban*. Cambridge: Polity.

Blok, Anders. 2016. 'Assembling Urban Riskscapes'. *City* 20(4): 602–618.

Blok, Anders and Ignacio Farías (eds.). 2016. *Urban Cosmopolitics: Agencements, Assemblies, Atmospheres*. New York: Routledge.

Callon, Michel. 1998. 'Introduction: The Embeddedness of Economic Markets in Economics'. In *The Laws of the Markets*, Michel Callon ed., pp. 1–57. Oxford: Blackwell.

Carr, E. Summerson, and Michael Lempert. 2016. 'Introduction: Pragmatics of Scale'. In *Scale: Discourse and Dimensions of Social Life*, E. Summerson Carr and Michael Lempert eds., pp. 1–21. Berkeley: University of California Press.

Farías, Ignacio. 2010. 'Introduction: Decentering the Object of Urban Studies'. In *Urban Assemblages: How Actor-Network Theory Changes Urban Studies*, Ignacio Farías and Thomas Bender, eds., pp. 1–24. New York: Routledge.

Farías, Ignacio and Thomas Bender (eds.) 2010. *Urban Assemblages: How Actor-Network Theory Changes Urban Studies*. New York: Routledge.

Farías, Ignacio and Anders Blok. 2016. 'Introducing Urban Cosmopolitics: Multiplicity and the Search for a Common World'. In *Urban Cosmopolitics: Agencements, Assemblies, Atmospheres*, Anders Blok and Ignacio Farías eds., pp. 1–22. New York: Routledge.

Gad, Christopher and Casper Bruun Jensen. 2010. 'On the Consequences of Post-ANT'. *Science, Technology & Human Values* 35(1): 55–80.

Hood, Christopher P. 2006. *Shinkansen: From Bullet Train to Symbol of Modern Japan*. London: Routledge.

Hull, Matthew S. 2012. *Government of Paper: The Materiality of Bureaucracy in Urban Pakistan*. Berkeley: University of California Press.

Kang T'aeho. 1996a. 'Hyŏnjae nosŏn Namsan kwa injŏp chayuroun tosi kaebal pulkanŭng, Kyŏngju arŭmtaum ŭn Kyŏngju tapge chik'il ttae tŏuk pit nanŭn kŏt'. *Kyŏngju Sinmun*, January 15: 8–9.

———. 1996b. 'Kyŏngbu kosokch'ŏldo Kyŏngju kugan ŭi munjejŏm mit haegyŏl pangan'. *Kyŏngju Paljŏn* 5: 54–65.

Kim, Samuel S. (ed.). 2000. *Korea's Globalization*. New York: Cambridge University Press.

Kongboch'ŏ. 1995. *Segyehwa*. Seoul: Kongboch'ŏ.

Kyŏngju Sinmun. 1996. ''Kosokch'ŏl' hapdong chosadan p'aengp'aeng han ŭigyŏn taerip'. *Kyŏngju Sinmun*, May 8: 2.

Latour, Bruno. 2005. *Reassembling the Social: An Introduction to Actor-Network-Theory*. New York: Oxford University Press.

Law, John. 2004. *After Method: Mess in Social Science Research*. London: Routledge.

Lepawsky, Josh, Grace Akese, Mostaem Billah, Creighton Conolly, and Chris McNabb. 2015. 'Composing Urban Orders from Rubbish Electronics: Cityness and the Site Multiple'. *International Journal of Urban and Regional Research* 39(2): 185–199.

Mol, Annemarie. 2002. *The Body Multiple: Ontology in Medical Practice*. Durham, NC: Duke University Press.

Mol, Annemarie and John Law. 1994. 'Regions, Networks and Fluids: Anaemia and Social Topology'. *Social Studies of Science* 24(4): 641–71.

Oppenheim, Robert. 2014. 'Thinking Through Place and Late Actor-Network-Theory Spatialities'. In *Objects and Materials: A Routledge Companion*, Penny Harvey, Eleanor Conlin Casella, Gillian Evans, Hannah Knox, Christine McLean, Elisabeth B. Silva, Nicholas Thoburn, and Kath Woodward eds., pp. 391–398. London: Routledge.

———. 2008. *Kyŏngju Things: Assembling Place*. Ann Arbor: University of Michigan Press.

———. 2007. 'Actor-network Theory and Anthropology after Science, Technology, and Society'. *Anthropological Theory* 7(4): 471–493.

Philips, Susan U. 2016. 'Balancing the Scales of Justice in Tonga'. In *Scale: Discourse and Dimensions of Social Life*, E. Summerson Carr and Michael Lempert eds., pp. 112–132. Berkeley: University of California Press.

Simone, AbdouMaliq. 2010. *City Life from Jakarta to Dakar: Movements at the Crossroads*. New York: Routledge.

Slater, Don and Tomas Ariztia. 2010. 'Scaling Asturias: Scaling Devices and Cultural Leverage'. In *Urban Assemblages: How Actor-Network Theory Changes Urban Studies*, Ignacio Farías and Thomas Bender eds., pp. 91–108. New York: Routledge.

Strathern, Marilyn. 2004. *Partial Connections*. Updated edition. Lanham, MD: AltaMira.

Ureta, Sebastian. 2014. 'The Shelter that Wasn't There: On the Politics of Co-ordinating Multiple Urban Assemblages in Santiago, Chile'. *Urban Studies* 51(2): 231 246.

Yi Sŏnpok. 1996. *Yi Sŏnpok kyosu ŭi kogohak iyagi*. Seoul: Gaseowon.

Yu Hongjun. 1993. *Na ŭi munhwa yusan tapsagi 1*. Seoul: Ch'angjak kwa pip'yŏng sa.

31

How can ANT trace slow-moving environmental harms as they become eventful political disruptions?

Kregg Hetherington

In early 2017, I went to the eastern frontier region of Paraguay to interview soybean farmers about their ongoing dispute over national agriculture regulations. Soy is not only the most extensive export crop in South America, it is one of the most economically significant crops globally, providing most of the world's animal feed, edible oils and preservatives, along with hundreds of industrial by-products (Du Bois et al. 2008). Modern, chemical-intensive, soy farming is by far the largest contributor to the Paraguayan economy, and over the last 20 years, it has been responsible for huge environmental and social changes, including deforestation, land concentration and the creation of a new wealthy class of farmers, most of them Brazilian migrants.[1] It is not surprising that soybeans should be one of the most controversial things in Paraguayan public discourse, what in actor–network theory is commonly called a 'matter of concern' (Latour 2004). But one of the things that struck me in my interviews was how often, when asked *why* soybeans had generated such widespread controversy, farmers made reference to a single event, the death of a young boy from pesticide poisoning 14 years previously.

I had heard about the case of Silvino Talavera from environmental justice activists in Paraguay, but it had been years since it had come up in those circles. And yet here were the soy farmers themselves still talking about it as a historical rupture in their relationship with soy and the state. After a court found Talavera's neighbour guilty of negligence in Silvino's death, these farmers believed that *they* were the victims in the case, which had pitted new social movements against them, created the regulatory landscape they were still fighting and even helped elect a president they despised.[2] Silvino had died of 'natural causes,' they told me (often with awkward caveats like 'that doesn't mean it wasn't a tragedy'), and belief to the contrary was a sign of both Paraguayan ignorance and corruption.

Silvino's death was a political event in the sense meant by Jacques Rancière (1999), in that it disrupted the way people made sense of the world. To most Paraguayans, prior to that, soy had seemed like one agrarian product among many, and the many pesticides used to grow it simply part of the farmer's toolkit. Afterwards, soy had become what Timothy Morton (2013) calls a 'hyperobject,' a massive, harmful thing, with limits that were hard to define, and relations that were hard to account for.[3] As such, it was also the source of endless concerns and controversies, and failed attempts to define the problems it created or to solve them with regulations.

This case is instructive for ANT for at least two reasons. On the one hand, it demonstrates how situations become eventful, and, by extension, how mere things become matters of concern. It suggests that attention to the temporality of such things helps us to see how publics are generated around environmental harms (Marres 2010). By highlighting some recent theoretical crossings between ANT and Jacques Rancière (1999), this essay helps to explain how much effort it takes to make certain things public, and helps to see the role of activists and criminal courts in that effort. If the history of environmental problems in the 20th century has taught us anything, it is that the creeping harms of late industrialism are very hard to make visible (cf. Fortun 2012). This essay suggests that ANT's recent engagements with contemporary environmental politics have much to tell us about why this is so. In doing so, it also offers ANT a way of heeding more carefully the delicate line that separates the uneventful from the eventful.

Eventfulness

In the last 15 years, Bruno Latour has made a forceful pitch to rethink the politics of ANT around what he calls *dingpolitics,* a form of democratic contention organised not only around ideas, deliberation or facts, but around what he calls 'matters of concern,' those lively things that have presented themselves as problems (Latour 2005). Starting with the *Politics of Nature* (2004a), much of this work is presented as a kind of recipe for radical democratic inclusiveness of the non-human, and increasingly a way for ANT to engage in 'ecologization.' The argument goes that since there can be no objective 'nature' to which we can refer unequivocally in environmental politics, and therefore no final experts or spokespeople for a singular universe, politics ought to be organised around a plurality of voices speaking for things. In other places, Latour (2004b) has called this a 'cosmopolitics,' building primarily on Isabelle Stengers' (2010) work. If traditional cosmopolitanism is the rejection of parochial politics among inhabitants of a singular cosmos, then for Stengers, cosmopolitics further rejects the parochial ontology of such a universe and asks us to be involved in the production of multiple worlds (see Savransky, this volume). Latour proposes a 'due process' for this cosmopolitics, inviting multiple representatives of matters of concern into metaphorical 'parliament of things' which attempts to 'compose a common world' (Harman 2014).

It is tempting to apply Latour's vision of multiplying dialogues to a situation like the Paraguayan soy boom, where different people have quite different senses of what soy is and what kinds of harms it perpetrates, particularly when one of the primary contenders is routinely disqualified from official politics. Peasants speaking an indigenous language, living on a frontier in a country that most consumers of soybeans can barely place on a map, are rarely listened to, particularly if their opinions don't sit comfortably within a standard political ontology. But ANT's rush to embrace new entities and connections is also, potentially, its blind spot. Latour's parliament of things, for instance, tends to understate the enormous difficulty of rendering something like soybeans public, to turn it into a matter of concern and, by extension, of allowing peasants to be its spokespeople (see Marres 2007). As Watson (2011) has argued, while Latour's parliament breaks down divides between the social and the natural, it tends to do so while (still, inadvertently) privileging the authority of science, and dismissing the kinds of oppositionality that is central to subaltern politics. This is why it is important to pay attention to those moments when new beings are admitted in public discussion, and see what sorts of spokespeople emerge alongside them.

A focus on events, in addition to things, is very much in keeping with Latour's conception of networks. Like Stengers, Latour borrows from Whitehead an underlying 'occasionalist' metaphysics, which sees propositions as 'occasions given to different entities to enter into contact' (Harman 2009: 79–82; Latour 1999: 141). Part of the charisma of Latourian ontology is the irrepressible profusion of relations, as new entities arise from such occasions all the time. But as Marilyn Strathern (1996) points out, actor-networks do not go on forever. They are, at different moments, 'cut' in the act of observation, interpretation and narration, and cuts are just as consequential as connections.[4] Events, I suggest, are one kind of cut: Occasions for new forms of relating which simultaneously produce a rupture in other relations. In doing so, they also produce a temporal distinction between a past and a future, and a specificity that allows them to be reiterated into the future (see Wagner-Pacifici 2017). Or, to follow Stengers (2015: 39), an 'event … brings the future that will inherit from it into communication with a past narrated differently.'

A brief discussion of a technical aspect of the modern soybean industry helps to make this clear. The Paraguayan soy boom really took off after 1999, when farmers began to plant a new variant known as Roundup Ready, the first commercially available genetically modified crop. The biotech industry that created them is built around the ability to patent living organisms that they have engineered. More precisely, though, biotech patents 'events,' the name for any identifiable gene mutation which reiterates itself through an organism's reproduction. Thus, Roundup Ready soybeans bear the genetic echo of a temporal rupture in the soybeans' genealogy, produced and patented by Monsanto engineers. If the event is thus partly a result of new sets of relations occasioned in the lab, property in the event is derived from further cutting of relations (to other soybeans, for instance, to previous genetic knowledge, and to the people working in the cafeteria at the lab) in order to isolate a patentable difference.

But genetic modification is eventful beyond the strictly genetic sense. When these new food crops appeared, they also produced knowledge controversies around the world, with environmentalists arguing that genetic modification posed risks to human health, to crop biodiversity, to independent farms or even to ecosystems broadly. In a series of cases brought before the Environmental Protection Agency in the United States, crop scientists argued that genetic modification did not substantially change the nature of the organisms in question, and therefore should not be subjected to further regulatory restrictions. Essentially, they argued that genetic modification may have produced a genetic event, sufficient for the recognition of property rights, but not an environmental, nor public health one, which would entail further public oversight (Schurman and Kelso 2005). Nonetheless, in other jurisdictions, the controversy over genetic engineering was successfully made eventful. Stengers herself names the struggle over GMOs as paradigmatically eventful, not just because it produced various regulatory changes within the European Union (EU), but because it 'made both scientific experts and … politicians think, as if a world of problems that they had never posed was becoming visible to them' (2015: 39).

I suggest that Rancière's (1999) notion of the political, as moments of 'disruption' or interruptions in a given distribution of the sensible, is a useful way to think about how we might identify such events.[5] For Rancière, novel associations are precisely *not given* in the way that Latour might have it, because specific distributions of bodies are held in place by regimes of visibility, or what Rancière refers to as 'police,' rather than politics. What Rancière allows us to see here is the effort that goes into bringing new beings into view.[6] Indeed, I would suggest that the event, which first allowed biotech companies to patent living organisms (a US supreme court decision known as Diamond vs. Chakrabarty, in 1980), is paradigmatically political in Rancière's sense, because it produced what he would call a 'scandal in thinking,' evidenced by the many denunciations of 'Frankenfoods' that followed (Whatmore 2002). In other words, the political event here is not the genetic event produced in the lab, but the court case which made certain kinds of genetic mutations into a form of capital. Those who benefited from their new rights to own property in genetic events then set about trying to police further disruptions to the distribution of these new organisms, disruptions that might have limited the profits they could derive from them.

As Alain Badiou (2005) has it, an event produces a rupture with the context that produces it. But Rancière takes this further, suggesting that events enable new languages (see Papastergiadis 2014), like Frankenfoods. Only then do occasions produce what Povinelli (2011) refers to as their own interpretive frame, making it possible to narrate the event as a separation of before and after (Wagner-Pacifici 2017). It's this second order of interpretation, and iteration that makes events necessarily a public affair. Here, Noortje Marres' (2010) provocative work on the assembly of publics is useful. Instead of understanding publics as abstract spaces of deliberation, she favours the notion of ad hoc groupings that coalesce around issues or objects which affect people in a way that requires systematic attention (ibid: 191). Publics, for Marres, are 'heterogenous assemblages' formed around any sort of harm that implicates a wide variety of actors for collective intervention (Marres 2007: 768).[7] Elaborating somewhat, the key here is that publics are not formed by people around things or relations per se, but by harms which implicate people, who then go on to reiterate those harms as events.

And this finally offers us a clue as to why criminal courts are such key sites of struggle for so many activist movements, despite the obvious problem that court cases are so rarely effective for redressing widely distributed harms, and, as an extension of the police, are conservative with regard to the existing distribution of things. What the Talavera case shows so well is that law, and criminal law in particular, is a potent site for the production of events. As Latour (2010) himself pointed out, law does not produce facts (in the way that science does) but rather judgements, which are made by linking a series of texts produced by a particular circumstance (some of which might be scientific) to a larger body of texts, including legislation and past judgements, that together constitute *existing law.* Internally, law changes through judgement events (precedents), similar to DNA mutations, as these are reiterated in future legal settings. But legal tinkering with eventfulness can also come to exceed law and the courtroom. Creating a discourse whereby particular circumstances come to fit, or not, a repertoire of criminal narratives, can produce a kind of public amplification – one powerful node along which Marres' heterogeneous assemblages come to be assembled. In that sense, the court's actual judgements may not matter as much as the way proposals assembled for the court contribute to the construction of an event.

The temporality of natural causes

At 7 am on January 2nd, 2003, Petrona Villasboa de Talavera sent her 11-year-old son Silvino out with his cousin to the local corner store. It was already very hot, and a key point in the day when both domestic and agrarian work was being carried out in anticipation of the midday heat. So when Silvino was riding his bike home along the dirt road, his neighbour, Hermann Schlender, was out spraying his soybeans. Schlender drove too close to the road; one of the wings on his sprayer passed over Silvino's head, dousing him in a mix of herbicide and insecticide. Later that night, the whole family fell ill with debilitating stomach cramps, vomiting and diarrhoea. Everyone recovered after a few days, except for Silvino, who was finally brought to a local health post on January 7th, where the doctor diagnosed him as having been intoxicated with organophosphates. He was dangerously dehydrated, but the health post had no saline solution with which to treat him. Over the course of the day, he was transferred to another health centre and finally to a regional hospital, where he died that afternoon.[8]

All of this happened in the epicentre of the soy boom, where migrant farmers from Brazil had been resettling to for two decades. Because soy is a high-intensity, mechanised crop, it also has dramatic local effects in places like eastern Paraguay that had previously been dominated by forests and small-scale peasant farms.[9] It was now a centre of new wealth and cutting-edge technology. But the transformation had excluded families like the Talaveras who had been living there much longer but could not afford the upfront costs of this kind of agriculture. One of the major controversies that followed from the case was around who bore responsibility for protecting families like his, living in small, permeable shacks, from soy's various risks to human health, particularly the widespread use of Roundup herbicide necessitated by Roundup Ready beans.

It is not that peasant families like Silvino's hadn't been complaining for years about soy farms, even decrying them as violent. But despite the comparatively rapid advance of the frontier, soy's violence is slow, in the sense meant by Rob Nixon (2011: 6): 'calamities that patiently dispense their devastation while remaining outside of our flickering attention spans.'[10] Soy's violence creeps up on you like the smell of Roundup wafting onto your patio just before breakfast. Or like the shift from precarious employment to chronic unemployment. For a population that has fought poverty for generations, that has endured brutal military and police repression, it is hard to get many people excited, to get them to notice even, the slow accumulation of rashes, headaches, cancers and birth defects that seem to cluster around soy fields. The same is true of the general, cumulative effects of poverty, which is also uneventful. Poor health and diet, labour insecurity, poor education, crumbling infrastructure, displacement, high rates of infection, injury, violence and mortality, all predate soy. These do not rise, as Elizabeth Povinelli (2011) puts it, to the level of event, because they are part of the cruddy and cumulative effects of abandonment. From an experiential point of view, the soy frontier primarily entails an intensification of these cruddy, cumulative conditions, and each rash and each day spent home from school are merely 'quasi-events' that 'confound response' (ibid: 144). It is not hard to relegate these effects to the background of rural experience. But this is why the death of a small boy is so potentially disruptive. On its own terms, deaths always produce cuts for someone. Separated out from the merely cumulative, the story of a death is easily repeated, and invites others to see it as eventful.

In the trial, this attribution of responsibility boiled down to the question of whether Silvino had died because of Schlender's actions, or because of 'natural causes.' In some ways, then, the trial recapitulated what Latour (1993) has called the 'modern constitution,' the ontological premise that human and natural causes belong to different realms. I argue that the invocation of natural causes is also an argument about time. Human agency has its own time – and for anyone to be found culpable of a criminal act (indeed, for a criminal act to exist at all), they need first be thought of *as events* rather than merely as iterations of an ongoing and unexceptional process called 'nature.'

In the trial, then, the invocation of 'natural causes' is a statement about uneventfulness, or, to use a language closer to Rancière's, it is the means of policing the reigning distribution of the sensible. This is why for the trial lawyers as well, the distinction between nature and poverty is hazy. The prosecution argued that the toxic exposure was negligible and that Silvino had died from 'malnutrition,' or perhaps food poisoning due to the house's unsanitary conditions, both of which were considered 'natural.' In his guilty ruling, the judge acknowledged that there were two kinds of causes for the death at play, but reversed their significance. 'We are not talking about a NORMAL boy, who is well-nourished and strong,' he wrote in the ruling, and for this reason one should be sceptical of the claims by industry spokespeople that the chemicals could not have killed him. For him, Silvino's poverty was a background condition that added to the responsibility of those spraying dangerous substances. What the defence called a natural cause, the judge turned into the context for Schlender's negligence, which stuck out as part of a particular event. This was the basis for the first ruling of guilt against Schlender.

Disqualification

If the court's institutional function in cases like this is to judge whether or not an event has occurred, the way it proceeds is primarily through disqualifying potential narratives. ANT readers are used to beginning their analysis from an injunction against disqualification. In his essay on facts and fetishes, for instance, Latour (1999) shows how the figure of the fetish functions as a form of accusation through which the fetishist is disqualified for making an ontological category error by ascribing divinity or agency to a mere thing. Stengers (2010) takes this further, showing that there is a direct relationship between the absence of non-humans in politics and the disqualification from political speech of those deemed non-rational. Operationalised in the criminal court, this kind of disqualification takes the shape of attacking the credibility of witnesses and the evidence that they produce.[11]

When the case against Schlender reached its second appeal, in 2005, the defence attacked the physical evidence that had been presented by the prosecution. In lurid detail, backed up by the explanations of 11 biochemists, the defence went after both the clinical assessment of local doctors and the various substances extracted from Silvino's corpse and from his surviving family members. None of the original lab reports could be trusted, they argued, because the materials had been mishandled from the outset. A full necropsy hadn't been performed on Silvino's body because the family didn't have the resources to do so right away, and as per Paraguayan custom, he had been buried within 24 hours of his death. Tissue samples were only taken, therefore, after the body had been in the ground for a week. The local doctor had written in his report that he suspected Silvino to have been intoxicated with organophosphates, a highly toxic class of insecticides which were not found on the farm. The same doctor (who had been unable to find saline solution the day of Silvino's death) had also apparently used saline solution to preserve tissue samples from the autopsy, which had further degraded the samples and undermined lab results. The defence mocked the doctor's incompetence both for having an unstable supply of saline solution in his clinic *and* for not being an expert pathologist. The family came in for even less subtle disqualification, as trace amounts of chemicals in their blood were attributed to the family's lack of personal hygiene, their inability to wash their clothes daily and inadequacy of their storage for household insecticides.

Silvino's body, and the household he lived in, was too poor to offer a reliable account of events, and so too was the minimal public health infrastructure of the town. Even the judges would later be accused of being parochial and corrupt. The true stakes of the 'natural cause' argument now become sharper. In the first instance, nature, including poverty, was a state of uneventfulness. Having lost this argument, the defence returned again, arguing that poverty made everyone and everything around Silvino's body incapable of bearing witness to that event. They were disqualified for living in conditions where expertise, or even reliable subjectivity, was impossible. Whatever the case, not only did poverty and ignorance kill Silvino, poverty and ignorance also made the network around Silvino incapable of speaking on his behalf.

This tells us a lot about why criminal courts, on their own terms, are not very good at dealing with economic inequality or environmental injustice. Povinelli's distinction, between the cruddy and the eventful, is not just a matter of temporal perception (this would be Nixon's argument), it is a partition of time that is held in place by the courts themselves. As Javier Auyero and Deborah Swistun (2009) argue in their remarkable book about lead contamination in an Argentine slum, these are common features of environmental justice cases. Poverty is the perfect alibi, because it not only distributes causes across a range of sites, it also fogs the accounting practices on which the court depends for disentangling one cause from another.

The judge who heard the second appeal showed just how uncertain any of the *facts* of the case remained. Sentencing Schlender to two years in prison for criminal negligence, he then immediately suspended the sentence and allowed Schlender to go home.[12] Silvino's family, and the now massive network of lawyers, activists and supporters gathered around them, declared this a compounding injustice. But injustice or not, they had accomplished what in retrospect was a far more impressive goal, and which all those supporters embodied: a new public had been born of disruption. More interesting than what went on in the trial, the facts presented, the expertise marshalled and the decisions reached, was everything that happened around the trial. Fourteen years later, no one I spoke to ever brought up Hermann Schlender's name, and few remembered Silvino's. But they could narrate the event in detail, and bristle about the new language it had created: *agroquímicos* (agrichemicals) had become *agrotóxicos* (agrotoxins), *biotecnología* (biotechnology) had become *transgénicos* (transgenics) and *productores* (producers) had become *sojeros* (soy farmers). And what assembled these objects, which together were (perhaps) capable of killing a boy who was not Schlender, but soy itself. The abiding slogan from the Talavera case is *la soja mata*, soy kills.

Disqualification as ethnographic opportunity

The literature on environmental justice is full of stories such as this one, written by activist academics drawn to the charismatic stories of communities rising up against corporate polluters. And yet as Auyero and Swistun (2009) point out, most environmental harm does not produce such disruptions. Instead, it further disqualifies those affected from speaking about their own marginality. Recent ethnographic work, sitting somewhere between ANT and political economy, has begun to correct this, providing us with careful portraits of the complexities of environmental harm, and the many lines that lie between disqualification and eventfulness (De la Cadena 2015; Li 2014; Lyons 2016; Tironi and Rodriguez-Giralt 2017). Above all, these sorts of studies have shown how difficult it is to isolate the relationships and to generate simple solutions. Soy is a perfect instantiation of this problem. It cannot be tried in a courtroom, isolated in a lab or regulated on the basis of inherent properties. It is not a proximate agent in Silvino's death nor in any other grave harm – the way, for instance, Schlender or Roundup or even Salmonella might be. But nor is it so diffuse as to be entirely dismissed as 'nature,' the way poverty is. Instead, it is somewhere in between, a 'hyperobject,' hard to delimit, but also clearly present, offering footholds for activists to make it eventful.

The same is true of an environmentally engaged *dingpolitics*, for which such environmental actors, and their awkward, stuttering temporality, offer an ethnographic opportunity. In contrast to earlier iterations of ANT, which contented themselves with tracing the occasions for the emergence of all sorts of new beings, this new project tells the stories of those beings that have already broken through disqualification to disrupt the distribution of the sensible. In doing so, it directly engages with the tyranny of those powerful ways of thinking that hold relations of violence in place by calling them 'natural.' That project need not hew to a particular ideological line, nor does it come with a guaranteed slate of answers. Instead, it does something more subtle, as criminal courts sometimes do, by amplifying disruptions already underway.

Notes

1 For an overview, see Hetherington 2013; Fogel and Riquelme 2005.
2 Fernando Lugo was Paraguay's first leftist president, elected from 2008 to 2012, on a promise, among other things, to improve environmental regulations and protections for the rural poor.
3 This is somewhat different than the formulation of object multiplicity more common in ANT (for example, Law 2004; Mol 2002). Soy is multiple as well (Hetherington 2014), but what is at stake in making soy public is less a struggle to define soy *differently*, as it is to be overwhelmed by its excessiveness.
4 Latour might object that the very act of cutting creates its own connections – that the repression of hybrids generates new ones (esp. Latour 1993). But in the sorts of struggles we are talking about, the question of who is trying to create connections and who is trying to cut them is consequential.
5 A similar line of argument has been taken up by Marisol de la Cadena (2015) and Marrero–Guillamon (2016).
6 Rancière's primary concern here is with human actions, but there's no reason why we needn't think of such disruptions beyond human agency (Bennett 2010: 105–107; Marrera-Guillamon 2016).
7 Marres follow's John Dewey's pragmatic conception of publics here.
8 For a full recounting, see https://kurtural.com/cicatrices/.
9 Although this effect can be found throughout South America, it is especially acute in Paraguay, where rural inequality is more pronounced and regulations less strict (Hetherington 2013).
10 As in Nixon's usage, I only evoke this general 'us' as a first approximation of the various tempos of environmental violence and the way they can be taken up and made perceptually available – or publicly recognisable – by political action.
11 This well-known tactic for discounting the knowledge of affected populations which Wynne (1991) called the 'cognitive deficit model' of public understanding of science.
12 'Suspension' meant transforming the sentence into a period of probation, conditional on his not reoffending. Since two years had already passed, the probation was deemed to have been served.

References

Auyero, Javier and Débora Alejandra Swistun. 2009. *Flammable: Environmental Suffering in an Argentine Shantytown*. New York: Oxford University Press.
Badiou, Alain. 2005. *Being and Event*. London: Continuum.
Bennett, Jane. 2010. *Vibrant Matter: A Political Ecology of Things*. Durham, NC: Duke University Press.
De la Cadena, Marisol. 2015. *Earth Beings: Ecologies of Practice Across Andean Worlds*. Durham, NC: Duke University Press.
Du Bois, Christine M., Chee Beng Tan, and Sidney Wilfred Mintz. 2008. *The World of Soy*, Food Series. Urbana: University of Illinois Press.
Fortun, Kim. 2012. 'Ethnography in Late Industrialism.' *Cultural Anthropology* 27(3): 446–464.
Harman, Graham. 2009. *Prince of Networks: Bruno Latour and Metaphysics*: Prahran, Vic.: Re.press, 2009.
———. 2014. 'Bruno Latour: Reassembling the Political.' London: Pluto Press.
Hetherington, Kregg. 2013. Beans Before the Law: Knowledge Practices, Responsibility, and the Paraguayan Soy Boom. *Cultural Anthropology* 28(1): 65–85.
———. 2014. Regular Soybeans: Translation and Framing in the Ontological Politics of a Coup. *Indiana Journal of Global Legal Studies* 21(1): 55–78.
Latour, Bruno. 1993. *We Have Never Been Modern*. Cambridge, MA: Harvard University Press.
———. 1999. *Pandora's Hope: Essays on the Reality of Science Studies*. Cambridge, MA: Harvard University Press.
———. 2004a. *Politics of Nature: How to Bring the Sciences into Democracy*. Cambridge, MA: Harvard University Press.

———. 2004b. Whose Cosmos, Which Cosmopolitics? *Common Knowledge* 10(3): 450–462.

———. 2005. 'From Realpolitik to Dingpolitik or How to Make Things Public.' pp. 14–41 in *Making Things Public: Atmospheres of Democracy*, edited by B. Latour and P. Weibel. Cambridge, MA: MIT Press.

———. 2010. *The Making of Law: An Ethnography of the Conseil D'etat*. Cambridge: Polity.

Law, John. 2004. *After Method: Mess in Social Science Research*. London: Routledge.

Li, Fabiana. 2014. *Unearthing Conflict: Corporate Mining, Activism, and Expertise in Peru*. Durham, NC: Duke University Press

Lyons, Kristina Marie. 2016. Decomposition as Life Politics: Soils, Selva, and Small Farmers under the Gun of the US–Colombia War on Drugs. *Cultural Anthropology* 31(1): 56–81.

Marrero-Guillamón, Isaac. 2016. The Politics and Aesthetics of Assembling: (Un)building the Common in Hackney Wick, London. In *Urban Cosmopolitics: Agencements, Assemblies, Atmospheres*, edited by A. Blok and I. Farias. 125–145

Marres, Noortje. 2007. 'The Issues Deserve More Credit: Pragmatist Contributions to the Study of Public Involvement in Controversy.' *Social Studies of Science* 37(5): 759–780.

Marres, Noortje. 2010. 'Frontstaging Nonhumans: Publicity as a Constraint on the Political Activity of Things.' in Sarah Whatmore and Bruce Braun (eds.) *Political Matter Technoscience, Democracy, and Public Life*. Minneapolis: University of Minnesota Press, pp. 177–210

Mol, Annemarie. 2002. *The Body Multiple: Ontology in Medical Practice, Science and Cultural Theory*. Durham, NC: Duke University Press.

Morton, Timothy. 2013. *Hyperobjects: Philosophy and Ecology After the End of the World*. Minneapolis: University of Minnesota Press.

Nixon, Rob. 2011. *Slow Violence and the Environmentalism of the Poor*. Cambridge, MA: Harvard University Press.

Papastergiadis, Nikos. 2014. A Breathing Space for Aesthetics and Politics: An Introduction to Jacques Rancière. *Theory, Culture & Society* 31(7–8): 5–26.

Povinelli, Elizabeth A. 2011. *Economies of Abandonment: Social Belonging and Endurance in Late Liberalism*. Durham, NC: Duke University Press.

Rancière, Jacques. 1999. *Disagreement: Politics and Philosophy*. Minneapolis: University of Minnesota Press.

Schurman, Rachel, and Dennis D. Kelso. 2003. *Engineering Trouble: Biotechnology and Its Discontents*. Berkeley: University of California Press.

Stengers, Isabelle. 2010. *Cosmopolitics*. Minneapolis: University of Minnesota Press.

———. 2015. *In Catastrophic Times: Resisting the Coming Barbarism*. London: Open Humanities Press.

Strathern, Marilyn. 1996. Cutting the Network. *Journal of the Royal Anthropological Institute* 2(3): 517–535.

Tironi, Manuel, and Israel Rodríguez-Giralt. 2017. Healing, Knowing, Enduring: Care and Politics in Damaged Worlds. *The Sociological Review* 65(2_suppl): 89–109.

Wagner-Pacifici, Robin. 2017. *What is an Event?* Chicago, IL: University of Chicago Press.

Watson, Matthew C. 2011. Cosmopolitics and the Subaltern: Problematizing Latour's Idea of the Commons. *Theory, Culture & Society* 28(3): 55–579.

Whatmore, Sarah. 2002. *Hybrid Geographies: Natures, Cultures, Spaces*. London: Sage.

Wynne, Brian. 1991. Knowledges in Context. *Science, Technology, & Human Values* 16 (1): 111–121.

32

Is ANT equally good in dealing with local, national and global natures?

Kristin Asdal

What is the size of an actor-networked nature?

The question indicated in this chapter's title immediately made me want to ask back: Is this not precisely the kind of question that actor-network theory (ANT) does not 'allow' for? That of questioning and troubling categories of scale like the local, the national and the global is core to what ANT is about. Put a little differently, to presuppose that there is something we could call local, national or global natures is to take for granted precisely what ANT would start questioning right from the start.

This side to ANT is not a hidden mystery. To the contrary, it is something those who have become renowned for inventing the theory have always been quite explicit about. Take, for instance, Bruno Latour and his 'The flat-earthers of social theory' from 1996 where the flat-earthers are precisely himself and colleagues:[1] 'Look at those flat-earthers! They do not have the modesty of interactionists. They claim to account for big things as well as for small talks, for agencies as well as for structures' (Latour, 1996, p. xii). But how then, we must ask, do these flat-earthers claim to account for these things? The answer is '[l]ocal interactions, plus objects and instruments, plus number of connections, plus accounting procedures! That's all there is in their Meccano' (Latour, 1996, p. xvi). The point then is not that everything *is* local, but how something may *become* larger – by way of instruments and connections.[2] Hence, instruments (as well as theories) are performative and produce size that was not there to begin with.

Even though the intervention from the flat-earthers of actor-network theorists was directed to sociologists in particular, Latour is perfectly right when it comes to the flat-earthers' immodesty. ANT is not a theory trying to encompass simply the social in a different way, but also nature and the natural.

As in Latour's account again:

> …if you wish to go further and shift from the social to natural sciences and start talking about the geography of Britain, the hydrography of France, or the meteorology of Germany, they [the flat-earthers] will answer by showing you geologists drawings maps of Britain; you will learn everything about instrument networks along French rivers;

337

and instead of the weather above Munich, they will list the satellites and weather stations and computers that 'make up' the meteorological charts and draw the polar fronts on your evening newspaper.

(Latour, 1996, p. xiii)

Hence, the performativity thesis goes for nature as well, that is, the size of nature is not simply there to begin with, but made by way of instruments and connections.

So we might conclude already here; from an ANT perspective, the perspective of 'the flat-earthers,' the aforementioned question is downright impossible to answer or, worse, not even the slightest interesting.

From describing to troubling actor-network theory

But maybe there is something interesting to this question after all? Perhaps precisely the aforementioned question might help us with not only *describing* ANT, but also with seriously engaging with troubling and questioning it? What I will be doing in the following is to keep carrying this initial question with me, first in relating to it more indirectly, and then later, towards the end of this contribution, to answer it more directly. However, in order to answer this question well, I suggest we trace what nature *does for* ANT, that is, what kinds of inquiries and questions it has allowed for.

My take on this will be to do this through different *versions*. I will attend closely to Bruno Latour's work from the late 1990s. This was the point in time where ANT proponents, notably Latour, started explicitly to relate to questions that had to do with ecology and environmentalism. Hence, this was not so much the nature that emerged in laboratories, but rather the natures that appeared *beyond* laboratories and became part of a politics of nature – that is, in political controversy. However, in order to grasp this move towards an explicit politics of nature, it is both needed and useful to read the move in light of earlier work and debates.

Allowing for nature: a baseline version of actor-network theory

To be clear, ANT was already very well equipped to do this move into politics of nature as ANT proponents had from its very beginnings criticised conventional social sciences for their ways of 'escaping' nature and all things 'natural' for the benefit of sticking solely to *social* categories. The turn and escape to 'the social' on the part of social sciences could be understood; it was reasoned, in light of the fight against positivism and the fear to be swallowed by an imperialist natural science in the late 19th century.

But the result of this fight, where the social sciences handed over 'nature' to the natural sciences for the benefit of claiming jurisdiction over the social, was that the world became divided into disciplines in ways which did not reflect how the social and the natural *in reality out there*, outside disciplinary borders, is very much entangled.[3] The point of departure for a follower of ANT would then be to acknowledge that society is composed just as much by the natural and the technical as by the social, and so to take this into account when *doing* social science.

Following this line of reasoning, one could argue that to keep doing social science without taking nature into account equals doing bad science, or put differently, misrepresenting the world as it actually plays out – in the wild of socio-material controversies.[4]

To be sure, there certainly is and never was full agreement over this, not even in science and technology studies (STS). The book 'Science as Practice and Culture,' edited by Andrew Pickering (1992) allows for a very interesting entry point which has come to go under the name of the chicken-debate. What would an 'allowing for nature' imply in practice? Would one not simply run the risk of losing all that which the tradition of the sociology of scientific knowledge had accomplished and now instead return to a version of positivism? That is, if one wanted to include nature and the natural in the social scientist's equation and explanation, how to do this without simply relying on the natural sciences which one had spent years finding means and approaches to interrogate and critically analyse by social science means?

Hence, one of the core objections to the ANT approach to nature from a sociology of scientific knowledge approach was that this would leave the social scientists to once again take what the natural sciences said about nature for granted. From another angle, meanwhile, critique came from feminist science studies, most notably as formulated by Donna Haraway (1997) along the line that 'nature' in the Latourian ANT version was a nature too much dominated by men and far from sufficiently lively.

To be clear, the notions I relied on earlier, namely 'bad science' and 'bad representations,' were never the actual wording by ANT scholars like Latour and others. But let me nevertheless use it here in order to highlight that there is indeed a moral to or in ANT. This 'moral' has to do with providing better descriptions and doing better science – and to facilitate a better politics of nature. I will return to this later on. For now, we can simply state that a version one or perhaps a baseline version of what ANT allows for is to take nature into account in our social, or socio-material, analyses.

Re-inventing nature empirically

One both influential and controversial contribution by Michel Callon (1986), which has come to go under the name of the 'scallop paper,' was in fact a recipe (coming even with its own vocabulary) for pursuing such a procedure, i.e. allowing for nature in social analysis. The recipe was about not taking only social forces and actors into account but also the forces and agency (including resistance) of non-humans, scallops in this case.

As such, we could perhaps answer in the affirmative also when it comes to the question if ANT is good at taking *local* natures into account in the social scientist's analysis and narrative. Because in one important and particular respect, the scallop paper does its nature–analysis in very much the same way as Latour (2007) would later argue was how actors engaged in environmental dispute, when we study them closely, always do. They always relate to nature in its concrete and, to use that word, local, specificity: It is always about *that* river, *those* trees, *this* pack of wolves. And in Callon's case, the scallops of St. Brieuc Bay.

I will return to this crucial nature-point later. But first we must ask again if it makes sense or does justice to ANT to employ the label *local* nature. As I touched upon already earlier, what ANT allows for is rather to pose the question of scale as *empirical* questions. For instance, how far does a specific 'local' nature go or how far does it reach? Moreover, and not the least important, what enables this 'nature' to travel in the first place? What are the means by which nature or non-human entities and organisms are equipped or enabled to move or to grow? Hence, importantly, when allowing for nature, it is not nature *as such* that is the entity ANT puts forward. In fact, it is quite problematic to equate ANT too quickly with any 'material turn.'

Not a material turn, but a turn to words *and* things, a 'nature-semiotics'

Scholars like Jane Bennett (2010) have been drawing on Latour in order to underpin and substantiate a material turn, a turn towards taking 'vital materiality' into account. However, the materiality and the nature in ANT are close to always a nature mediated by instruments or devices. It is a nature that is equipped to move and to travel by way of a heterogeneous set of devices, such as graphs, maps, computers and documents.

ANT is a *material-semiotic* turn which does not live by any opposition between the material, on the one hand, and the linguistic, discursive or semiotic, on the other. As is well-known, ANT borrowed enormously, not so much from sociology as from the humanities and literary theory, more precisely a theory of semiotics, but also book-history (see e.g. Asdal and Jordheim 2018; Law 1999). But it did not do so in order to exchange the semiotic for the material, to exchange things for words. As Latour, for instance, writes in *Pandora's Hope* (1999):

> The old settlement started from a gap between words and the world, and then tried to construct a tiny footbridge over this chasm through a risky correspondence between what were understood as totally different ontological domains: language and nature. I want to show that there is neither correspondence, no gaps, nor even two distinct ontological claims but an entirely different phenomenon, circulating reference.

The point then is not to put two domains of language and nature in opposition, but rather to investigate how nature is enabled to move, to circulate, by way of, for instance, words inscribed into documents, maps, digital devices and so on. In doing this move, ANT does not only ask us to take nature into account, but also the means by which nature is taken into account and made available to us, as, for instance, by way of the scientific lab. The nature in Latour or the initial ANT version was always a nature constructed, represented and modified. It is a nature that is made real by way of its representations. The latter is the particular version of realism (if we would like to use that word) which we can take from ANT.

So the interesting question here, from an ANT perspective, is not so much if nature is local, national or global but the extent to and the means by which nature is enabled to move from one locale to another locale and the means, quite precisely, by which nature is made available to us. For this purpose, what we could call the 'nature-semiotics' offered to us by ANT is, I would argue, indeed quite promising.

Doing away with a global or all-encompassing nature?

But now we must return to where I initially promised to start, namely with Latour and his book *Politiques de la Nature* (2004 [1999] but see also Latour 2007). One way of describing this work is in fact that it was a project for doing away with any such thing we could call a global nature. The kind of nature taken as a global whole, an all-encompassing reality or system of interconnections. As such, it would be fair to say that even if scale to ANT is made into an empirical question, that of dealing with *global* nature is something ANT must be really bad at!

Let me explain: As long as ANT scholars were simply talking about scallops, or whatever single species we could think of, this move to encompass non-humans could indeed be irritating and provoking to social scientists and others. Hence, there was potentially a problem with too many enemies. One the one hand, those who read ANT as a way of giving subject status and position to non-humans and on the other hand, those who took it to offer nothing but a realist, positivist trap of taking natural science and its facts and objects for granted.

Yet, there was also a problem with (potentially) the wrong kind of friends. In particular, radical or 'deep ecological' environmentalists who would gladly welcome nature into the equation and argue that Nature, capital N, was more important to preserve than single humans and the will and desires of the human population. Hence, again, there is this issue of normativity and a *politics* of nature to take into consideration and account for. What if a Nature-global was being put up against the social? And who was, eventually, to speak on this Nature's behalf? A group of experts, a technocracy, at the cost of ordinary democracy? Here, it is worthwhile reminding that Latour's book also has a subtitle: How to bring the sciences (read: nature) into democracy.

The assumption that nature will cut short democracy

At this point, to Latour, the precondition for including nature in the equation became, so to speak, to eradicate capital-N nature. The argument is substantiated with empirical findings. Arguably, when people, environmentalists for that matter, talk about their attachment to nature, this is always in the concrete. They talk about *this* specific river, *these* Platan-trees in the midst of Paris, *this* park of elephants in Amboseli ... Hence, also empirically, the argument goes, nature is, if not *local* in character, then at least concrete, situated and specific.

Hence, in trying to answer our question, we must have an eye for how the demand for taking nature into account (as was indeed the demand from ANT) was simultaneously a demand to do this in a quite specific way. This is related to how such an inclusion was also part of an argument and indeed a critique on ANT's part. A critique of Nature capital n, which would, if let in, rather than becoming *integral* to democracy, cut short politics and the political process. In short, the way in which nature was let into the equation by ANT was simultaneously as part of a democratising and politicising argument. 'Nature' was conceived of as a determinist anti-political entity which had to be bracketed, indeed taken apart and away, for the benefit of nature-entities in their (so to speak) 'uniquenesses,' 'richnesses,' 'heterogeneities' and singularities. Only then could nature, which had now been transformed into plural 'natures,' be taken into democracy.

Obviously, we need to see this move as a truly classic STS argument: An argument in favour of politicising and democratising the sciences at the expense of expert rule and technocratic government. But we must also see it in line with ANT more specifically: When taking 'nature' into account, this was not to be accepted as an entity above and beyond humanity, but rather an entity to be studied in its imbroglios with technologies and humans. So does this then imply that ANT is not good at dealing with global natures? The answer must indeed be in the affirmative, again, if we equal Nature capital n with the global. Or, more precisely, it is not so much bad at it, as it is downright against it.

At this point, you, our reader, might already have protested: The aforementioned conclusion comes a little bit too easily! So let me rephrase this just a bit: Dealing with 'nature' in an ANT manner is a laborious affair. To do so demands that one actually traces and demonstrates how (1) an assumingly global nature comes into being – hence not takes it at face value but demonstrates how such versions of nature are produced in the first place, and (2) how humans (for instance) are involved, entangled with and attached to a global nature.

So far, we have treated our question regarding what ANT is good at when it comes to nature (be it local, national or global) very much in line with earlier work by ANT scholars and Bruno Latour most notably. However, there is in fact a quite important thing to note here. While the laboratory studies through which ANT was very much developed and employed (for instance, Latour, 1987, 1988, Latour and Woolgar, 1986, Law, 1994) were detailed empirical and ethnographic studies, this is not the case when it comes to the *Politics of Nature* book of Bruno Latour. In contrast to many of Latour's and companions famous and earlier studies, this one is not based upon comprehensive ethnographic studies. In fact, it does not so much study neither nature nor politics in any extensive detail. This is actually an interesting weakness of the book, as the book stands out as more of an overriding philosophical argument than a broad empirical study (see also Jensen, 2006). In doing this, it also very much presupposes how nature 'behaves' in politics. And the presupposition is that Nature capital n cut shorts politics and acts in a determinist way.

How does nature behave in democracy?

So let us return again to our initial question: Is ANT equally good in dealing with local, national and global natures? Well, the problem as I have now pointed out is that ANT in the version spearheaded in Latour's book treated natures as if there *is* only local, that is, specific, concrete and 'individual' nature-objects. It is only in environmentalist's theories, Latour argues, that there is a Nature capital n, and *if* there was a Nature capital n in *practice*, this would be a threatening de-politicising entity, impossible to make integral to democracy.

But surely, nothing prevents us from working with nature in an ANT manner and demonstrating that there is more to nature than scallops and elephants and so on, and, that there are indeed 'nature-wholes' in practice, too, that is natures that go beyond the single and the more easily situated versions. We can and should trace how such nature-wholes are scientifically, bureaucratically and politically made (Asdal, 2008). In fact, when pursuing such a task, the argument can be justified that Latour was wrong in his argument that the natures we are dealing with are simply local (to use that word then) in character. Moreover, 'nature-wholes,' versions of Capital-N nature, do not work in a determinist way and do not de-politicise and cut short politics the way it was supposed. In fact, is not this precisely what Latour (2017 [15]) more recently have been realising (in a dual sense) in and via the figure of 'GAIA'?[5]

Becoming good at allowing for nature

The space for words is too limited to delve into the GAIA-theory and the cosmopolitics of Latour in this intervention (but see e.g. Blok and Farias 2016). Let me rather stay with the question I was asked to answer. Or perhaps it would be a good idea to modify the question that I was asked to answer. Because a different way of asking the question is perhaps to stay where I began: What is it, when it comes to nature, that ANT allows for and can assist with? Then, answering the question would be much easier: ANT definitely allows for and may assist us in taking nature into account in analysis. I would even argue that more than allowing for, it demands us to be inventive and to *become* good at doing precisely that. Here, the question is not so much if the nature *is* local, national or global, but rather what local, national or global natures do, and how they come to 'be' their size?

And interestingly, there are a couple of things that ANT has been very good at – and which is particularly important, I would argue, in discourses about nature. ANT has been particularly good at tracing and demonstrating *the means* by which nature-objects come into being, the way they are being realised as well as what it takes to *achieve* agency. One of the ways in which this has been put into words (apart from throughout the series of laboratory studies) is Latour's way of pointing out how nature is 'instrumentized' (Latour in Asdal and Ween, 2014) – that is, how it is enabled by instruments in order to emerge, become visible and realised. Also in this respect, it is a slight irony how ANT continuously is made part of the so-called material turn and the argument that non-humans *have*, as opposed to *achieve*, agency.

There is no easy turn to nature, be it local, national or global, with ANT. ANT is actually *very bad* at turning directly to the material. Which, to my mind, is one of its merits. Nevertheless, the 'translational' and the material-semiotic which I delved into earlier – that natures, for instance, often are coupled with words, enter into our forms of life by way of documents, via the digital, by particular conceptual formations, via laboratory equipment, staining-techniques and so on – is too easily forgotten in favour of the apparently lively and 'into your face' materiality.

A nature modified

But then, what was the question again – and did I actually answer it? I mean, what did the editors mean by their question: Is ANT good at *dealing with…*? So far, I have made my task easy, taking this expression to mean more or less the same as *relating to* or *taking into account*. In what other manner might ANT, whether local, national or global, be said to be good at *dealing with* natures? This warrants us to address the tricky question regarding how ANT relates to political theory and political practice.

In fact, there seems to be an underlying ideal, sometimes perhaps an explicit ideal, in ANT, that nature is something we can and should deal with, in a deliberative manner, a manner of reasoned, democratic dialogue. And that through such deliberative manners we will, eventually, come to agreement with which versions of nature to take into account, account for, bring into the collective and so forth. Ironically however, tracing how nature is already being taken into account and how nature is already part of ordinary institutions such as parliaments and bureaucracies have been done by ANT scholars only to a much lesser degree.

Interestingly, it is as if ANT has been much better at dealing with arguments *around and about* politics of nature, than in fact *doing* politics of nature – in other words, tracing the ways in which political arrangements, ministries, parliaments, councils, bureaucratic offices do politics of and with nature, be it local, national or global in character. In order to be good at this, we might benefit more from borrowing the methods by which ANT scholars pursued their laboratory studies than the reasoning around how nature behaves in politics and ordinary political institutions (Asdal and Hobæk, 2016, see also Danyi this volume, as well as, for instance, Barry, 2001 and Marres, 2012).

How are policies and politics of nature in such arrangements instrumentised and realised? In pursuing this task, we might come to the conclusion that Nature does not play the hegemonic part in politics as ANT initially reckoned it did. Maybe Nature, even if talked about as a unitary thing, more acts as a starting point for the political discussion rather than as the endpoint? What if nature is constantly modified by other entities, such as the economy? (Asdal, 2008). Hence, more studies might end up modifying ANT. But then again, it is not really 'ANT' but *us* that need to be good at pursuing such tasks.

So maybe one lesson when it comes to ANT ought to be like the one ANT scholars used to apply on environmentalists, philosophers and scientists alike: Do not listen to what the classic ANT scholars *say* about nature/s (be it local, national or global). Follow their practices and the methods they used in their laboratory studies and let these inspire us in our ways to new sites, such as parliaments and bureaucracies and offices.

Notes

1 The contribution is a foreword to the book edited by Michael Power (1996): Accounting and Science: Natural Inquiry and Commercial Reason. Cambridge Studies in Management 1996.

2 For this concern with that which seeks to become larger, see also interestingly Bruno Latour's postscript Irreductions in *The Pasteurization of France* (Latour, 1988).

3 See, for instance, Bruno Latour (1993). We have never been Modern. For the debate and position in relation to positivism and the critique of positivism and the actor-network theory response, see Kristin Asdal (1995).

4 See in particular the three contributions in the book by H.M. Collins and Steven Yearley (1992), Steve Woolgar (1992) and Michel Callon and Bruno Latour (1992).
5 See also the very interesting interview of Latour in *New York Times* by Ava Kofman (2018).

References

Asdal, K. (2008). Enacting things through numbers: Taking nature into account/ing. *Geoforum*, 39(1), pp. 123–132.

Asdal, K. and Ween, G. (2014). Writing nature. *Nordic Journal of Science and Technology Studies*, 2(1), pp. 4–10.

Asdal, K. and Jordheim, H. (2018) Texts on the move. *History and Theory*, 57(1), pp. 56–74.

Barry, A. (2011) *Political Machines: Governing a Technological Society*. London: Athlone Press.

Bennett, J. (2010) *Vibrant Matter: A Political Ecology of Things*. Durham, NC: Duke University Press.

Blok, A. and Farias, I., (2016) *Urban Cosmpolitics. Agencements, Assemblies, Atmospheres*. New York: Routledge.

Callon, M. (1986). Some elements of a sociology of translation: Domestication of the scallops and the fishermen of Saint Brieuc Bay. In: Law, J., ed., *Power, action and belief: A new sociology of knowledge?* Sociological Review Monograph, 32, London: Routledge & Kegan Paul, pp. 196–233.

Callon, M. and Latour, B. (1992) Don't throw the baby out with the bath school. Epistemological chicken. In Pickering, A. ed., *Science as practice and culture*. Chicago, IL: The University of Chicago Press, pp. 343–368.

Collins. H.M and Yearley, S. (1992) Epistemological chicken. In Pickering, A. ed., *Science as practice and culture*. Chicago, IL: The University of Chicago Press, pp. 301–326.

Haraway, D. (1997) *Modest_Witness-Second_Millennium-FemaleMan_Meets_OncoMouse*. New York: Routledge.

Jensen, C. B. (2006). Experimenting with political ecology: Bruno Latour, Politics of nature: how to bring the sciences into democracy. *Human Studies*, 29(1), pp. 107–122.

Kofman, A. (2018, Oct. 25) Bruno Latour. The Post-Truth philosopher, Mounts a Defence of Science. In: *New York Times Magazine*.

Latour, B. (1987). *Science in action. How to follow scientists and engineers through society*, Cambridge, MA: Harvard University Press, 288 pp.

Latour, B. (1988). *The pasteurization of France*. Cambridge, MA: Harvard University Press. 292 pp.

Latour, B. (1993). *We have never been modern*. Cambridge, MA: Harvard University Press. 168 pp.

Latour, B. (1996). The flat-earthers of social theory. In: Power, M., ed., *Accounting and science: Natural inquiry and commercial reason*. Cambridge: Cambridge University Press, pp. xi–xviii.

Latour, B. (1999). *Pandora's hope. Essays on the reality of science studies*, Cambridge, MA: Harvard University Press, 336 pp.

Latour, B. (2004 [1999]). *Politics of nature. How to bring the sciences into democracy*. Cambridge, MA: Harvard University Press, 320 pp.

Latour, B. (2007) To modernise or to ecologise. That is the question. In: Asdal K., Moser I and Brenna, B. eds. *Technoscience. The politics of interventions*, Oslo: Unipub, pp. 249–272.

Latour, B. and Woolgar, S. (1986). *Laboratory life: the construction of scientific facts*. Second edition. Princeton, NJ: Princeton University Press, 293 pp.

Law, J. (1994). *Organising modernity: Social ordering and social theory*. Oxford and Cambridge, MA: Blackwell, 228 pp.

Law, J. (1999) After ANT: complexity, naming and topology. In: Law, J. and Hassard, J. eds. *Actor network theory and after*. Oxford: Blackwell Publishers/The Sociological Review, pp. 1–14.

Marres, N. (2012) *Material participation: Technology, the environment and everyday publics*. Basingstoke: Palgrave Macmillan.

Pickering, A. ed., (1992). *Science as practice and culture*. Chicago, IL: The University of Chicago Press.

Woolgar. S. (1992) Some remarks about positionism: A reply to Collins and Yearley. In: Pickering, A. ed., *Science as practice and culture*. Chicago, IL: The University of Chicago Press, pp. 327–342.

33

What happens to ANT, and its emphasis on the socio-material grounding of the social, in digital sociology?

Carolin Gerlitz and Esther Weltevrede

Data-intensive platform media bring up but also reconfigure the question of the socio-material grounding of the social once again. Unlike popular understandings, actor–network theory (ANT) maintains that digital platforms offer 'a more material way of looking at what happens in Society' (Latour, 1998 in Venturini et al., 2017, p. 3), since it materialises social interactions. Viewed from the perspective of digital sociology, platforms take on at least a threefold role regarding the social (Marres, 2017) – as site for the distributed accomplishment of the social, as methodological access point to data about the social and as infrastructures for participatory research and outreach.

In this chapter, we explore how recent engagements with digital platforms in digital sociology and related areas does more than just vindicate ANT's outlook (Latour et al., 2012) of a flat socio-material account of the social without inbuilt levels of 'micro' and 'macro.' Platforms are characterised by providing computational infrastructures (Bogost and Montfort, 2009) that enable heterogeneous stakeholders including users, developers, partners, organisations, companies and others to participate in their data and features (Gillespie, 2010). In this context, we argue, social media platforms reconfigure who or what can count as an actor, what counts as social and what as material, and, we argue, render these ambivalences very much a question of method, thereby posing new challenges to ANT. We take up ANT's empirical principle to 'follow the actors' and draw on an investigation of automated platform engagement – that is the use of bots, scripts, cross-syndication or auto-created content – to revisit fundamental principles of ANT. What or who is an actor? The platform? The user account? Third-party software? The issue?

The question of concealment of actors is not new to ANT and there was a brief period during which sociologist believed that social media may have solved this problem through their pre-structured profiles, actions and data formats (Savage and Burrows, 2007). Recent research and particularly the rise of bots and automation have shown that the problem is reintroduced again, this time in a medium-specific way, as platforms do not unveil or do not know if one is dealing with human, bot or half-automated use practices and are characterised by missing, nested or invisible traces, obfuscated infrastructures, automated usage and commensurating data-points (Gerlitz and Rieder, 2018). This does not only open up the question what constitutes an actor, but also the delineation between the human and the non-human and the affordances of 'following' as a methodological principle. What or whom can we follow, that is, trace from one situation, relation or context to the next one? We suggest that accounting for the non-human/human configuration is increasingly a question of method and of infrastructure in the context of platforms. By revisiting discussions of how methodological alignment and interface method may require to work with and against platform configurations and their infrastructural affordances (Marres and Weltevrede, 2013; Marres and Gerlitz, 2015; Weltevrede, 2016), we suggest that platform-based methodologies may involve different, unexpected and more difficult manoeuvres than straightforward acts of following.

Digital sociology and the socio-material grounding of the social

The transforming capacities that computational and digital media technologies bring to the study of the social have been addressed early on in ANT work. In a first step, we draw on a few selected interventions to arrive at the current state of digital sociology.

Most authors initially focused on the digital as a seemingly bigger and better access to social data, which offers direct, transactional data (Savage and Burrows, 2007), allows to trace (issue) networks (Marres and Rogers, 2005) or vindicates a flat ontology between actors and networks (Latour et al., 2012). Latour et al. (ibid), for instance, suggest that the socio-materiality of online and particularly platform media finally allows to experience and realise a variety of more abstract ANT claims, most notably the seamless transition between actors, relations and networks. From the authors' perspective, such flat ontologies – or, more precisely, '1.5 level' ontologies, in that the digital allows for shifting aggregates to emerge bottom-up, so to speak – even seemed to be hard-wired into the web, through hyperlink-relations or user profiles. Digital media are supposed to produce and present data in such ways that researchers can travel – or 'click' (ibid, p. 599) – from actors to relating attributes to networks and back without proclaiming a higher structural level as 'there is no whole superior to the parts' (ibid, p. 600).

Take the example of the social media platform LinkedIn: Users are characterised by the organisations they worked for, which are themselves characterised by the other users who worked there or endorsed them. The technicity of platforms, Latour et al. continue, lures researchers into following the actor by clicking on their profiles and following their associations, attributes and unfolding networks. It is through digital methods and tools that such experience can be transformed into scaled-up issue networks which abstain from reintroducing hierarchical distinctions between actors and networks. Even though the role of method is emphasised, there is, however, little methodological reflexivity in this 2012 Latour et al.'s account: Much like with ANT in general (Latour, 2005), the question of which relations and attributes can be accessed and which remain inaccessible in platform back ends is bracketed out; and how platform data have been accomplished in the first place is seemingly of little concern to the authors.

Gradually, attention amongst ANT-influenced scholars shifted towards the work that platforms and research devices do in enabling and rendering the social methodologically accessible: The idea of following the actor was supplemented by following the medium and the development of medium-specific research devices (Rogers, 2013). Digital media surfaced more prominently as methodological access point and research objects themselves – leading to forms of medium research – and the ways digital and methodological devices inform research were problematised and reflected upon (Ruppert et al., 2013; Marres, 2012). Drawing on a long-standing interest in the role of the research apparatus as an active agent in the production of knowledge, Ruppert et al. (2013) stress that digital platforms 'are both the material of social lives and form part of many of the apparatuses for knowing those lives' (2013, p. 24). As such, they inform both what modes of sociality can be realised and how they can be known.

Following Ruppert et al. (2013), digital devices collapse action and datafication, and produce transactional data which provide heterogeneous and granular accounts of actors. Transactional platform data allow to engage with what seems to be 'whole populations' that are non-coherent, dynamic and mobile. While the idea of transactional data promises access to the social, it also complicates it. Digital devices, Ruppert et al. continue, generate data or 'inscriptions' which are assembled into specifically configured 'research apparatuses,' together with other tools to extract, analyse and visualise these data. As a result, they enable cascades of inscriptions, in which data are translated into new formats and combinations. This, however, also increasingly makes previous inscriptions difficult to retrace. Although Ruppert et al. set out to move away from using the digital as mere access to the social, and towards a reflexive deployment of its inscriptions, they do not fully attend to how that data are accomplished and the entanglements between the human and the non-human that are enabled, pushed or made impossible through platform data.

In her account of digital sociology (2017), Marres draws these questions together by claiming for a 'device-aware' sociology (2017, p. 114), which 'foregrounds the computational dimension of social enquiry as well as social life' (2017, p. 39). Making sociology 'device-aware,' Marres claims, is largely a concern of method, and raises questions about whether digital media have in fact introduced new methods or not. Yes, they have, suggests Richard Rogers (2013), as they offer 'natively digital' data-points such as links, likes or tweets which afford the development of new methods and tools. No, they have not, others claim (Abbott, 2011; Venturini et al., 2012), as digital media come with established concepts and methods of the social build into them which they simply allow to operate on a larger scale. Hyperlink analysis, from this perspective, builds on the traditions of predigital network analysis and just allows to scale it from micro-ethnographies to aggregated overviews spanning across platforms (Venturini et al., 2012).

Marres and others intervene in this debate by asking how inbuilt methods interfere in research in new ways. Digital platforms may come with specific analytics built into them (Marres and Weltevrede, 2013; Marres and Gerlitz, 2015; Weltevrede, 2016), some of which are fairly straightforward, such as the bias towards positive affect on Facebook or the interest in trends and popularity on Twitter, whilst others are more difficult to detect. Such bias of data may not be specific to digital platforms (Venturini et al., 2017); however, they still challenge the 'sovereignty' of the researcher (Marres, 2017, p. 95) as the methods and concepts pushed by the medium may not be in line with one's own. From this perspective, Marres continues, simply following the actors (in classic ANT vein) is being complicated and she instead invites researchers to gain a sensibility for how platforms and research tools inform research, and what account of sociality they produce or make traceable (2017, p. 97).

Interface methods (Marres and Gerlitz, 2015) are methods that pay attention to such 'methodological uncanny' which may arise when the concepts and methods of the devices do not align neatly with those of the researchers. These methods explore different ways how research questions and the bias built into devices can be made to 'interface,' by following platforms, but also working around them (for instance, through means of data cleaning) or even working against them. Attention to the socio-material grounding of the social thus becomes a question of method, which could either focus more on the medium (by following its bias) or the social (by aligning differently to media bias) (Marres and Weltevrede, 2013). The social cannot simply be described by tracing associations in digital platforms, as such descriptions (a) remain inattentive to how these relations have been (infrastructurally) accomplished and (b) are difficult to accomplish when scaling up to tens of thousands of tweets, posts or network relations (Marres, 2017, p. 132). Digital sociology thus challenges ANT to move from following the actor (that is platforms, their users and their predefined data formats) to reconfiguring the empirical apparatus in ways that explicate, deploy or work against medium biases.

In their latest publication, Latour, Venturini and others (2017) acknowledge the limits of simply tracing actors by recalling their past engagement with digital media. Admittedly, they fell for the promise of preformatted digital data as opportunity to scale ANT up, only to be confronted with the bias built into the data. Venturini et al. thus distinguish between traces ('all the inscriptions produced by digital devices') and data ('inscriptions having undergone the cleaning and refining necessary to make them useful knowledge objects') (2017, pp. 2–3). Doing digital ANT is no longer presented as following (or clicking) data traces in digital media, but as a methodological accomplishment that involves turning traces into data and making them accessible in research infrastructures. In this sense, Venturini et al. claim that the conceptual flattening of actors and networks is not build into the medium but needs to be accomplished on the methodological level.

Whilst the notion of straightforward acts of following actors (and their networks) as well as representational data has thus been put to question, we would like to think these interventions further by problematising the distinction between actor and network as well as the human and not-only-human anew. We do so by drawing on a research project that explores automation in digital platforms.

Social media automation or the not-only-human of digital platforms

Automated use of social media platforms has advanced into the subject of both academic interest and public deliberation (Ferrara et al., 2016). Different from manual engagement with platform features, automation involves all forms of software-supported platform practices, ranging from bots, cross-syndication (automatically posting from one platform to another), algorithmic content selection or creation, sensor-triggered engagement and other scripts. These practices are usually realised with the help of third-party apps that directly connect to platforms via their application programming interface (API) and are regulated by platform policies. Automation is often associated with spam, trolling or manipulation of trending topics, but can also entail scripts to automatically promote new posts on news sites or blogs in social media. Especially in the aftermath of recent elections in the US and the Brexit vote, automation has been criticised to manipulate public opinion by pushing niche positions, spreading fake news, trolling users or altering the tone of debates through affective politics (Woolley and Howard, 2016).

These discussions often reduce the diverse spectrum of automation to the phenomenon of bots, that is, fully automated robots, as a platform-specific and rapidly proliferating phenomenon. Bots are presented as the unwanted non-human 'other,' technically complex and malicious, capable of political and social manipulation. However, on a technical level, bots are not too dissimilar from the software and scripts social media managers and publishers have been using on a daily basis over the last decade: Social media management software such as Dlvr.it or HootSuite provide the software infrastructures that assist users in writing and scheduling social media posts, make content suggestions, allow cross-posting on multiple platforms and auto-replying to customer requests. These practices are now largely associated with bots but are part and parcel of software supporting social media professionals and in the latter case explicitly invited by platforms. Furthermore, an increasing amount of platform users do not engage with Twitter, Facebook or Instagram via their official web and app interfaces, but use third-party software called social media 'clients,' 'sources' or 'apps' that are built on top of platforms, connect directly to their databases and may offer additional functionalities as listed earlier (Gerlitz and Rieder, 2018).

The question is what counts as an actor or a network in this context. Is the automated user account an important actor? Or should the emphasis rather be put on the human user that operates connected user accounts? And is the total amount of automated tweets that serve one specific purpose, such as pushing a certain issue agenda, the preferred network? Or should we rather focus on the third-party apps that enable forms of automated engagement? Following from this, how to attend to the human and non-human configuration in either of these perspectives? In what follows, we ask how digital platforms reconfigure human/non-human dichotomies into a more complex continuum and therefore challenge certain methodological assumptions underpinning both ANT and digital methods.

Departing from an investment in human/non-human symmetry, it is problematic on several levels to limit automation to a sole focus on bots. First, this fairly narrow category reduces what should instead be seen as a spectrum of automation. Bot detection methods follow the data formats delineated by the medium and largely aim to detect automation on account level, by studying profile information, linking patterns, rhythms of activity, follower/followee ratios, style of writing or a combination of these (Ferrara et al., 2016). Such approaches often lead to false positives as it proves difficult, for instance, to distinguish between bots and manual accounts tweeting with a very high frequency. Secondly, such a narrow perspective renders invisible many modes of automation practices including those that focused on the redistribution of information, accounts that mix manual and software-assisted tweeting (for instance, through Tweet-buttons), as well as forms of cross-syndication which translate manual content from one platform into the data formats of another. While debates often focus on the actors (i.e. bots), what is more relevant are automation practices and accomplishments. However, these are being obfuscated by the platform. Automation is therefore highly unstable and hard to disentangle from 'human activity' due to obfuscation by the platform infrastructure.

This leads us to the third point: The reduction of actors to bots suggests that manual and seemingly social practices can be disentangled from automated and material-technical practices. It proposes that the material inscriptions of automation software can be carved out, or at least be named, and supports a representational approach to platform data. In that sense, the focus on bots attempts to stabilise a contested, unstable phenomenon. The nuanced automation practices, original data-sources and underlying technicity that feed into platforms, however, get largely bracketed out as platform infrastructures and especially APIs often obfuscate where and from whom data came from. If we 'follow the actors,' we would be leaning to clean data from all signs of automation to get to human actors. If we 'follow the medium' – understood as the platform API-preferred data collection and analytics – we would remain inattentive to the wide spectrum of human and non-human accomplishment in the data. By contrast, to maintain a sensibility for the socio-material accomplishment of the social, we cannot hold on to purified accounts of the social but need to consider various degrees of automation as part of the social.

Methodological manoeuvres for tracing social media automation

In our work, we explore these issues by tracing how automation is technically accomplished through third-party apps, how it is subject to multiple nested infrastructures and works with, through and against the concealment of these infrastructures. In what follows, we experiment with different methodological manoeuvres of following and working with and against actors and medium that may allow us to gain a different sensibility for the fabrication of the social in the context of Twitter. We do so by selecting a specific starting point to account for the socio-material accomplishment of the social on Twitter, namely the apps from which accounts tweet.

These apps make use of the Twitter API infrastructure to directly input and output selected data to and from Twitter and are called 'sources' in Twitter's developer documentation.[1] The platform offers a variety of sources for data input and output themselves, most notably the Twitter.com web interface and specific apps for smartphones and tablets, but the majority of sources are developed by third parties. These are crucial for the implementation of custom scripts and software that support automation practices, rely on Twitter's technical infrastructure and are subject to its regulation. Twitter has invited for a multiplication of sources from the launch of the platform; however, it follows mixed politics regarding automation which is both encouraged in some cases and contested in others within the platform.

During its first years, Twitter encouraged third-party development of apps without many restrictions on automation, before successively implementing restrictions on different levels. The most notable cut was implemented in 2012 under the category of the Developer Rules of the Road in combination with API updates when Twitter introduced a stricter authentication system (before developers could use the API anonymously) and rate limits to API calls, which ruled out anonymous and high-volume automation.[2] Now, automation is largely regulated through Twitter's Automation Rules, which explicitly encourage creative experimentations, marketing, automated direct messaging and automated content sharing.[3] Even though developers can no longer operate anonymously, creating an app through the Application Management remained fairly easy (Gerlitz and Rieder, 2018) up until July 2018, when Twitter introduced a more strict app application process.[4] The distinction between promoted and prohibited behaviour is not always entirely clear. For example, in the Twitter's Automation Rules, the platform forbids activities that can be categorised as spam, misleading, conflicting with privacy regulations, exceeding rate limits and specific actions like doubling accounts, auto-posting to trending topics, which require a situated evaluation. In short, the type and degree of possible automated practices are being co-produced between third-party software, the materiality of the platform API and its policy rules.

There are several methodological obstacles to attend to when approaching automation via the source metric which complicate straightforward ideas of 'following' traces. The first obstacle is that data on the source of a tweet are largely not accessible in the interface of official Twitter clients as the platform does not straightforwardly make infrastructures build on top of the platform visible to its private users. In order to arrive at the source of a tweet, one needs to access the attributes of tweets via the Twitter API. Knowing the source, however, does not come with knowledge about its automation features and practices, which need to be explored differently. Second, automation can be obfuscated and rendered untraceable on source level – Tweet Button activity, for instance, appears in the database as coming from the official Twitter Web Client (Gerlitz and Rieder, 2018) – but also through bought interactivity and engagement offered by social media marketing services. Developers might resort to these tactics because Twitter has doubled down efforts to identify and block un-wanted automation, amongst other things by focusing on the source.[5] Although automation is an integral part of the platform infrastructure – our research interest thus aligns with the medium's configuration – methodologically the analytic affordances of the platform infra-structure do not privilege an easy pathway to follow the continuum between automation actors and their networks. It is through this nesting of infrastructures of platforms and apps that an alternative positioning is needed, which both follows and works against the medium. It 'follows the medium' by making use of the materialisation of automation in the platform infrastructure; however, it goes *against medium method* by circumventing the preferred way the platform affords to follow traces and thus develops shortcuts and workarounds to render the platform ontology flat(ter).

In the data set used for this research,[6] we explore tweets from the week running up to the Brexit referendum[7] as this event is considered particularly contested and prone to automation (Howard and Kollanyi, 2016). Within our data set, we take the top 100 most used sources[8] and identify their affordances for automation. To gain more information, we draw on the websites provided by developers when registering an app (this means that information is not always available), create accounts to analyse the interfaces of the third-party app (if available, some software does not have a public-facing website) and/or explore their tweet syntax, informing a bottom-up characterisation of automation attributes. We do not use sources to merely identify automation, but to contribute to a more fine-grained language to address the spectrum of automation on social media and to reflect on the entanglement between the human and non-human configuration.

In accounting for automation attributes and the human/non-human configuration they enable, we differentiated three partly overlapping levels, namely degree of automation, type of source (and what practices it enables) and its main (automation) functionalities. Figure 33.1 hierarchically visualises the degrees of automation and their types. The degree is understood

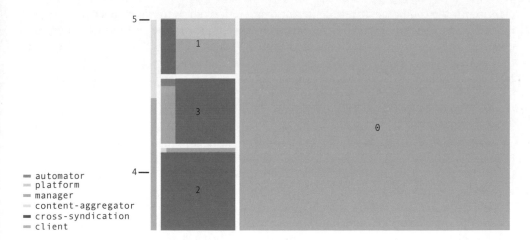

Figure 33.1 Spectrum of automation

Differentiation based on the top sources in the Brexit Twitter data set. The tree map visualises the hierar-chy of sources in the Brexit data set, 17–23 June 2016. The numbers 0–5 indicate sources categorised by degree of automation, the size of the parts is scaled by the number of tweets sent from a source and the colour of the parts indicates the type of source. Visualisation made with RAWGraphs.io.

as the spectrum of handing over calculative capacities to software and allowing it to find, select and post content on one's behalf. The distribution of sources indicates that the majority of the top 100 sources are classified as degree 0 of automation, which refers to manual tweeting – a surprising finding in the Brexit data set. However, our approach cannot account for masked automation as discussed earlier. The only type of source in this category are social media clients, which we defined as external software offering largely the same functionalities as Twitter, allowing private users to read/post content via external interfaces. In some cases, clients allow tweeting from multiple accounts or to operate multiple platforms.

Degree 1 introduces blended forms of automation by allowing users to (semi-)automatically repost or schedule their *own* content from other sources. The most recurring type in this category is cross-syndication, which we define as the reposting of existing content defined by the user, including cross-platform posting, buttons and automation recipes. Sources in degree 2 add an additional level of automation as they facilitate the (semi-)automatic reposting of content of others, often based on recipes, cross-syndication or topical selection. The most recurring type in this category is again cross-syndication with sources focusing on the reposting of *external* content of *other accounts* from the web or social media, including, for instance, reposting rss-feeds. Sources in degree 3 introduce significant, but often rudimentary, automation by offering automated interactions, such as auto-direct message, auto-reply or auto-retweet. Types of sources in this category are more diverse, including clients, cross-syndication, but also introducing two new types: automators and managers. Automators include sources which only function on automating Twitter action and interaction, often in a rudimentary way, such as auto-retweeting. Managers are addressed at professional users and largely offer advanced functionalities such as collaborative tweeting, multi-accounts, advanced analytics, (algorithmic) content-recommendation and scheduling.

Sources of degree 4 introduce auto-created or recommended content based on (algorithmic) selection or recommendations by the source. Managers and content aggregators are prominent in this category, handing over more and more calculative and agentive capacities to software. They can include more complex forms of automation such as machine learning or algorithmically composed tweets. Different than cross-syndicators which operate based on user selection, content aggregators directly offer users' topics, tweets or links for automated posting. Only one source is categorised as degree 5 – sources are included in this degree if they can auto-post auto-created content, such as data from sensors, external databases or the Internet of Things (IoT), taking over the process of content selection and posting entirely. The source in this category, for instance, offers the functionality to automatically create accounts and creates news-based content to tweet from these accounts.

In addition to the degree and type of automation, we differentiate the sources based on the automation functionalities they offer. In Figure 33.2, the dendrogram visualises the spectrum of functionalities of sources hierarchically by degree of automation and types of sources, and highlights how automation functionalities of sources increasingly gain complexity in the higher degrees of automation. For example, a client in degree 0 obviously offers little automation functionality besides allowing users to access multiple accounts. Clients in degree 1 typically offer additional functionalities such as cross-syndicating and scheduling posts. Clients in degree 3 offer content-recommendation, advanced analytics and automation of core Twitter functionalities such as auto-retweet, auto-like and auto-direct message.

In this sense, the visualisation supports the differentiation in degree of automation and shows how automation is involved in the redistribution of both calculative and agentive capacities – as automation software not only takes over the act of posting, retweeting or replying, but also the identification of content based on search, algorithmic recommendations

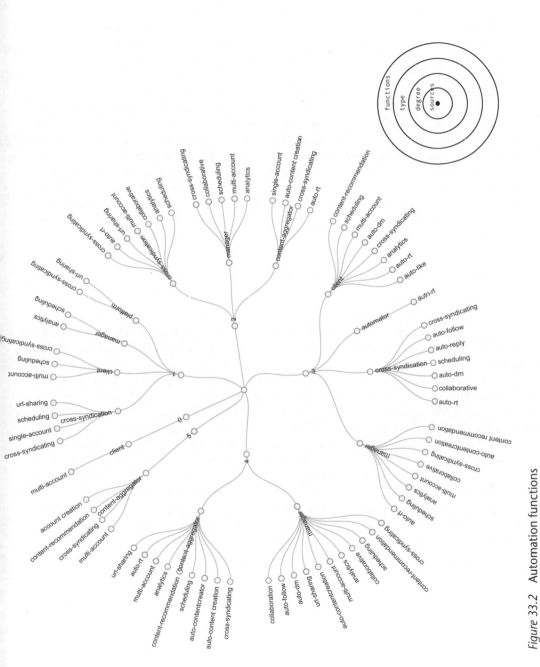

Figure 33.2 Automation functions

Differentiation based on the top sources in the Brexit Twitter data set. The dendrogram visualises the hierarchy of sources, degrees of automation, types of sources and their functions in the Brexit data set, 17–23 June 2016. Visualisation made with RAWGraphs.io.

or machine learning. Furthermore, these proliferating functionalities complicate the notion of the individual account. In digital research, platform accounts are often considered as pointing to one individual user. A variety of sources, however, allow users to run several accounts (see function 'multi-account') or collaboratively handle a single account (see function 'collaborative'). In combination with other functions especially offered by managers – such as content-recommendation and scheduling – manual and automated tweeting can be blended in the context of a single account and the notion that accounts equal individuals is put to question.

Our categorisation exercise offers first pointers for a more nuanced account of automation, which is (a) not limited to account level, (b) differentiates between own and external content, (c) distinguishes automated content production, selection and interaction, and (d) shows the variety of ways in which manual tweeting can be software-supported or translated through source functionalities, which themselves are part and parcel of the material infrastructure. To account for the entanglement between human and machine as practice requires methodological manoeuvres beyond simply following traces, as the seemingly ready-made data formats obfuscate their heterogeneous accomplishment. The case of automation also shows that there is no single medium to follow, as platforms are set up as infrastructural, socio-material assemblies that enable but do not easily explicate the relation-making of their data.

The socio-material accomplishment that leaves no traces

We would like to conclude by drawing out a series of interventions we think digital sociology offers to ANT, using our interest in distributed platform infrastructures to push the three roles platforms play in digital sociology (Marres, 2017) further and ask: How does one do digital sociology in ways that are attentive to the role of the digital as site, access and engagement of the social?

Firstly, digital sociology and platform-sensitive engagement with the digital respecify ANT's established idea of the continuum between the actor and the network (Venturini et al., 2017). The issue of automation has shown that research will have to accommodate a variety of digital part-humans part-machines which, however, cannot (only) be treated as type or actor but as *practice*, that is dynamic, situated and emerging in relation to the material-technical affordances of the medium. As shown in the case of Twitter, an account does not necessarily represent a human user, as it is accomplished in distributed and situated ways, just as a tweet is not a tweet, commonly understood as a uniquely typed post. Both occupy a continuum between manual and fully automated activity, which is supported by distributed infrastructures of support software. The notion of the continuous and relational monad (Latour et al., 2012) is useful when approaching (semi)automated activity; however, it is not possible to seamlessly follow them, as often automation does not translate into visible (or even clickable) traces.

Secondly, the infrastructural concealment of platforms complicates the distinction between what Venturini et al. (2017) consider as trace, namely inscriptions by digital media, and data, that is, traces readied for research through the methodological apparatus. What appears as trace or inscription in the context of the platform is often a highly composite measure assembling heterogeneous input sources and has been subject to cascades of inscription (Ruppert et al., 2012) by various stakeholders such as developers or users which have been rendered invisible. Instead of stabilising platform data as demarcated by the medium, digital sociology needs to revisit distinctions between traces and data, their accomplishment and their alignment with one's research objectives (Weltevrede, 2016). Platform media thus challenge ANT to not only pay attention to existing but also to missing, nested or no longer visible traces, most notably in the context of black-boxed dataflows. As platforms constantly recombine, commensurate and circulate data through their related infrastructures (Beer, 2017), a sole focus on flat ontologies and associative relations of actors needs to be supplemented by a critical investigation of what counts as platform data, and how it has been accomplished and stabilised. Here, both ANT and digital sociology would benefit from engaging more carefully with an infrastructural perspective which draws attention to exactly such black-boxed, distributed accomplishment. From this perspective, ANT's notion of 'following' as a methodological principle advances from a seamless movement to a situated methodological configuration which may involve cuts, jumps and fissures.

Thirdly, the platform perspective pushes the notion of interface methods (Marres and Gerlitz, 2015) further, as bias is not limited to the capacity of a platform to pre-structure its data for preconceived analytical aims (positive affect on Facebook, popularity measures in Twitter) – but subject to 'in actu' realisation through different stakeholders and their practices on a platform. What does it mean, if users, companies, organisations and bots all use the platform with a different idea of sociality (promotional, deceptive, affective, etc.) in mind? Whilst the data capture mechanisms of platforms may come with some forms of bias built into them, the way they are appropriated, negotiated and repurposed by their stakeholders may introduce yet other, potentially even more heterogeneous bias. In the context of distributed platform media, interface methods thus need to negotiate and configure these multiple dimensions of bias.

To return one final time to Marres (2017), when taken together this means that methodological alignment may require to work against some actors and biases, whilst allowing in others. Furthermore, medium research and social research occupy a continuum in which they challenge the disciplinary boundaries between sociology and media studies. It is not only in 'social research' that attention to the socio-material grounding of the social becomes a question of configuring method. Also 'medium research' can open up questions of alignment to media bias, following the preferred configurations of the medium to study bias effects, or by aligning differently the methods built into the medium – following the medium against the medium.

Notes

1 Twitter defines the source metric as 'Utility used to post the Tweet, as an HTML-formatted string. Tweets from the Twitter website have a source value of web.' See https://dev.twitter.com/overview/api/tweets (last opened April 3, 2019).
2 See https://blog.twitter.com/developer/en_us/a/2012/changes-coming-to-twitter-api.html (last opened April 3, 2019).
3 See https://support.twitter.com/articles/76915 (last opened April 3, 2019).
4 See https://blog.twitter.com/developer/en_us/topics/tools/2018/new-developer-requirements-to-protect-our-platform.html (last opened Nov 12 2018).
5 See https://blog.twitter.com/official/en_us/topics/company/2017/Our-Approach-Bots-Misinformation.html (Last opened April 3, 2019).
6 A special thanks to all members of our Digital Methods Summer School 2018 research project group entitled 'Method Maps: Visualising Automation': Anne Helmond, Fernando van der Vlist, Mace Ojala, Laetitia Della Bianca, Karmijn van de Oudeweetering, Cindy Krassen, Daniela van Geenen, Lisa-Maria van Klaveren, Angeles Briones, Iulia Coanda, Liping Liu, Emilija Jokubauskaite, Ece Elbeyi, Gabriela Sued, Enedina, Eloy Caloca Lafont.
7 The data set contains tweets mentioning 'brexit' from 17.-23.06.2016 and consists of 2,914,134 tweets by 786,494 users.
8 The data set features tweets from 8,302 distinct sources. 2,818,718 tweets are sent from the top 100 sources, which is roughly 97% of the tweets in the data set.

References

Abbott, A. (2011) Googles of the past: Do keywords really matter? Annual Lecture of the Department of Sociology, Goldsmiths, 15 March.

Beer, D. (2017) *The social power of algorithms*. Taylor & Francis.

Bogost, I. and Montfort, N. (2009) 'Platform studies: Frequently questioned answers,' *UC Irvine: Digital Arts and Culture*. Available at: https://escholarship.org/uc/item/01r0k9br.

Ferrara, E. et al. (2016) 'The rise of social bots', *Communications of the ACM*, 59(7), pp. 96–104.

Gerlitz, C. and Rieder, B. (2018) 'Tweets are not created equal. Investigating Twitter's Client Ecosystem', *International journal of communication,* 12 (2018), pp. 528–547.

Gillespie, T. (2010) 'The politics of "platforms"', *New Media & Society*, 12(3), pp. 347–364.

Howard, P. N. and Kollanyi, B. (2016) 'Bots, #strongerin, and #brexit: Computational propaganda during the uk-eu referendum', Working Paper 2016.1. Oxford, UK: Project on Computational Propaganda.

Latour, B. (2005) *Reassembling the social: An introduction to actor-network-theory*. Oxford University Press.

Latour, B. et al. (2012) '"The whole is always smaller than its parts"–a digital test of Gabriel Tardes' monads', *The British journal of sociology*, 63(4), pp. 590–615.

Marres, N. (2012) 'The redistribution of methods: On intervention in digital social research, broadly conceived', *The Sociological Review*, 60(1_suppl), pp. 139–165.

Marres, N. (2017) *Digital sociology: The reinvention of social research*. John Wiley & Sons.

Marres, N. and Gerlitz, C. (2015) 'Interface methods: renegotiating relations between digital social research, STS and sociology', *The Sociological Review*, 64(1), pp. 21–46.

Marres, N. and Rogers, R. (2005) 'Recipe for tracing the fate of issues and their publics on the web,' in B. Latour and P. Weibel (eds) *Making Things Public*. MIT Press, pp. 922–35.

Marres, N. and Weltevrede, E. (2013) 'Scraping the social? Issues in live social research', *Journal of Cultural Economy*, 6(3), pp. 313–335.

Rogers, R. (2013) *Digital methods*. MIT Press.

Ruppert, E., Law, J. and Savage, M. (2013) 'Reassembling social science methods: The challenge of digital devices', *Theory, Culture & Society*, 30(4), pp. 22–46.

Savage, M. and Burrows, R. (2007) 'The coming crisis of empirical sociology', *Sociology*, 41(5), pp. 885–899.

Venturini, T. et al. (2017) 'An unexpected journey: A few lessons from sciences Po médialab's experience', *Big Data & Society*, 4(2).

Weltevrede, E. (2016) *Repurposing digital methods: The research affordances of platforms and engines*. PhD Thesis. University of Amsterdam.

Woolley, S. C. and Howard, P. N. (2016) 'Automation, algorithms, and politics| political communication, computational propaganda, and autonomous agents – introduction', *International Journal of Communication*, 10, p. 9.

Section 6

The uses of ANT for public–professional engagement

Celia Roberts, Anders Blok and Ignacio Farías

What happens when academics and other public intellectuals take ANT out into the world in an attempt to make a deliberate difference and/or to set up some kind of social experiment? In this section, authors narrate and analyse their experiences of working with ANT and closely related concepts and approaches in a diverse range of places and spaces. Coming from a variety of intellectual and professional backgrounds, these authors engage ANT to shape political actions, environmental interventions and design processes, even, astonishingly, to run a public hospital! In each case, the chapters describe how ANT travels, where and how it works or gains traction, and where it stumbles or needs development. As a group, the chapters provide strong testimony to creative intellectual work and energetic engagement in the problems of contemporary life: In our view, they are inspiring, even exhilarating.

As with all the chapters in this collection, the authors in this section present personal, arguably eclectic, versions of ANT. From our perspective, this is completely as it should be (perhaps even more so in this section than elsewhere). Here, ANT operates as a kind of tool kit or set of resources, rather than a totalising worldview or methodological approach. (Indeed, perhaps we can conclude, finally, that this is how ANT works best.) We could have filled a whole book with chapters like these – such a text would be a brilliant sequel or companion to this volume. Our choice of authors for the current book focussed on diversity of topics and approaches, as well as national contexts. As ever, we also chose people's whose work we really like! As with all the sections, then, this group of chapters is intended to be inspirational and provocative rather than exhaustive.

In terms of geography, Section 6 takes us around the globe: From Japan, to Chile, Spain and the UK. We travel to famous cities – London, Barcelona; to smaller towns – Puerto Montt and Saitama; and to the Cumbrian fells and lakes. Here, we encounter many challenging materialities: Overflowing cow shit, burst hospital water pipes, radiation-affected vegetables, electricity-monitoring devices and wheelchairs trying to mount kerbs. We are introduced to people working in, with and against these materialities: Farmers collecting and redistributing slurry; labourers trying to fix the pipes; activists and consumers trying to decide which vegetables are safe to eat; designers making new monitoring devices and disability activists building and testing ramps. In each case, our authors contend, ANT helps not only to understand what's going on intellectually (to provide an interesting analysis) but actually to make things happen differently.

Sometimes, this works by facilitating new conversations and by opening up lines of thought and action that may otherwise have remained unclear. For disability activists in Barcelona, for example, ANT became a resonant framing for thinking about relations between bodies, technologies and environments. For the ANT researchers involved, working in and with activists also hugely enlivened their own understandings of these issues. Tomás Sánchez Criado and Israel Rodríguez-Giralt suggest the term 'joint problem-making' to describe the ways in which researchers and activists can work together to make a difference to the world and to ANT. For the farmers, consumers and activists of Kashiwa trying to work out what to do about irradiated food in post-Fukushima conditions, ANT has helped to make sense of scientific information, to build a workable understanding of the dangers of food grown or harvested in radiation-exposed conditions, and to reach forms of consensus about how to proceed. Shuhei Kimura and Kohei Inose argue that ANT has helped them, both as anthropologists and as concerned citizens, to 'gift an ethnography-based story' to their compatriots in order to help everyone live better with the enduring effects of the 3.11 disaster. In both cases, ANT is not a theory or method that can be simply applied to the work of activists but rather is a productive material-semiotic practice that stimulates questioning, provokes alternative visions and facilitates productive storytelling.

On other occasions, ANT has been used to establish particular kinds of forum. Following Latour's work on the politics of Nature and Dingpolitik, as well as Callon's *Acting in an Uncertain World*, Emma Cardwell and Claire Waterton describe and explore the capacities and challenges of hybrid forums or knowledge collectives in addressing environmental and farming-related controversies. Reporting on two extensive empirical studies – of algae in the Lake District's Loweswater, and of slurry on UK beef farms – they argue that ANT-inspired knowledge collectives can create spaces to say and do things that might otherwise remain unsaid and undone. Engaging with recent STS work on matters of care, they also explore what remains ungraspable in these collectives, and start to think about the significance of such challenges for ANT-inspired involvement in environmental politics.

For Alex Wilkie and colleagues, ANT provides a similarly material resource. ANT is in the room when the 'Sustainability Invention and Energy-demand Reduction: Co-Designing Communities and Practice' research group engages in design practices to assess the limitations of existing electricity-usage monitors and to build more interesting ones. Wilkie's chapter outlines and explains this process, introducing some fascinating and important concepts (retroscription and procomposition) to describe the resultant practices

> that, on the one hand involves an active participation in lived reality of using a smart energy monitor and transferring this experience, by way of visualisation, into a diagram of use in which the efficacy of the appliance is laid bare.

Designing the new Energy Babble, Wilkie argues, shows that ANT can be used to do so much more than generate clever description: It can be enrolled to make literally shape collectives of people, environments and things. Indeed, each of these chapters shows how bringing ANT to the table, allowing it to flourish in spaces of deliberate intervention and experimentation, facilitates imaginative and non-normative ways of working with and through such assemblages.

But who could have thought that ANT might be useful for actually running a public hospital? In his wonderfully evocative piece, Hospital Director and ANT scholar Yuri

Carvajal Bañados shows us how. For Carvajal, working with ANT and related theories is a double-edged sword —necessary and clarifying but also politically risky. Indeed, during the drafting of his chapter – a drawn-out process for most of the authors and the editors of this book – Yuri stopped being the Hospital Director. Nonetheless, he concludes that 'ANT worked for the Puerto Montt Hospital as a valuable enzyme.' Concluding the section and thus the book, this sentence can perhaps work as a metaphor for ANT more generally: Enzymes are necessary, vitalising catalysts that are part of complex systems of action and reaction. They cannot work alone, but form lively connections to a wide range of other entities and systems.

34

Can ANT be a form of activism?

Tomás Sánchez Criado and Israel Rodríguez-Giralt

Ever since the 2011's *indignados* or '15-M' (May 15th) movement started – occurring after the occupation of Spain's main city squares, and later on developing into interesting forms of urban activism – the personal and academic life of many of us in Spain has turned into a perpetual unfolding of particular protests and activist struggles, with a lasting impact on our modes of doing research. One good example of this was Tomás' ethnographic involvement, since 2012, in an open design initiative called *En torno a la silla* (ETS). This initiative comprised a few people who had met in the 'functional diversity commission' of the 15-M encampments in Barcelona,[1] where different wheelchair-using activists of the Independent Living Forum had come across other people – all of them long-standing participants in several activist groups and initiatives and some of them with design and making skills – and started doing things together in a 'free culture' atmosphere. After suffering from the inaccessible conditions of the precarious places they started to occupy and share, many of those participants decided to intervene in these spaces.

For instance, in the summer of 2011, before ETS came into being, a material intervention was undertaken in the yearly Gràcia district festivities, aiming to reclaim the accessibility lost during the festival due to the emplacement of street decorations and concert stages. At the request of the Festivities Council of the Fraternitat Street, where Sebastián Ledesma (one of the members of the functional diversity commission) lived, singular yellow ramps (Figure 34.1) showing the black and yellow inscription 'revolution will be accessible or it won't be' were produced and displayed. After this first attempt, the people who would later initiate ETS – Antonio Centeno, an independent living activist, Alida Díaz, architect, and Rai Vilatovà, anthropologist and craftsperson – sought to join forces to co-design a free/open kit 'to activate the wheelchairs' environments' otherwise, comprising a portable wheelchair ramp and other gadgets transforming the wheelchair,[2] a kit whose main results would be digitally documented and openly distributed so that anyone could draw inspiration from it. This was not an act of charity by do-gooders, but rather an experimental attempt to prolong a nascent friendship between specific 'disabled' and 'non-disabled' people by technical and material means. Tomás later joined the project when they were beginning to prototype these elements, and in the practice-oriented and engaged context of everyone doing something for the common project, he made his ethnographic skills available, acting as an audiovisual 'documenter' of the

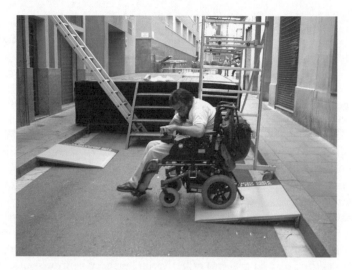

Figure 34.1 Sebastián Ledesma testing the ramps at Fraternitat Street. Picture CC BY Functional Diversity Barcelona commission (August 2011, used with permission)

collective. Several other people came and went in the following years, devoting great amounts of time to the audiovisual documentation, the open prototyping, the public reflection and/or the organisation of workshops and events to show and discuss what it meant to act in this way.

All of this was a common way of 'acting' popularised at the time of the 15-M. Indeed, ETS could be said to be part and parcel of a singular activist ecology, an heir of the particular 'collaborative turn' that the 15-M enacted. As various scholars have tried to convey (Arribas, 2018; Moreno-Caballud, 2015), 15-M's main form of collective action pivoted around not only just a wild experimentation with 'artivist' initiatives – mostly against austerity measures, political corruption and technocratic forms of knowledge production. It also showed a deep concern with enacting a politics of listening and of the 'liberation' and documentation of knowledges in the wake of a broad 'free culture' ethos (Corsín & Estalella, 2016; Martín Sáinz de los Terreros, 2018; Postill, 2018). In a time of great collective effervescence, this '15-M' way of doing politics also shaped forms of research grounded on an ethics of companionship and an experimentation with the modes of joining together different people, knowledges and kinds of worlding.

In this chapter, we would like to *think with* the ETS collective, and in particular the research engagement afforded by their intense social and material explorations in the environmental intervention and remaking of wheelchair users and their surroundings. We will characterise this particular form of research activism as 'joint problem-making': Comprising a series of social and material interventions to problematise, transform and account for the worlds being produced together with others. Our main aspiration in doing this would be to describe the impact it had on us as researchers: Or, to be more specific, on our ways of engaging ethnographically, and to consider how this might inspire the 'experimentally collaborative' or 'activated' ways in which ANT researchers might engage in other activist ecologies.

ANT research on technoscientific activisms

But before attempting to delve into this, allow us to briefly outline some of the most important ANT-related works dealing with activist initiatives and social movements. Already in the 1990s, ANT had attempted a certain redefinition of 'collective action' beyond human-centric accounts and a means-ends understanding of action that was fundamental in opening up an agenda around technoscientific activism, charting *hybrid collectifs* and their 'multiform kinds of agency' (Callon & Law, 1997: 113). On the one hand, for ANT accounts, action was framed as 'collective': Meaning that agency is to be described as a distributed and ma-terially heterogeneous process, imputable to particular entities only after the fact. This par-ticular stance allowed the sociology and anthropology of social movements to expand their reach, paying attention to the role of more-than-human actors in particular forms of activ-ism (Rodríguez-Giralt, 2011). On the other hand, ANT-related scholars became interested in going beyond an instrumental understanding of action (Latour, 2002): Paying attention to the thorny issue of mediations or the multifaceted role of 'attachments' in these *hybrid collectifs* (Gomart & Hennion, 1999) and how they prod particular forms of doing, either in action or in passion, in particular events. In fact, continuing their early analysis of the very particular 'power of reason' (Latour, 1987), ANT-related scholars took issue with how forms of technoscientific activism challenge that power in particular ways. That is, they sought to engage in understanding particular forms of 'activating' technoscientific issues, be it through expanding given matters of concern and 'truth politics' (e.g. addressing experiences not con-sidered), or disputing particular arrangements or *agencements* that make *hybrid collectifs*, such as markets (Callon, 2008).

In *Acting in an Uncertain World*, Michel Callon and associates (2011) explored the boundaries and challenges of the mode of knowledge production known as 'secluded research.' Building on Latour's (1987) seminal explorations on how laboratories gain power through gathering inscriptions in centres of calculation to 'act at a distance,' they describe the particular op-eration that takes place in laboratories as one of 'Translation' (capitalised in the original): A threefold process that goes from a transportation of the world into the miniature space of the lab through the accumulation of inscriptions (translation 1), to the many manipulations and calculations of a research collective in that space until the phenomenon is controlled or 'black boxed' (translation 2), and can then be brought back into the world as a stable entity potentially changing existent relations (translation 3). However, as they suggest, in the last decades, a whole gamut of 'hybrid forums' – including both experts and laypersons – of the most varied kinds (e.g. environmentalist or embodied health collectives) have been putting secluded research in crisis (see Waterton and Cardwell, this volume), generating many avenues for the renewal of the ancient STS project of 'technical democracy.' Indeed, their main proposal was to open up a research programme on what they called 'research in the wild' (Callon & Rabeharisoa, 2003), a term used to describe the many different practices of laypersons seeking to 'activate' technoscientific practice in particular ways, potentially democratising the who and how of technoscience. One of the most interesting contribu-tions of these works is its singling out of the role of 'emergent concerned groups' (Callon & Rabeharisoa, 2008): That is, those who audaciously search to find ways to transform the market and/or technoscientific arrangements that leave them behind or cause them to suffer from particular unpredicted overflows.

Indeed, this was ANT researchers' main contribution to broader discussions in STS around activism and social movements (Hess et al., 2008). Interestingly, these discussions have also been important for the consolidation of STS:[3] Indeed, this has become a research domain probing into the processes by which scientific truth and credibility, or technical objects, are publicly opened up for debate, scrutiny, control or co-construction in different narratives in which activism becomes a democratising agent, an 'activator' of alternative technoscientific practices.[4] In turn, these reflections have opened up interesting debates with regard to: (a) The normative horizons of STS practice and how activism may contribute to the co-construction of science and technology and to public control and participation in the development of science and technology (Sismondo, 2008); and (b) the different meanings of politics and democracy, and what STS could do to better articulate science, technology and democracy (Latour, 2007; Marres, 2007). However, in most of ANT and STS literature, the role of the analyst is left untouched: Most works present themselves as a study *of* activism, not studies *in* and *through* activism. In contrast, we want to ask: How does the activation, criticism, contestation and dispute of technoscientific knowledges and arrangements 'activate' (Rodríguez-Giralt et al., 2018) our own research practices or us as researchers?

Becoming 'activated' as researchers

But how to open up a reflection against our own very 'dogmatic negations,' to use Stengers and Pignarre's (2011) terms? That is, how to resist the impinging professionalisation of ANT as a 'too comfortable,' 'self-complacent' or 'undisputable' research space? Perhaps we might draw inspiration from two research programmes in STS that have been developing more explicit 'materialist-interventionist' versions of what an ANT-related mode of engagement might look like: (1) The debate around action-oriented research or the role of 'situated interventions'; and (2) the debates around 'matters of care.'

In recent years, a debate has arisen in STS around how to develop a more 'action-oriented' type of engagement in the worlds we study, and the practical ways in which such more socially and politically committed interventions might take place. We specially refer here to the work of Teun Zuiderent-Jerak and colleagues (Zuiderent-Jerak & Jensen, 2009; Zuiderent-Jerak, 2015). In contrast to some calls for a transformation of STS foregrounding, a 'realistic' (Bal et al., 2004) or an 'engaged' (Hamlett, 2003) political *ethos*, theirs stands out because of a desire to develop a situated stance: One where normativities are addressed as part of the reflexivity displayed in different types of experimental modes of intervention beyond textual/discursive means. This is not, for sure, a new relativist agenda but rather an approach to specificities: A descriptive and reflexive attitude of exploring

> the different forms interventions can take [...] in a hybrid space, in which many agents constantly negotiate and influence each other, in order to achieve multiple conflicting goals. A picture emerges of intervention as mutual betrayal, or, more positively, as a process of artful contamination whereby actors spread their agendas, ideas and aspirations.
>
> *(Zuiderent-Jerak & Jensen, 2009: 230–231)*

Another inspiring materialist-interventionist agenda we could draw from to rethink ANT modes of engagement stems from the critical reflections by several feminist technoscience scholars around care. Feminist debates in STS not only have complicated and unsettled care as an analytic (Martin, Myers and Viseu, 2015; Mol, 2008; Jerak-Zuiderent, this volume; Puig de la Bellacasa, 2017), but have also sought to foreground the impact these vocabularies and reflections might have on our role as analysts. For instance, in Puig de la Bellacasa's (2011) parlance, 'care' ought to be considered as, simultaneously, an 'affective state,' a 'material/vital doing' and an 'ethico-political obligation.' Interestingly for our purposes, she is perhaps one of the most vocal in inviting us to *think with* the collectives and issues we work with as a matter of care. As she puts it: 'transforming things into matters of care is a way of relating to them, of inevitably becoming affected by them, and of modifying their potential to affect others' (Puig de la Bellacasa, 2017: 64).

This affective obligation should then lead scholars to address carefully those collectives whose particular ecology of practices makes them more vulnerable with regard to particular hegemonic forms of knowledge production and politics. This involves finding ways of 'speaking well' about the practices we study: That is, characterising the mode of action of its practitioners without disregarding what matters to them (see López, this volume), but also foregrounding the frictional relations we may have with them. In some occasions, as it happened with our independent-living friends, a careful attitude might entail finding alternative vocabularies to the one around care – highly disputed because of its perceived connotations, implying their treatment as passive objects or recipients of care (see Shakespeare, 2006).

When brought to the study of activist groups, these series of concerns may be translated into a commitment to approach such groups not only as 'recipients of the academic gaze' but also as 'knowledge-making agents in their own right,' as several works in the literature of social movements have stated for years (Casas–Cortés et al., 2008; Melucci, 1989; Rabeharisoa et al. 2014). This sensitivity has also helped to articulate a wealth of modes of research engagement that open themselves to activism: such as militant research, feminist research, emancipatory research, indigenous and decolonial research, participatory action research, etc. We believe that all of these lines of inquiry, foregrounding a problematisation around intervention, and in allowing us to think *with* and *from* activism as a mode of research pave an interesting way for ANT to draw inspiration and to carefully explore the ways in which activism might change our own research practices (Rodríguez et al., 2018). To expand on this, allow us to go back to the case of ETS.

Engaging in 'joint problem-making'

In late 2012, Tomás began curating the social media of the ETS collective – notably, the blog and social networks – taking minutes, and helping in the coordination of the process of documentation, crafting tutorials and how-to guides, and preparing meetings and public presentations. This mode of ethnographic engagement had lasting effects not only in the particular ways of undertaking research, but also in what the collective was doing. In fact, their free/open design practice foregrounded different practices of forms of 'relating' – forging relations with others and accounting for them – that were absolutely central in their design/ interventive practice. Even as documentation played a fundamental role, it was taken as a driver of non-predicative forms of knowledge (Criado & Cereceda, 2016): That is, as a form of showing others our efforts to make do, either to inspire them or so that they could relate to us and start newer dialogues, rather than as a form of evidence (or, if at all, as evidence that another life could be possible).

Actually, as was the case in other developments and alliances derived both from the 15-M encounters (see García-Santesmases et al., 2017) and from the Spanish independent living modes of doing 'emancipatory networks of knowledge' (Centeno, 2009), ETS can be considered an interesting exploration in collaborative modes that related to particular forms of engagement and co-activation around research.[5] Ours was a paroxysmal state of ongoing convulsion, sadness and suffering, but also full of moments of sheer joy and exaltation with many being drawn into calls for action, feeling moved and 'taken' by different and previously incompatible kinds of collective explorations: From actions around the monstrous problem of mortgages and evictions to the expansion of DIY practices and collective architecture; from a renewal of movements searching for a re-democratisation of institutions and decision-making or even for the transformation of political topologies (e.g. Catalan independence movements or the creation of new political platforms, but also emerging trans-feminist, environmentalist, decolonisation and independent-living movements) to street-level democratic initiatives in social (i.e. anarchist-related, commons-led) spaces. This, of course, is not particularly new if one pays attention to the impact of, say, feminist or disability rights activist agendas in different strands of post-ANT research. But, in our case, it decisively contributed to activate us as ANT-inspired researchers:[6] Something we felt when engaging with such an open and 'self-experimental' ethos to create alternative research relations and to collaboratively forge other tales or accounts.[7]

The 15-M's political effervescence produced alternative knowledges that impacted on established forms of expertise and markets (cf. Roelvink et al., 2015). But in there we also witnessed a repertoire of activism going beyond the classic ANT concern with the 'translation of knowledge' (*research in the wild*) or the 'modulation of markets' (*concerned groups*). The 15-M, in turn, created the conditions of what could be called *wild research*: That is, a form of activism that entailed different forms of gathering together laypeople, social scientists, long-time activists and designers in exploratory practices in and through where all involved felt activated to engage in alternative and always collaborative forms of research. This meant avoiding ready-made solutions and explanations, giving centre stage to the ones usually left aside, and promoting an innovative exploration of arranging worlds together with many others having the experience, capability and the means to describe and transform them in new and creative ways.

Rather than a desire to produce 'situated interventions,' it was this collaborative intensity that 'activated' us, as researchers, towards other modes of moving and activating research, and through which we also put others in motion.[8] This was also what activated our particular modes of 'speaking well' of these activist practices (cf. Rodríguez-Giralt et al., 2018), something we attempted in and through documentary and academic activities, but also engaging in research interventions that facilitated particular forms of what we could call, following our epistemic partners' modes of relating, 'joint problem-making' (Criado & Rodríguez-Giralt, 2016): That is, ways of problematising – both materially transforming and conceptually accounting – their own worlds together with others (a friend from ETS joyfully called this a practice of 'adding up our fates,' which is probably a more appropriate register than the one around care). In fact, 'joint problem-making' also became a particular mode in which Tomás engaged ethnographically in activities of note-taking (Criado, 2018); ETS's documentation being not just a note-taking device 'from the field' – for ETS members to narrate their undertakings – but a fundamental space in the articulation of that very field as a joint research space for all involved. These forms of joint problem-making also developed into manifold 'experimentally collaborative' research endeavours (Criado & Estalella, 2018), when we hosted events where our friends and colleagues could present their ideas in public, and where different approaches to DIY could be shown and demonstrated; or in bringing our counterparts so that they could present on an equal footage in academic venues; or in co-writing texts, open design tutorials and blog posts with reflections about our experiences in common.

Conclusions

In this text, we have explored what happens to ANT when we engage in activist ecologies. We have done this through the particular reflection of our involvement in a particular instance of the 15-M and independent living movements in Spain, and how they affected or, rather, 'activated' alternative modes of engaging in these particular ecologies. We believe ETS to be a good case to think with about the different issues arising when trying to experiment with more collaborative, engaged and action-oriented ways of doing ANT research. In particular, we have expanded on a particular mode of ethnographic 'joint problem-making' as a concrete strategy to do this. But this is far from the only possibility in which ANT's research modes could be experimented upon, if not repurposed, when engaging in activist practices. In what ways could ANT be reshaped when engaging in other activist settings? Or, rather, can ANT be a form of activism?

In pointing out all this, our hope would be that ANT could become a more open and nonconformist research space: An 'activated' practice, problematising in newer ways the relationship between description and action, exploring the manifold ways of being an analyst or a researcher that might be available when engaging in these settings. This does not necessarily mean a call to engage in normative or programmatic interventionist agendas. On the contrary, it might mean developing a more careful, curious and situated *ethos*, whereby newer understandings of normativity and the politics of research, including experimental forms of joint problem-making, could be developed through particular ways of relating to or, rather, 'becoming activated' as ANT researchers in, activist ecologies.

Notes

1 Stimulated by the inclusive and participatory atmosphere of 15-M movement, several 'functionally diverse people' – most with activist backgrounds in the Spanish Independent Living Movement – congregated in the Spanish squares and created commissions of 'functional diversity' (Arenas & Pié, 2014). These commissions created the conditions for engaging in lively discussions about the several challenges faced by the functionally diverse and the empowering dynamics taking place in the squares.

2 For a story of the ramp's open prototyping, see Criado et al. (2016) where we detail how the 'inclined plane' of the ramp was an experimental space, putting to a test given care arrangements, and allowing tinkering with alternative ones.

3 For instance, Cozzens (1993) reflected on whether STS could be considered a social movement in itself.

4 A classic reference of STS work on technoscientific activism is, perhaps, Epstein's (1996) on AIDS and the 'impure science' of ACT-UP interventions. But other interesting have arisen in the last decades, such as 'embodied health movements' (Brown, 1997), 'evidence-based activism' (Rabeharisoa et al., 2014) or 'technoscience otherwise' (Murphy, 2012). Besides, activism and social movements have been at the background of relevant moves in STS ways of addressing public engagement in science and technology, from the work on 'civic epistemologies' (Jasanoff, 2005) to the more recent emphasis on 'material publics' (Marres & Lezaun, 2011).

5 At that time of the main 15-M mobilisations, we were both starting a collective research project called EXPDEM: 'Political Action of Groups Concerned with the Promotion of Independent-Living in Spain' (funded by the Spanish Ministry of Science and Competitiveness grant CSO2011-29749-C02-02). In an explicit gesture towards forms of 'emancipatory research' (Oliver, 1992), the EXPDEM project was conceived from the onset including different activist members as part of its advisory board and as co-researchers (Antonio, one of ETS's members being one of them). With this, we aimed to prevent researchers from 'speaking for the other' (cf. Ruby, 1992) and to create instances of friction and shared reflection.

6 These issues were articulated in *RedesCTS*, the STS space/network where many ANT-oriented researchers in the Spanish context regularly met. For a polyphonic account of it, check the collective video, produced as a report of the 2013 ¿Y si no me lo creo? | *What if I don't buy it?* meeting in Barcelona, available at https://redescts.wordpress.com/archive/

7 Indeed, in this we also felt the inspiration of the 15-M movement, which had entailed in a brutal explosion of forms of documentary 'open-sourcing' – in digital platforms, booklets, zines, blogs or how-to guides – whereby our cities were turned into 'infrastructures of apprenticeship' (cf. Corsín & Estalella, 2016).

8 After disassembling the notion of the social and interrogating the concept of movement, Latour has recently rearticulated the notion of social movement to put forward a politics that *moves* individuals into *active* engagement and dialogue (Latour et al., 2018).

References

Arenas, M., & Pié, A. (2014) Las comisiones de diversidad funcional en el 15M español: Poner el cuerpo en el espacio público. *Política y Sociedad*, *51*(1), 227–245.

Arribas Lozano, A. (2018). Knowledge co-production with social movement networks. Redefining grassroots politics, rethinking research. *Social Movement Studies*, doi: 10.1080/14742837.2018.1457521

Bal, R., Hendriks, R. & Bijker, W. (2004) 'Get real!' From scholarly work to recommendations: sailing between Scylla and Charibdis. Paper presented at 4S/EASST Conference, Paris, 26–28 August.

Brown, P. (1997). Popular epidemiology revisited. *Current Sociology*, *45*(3), 137–156.

Callon, M. (2008). Economic markets and the rise of interactive agencements: from prosthetic agencies to habilitated agencies. In T. Pinch & R. Swedberg (Eds.), *Living in a Material World: Economic Sociology meets Science and Technology Studies* (pp. 29–56). Cambridge, MA: MIT Press.

Callon, M., & Law, J. (1997). Agency and the Hybrid Collectif. In B. Herrnstein Smith & A. Plotnitsky (Eds.), *Mathematics, Science and Postclassical Theory* (pp. 95–117). Durham, NC: Duke University Press.

Callon, M., & Rabeharisoa, V. (2003). Research "in the wild" and the shaping of new social identities. *Technology in Society*, *25*, 193–204.

Callon, M., & Rabeharisoa, V. (2008). The growing engagement of emergent concerned groups in political and economic life: lessons from the French association of neuromuscular disease patients. *Science, Technology & Human Values*, *33*(2), 230–261.

Callon, M., Lascoumes, P., & Barthe, Y. (2011). *Acting in an Uncertain World: An Essay on Technical Democracy*. Cambridge, MA: MIT Press.

Casas-Cortés, M. I., Osterweil, M., & Powell, D. (2008). Blurring boundaries: recognizing knowledge-practices in the study of social movements. *Anthropological Quarterly*, *81*(1), 17–58.

Centeno, A. (2009). Redes de Conocimiento Emancipador. Taken from his blog, http://antoniocenteno.blogspot.com/2009/02/redes-de-conocimiento-emancipador.html

Corsín, A., & Estalella, A. (2016). Ecologies in beta: the city as infrastructure of apprenticeships. In C. B. Jensen, A. Morita, & P. Harvey (Eds.), *Infrastructures and Social Complexity: A Companion* (pp. 141–156). London: Routledge.

Cozzens, S. E. (1993). Whose movement? STS and social justice. *Science, Technology, & Human Values*, *18*(3), 275–277.

Criado, T. S. (2018). Note-taking: a 'fieldwork device' duplex. In *Allegra Lab*, http://allegralaboratory.net/note-taking-a-fieldwork-device-duplex-collex/

Criado, T. S., & Cereceda, M. (2016). Urban accessibility issues: Techno-scientific democratizations at the documentation interface. *City*, *20*(4), 619–636.

Criado, T. S., & Estalella, A. (2018). Introduction: Experimental collaborations. In A. Estalella & T. S. Criado (Eds.), *Experimental Collaborations: Ethnography through Fieldwork Devices* (pp. 1–30). New York/Oxford: Berghahn.

Criado, T. S., & Rodríguez-Giralt, I. (2016). Caring through design? En torno a la silla and the 'joint problem-making' of technical aids. In C. Bates, R. Imrie, & K. Kullman (Eds.), *Care and Design: Bodies, Buildings, Cities*. (pp. 198–218). London: Wiley-Blackwell.

Criado, T. S., Rodríguez-Giralt, I., & Mencaroni, A. (2016). Care in the (critical) making. Open prototyping, or the radicalisation of independent-living politics. *ALTER - European Journal of Disability Research*, *10*, 24–39.

García-Santesmases Fernández, A., Vergés Bosch, N., & Almeda Samaranch, E. (2017). "From alliance to trust": Constructing Crip-Queer Intimacies. *Journal of Gender Studies*, *26*(3), 269–281.

Gomart, E., & Hennion, A. (1999). A sociology of attachment: music amateurs, drug users. In J. Law & J. Hassard (Eds.), *Actor-Network Theory and After* (pp. 220–247). Oxford: Blackwell.

Hamlett, P.W. (2003) Technology theory and deliberative democracy. *Science, Technology, & Human Values*, *28*(1), 112–140.

Hess, D., Breyman, S., Campbell, N., & Martin, B. (2008). Science, technology, and social movements. In E. J. Hackett, O. Amsterdamska, M. Lynch, & J. Wajcman (Eds.), *The Handbook of Science and Technology Studies, Third Edition* (pp. 473–498). Cambridge, MA: MIT Press.

Jasanoff, S. (2005). *Designs on Nature: Science and Democracy in Europe and the United States*. Princeton, NJ: Princeton University Press.

Latour, B. (1987). *Science in Action: How to Follow Scientists and Engineers Through Society*. Cambridge, MA: Harvard University Press.

Latour, B. (2002). Morality and Technology. The End of the Means. *Theory, Culture & Society*, *19*(5–6), 247–260.

Latour, B. (2007). Turning Around Politics: A Note on Gerard de Vries' Paper. *Social Studies of Science*, *37*(5), 811–820.

Latour, B., Milstein, D., Marrero-Guillamón, I., & Rodríguez-Giralt, I. (2018). Down to earth social movements: An interview with Bruno Latour. *Social Movement Studies*, doi: 10.1080/14742837.2018.1459298

Marres, N. (2007). The issues deserve more credit: pragmatist contributions to the study of public involvement in controversy. *Social Studies of Science*, *37*(5), 759–780.

Marres, N., & Lezaun, J. (2011). Materials and devices of the public: an introduction. *Economy and Society*, *40*(4), 489–509.

Martin, A., Myers, N., & Viseu, A. (2015). The politics of care in technoscience. *Social Studies of Science*, *45*(5), 625–641.

Martín Sáinz de los Terreros, J. (2018). Welcoming sound: the case of a noise complaint in the weekly assembly of el Campo de Cebada. *Social Movement Studies*, doi: 10.1080/14742837.2018.1456328

Melucci, A. (1989). *Nomads of the Present. Social Movements and Individual Needs in Contemporary Society.* London: Hutchinson.

Mol, A. (2008). *The Logic of Care: Health and the Problem of Patient Choice.* London: Routledge.

Moreno-Caballud, L. (2015). *Cultures of Anyone: Studies on Cultural Democratization in the Spanish Neoliberal Crisis.* Liverpool: Liverpool University Press.

Murphy, M. (2012). *Seizing the Means of Reproduction: Entanglements of Feminism, Health, and Technoscience.* Durham, NC: Duke University Press.

Oliver, M. (1992). Changing the social relations of research production? *Disability, Handicap & Society*, *7*(2), 101–114.

Postill, J. (2018). *The Rise of Nerd Politics: Digital Activism and Political Change.* London: Pluto Press.

Puig de la Bellacasa, M. (2011). Matters of care in technoscience: Assembling neglected things. *Social Studies of Science*, *41*(1), 85–106.

Puig de la Bellacasa, M. (2017). *Matters of Care: Speculative Ethics for a More Than Human World.* Minneapolis: Minnesota University Press.

Rabeharisoa, V., Moreira, T., & Akrich, M. (2014). Evidence-based activism: Patients', users' and activists' groups in knowledge society. *BioSocieties*, *9*(2), 111–128.

Rodríguez-Giralt, I. (2011). Social movements as actor-networks: Prospects for a symmetrical approach to Doñana's environmentalist protests. *Convergencia*, *18*(56), 13–35.

Rodríguez-Giralt, I., Marrero-Guillamón, I., & Milstein, D. (2018). Reassembling activism, activating assemblages: An introduction. *Social Movement Studies*, doi: 10.1080/14742837.2018.1459299

Roelvink, G., St. Martin, K., & Gibson-Graham, J. K. (Eds.). (2015). *Making Other Worlds Possible: Performing Diverse Economies.* Minneapolis: Minnesota University Press.

Ruby, J. (1992). Speaking for, speaking about, speaking with, or speaking alongside: An anthropological and documentary dilemma. *Journal of Film and Video*, *44*(1–2), 42–66.

Shakespeare, T. (2006). *Disability Rights and Wrongs.* London: Routledge.

Sismondo, S. (2008) Science and technology studies and an engaged program. In E. J. Hackett, O. Amsterdamska, M. Lynch, & J. Wajcman (Eds.), *The Handbook of Science and Technology Studies, Third Edition* (pp. 13–31). Cambridge, MA: MIT Press.

Stengers, I. & Pignarre, P. (2011) *Capitalist Sorcery: Breaking the Spell.* New York: Palgrave Macmillan.

Zuiderent-Jerak, T. (2015). *Situated Intervention: Sociological Experiments in Health Care.* Cambridge, MA: MIT Press.

Zuiderent-Jerak, T., & Jensen, C. B. (2009). Editorial introduction: Unpacking "Intervention" in science and technology studies. *Science as Culture*, *16*(October 2012), 37–41.

35

How has ANT been helpful for public anthropologists after the 3.11 disaster in Japan?

Shuhei Kimura and Kohei Inose

After the 3.11 disaster hit Japan

At 14:46 JST on March 11, 2011, an earthquake of magnitude 9 occurred off the Pacific coast of Japan. It led to gigantic tsunamis, which swallowed a wide range of the coastal landscape of the country, and then caused explosions and meltdowns at three reactors of Tokyo Electric Power Company (TEPCO)'s Fukushima Dai-ichi Nuclear Power Plant (F-1). This triple disaster killed more than 15,000 people mainly in the Tohoku region and, by releasing massive amounts of radiation (the second level 7 accident on the International Nuclear Event Scale after Chernobyl), damaged the planet in a multifaceted way. With its direct and indirect damage, the 3.11 disaster is one of the worst in the Anthropocene.

Not surprisingly, it is highly troublesome to demarcate the disaster. As historians have stressed, March 11 is not the date when it was created: This accident has deep roots in Japan's history of modernisation. Although it was the accident of F-1 that brought the dame to the fore, it is severely difficult to cut the network of the responsibility for it, legally or scientifically (Strathern 1996). A comment from the chairperson of the independent investigation commission in the National Diet is symptomatic: The fundamental causes of the nuclear accident 'are to be found in the ingrained conventions of Japanese culture' (The National Diet of Japan Fukushima Nuclear Accident Independent Investigation Commission 2012). Also, it should be noted, the 3.11 disaster is still unfolding. Most of the affected communities have not recovered. One hundred and thirty-six people who died in 2016, for example, were officially recognised as 'disaster-related deaths,' that is indirectly caused by the 3.11 disaster. Moreover, its long-life radiological impacts on humans and non-humans on the planet are largely still uncertain.

Despite this, the political actions to make the disaster a thing of the past are compelling in Japan. Just nine months after the explosion, Prime Minister Noda (DPJ)[1] declared officially the 'convergence' (*shūsoku*, a bureaucratic, euphemistic expression describing the demise) of the Fukushima accident. In October 2013, hoping to host the Olympic Games in 2020, Prime Minister Abe (LDP) announced to the world that the Fukushima radioactive water was 'under control.' Domestically, the government strongly encourages evacuees to return home according to their plan, ignoring dissenting voices.

Against the government's bureaucratism, the momentum of social change, a temporary alignment of diverse groups under a banner of 'anti-nukes,' could not bear fruit. Gradually, conflicts between different knowledge and agendas, small and large, came to the surface. Intense debates about the prevention of health risk, energy policy, what people outside Fukushima should do and so on made citizens weary. While some kept resisting the government's high-handed reactions, others overcame (or threw away) their political distrust through the 'change' of the government at the first national election after 3.11, ironically, from DPJ to LDP.[2] In terms of political atmosphere, Japan quickly seems to have returned to the same state before the accident, or, dare we say, become worse. Under the dominating, 'restorative' LDP administration, individuals become nervous about being seen as a 'politicized' person, who openly expresses their opinions on 'Fukushima' or other sensitive social issues. Public discussions about 'Fukushima' are thus becoming less visible, and different perspectives on the issue are left incommensurate. Each camp gains ground from different information and knowledge from various sources, factual or fake, and evaluates others' words and deeds from their (inflexible) standpoint. They are like different ontologies: While there *is* health damage from low-dose exposure in a world, there is *no such thing* in another. In this world(s), coordination between them (Mol 2003) can no longer be expected.

In this chapter, we retrace emerging networked engagements with the sociopolitical state of things triggered by the 3.11 disaster in Japan.[3] It is not an objective study of radiation after the accident; it looks more like storytelling. A key task of public anthropologist is, we believe, to gift a slow, ethnography-based story to the public to cope with the given reality, and hopefully, to help bring forth a better reality. As shown later, our approach benefits from ANT's basic ideas: A fact is realised with a network of human and non-human actors; non-humans can play as an important role as humans; the reality of a fact, the identity of an actor or the meaning of the event is relational to its association with other (Latour 1987, 2005). In this chapter, we do not pursue 'normative' actions of ANT to open the black box of a fact and to follow the process of its closure. Instead, we attempt to display a network's potential to expand as a way to resist the hasty closing of the black box.

Repairing the reality

In the aftermath of the earthquake, tsunami and the nuclear accident, major parts of Japan's social infrastructures of information, transport and energy were temporally paralysed, splitting the country into (roughly) three categories with significantly different experiences of the early phase of the disaster: Those directly affected by the earthquake, tsunami and/or the radiation leak; those who felt their lives were in danger of possible radiation exposure and those who could manage to live ordinary life away from the epicentre geographically and/or psychologically. Both of the authors of this chapter fell into the second category. For several weeks after March 11, we were so frustrated, prevented from knowing what was really going on. Tense reports and miscellaneous predictions (optimistic and pessimistic, realistic and exaggerated) regarding the Fukushima power plants were continuously aired. Without adequate measuring devices, radiation is silent and invisible. Even with them, it is highly difficult to predict its behaviour and consequences.[4] We felt as if our reality dissolved in the extreme uncertainty.

The initial reactions were to 'repair' this collapsed reality by visualising what was going on.[5] Morita, Blok and Kimura (2013) articulated an ad hoc public which assembled a civic radiation monitoring map by using freely available software. On March 13, a video director in central Japan called via twitter for people's collaboration in building a Google Map covering radiation monitoring posts all over the country.[6] The map soon attracted followers, over time becoming one of the most important tools for citizens concerned by radiation risks to take stock of the situation. The map came to serve as the infrastructural node of an emerging sociotechnical network, formed through the ad hoc collaboration of 'amateur' citizens with no formal scientific credentials, but with practical knowledge gained through collective learning processes.[7] This civic map gradually came to structure what people 'saw' in their immediate surroundings (Fortun 2004) and came to play an important role in supporting people's assessments and actions. In other words, it became an important 'cartographic' tool for citizens to restore some sense of *grip* on an otherwise fractured reality. With the map, people gained a (temporary) means not only for knowing the real-time radiation levels in (or near) one's locality, but also for assessing the plausibility and credibility of ('reassuring') official press releases and ('catastrophic') circulating rumours.

While with his colleagues Kimura followed an ad hoc, online network, Inose describes his own attempts to 'repair the reality' on his 'welfare farm (fukushi nōen)' (2015). As a part of a sustained social movement concerning disability in Inose's hometown of Saitama,[8] a relatively big city 25 km north of Tokyo, a group in which he participates has run a 'welfare farm' where disabled people, volunteers, students, children and parents, and farmers have worked together for about a decade. By exposing their vegetables to radiation, the 3.11 disaster endangered the group's activity. On March 15, a monitoring post in Saitama measured a radioactivity level of 1 μSv/hr, 40 times higher than usual. They were unprepared since Saitama is more than 200 km away from F-1 and got worried whether their vegetables were safe or not. Although a young researcher of soil science and volunteer named Ishii shared with other members his knowledge about the radiation, it was too abstract and complicated for them to be reassured. Some even distrusted his 'scientific,' numerical explanation.

Yet, Ishii did not give up. He reviewed the relevant literature and participated in various activities such as a visit to Chernobyl organised by the agricultural cooperative of Fukushima. Finally, he launched an experiment on the transfer of radioactive materials from the soil to a plant in a town in Fukushima. His experiment was, retrospectively speaking, a re-examination of how multiple actors, human and non-human, are related to each other in the complex process called agriculture, and a struggle to reconnect the relationships in a more resilient way. Consequently, it led to his (re)discoveries including the effectiveness of traditional mix cropping in keeping lower the radiation dose of the crops. What is crucial is that he could share the findings of his experiments with other members of the welfare farm this time, by providing it as a story, where materials and places otherwise alien – vegetables in the welfare farm, caesium, traditional farming, Saitama, Fukushima and Chernobyl – meet. In a sense, Ishii gifted his fellows an actor-network story! Resonating with their own story of the disability movement in Saitama, his story touched the group members, providing them with a touch of reality and clue how to live with it.

Confronted with the harsh 'fact' of nuclear accident, some might have thought that the intellectual game of ANT was over. Yet as demonstrated earlier, with ANT, we can capture the emergence of local efforts of assembling human and non-human actors that 'patch' a collapsed reality with a temporary tool (map or story) created by themselves.

Remaking a reality[9]

Let's turn to our next story. Kashiwa is another commuter town 50 km northeast of Tokyo. As with Saitama, a rapid influx of newcomers after the Second World War created friction with natives. To mediate this and to animate the city after the burst bubble economy, in the late 1990s, the local chamber of commerce held a town planning course (*machi zukuri juku*) for young residents. A mixed group of newcomers, sons of local farmers, owners of the shops at the mall and so on joined in. When the course was completed, this group formed a voluntary organisation called 'Street Breakers' (hereafter SB) and started hosting social events to revitalise the central district of the city. Among their successful projects are those of supporting street musicians and managing a monthly street market of local goods. Along with the street market, local vegetables gained popularity with local restaurants as well as health-conscious Kashiwans. The concept of local production for local consumption (*chisan chishō*) was gradually taking root in Kashiwa.

The 3.11 disaster ruined this optimistic mood. At the early stages, people in Kashiwa, more than 200 km away from the F-1, kept watching. Yet in April, unexpectedly, it turned out that there were many hot spots of radiation in the area including Kashiwa. Repeated again and again in media coverage, the name of Kashiwa became a symbol of hot spots away from Fukushima. Although many Kashiwans felt disturbed, their concerns and reactions were not identical: Mothers were anxious about their children's health; farmers about the shipment of their vegetables; municipal officers about population outflow and so on. Worried mothers promptly launched a campaign for collecting signatures to make the city government prevent schoolchildren's radiation exposure, asking them, as part of increased food safety, not to use local vegetables for school lunches (cf. Sternsdorff-Cisterna 2015). Their action made the farmers feel abandoned because the aware mothers had been their best customers. In contrast, farmers could not take immediate action. Individual farmers hesitated to calculate the radiation levels of their land or vegetables, since a high score would constitute, they supposed, objective evidence to stop the shipment of overall agricultural products from Kashiwa. Therefore, they attempted to keep the score 'unknown' (cf. Petryna 2002), which, in turn, disturbed mothers. In this tense situation, the municipal government did not actively intervene 'to stay out of trouble.'[10]

As a heterogeneous civic group, SB gradually took actions to change the stressful, hostile atmosphere. At the end of July, 2011, they hosted a closed roundtable, inviting concerned citizens. A small group of people including local farmers of direct sales, consumers-residents including the signature campaigners, and the owner of a farmer's market, sat down at the discussion table. Yet, not surprisingly, in an oppressive air, everyone attempted to probe others participants' thoughts without offering their own. The only thing the first roundtable could decide on was its general rules: To hold monthly meetings, not to discard any opinions and to adopt the rule of unanimity for future actions.

SB then invited Takahashi, manager of an IT company, to join the roundtable. They thought he would be interested in it, knowing he had personally imported dosimeters. As a business person, his position was different from the farmers or consumers, but he was also a concerned resident attached to Kashiwa. At the second roundtable, his dosimeters attracted both sides' interest: He (or his instruments) came to mediate between them. During the summer of 2011, he devoted himself to mastering these devices, resulting in launching a private dosimetric service for foods at a low price in October.

In November, Igarashi, a sociologist member of SB, shared with local farmers the results of a consumer questionnaire about Kashiwa vegetables that he had conducted with the parents of preschool children. It suggested that if the farmers show diligent effort to provide appropriate information about radiation, health-conscious consumers in Kashiwa would be strong supporters of local agriculture. His findings gave them a glimmer of hope. The atmosphere of the roundtable improved gradually and mutual trust among the members was fostered.

To reconstruct face-to-face relationship between consumers and farmers in Kashiwa, SB embarked on a dose measurement project. It is easy to predict that a design for a meaningful test for both farmers and consumers is extremely difficult. If air dose can be measured to some degree by a portable instrument on site, the testing of soil and crops requires a sample inspection by a large machine in a laboratory. While a rigorous test (such as an inspection of every crop in a laboratory) would require too much time and money, a simple test on site is not accurate enough to convince consumers that the crops are safe. In addition, the total mechanism of radiation transfer between the soil, water and crop was too complex for SB members to identify. As a trial, they measured the variation of the air dose in several spots in one crop field and compared the scores with the radiation level of crops grown there, and found that there was no direct relation.

After a long discussion, they decided to inspect the soil dose, as a substitute for total inspection, accepting that there was a rough correlation between the radiation level of crops and the soil (not the air) where it grew. Another breakthrough was made by Takahashi's comment: 'LB200 is portable and we can use it for on-site inspection.' By then, he had obtained different types of dosimeters. LB200 was one of them. Although it was a rather simple measuring device running on an electric battery, Takahashi realised, the LB2000 could be effective in Kashiwa in determining the location of relatively contaminated soil. Yet, no one except Takahashi thought that a 25-kilogram instrument was portable. Thus, SB established a two-step test of radiation level. First, they would measure on site the radiation level of the soil sample obtained from the five points (four corners and the centre) of a crop field with the 'portable' instrument with moderate accuracy. After finding the highest radiation point of the field, they would then rigorously test the radiation level of the crop taken from that point in a laboratory. They also independently established the standard for shipment: less than 20 bq/kg.[11] What was crucial was that it was not a value decided by outsiders like scientists or bureaucrats, but the result of their – a diverse group of Kashiwans' – own discussions. It was a private but deliberate and sharable standard. 20 bq/kg is a criterion acceptable for most (if not all) health-conscious consumers, settable as a realistic goal for local farmers, and a detectable amount that is both technically and financially feasible for the SB.

The members of the roundtable named this two-step inspection as 'My Farmer Method' (*mai nouka houshiki*), suggesting the importance of building personal connections between farmers and consumers. As they released the results of the inspections publicly via the Internet, the anxiety in Kashiwa concerning local vegetables was alleviated. Additionally, a substandard soil or vegetable no longer meant merely the presence of urgent risk. Instead, it offered a tangible, viable target that encourages Kashiwans' collaboration in taking countermeasures. Obviously, such an inspection never ameliorates the damage from the 3.11 disaster, nor makes clear the entire mechanism of the radioactive damage. However, through the pragmatic and democratic process to implement the two-step inspection, Kashiwans could redirect their attention from present uncertainties into their liveable future.

The spontaneous 'citizen science' emerging from a commuter town suggests an 'art of living' with uncertainty (cf. Tsing 2015), if not a solution for an Anthropocenic disaster. Here, ANT's ideas help us elucidate what is going on and narrate it as a story. With time and care, SB has reassembled human and non-human actors, based on their attachment to the town. Tools, numbers and affect mediate their collaboration. Scientific knowledge is vital for its development, but never eliminates uncertainty related to radiation. Personal stories are fragile, but can create a temporal alliance among heterogeneous actors, bracketing uncertainty. SB was willing to open the network to new actors and to explore potential of each actor. The capability of the actors is relational: For example, in the network, Takahashi is not merely a manager of an IT company nor is LB2000 merely a dosimeter running on an electric battery. In this way, Kashiwan network expands, bringing change in their perspective regarding the landscape they are facing with (cf. Waterson and Cardwell, this volume).

Towards a more public network

As time passes by, visible traces of the 3.11 disaster are rapidly covered or erased, sometimes by force. The Anti-Nuclear Tents, set up by citizens in September 2011 in front of Ministry of Economy, Trade and Industry, have served as a symbol of anti-nuclear movement for years. At midnight on a Sunday in 2016, they were removed forcibly but silently by the government, without citizen's stiff resistance and criticism. The removal was symbolic in two senses: It embodied state actions to 'close the black box' or to make the disaster a thing of the past; and on the other hand, it epitomised the feeling of weariness among citizens these days. Now, the majority of Japanese people are for the government, passively or actively, as LDP's consecutive election victory suggests. As we explained earlier, public discussions on 'political' issues are becoming less visible. Resonating with neo-liberalism, difficulties related to the nuclear disaster are being considered personal rather than public issues. While thousands of people are still living as evacuees, Fukushima City is enjoying an economic boom generated by reconstruction projects.

In this situation, empowerment of the affected people is vital and necessary but yet not enough. The present (and perpetual) challenge for public anthropologists is to reach the public, who keep a silent distance from 'political' issues. For this purpose, with help from ANT, we follow the sequels to the stories we have narrated, attending to the importance of the relationship generated through eating. In fact, food has been a target of heated controversies after the 3.11 disaster. Food is vital for and, in our context, embodies the risks we take in our lives. As the above-mentioned cases suggest, foods unavoidably take part in the complex interactions between humans and non-humans, and are simultaneously themselves the outcome of such interactions (Mol 2008). Note that our emphasis is not on the binary relationship of predator/game or friend/foe that food may generate. Instead, we explore the physiological and moral relationship that would be created by the act of eating together, offering or sharing food, a relationship which can lead us to share a sense of living in the same world (cf. Tsing 2015). As Sternsdorff-Cisterna writes, 'It is an intimate act of ingestion, which connects the body to the outside world and connects givers and recipients of sustenance in affective relationship' (Sternsdorff-Cisterna 2015:464).[12]

During his struggle to reduce radiation in the soil, Ishii from Saitama encountered Igarashi from Kashiwa. Although neither of them were farmers, they had a deep concern about the crops they grew. Then, Komatsu, a young journalist from Iwaki, Fukushima, a coastal town near F-1, contacted Igarashi. When the disaster hit, Komatsu returned to his hometown and, as a spokesperson for a local fish product company, started transmitting 'on-site' information about fish, mainly via the Internet. In the subsequent years, fish and the fishery in Iwaki were endangered by not only radiation but also uncertainties it entails. Ironically, Komatsu's effort to provide accurate information was often accused of his 'biased' view via social media by those acutely aware of health damage or those against nuclear power. In the fragmented space of social media, it is difficult to bridge gaps between different already-formed circles. He changed his strategy.

Collaborating with Igarashi and local veterinarians working for the prefectural aquarium of Fukushima, Komatsu launched a project, The Sea Laboratory (*Umi rabo*) and The See Laboratory (*Shirabe rabo*) in 2015. They go out to the sea of Fukushima to catch fish and obtain marine soil, and then inspect their radiation level. He encourages local consumers and anglers to join. Thus on the boat of The Sea Laboratory is always a heterogeneous group. In The See Laboratory, a monthly public event they hold at the aquarium, the veterinarians report the result of inspection (although the values vary among species, they are generally low and show a steady decrease over time) and to eat together the fish they caught. In face of delicious-looking fish, people typically reach out their hand, regardless of their view on radiation. 'What convinces people is neither reason nor logic,' Komatsu assures, 'but pleasant taste.' Eating together fosters friendly relationships among professionals, citizens and fish. Pleasantness is crucial here, like Kashiwans' attachment to their hometown. Certainly, emphasis on taste risks descending into emotional propaganda, or naturalises the difference of social class (Bourdieu 1987). In this respect, of great importance is that, with a 'scientific' name, Komatsu builds his project around scientific research.

As a former journalist displaying his capabilities by conducting the radiation tests and holding social events, Komatsu associates with multiple actors, and changes the atmosphere of the politics slowly. In 2016, Komatsu started a second project, 'Place of Fish' (*Sakana no ba*), a monthly event to get people together to eat local fish and drink local *sake* in Iwaki. This project was inspired by Ohashi, whom Igarashi introduced to Komatsu. A former would-be theatre actor and graduate student who studied anthropology with Inose and Kimura, Ohashi runs a small bar located in a busy shopping street in Ueno, Tokyo. No, he does not run a bar but creates a stage, Ohashi tells us, where a heterogeneous group of actors, human and non-human, including seafood and *sake* from tsunami-affected areas, interact. With sympathetic interaction, a network of humans and non-humans expands. When a network expands, the actors unfold their potential accordingly, which, in turn, help the network incorporate new actors.

Conclusion

In this chapter, we have narrated emerging networked actions after the 3.11 disaster. As our cases show, in the network, the distinction of the actor (activist) and the observer can easily dissolve – a student of ANT becomes naturally a public anthropologist (cf. Criado and Rodriguez-Girault, this volume). And by expanding a network continuously, as Fortun (2001) suggests, we can keep things unsettled and resist the hasty closing of the black box.

In the beginning, we asked: How has ANT been helpful for public anthropologists after the 3.11 disaster in Japan? ANT helps us in several ways: To attend to local, down-to-earth activities as well as large-scale movements; to follow what was going on there; to narrate them not as a 'hero legend' nor as 'best practice' transferable to other sites, but as a story that affects the readers; and to understand the dynamism of how a network and its actors transform each other through interaction. Above all, in the aftermath of a catastrophe, ANT reminds us that writing is a way to contribute to bring forth a better reality.

Notes

1 Democratic Party (DPJ) was a seemingly reformist party, which existed from 1998 to 2016. Liberal Democratic Party (LDP) is conservative, the ruling party for most of the years since 1955 in Japan.
2 DPJ took the reins of government in 2009, criticising LDP government's political corruption. However, mass enthusiasm for 'change' turned quickly into frustration because of DPJ's indecisiveness. Although DPJ reformed little the governmental regime LDP had elaborated, LDP regained the reins in 2012, blaming the DPJ government for all the problems related to the 3.11 disaster.

3 Since space is limited, we have to omit a variety of the actions in and out of Japan, in which anthropologists are taking part (Gill et al. 2013; Naito et al. 2014; Manabe 2015).

4 In fact, real-time macro-behaviour was simulated and visualised by SPEEDI (System for Prediction of Environmental Emergency Dose Information), developed by The Nuclear Safety Division of the Ministry of Education, Culture, Sports, Science and Technology (MEXT). However, thousands of sheets of the simulation maps produced by SPEEDI were not used for evacuation nor even published for about two months.

5 Many attempts to contextualize the disaster or predict its future course based on prior cases were also made (but largely from those in the third category).

6 In the course of events, several experts – including atomic physicist Ryugo Hayano and volcanologist Hayakawa Yukio – transmitted technical knowledge via their twitter accounts. As a 'layperson,' however, the video director did not aim to transmit his own knowledge but instead used his account to build collaboration on the civic radiation monitoring map.

7 'Amateur' obviously carries no derogatory connotations here, but simply points to the fact that engaged citizens were not 'professional' (i.e. trained) radiation monitoring experts.

8 During the rapid economic growth of Japan after the Second World War, Saitama developed as a commuter town of Tokyo (cf. Robertson 1991). In the dynamic interaction between natives and newcomers, disability emerged as a critical social issue. In the 1970s, a group of people with and without disabilities launched an action to make the society more inclusive. They resisted local practices of shutting family members with disability in their houses and acted against national education policy separating classes for disabled students and able-bodied students.

9 This section is written based on Igarashi et al. 2012.

10 At around the same time as the roundtable, the mayor of Kashiwa expressed his opinion on his personal blog, which was heavily criticized since it revealed his insufficient understanding of radiation. He apologized and promised to address the radiation issue seriously. Subsequently, the municipal office started to calculate the radiation of local vegetables.

11 National standard prescribed by Ministry of Health, Labor and Welfare in 2012 is 100bq/kg (cecium-137).

12 Foods also interact with radiation and, as a result, can transform at a molecular level. Such a micro level transformation can cause unexpected effects on living organisms, and the relationship among them. Thus, eating (contaminated) foods would pose additional questions such as: Is an individual human or non-human actor to be treated as an actor, or as a collection of actors (i.e. genes)? How can we follow them? By taking seriously radioactive substance as an actor, what alternative political perspective will be opened?

References

Bourdieu, Pierre 1987. *Distinction: A Social Critique of the Judgement of Taste.* Richard Nice (trans.), Harvard University Press.

Fortun, Kim 2001. *Advocacy after Bhopal: Environmentalism, Disaster, New Global Orders.* The University of Chicago Press.

——— 2004. Environmental Information Systems as Appropriate Technology. *Design Issues* 20(3): 54–65.

Gill, Tom, Brigitte Steger, and David H. Slater (eds.) 2013. *Japan Copes with Calamity: Ethnographies of the Earthquake, Tsunami and Nuclear Disasters of March 2011.* Peter Lang.

Igarashi, Yasumasa et al. 2012. *Minna de kimeta 'anshin' no katachi: posuto 3.11 no 'jisan jishō' o sagashita Kashiwa no ichinen* [Determining 'Safety' Together: One Year in the 'Eat Local' Movement in Post-3.11 Kashiwa]. Aki shobō.

Inose, Kohei 2015. Living with Uncertainty: Public Anthropology and Radioactive Contamination. *Japanese Review of Cultural Anthropology* 15: 141–150.

Latour, Bruno 1987. *Science in Action: How to Follow Scientists and Engineers through Society.* Harvard University Press.

——— 2005. *Reassembling the Social: An Introduction to Actor-Network-Theory.* Oxford University Press.

Manabe, Noriko 2015. *The Revolution Will Not Be Televised: Protest Music after Fukushima.* Oxford University Press.

Mol, Annemarie 2003. *The Body Multiple: Ontology in Medical Practice.* Duke University Press.

Morita, Atsuro, Anders Blok, and Shuhei Kimura 2013. Environmental Infrastructures of Emergency: The Formation of a Civic Radiation Monitoring Map during the Fukushima Disaster. Richard Hindmarsh (ed.) *Nuclear Disaster at Fukushima Daiichi: Social, Political and Environmental Issues*, pp. 78–96. Routledge.

Naito, Daisuke, Ryan Sayre, Heather Swanson, and Satsuki Takahashi (eds.) 2014. *To See Once More the Stars: Living in a Post-Fukushima World*, New Pacific Press.

Petryna, Adriana 2002. *Life Exposed: Biological Citizens after Chernobyl*. Princeton University Press.

Robertson, Jenifer, 1991. *Native and Newcomer: Making and Remaking a Japanese City*. California University Press.

Sternsdorff-Cisterna, Nicolas 2015. Food after Fukushima: Risk and Scientific Citizenship in Japan. *American Anthropologist* 117(3): 455–467.

Strathern, Marilyn 1996. Cutting the Network. *The Journal of the Royal Anthropological Institute* 2(3): 517–535.

The National Diet of Japan Fukushima Nuclear Accident Independent Investigation Commission 2012. *The official report of The Fukushima Nuclear Accident Independent Investigation Commission*, The National Diet of Japan.

Tsing, Anna Lowenhaupt 2015. *The Mushroom at the End of the World: On the Possibility of Life in Capitalist Ruins*. Princeton University Press.

36

How to move beyond the dialogism of the 'Parliament of Things' and the 'Hybrid Forum' when rethinking participatory experiments with ANT?

Emma Cardwell and Claire Waterton

Section 1: Thinking about participation with ANT

In *Reassembling the Social*, Bruno Latour (2005a) describes the defining moment of ANT as that when three social theorists (John Law, Michel Callon and himself) each managed to persuade others that three objects formerly understood to be non-social (respectively: reefs, scallops and microbes) could be described as *associated* with social entities.[1] The concept of association inspired detailed redescriptions of such objects, in ways which showed that 'they *make* others do unexpected things' (2005a: 106). The work of the analyst, as these authors saw it, was to get closer to entities that were traditionally seen as non-social, follow them and trace, in minute detail, their relations, translations and mediations. In order to do this, the very meaning of the social and the natural would need to be 'dissolved simultaneously' (Latour 2005a: 109).

These early iterations of ANT worked primarily to produce better accounts of how *scientific* knowledge was made: The socio-material practices that went into creating indisputable 'facts' which were accepted as existing 'out there' in an ontological world that existed independent of human agency. This core orientation has significant implications for experiments in participation. In a world where scientific and political controversies are often inseparable (e.g. climate change, GM foods, avian flu, tobacco, human reproduction, etc.), attempts to unsettle 'natural' facts are vital to the tricky task of participating in contemporary democratic politics.

In this chapter, we will outline the basics of an ANT approach to participation. We begin with a discussion of three key texts: Callon, Lascoume and Barthe's (2009) *Acting in an Uncertain World*, Latour's (2004) *Politics of Nature* and Latour's chapter (2005b) *From Realpolitik to Dingpolitik* (published in Weibel and Latour's 2005b *Making Things Public: Atmospheres of Democracy*). These texts spark readers to reimagine the spaces, procedures and philosophies in play around technoscientific controversies in public life. The insights of ANT, their authors believed, could address the proliferation of such controversies, and contribute to better democratising democracy. Through subtly different but overlapping argumentation, they advocated:

-   intervening in technical and scientific knowledge-making;
-   opening out the idea of expertise through the championing of different understandings;
-   reconfiguring lay and expert knowledges in research;
-   building new evidence;
-   creating 'spaces' or 'collectives' where people and things might gather together to discuss, differ and dispute; and
-   supporting an explicitly experimental mode of engagement around controversy.

In each of the texts, the authors established specific ideas about how to achieve the aforementioned. We outline these in section 1. Then, in section 2, we explore the performance of these ideas 'in the wild' through an experimental participatory collective formed around blue-green algae in a lake, Loweswater, in Cumbria, Northern England. After that, in sections 3 and 4, we query what a subsequent project, which looks at how farmers make decisions about the management of cattle slurry,[2] can tell us about moving beyond these approaches in rethinking participatory experiments in the spirit of ANT.

Science in the wild: hybrid forums

We begin our whirlwind tour of an ANT approach to participation with the ideas of Callon, Lascoumes and Barthe (2009) in *Acting in an Uncertain World*. Here, Callon and colleagues advocate for widening participation in the socio-material construction of scientific knowledge via 'hybrid forums.' Participation is seen by these authors as an appropriate response to ontological uncertainty and scientific controversy. They advocate a 'technical democracy' – one where political institutions are expanded and improved, to open out controversies through inclusive, public collaborations in knowledge-making and research.

A *hybrid forum* (henceforth HyFo) is a group that comes together to address a problem. The HyFo is *hybrid* both because the members are heterogeneous (scientists; politicians; interested laypersons), and the problem is addressed in many different domains (technological; ethical; economic). It is a *forum* because it creates an open, public space for debate, from which no one (and no subject) is excluded. A HyFo's members explicitly recognise that knowledge creation – even knowledge creation concerning the natural world – is inherently a political act. That is, it involves a politics performed through (public) dialogue, rather than delegation to political professionals or elected officials. HyFos are therefore based on collective experimentation and learning. They scramble the division between laypersons and officially appointed experts. This makes the HyFo a 'normatively oriented space' and one that has specific 'dialogic procedures' (Callon et al. 2009: 161).

Both the socio-material construction of science and the political implications of this construction are explicitly recognised in the HyFo, and participation in knowledge creation is extended beyond the milieu of the expert. Science leaves the protected space of the laboratory and goes 'into the wild.' Clear-cut decisions – the arrival at a final and absolute truth – are not necessarily the desired outcome of the HyFo; instead, a favourable outcome is an ongoing 'series of rendezvous' (2009: 223), repeated meetings that make provisional decisions, open to change as information, circumstances or priorities change. In *Acting in an Uncertain World*, Callon et al. lay out guidelines for how the successful HyFo should be both executed and evaluated.

The participatory non-human: the Parliament of Things

Further contributions to ANT participatory thinking are found in Latour's (2004) book *Politics of Nature* and in extensions of his arguments there in the essay 'From Realpolitik to Dingpolitik or how to make things public,' written for the exhibition catalogue *Making Things Public* (Weibel and Latour 2005). In *Politics of Nature*, Latour proposes a 'new constitution' to do away with the 'illicit' divide of nature (as represented by Science – note the capital 'S') and society (aka Politics). This new constitution would allow participants in democracies to work towards a 'common world,' not split into society and nature, where both humans (society, the realm of politics) and non-human 'things' (nature, the realm of science) are afforded due political process. This, in other words, is a vision of an enlarged democracy in which the hybridity of nature-society is recognised. In his later elaboration of this argument, Latour suggested that 'Dings,' where 'the Ding designates *both those who assemble* because they are concerned *as well as what causes* their concerns and divisions,' needed to be brought back into politics (2005b: 8, our emphasis). Dings, Latour argued, should become the centre of attention in what he called an 'object-oriented democracy' (2005b: 8).

Dingpolitik, or the 'parliament of things,' involves a vital move whereby participants in democracies start to work around 'matters of concern' rather than 'matters of fact.' For Latour, a 'matter of fact' (henceforth MoFact) is firmly linked to a particular conception of 'Scientific knowledge.' Such 'Scientific knowledge' – an essential part of the post-enlightenment 'modern constitution' (Latour, 1993) – works by defining an objective nature. Latour deliberately gives this kind of 'Science' a capital 'S' to symbolise its power and to signal the way it closes down dissent and takes the place of politics in conventional decision-making fora. Latour suggests that such MoFact retain the power to render matter 'mute' and to make nature 'incontestable' (Latour, 2004:10). MoFact, in other words, work to stifle and deny politics, even in situations of controversy.

To address this, Latour puts forward a new concept: 'matters of concern'. To conceptualise things as matters of concern (henceforth MoConcern) is to allow for the transformation of MoFact, impotent as they are in their Scientific form, into participatory 'things' that are lively and have agency.

Latour's proposition is that new collectives are needed to allow us to move away from Science towards a situation where – similar to within the HyFo – both lay and expert practitioners might engage in the making of knowledge. These new collectives would not be silenced by indisputable scientific 'facts.' Rather, they would become politically active in MoConcern. Non-human 'things' would play a vital role (Latour 2004: 69), and be recognised as having the power to trigger 'new occasions to passionately differ and dispute' (Latour 2005b: 5).

In the Parliament of Things (PoT), participants are not required (as Scientists are) to leave their attachments, passions and weaknesses at the door in favour of an imagined objectivity. All scientific and political participants of collectives have the unavoidable 'disabilities' of weakness, passion and attachment, Latour argues; it is impossible to adopt a 'view from nowhere' (Nagel 1986), or speak on behalf of the world objectively. We are all inherently 'politically challenged' and democracy demands we recognise this. Latour rhetorically asks, 'Are we not, on the whole, totally disabled?' (Latour 2005b: 20), and urges participants to embrace this 'disability,' rather than chase the ideal of becoming perfectly eloquent, enlightened and objective disembodied thinkers.

Taken together, *Acting in an Uncertain World*, *Politics of Nature* and *Making Things Public* advocate a more open model for doing science, which is inclusive not just of a wider *human* demos, but also the vitality of the non-human material world – recognising that *matter matters* – in the production of scientific knowledge. Furthermore, all participants in the new knowledge-making collectives should acknowledge their own 'constitutive attachments' (Callon et al. 2009: 265) or 'disabilities' (Latour 2005b: 20) as essential to the work of democratising democracy.

So what does this mean in practice? In the next section, we describe how several simple principles drawn from these ANT ideas were adapted and utilised 'in the wild' (Callon et al 2009: 104), that is, within a specific participatory collective that emerged in Loweswater, Cumbria, in the north of England.

Section 2: ANT in the wild: a participatory experiment in Loweswater

In 2004, a group of three ecologists working with farmers around a lake that was exhibiting deteriorating water quality invited two sociologists of science (Claire Waterton and Jake Morris) to join their team. At that time, the Loweswater Lake was subject to increasingly frequent and unpredictable 'blooms' of cyanobacteria, or 'blue-green algae.' Cyanobacteria are organisms that depend on, and are 'limited by,' high nutrient loads within lake water, typically attributed to farm run-off of fertiliser or animal excrement. These algae can be poisonous to animals and sometimes humans, and the deteriorating quality of the lake, situated within the popular Lake District National Park, was becoming an issue of public concern.

The sociologists learned that, from the early 2000s onwards, scientists and environmental agencies had been monitoring the lake water using specified sampling methods to detect nutrients and to measure water flow. The data produced by this sampling would feed into a nutrient budget for the lake and contribute to a model that would enhance understanding of its thriving algal populations.

Asking the ecologists about the water quality sampling they had been doing, the sociologists heard about the use of standard techniques, the importance of replicability and the necessity that sampling was 'do-able' for scientists. Asking farmers about water quality and the possibility of monitoring pollution, however, brought up very different issues. Firstly, farmers questioned the possibility of deriving an accurate picture of catchment processes from samples that are taken only once every month. Secondly, farmers asserted that the ecologists' depth gauge was too far upstream of the lake to take into account the nutrient loading that four tributary streams were contributing. Thirdly, farmers spoke openly of the politics of monitoring. They knew that some of the issues in the lake concerned leaky farm infrastructures, but at the same time they felt that, for reasons of sensitivity, neither they nor the ecologists could realistically monitor these: '*You couldn't do samples outside of everyone's slurry tanks…*'

The controversy not only involved the politics of measurement. Whilst scientists were monitoring water quality under the observant eyes of farmers and local residents, cyanobacteria were 'blooming' with increasing frequency on the surface of the lake. When this happened, the surface of lake turned a lurid green and Loweswater farmers acutely felt the blame of those witnessing the unsightly blooms. They also felt the blame of the regulatory authorities – for example, the National Trust who owned the lake,[3] the Lake District National Park Authority charged with preserving the beauty and health of the Cumbrian lakes and fells, and the Environment Agency (EA) responsible for good water quality nationally.

This uncomfortable atmosphere of blame, rising in concert with the appearance of blue-green algae on the surface of the lake, was oddly associated with one of the most taken-for-granted trends in the UK countryside since the mid-19th century: The ongoing 'agricultural improvements' since the invention and widespread uptake of technologies such as field drains, mechanised farm machinery, artificial fertilisers and so on. Such improvements allowed for increases in the acreage of improved grassland on farms, increased livestock numbers per farm and an increase in the yields of pastures. What the algae began to vividly evoke, by the early 2000s, it was a sense of the complex and unintended ecological and social consequences of these agricultural improvements which, in turn, engendered a tangible atmosphere of dismay, anger, stigma and shame.

If a participatory forum was going to be developed around the issue of the blue-green algae at Loweswater, it needed to acknowledge entrenched positions involving hierarchies of knowledge and feelings of stigma and blame, and to provide a way to understand the human, non-human, economic, social, cultural, historical, emotional and political complexities in which all participants were implicated.

A new collective

The sociologists decided that a good way to respond to this situation might be to use some of the ANT participatory ideas outlined earlier. A new 'Loweswater Knowledge Collective' was proposed. This would be a PoT where it would be accepted that:

- nature is not self-evident;
- knowledge and expertise have to be debated;
- uncertainty is the main condition humans are in (rather than a condition of having knowledge);
- what is important is the creation of connections between people and things;
- doubt and questioning are extended to all our representations.

These principles were taken on board by participants from the very first full meeting of the new collective, which participants renamed the Loweswater Care Project (LCP) in June 2008.[4] Thereafter, evening meetings were held every two months, until December 2010. At each meeting, questions – about nature, knowledge, expertise, representations, uncertainty, etc. – were raised, and talks and investigations proposed by participants were given and discussed. Wide-ranging issues were brought into the forum, including: The spawning grounds of brown trout; the use of household detergents in the valley; a recent piece of legislation, the European Water Framework Directive; the maintenance of the banks of streams around the lake; the existence of well-functioning and less well-functioning septic tanks; changing rainfall patterns; changing patterns of, and futures for, farming in Loweswater; and unexpected ecological relations in the de-oxygenated water of the lake.

The simple principles referred to above were used to appraise and interrogate the large quantities of new information coming into the collective. These principles directly questioned the 'natural' and the 'factual,' thereby transforming MoFact into MoConcern. They allowed all participants to inquire, contest and unsettle established facts as a *bona fide* mode of engagement, freeing them from the previous necessity to assign responsibility and blame.

Participants became inquisitive: Nature and natural processes were no longer taken as self-evident; all claims to knowledge and expertise had to be debated; doubt and questioning, including doubt of scientific methods and monitoring, but also of lay claims and counterclaims, were encouraged and nurtured. The principles also helped the collective think about fact making and knowledge in more provisional ways, accepting the inevitability of uncertainty. They helped the collective build connections and create its own knowledge through a number of research investigations designed by participants themselves.

For example, drawing on the idea that 'nature is not self-evident,' participants invited a representative of the UK State's agency for water quality – the EA – to explain the classification of 'moderate ecological status' that had been designated for Loweswater. The designation of 'moderate' as opposed to 'good' ecological status implied that the lake has a water quality 'problem.' This invitation highlighted participants' weaknesses and attachments; they admitted that they had little insight into the way in which this complex classification had been drawn up, and were also aware of the threat that 'moderate ecological status' implied for farmers, as breaching environmental regulations carried punitive financial consequences.

The EA representative came to an evening meeting and described the way in which moderate status had been designated. She informed participants that four pieces of scientific evidence (regarding oxygen, plant life, algae and fish in the lake) were needed to make the designation. But as she came to the end of her talk, she admitted that the classification of 'moderate status' for Loweswater had been established with only three pieces of data, as the data on fish populations were missing. A heated debate ensued. A farmer intervened:

Farmer: '*Sounds like guesswork science to me...*'
The EA representative was taken aback: '*Sorry?*'
The farmer reiterated: '*Sounds like guesswork science to me...If one of these things [the four pieces of evidence] can knock it back to 'poor' and you can't count the fish, you haven't a hope, if fish is one of them!*'

The farmer questioned the very basis on which the official designation had been given. Many participants in the room backed his reasoning. Not only the farmers, but also the scientists, were vulnerable now, with the EA representative acknowledging that there were 'problems with the fish classification tool we are using.' The 'constitutive attachments' of different participants were visible to all. Following the farmer's questioning, participants learned, through fraught and difficult discussion, exactly what combination of nature and artifice was involved in designating the ecological status of Loweswater. In addition, investigating water quality as a MoConcern, as they did here, the collective also embraced the realm of emotion, feelings and affect.

Affect is a complex term, but here denotes the ways 'feelings' or 'emotions' are inherently and pre-personally immanent in the material world (Massumi 2002; McCormack 2003, 2010: 643). Affect is important. Discussions that took place in the collective were argumentative, passionate and moving. They exposed not only the vulnerabilities and attachments of participants but also their intelligence, inquisitiveness, creativity and sensitivity, and the emotional intensity of the forum had material consequences for the collective itself, the 'things' at its centre and the way in which its relationships were made.

Section 3: An omission in the parliament: can we learn to worry about slurry through theories of care?

The interactions of the LCP bring up an issue of vital importance for the dialogic experiment of the HyFo and PoT. Namely, that the attachments, weaknesses and passions that Latour encourages us to embrace on our entry into the parliament, just like things themselves, have agency. Embracing 'disability' is not simply a process of drily accepting that we all have bounded rationality. Disabilities have affective and relational consequences, and produce and demand particular configurations of labour and sensitivity from all members of the collective, having indelible effects on the potentiality of directions the collective may take.

Think, for example, of the unwillingness (mentioned in section 2) to monitor farms' individual slurry tanks, despite the widespread suspicion that these were important sources of nutrient pollution. Slurry – the liquid mixture of cattle excrement and water that arises as a result of certain kinds of cattle housing on beef and dairy farms – was considered to be one of several contributing factors to water enrichment in Loweswater, as it contains significant amounts of the nutrient phosphorus. However, the LCP never worked out precisely how to formulate questions around slurry, or how to investigate slurry production, storage and use as a fertiliser in the catchment. Within the pre-existing affective atmospheres of Loweswater, the toxic culture of blame that had been wrapped around the 'objective' presentation of water quality played an important part here. The desire of participants to move away from these negative relations, and the need to rebuild trust, played a significant role in the unfolding of the forum and the democratic decisions it made.

This sense of affect steered how some issues were, and others not, opened out for scrutiny in the LCP, because of the difficult affectual atmospheres they contained. As one ecologist participant reflected:

> It may be that there are things…like slurry tank management, things that we haven't directly challenged in this project that are a problem for some of the farmers. And you know, that is an issue.[5]

The forum, it seemed, could not apprehend slurry without re-energising the pernicious spectre of blame that would shut down discussion entirely.

Encompassing affect: from matters of concern to matters of care

How can we make sense of our finding that disability mattered in Loweswater and that participants' vulnerabilities had an effect? That attachments, weaknesses and passions didn't just allow for open debate, but drove debate in certain directions at the expense of others?

A useful resource for thinking with this is Maria Puig de la Bellacasa's (2017) *Matters of Care: Speculative Ethics in More Than Human Worlds*. Here, Puig de la Bellacasa proposes that we extend the movement from MoFact to MoConcern by further conceptualising controversies as *matters of care*. A matter of care (MoCare) weaves Latour's MoConcern into the rich history of thinking about care in feminist theory.

To care, Puig de la Bellacasa reminds us, is a verb; it is 'a necessary practice, a life-sustaining activity, an everyday constraint' (2017: 160). Care is unavoidable, an inherent part of the necessary yet mostly dismissed labours of the everyday maintenance of life. Care circulates unequally, and is not innocent: Its labours fall more heavily on some shoulders, be they human or non-human, than others (the night worker, for example, who cleans the academic's office, or the worms in soil that allow us to grow food). Caring for one thing can mean killing something else (as we dig up weeds in caring for gardens). As care is *always* happening, the important question we must ask is 'What worlds are being maintained and at the expenses of which others?' (Puig de la Bellacasa 2017: 44; see also Haraway 1994). Nothing holds together without a more-than-human chain of care, and those relations are deeply situated.

Circling back to Loweswater, can we understand the failure to open out the issue of slurry in the collective as a matter of care? Within the context of a historical situation of toxic blame, we can see participants attempting to balance care for the lake's environmental status with care for local farmers, so long villainised by Science. The need for the group to care for other things (such as farmers' emotional well-being) in order to ensure the parliament's maintenance and survival worked to render slurry 'mute' as a 'Ding.' Did this 'disable' the collective in the act of caring for the polluted lake? Does that mean the forum was ineffective? In the short-term timescale of 'traditional' political decision-making, yes. But as Callon et al. (2009) take care to mention, hybrid fora and their attachments and entanglements take time and are iterative. The LCP developed, in time, into another community-run participatory forum, the Loweswater Care *Programme*, working in partnership with the West Cumbria Rivers Trust. As part of this, significant progress was made on tacking 'point source pollution' from leaky slurry tanks in the valley.[6]

A parliament for slurry: fact, concern or care?

Slurry, then, was a troublesome 'Ding' in Loweswater, and in 2016, the current authors had the opportunity to encounter it again, via a call made by a partnership organisation, SARIC (the Sustainable Agriculture Research and Innovation Club), to study (deep breath) 'the provision of decision-support on organic slurry storage and treatment techniques to enhance nutrient use efficiencies.'

The assumption of the research call was that farmers needed to take better heed of scientific knowledge in their slurry management decisions. This was very much embedded in the 'modern constitution' of Nature/Science that Latour has critiqued. The 'issue' to be 'solved' was not slurry as a material entity, or the nature–culture practices entangled within it, but farmers' 'inefficient' and unscientific decisions, the 'fix' for which – a 'decision support' tool – was decidedly technical.

As we got to know the material-semiotic landscape in which the research project was positioned, it became apparent that a semi-stable assemblage existed around the slurry controversy. This consisted of many MoFact: Devices for establishing water quality status (such as those we met in Loweswater), agronomic science determining crop nutrient uptake, and policy guiding and enforcing farmers' activities. Importantly, this assemblage also included a huge number of experts (scientists, policymakers and agricultural advisors) labouring to improve the status of freshwaters in the UK, to care for the environment.

However, this assemblage was not achieving its intended aim, so tweaking was considered necessary. As one soil scientist put it: '*There are really good decision support tools out there, but a significant number of people don't engage…We need to identify the barriers to engaging with this information*.' Within this Science-based chain of care, the 'problem farmer' was the weak link, making poor decisions with slurry on farm. If the problem farmer would change, the assemblage could achieve their two priorities of maximising agricultural output and achieving 'good' environmental status. Our expected role as social scientists in the proposed project was to produce the knowledge necessary to manage this 'problem farmer,' in the same way that Science allows us to manage Nature.

Of course, this positioning was deeply problematic, particularly from an ethical point of view. We therefore attempted to open up space within the research (which we called 'Slurry-Max') to let farmers – and slurry – speak.[7] This process revealed a very different reality to that of the 'problem farmer' not understanding their 'valuable resource.' We learned that slurry was unpredictable, powerful and dangerous. There was usually too much of it when wet weather and ground conditions made it a pollutant, and not enough when they allowed it to be a fertiliser. Its qualities were variable, and didn't fit neatly into the 'book values' denoted by agronomic science (the averages that make up official estimates of the nutrient content of slurry, which had huge standard deviations). Farmers, far from being uninformed, understood this well, giving considered reasons as to why they couldn't follow the scientific advice. In short, 'the thing' of slurry, voiced through Slurry-Max's empirical work with the material itself and with those that were working daily with it,[8] turned out to be very different from the concept of slurry that was embedded within the science-policy assemblage and the SARIC research call.

Section 4: Democracy in the wild

How could ANT thinking on participation help the researchers in this scenario? It was clear from the beginning that we would not have the freedom to establish a slurry HyFo or PoT. And even had we wanted to, the huge national scale of the 'nutrient management' science-policy assemblage would make that problematic. 'Slurry' as a 'thing' at this scale is too much, too diffused and too varied to truly be granted due process – the 'Ding' loses the advantage of a position in the PoT. The HyFo and PoT, we found, can be difficult to achieve within the pre-existing assemblages – operating at pre-existing scales – which maintain our ongoing worlds.

So, must we give up on the idea of participation under circumstances where the democratic conditions of the HyFo or PoT cannot be achieved, and layperson 'publics' and non-human things are not procedurally afforded due process? Should the Slurry-Max researchers have turned away from ANT, and provided the assemblage with the means to 'tame' the problem farmer, as they had been asked to do?

Of course, the answer is no. But how can we move beyond the privileged space of the HyFo and PoT, to create the conditions for an expanded 'technical democracy' in the wild? We can find some answer to this by returning to *Matters of Care* (2017). Puig de la Bellacasa stresses that, unlike the HyFo and PoT, care is not a 'recipe' for doing our encounters (2017: 90), but rather a trope that we can use to think through a non-normative politics of knowledge. She writes that this is 'always specific; it cannot be enacted by a priori moral disposition, nor an epistemic stance, nor a set of applied techniques, nor elicited as abstract effect' (ibid: 90).

In the messy, imperfect fora in which most contemporary politics play out, such a standpoint can aid us in seeking a better democracy. If, as Puig de la Bellacasa suggests, we think of care as the ongoing, unavoidable labour whereby some worlds are maintained at the expenses of others, it is, of course, inevitable that we are constantly, in every act, *doing* politics. Considering this, we should perhaps move beyond thinking of participation as procedural politics, choreographed within bespoke 'hybrid fora,' and instead consider it as embedded within ongoing patterns of *taking part*. This implies that participation becomes a kind of unevenly distributed process of inclusion and exclusion, choice, care and labour, which is inherent to all our collective activities.

Applying this to the problem of slurry management, we can see that the proposed SARIC project expected the most of, and listened the least to, two actors: Small farmers and slurry. Our concern as social scientists was then to attempt to 'democratise democracy' by amplifying the voices of these actors in the fora of the assemblage. In Slurry-Max, the researchers attempted to achieve this by close interrogation of the existing assemblage of care around nutrient management, tracing which participants – both human and non-human – were in danger of being objectified or silenced, or bearing an unreasonable burden of care. Then, we attempted in our own actions of world-making to respectfully ameliorate these conditions, drawing attention to which worlds were being supported and made, and at the expenses of which others.

In this way, we acknowledge that we are acting in a world where fully democratic cosmopolitical parliaments are not always possible – or at least, are always inherently bounded. We suggest that one way forward could be to think beyond participation as a discrete idea – or as something that can absolutely and procedurally be 'achieved' – and instead, attempt to live, as world-makers, with inclusionary care at the centre of our doings.

Politics in the contemporary world can sometimes seem a million miles from the ideal of the more-than-human democracy of the PoT. But that doesn't mean we have to give up on the work of democratising democracy. In a parliament in which we are all partial and disabled, it's vital to remember that this means all participants require care. By thinking about participation as an active process of taking care within the parliaments we are already part of, we can work towards participatory futures beyond the privileged space of a HyFo, and, disabled as we are, strive for a better world for all.

Acknowledgements

The research reported in this paper was supported by two research grants: (1) Rural Economy and Land Use (RELU) project, 'Understanding and Acting in Loweswater: A Community Approach to Catchment Management,' RES-229-25-0008, 2007–2010; (2) NERC/SARIC (Sustainable Agriculture Research and Innovation Club) Translation Research Project, 'Slurry-Max: Holistic decision-support system for slurry storage and treatment techniques for maximum nutrient use efficiencies,' NE/P007902/1, 2016–2018.

We would like to thank Celia Roberts for her editorial support. We are also grateful to the organisers and participants of the ANT Handbook workshop held in Berlin in September 2017 for their help with the framing of this chapter.

Notes

1 The three studies are: Law (1984), Callon (1984) and Latour (1984). Les Microbes Guerre Et Paix; Suivi de, Irréductions (see p. 106 of Latour 2005).
2 Slurry is 'faeces and urine produced by housed livestock, usually mixed with some bedding material and some water during management to give a liquid manure with a dry matter content in the range from about 1%–10%' (Pain and Menzi 2003).
3 Technically, the National Trust owned the lakebed, not the water within it.
4 Participants voted to change the name of the collective to the 'Loweswater Care Project' (LCP). See www.lancaster.ac.uk/fass/projects/loweswater/noticeboard.htm for minutes of meetings of the LCP.
5 Ecologist Lisa Norton reflecting on the achievements and gaps of the Loweswater Care Project at the workshop, 'New Forms of Participatory Environmental Governance: Experiences and Challenges from Loweswater, Cumbria' Penrith, 3rd December 2010. www.lancaster.ac.uk/fass/projects/loweswater/noticeboard.htm.
6 https://westcumbriariverstrust.org/projects/the-loweswater-care-programme.
7 http://wp.lancs.ac.uk/slurry-max/.
8 Empirical research included: 11 in-depth farmer interviews, conducted on farm; three in-depth farmer interviews, conducted by telephone; focus groups with around 100 agriculture students in four agricultural colleges in England, Wales, Northern Ireland and Scotland; face-to-face survey questionnaires with 84 farmers at auction marts and agricultural shows in England, Wales and Northern Ireland; face-to-face survey questionnaires with 18 agriculture students in Scotland.

References

Callon, M., 1984. Some elements of a sociology of translation: domestication of the scallops and the fishermen of St Brieuc Bay. *The Sociological Review, 32*(1), pp.196–233.

Callon, M., Lascoumes, P. and Barthe, Y., 2009. *Acting in an uncertain world: An essay on technical democracy.* Cambridge, MA: MIT Press.

Haraway, D.J., 1994. A game of cat's cradle: science studies, feminist theory, cultural studies. *Configurations, 2*(1), pp.59–71.

Latour, B., 1984. *Les microbes, guerre et paix, suivi de Irréductions.* Paris: A.-M. Metailié La Découverte.

Latour, B., 1993. *We have never been modern.* Cambridge, MA: Harvard University Press.

Latour, B., 2004. *Politics of nature: how to bring the sciences into democracy.* Cambridge, MA: Harvard University Press.

Latour, B., 2005a. *Reassembling the social: an introduction to actor-network-theory.* Oxford University Press.

Latour, B., 2005b. From realpolitik to dingpolitik; Or, how to make things public. In: Latour, B, Weibel, P. (eds) *Making Things Public: Atmospheres of Democracy.* Cambridge, MA: MIT Press, pp. 14–43.

Law, J., 1984. On the methods of long-distance control: vessels, navigation and the Portuguese route to India. *The Sociological Review, 32*(1), pp.234–263.

Massumi, B., 2002. *Parables for the virtual: movement, affect, sensation.* Duke University Press.

McCormack, D.P., 2003. An event of geographical ethics in spaces of affect. *Transactions of the Institute of British Geographers, 28*(4), pp.488–507.

McCormack, D.P., 2010. Remotely sensing affective afterlives: the spectral geographies of material remains. *Annals of the Association of American Geographers, 100*(3), pp.640–654.

Nagel, T. 1986. *The view from nowhere.* Oxford University Press.

Pain, B. and Menzi, H., 2003. Glossary of terms on livestock manure management 2003. *RAMIRAN Network,* p.59.

Puig de La Bellacasa, M., 2017. *Matters of Care. Speculative Ethics in More Than Human Worlds.* University of Minessota Press.

37

How well does ANT equip designers for socio-material speculations?

Alex Wilkie

In conversation with Claire Parnet, Gilles Deleuze (2002: 2) evokes the image of the becoming of the orchid and the wasp: A 'double capture' whereby, and momentarily, each becomes a function of the other's doings. In considering actor-network theory and design practice, this image is appealing as it can be understood to stage two ways by which their practices can be seen to converge. In the first, each discipline captures something from the other. For orthodox scholars of ANT design becomes another empirical domain to describe. For designers, ANT becomes another explanatory resource with which to capture the social and warrant the arrival of the new. A second way of relating, recalling Isabelle Stengers' 'reciprocal capture,' a variant of double capture, opens up the possibility of the emergence of shared practices of empirical socio-material speculation where adding designed propositions to collectives brings into being new prospects and capacities to act. In what follows, and after a brief discussion of the possibility of ANT and design as a constructivist proposition, I outline how ANT can be understood to support design practitioners. I then draw on the case of an interdisciplinary project, involving designers and STS practitioners, to discuss how a workshop for exploring the use of smart energy monitors – and resourcing the design and deployment of an interactive research device amongst UK-based energy communities – involved practices of retroscription and procomposition. Here, retroscription characterises the visual and analytic evocation of the lived and situated experience of using smart monitors, whereas procomposition involves the mobilisation and re-patterning of retroscriptions into novel speculative compossibilities.

Reciprocal capture and speculative questions

For Stengers (2010: 36),

> we can speak of reciprocal capture whenever a dual process of identity construction is produced: regardless of the manner, and usually in ways that are completely different, identities that coinvent one another each integrate a reference to the other for their own benefit.

In her diagnosis of modern science, reciprocal capture provides a way of understanding the upshots of complex and dynamic relations between located scientific/research (knowledge) practices, in what she dubs an 'ecology of practices' (ibid. 32) where the interdependent combination of the political and the scientific produces new modes of existence and evaluation. In the context of science, reciprocal capture describes how novel (scientific) practices (and entities) arise out of the becoming together, rather than coming together, of other practices and how such practices present themselves to one another.[1] The term practice is noteworthy, here, as it underscores the inherent immanence of research, knowledge and its (human and more-than-human) participants. Perhaps of more importance is Stengers' insistence on the specificity – or constraints – of practices, where, rather than assuming an equivalence between science and technology (e.g. technoscience), but not discounting their interrelation, the question or challenge, is to distinguish the particular types of creation they make possible and their modes of existence. Thus, the subatomic neutrino particle has a dual existence as an entity inhabiting the reality of quantum physics and as an active member of scientific experiments. It is produced by and the producer of practices, simultaneously objective and constructed.

Reciprocal capture therefore invites the rethinking of the confrontation of ANT, in its various invocations, and design – or any social science and artistic discipline for that matter – not where the problems, constraints and products of the respective practices remain somewhat indifferent and untroubled by their encounter (the wasp and the orchid return to carrying out their business) – but as special kind of double capture vectorising the creation of new practices which are, themselves, capable of creating novel entities, assemblages, practices and speculative possibilities. Stengers provides a caution, however, in that the adaptation of the term to non-scientific practices, such as ANT and design, requires thinking through the specificity of these knowledge practices. We must actively 'relate knowledge production to the question it tries to answer' (2008: 92) and this includes the constructs of particular practices – what and with what they shape and how they shape epistemic things. The becoming together of particular practices will necessarily create their own obligations and constraints, their own modes of creation and existence.

Clearly, practitioners of design and ANT are posing questions different to those of experimental physicists. If the interrelated notions of reciprocal capture and the ecology of practices operate as tools to think with practitioners without losing the specificity of their practices, then these tools demand that we use them reflexively to think through the becoming-together of ANT and design. This, in part, echoes Marilyn Strathern's (1992: 10) adage 'it matters what ideas one uses to think other ideas (with),' which might be rephrased as 'it matters what matters we use to think other matters with….' In the context of the fusing of ANT and design, it matters what concepts, devices and techniques we use to think and make with and what assumptions they carry into research. In this becoming, ANT unequivocally loses any kind of pretension to neutrality towards the compositions that it dramatises – a caveat that should bear on the optics of visualising controversies (Venturini et al. 2017) – arguably an orthodox version ANT enhanced by information design.

This insistence on the specificity and interplay of research practices invites constructivist and speculative questions of the conjoining of ANT and design – about the kinds of knowledge practices it engenders, the devices it crafts, their becomings, compossibilities and how they are answerable to researchers, practitioners and community members, etc. This, then, involves approaching each research problem with the question of what it is composed of and what this composition is capable of, i.e. the compossibility immanent to the question at hand.[2]

Asymmetries of practice

It never ceases to mystify colleagues whenever, in answer to being asked my profession, I cagily admit to being both a 'designer' and 'social scientist' – 'STS scholar' being especially esoteric. 'Strange combination!' is the usual response, suggesting an inexplicable mixture of practices and knowledge. The more I examine the two disciplines, however, the more they seemingly go hand in glove. The complex of sub-disciplines and genres that comprise design has a long and elaborate history of engagement with the social sciences and vice versa. It is impossible to do justice to these entanglements and vectors, suffice to say multiple shared genealogies exist. Some, admittedly partial, examples, however, help to point towards the various co-becomings of two contemporaneous disciplines. In some cases, the two are ostensibly indistinguishable, such as the foundational graphic isotypes of Otto Neurath (for overviews, see Cartwright et al. 2008, Neurath and Cohen 2012) or various variants of 'participatory' design (Ehn 2017), including its more recent engagement with ANT (Binder et al. 2011). The related fields of Human–Computer Interaction (HCI) Design and Computer-Supported Cooperative Work (CSCW) have both emerged, in part, by way of encounters with ethnography (Button 2000) and ethnomethodology (Suchman 2006) and intricate crossovers and co-developments with approaches associated with STS scholars (Cooper and Bowers 1995, Jensen 2001, Star 1995, Suchman 1999). The aforementioned also testifies to how the social sciences are also routinely mobilised in industrial and commercial design practices (Wilkie 2010) as well as academic settings. More recently, adaptations of design and social science are at play in relatively new fields. On the one hand, what is characterised as speculative and critical design (Dunne and Raby 2013) has been captivated by the public understanding of and engagement with science and technology (Kerridge 2015) whilst on the other, recent innovations such as service design (Kimbell 2009) and the so-called 'social design' (Armstrong et al. 2014), which, amongst other enterprises, provide consultancy for non-profit, voluntary and non-governmental as well as government policy organisations.

This brief summary serves to overturn the view that traffic between ANT and design is exceptional. It is anything but. In this potted light, it is just another confrontation that dramatises difference questions, issues and problems, or, to paraphrase Barry et al. (2008), it stages another transformation in the epistemic and ontological imagination of design, where sociocultural research effects a change in how sociality is conceived and acted upon/within and vice versa. Furthermore, what is readily apparent from the aforementioned are the rhizomatic relations between social science and design, and the ecology of practices that this invariably includes. This, then, points to complex (a)symmetrical – or non-metrical – relations between the diverse multiplicity of practices of the two domains as a dynamic heterogeneous patterning that bears immanent equivalence and difference. It also suggests that the interrelations between design and ANT undergo continual ordering and reordering as new practices are folded into prior commitments.

Within this ongoing multiplicity of practices, there are various instances where ANT has been implicitly and explicitly taken up by designers for purposes of design pedagogy (Wilkie et al. 2018), to write accounts of design practice and designs (Calvillo González 2014, Wilkie 2010, Jaques Forthcoming), to reimagine and recompose participatory design practices and user-involvement, in, for example, healthcare (Storni 2015), to develop experimental research devices (Wilkie et al. 2015) as well as efforts to infuse design into the milieu of policy innovation (Kimbell 2015). DensityDesign, a research laboratory based in the Design Department of Design at the Politecnico di Milan, is both indicative of the organisational embedding of interdisciplinary engagements between design and ANT as well as a particular coalescence of interests between sociological research into big data, controversy mapping and communication design (Valsecchi et al.). As part of the MACOSPOL (Mapping Controversies on Science for Politics) project, led by Bruno Latour, DensityDesign was also involved in the development of digital 'tools' for tracing public engagement with science and technology. The preoccupation of STS researchers with the design of digital tools for controversy analysis can, arguably, be traced back to the 'Web Geographies' project, a collaboration between Science Dynamics at the University of Amsterdam and members of the Computer Related Design Department at the Royal College of Art, London, which developed into govcom.org (Rogers 2000).

The multiplicity of practices that comes to light given the aforementioned could be neatly explained (away) as having immanent epistemological, methodological and ontological valencies. Thus, and for instance, forms of user-centred and participatory design enlarge their respective collectives to include both human and more-than-human participants and reconsider the kinds of situated ontological politics (Berg 1998) they collude in and enact. New techniques, or methods, for identifying actors and rendering collectives and the composition of issues and controversies are devised by, for instance, activist and policy agencies alike. Finally, it proposes a redistribution of epistemic practices amongst those involved or implicated (Clarke 1998) in processes involving design, such as designer–scientist–laboratory engagements, DIY hackathons, activist workshops (see Criado & Rodríguez-Giralt, this volume) or probe workshops. One take on this (Latour 2008) figures the designer as a 'cautious Prometheus' whose artfulness, craft and skills are pressganged into the (STS) task of rendering political ecologies (ibid. 13). Implicit to this, however, is the premise that such practices primarily involve retrospective optics – scrutinising the present and/or past rather than animating the possible. More explicit is the subsuming of design to the obligations and requirements of STS – in such practices, designers equip ANT scholars with renewed technological and visual capacities, devices and techniques. Furthermore, it palpably risks reviving the optical illusion of the neutrality of ANT as the visualisation from nowhere where its practices remain withdrawn from view. Rather than appease the actualist obligations of ANT in which diagnosis is the principle commitment, design practices demand a shift to not only envisioning collectives and their composition or embellishing the specific modes of existence of objects but exploring the possibilities of collectives, their composition and the modes of existence of novel entities. This is precisely the risk or the exigencies of design practice, i.e. the potential to move from an actualist to a speculative constructivism (Wilkie et al. 2017). At base, this means adding propositions to a collective, such as that made by Scandinavian participatory designers in 1982 who inserted an innocuous cardboard box, with 'DESK TOP LASER PRINTER' handwritten on its side, into the workplace of newspaper typographers and journalists in order to reimagine the proofing machine and thus, their cooperative working practices (Ehn and Kyng 1991), or the deployment of an interactive radio-like device amongst local communities endeavouring to face down climate change by way of energy-demand reduction. In both, the insertion into, and this addition to, a collective explores its concrete possibilities to change and develop new practices.

In what follows, I elaborate on this additive version of engagement carried out in relation to local community efforts to effect energy-demand reduction in the UK. In this approach, design demands that ANT becomes a procompositional practice. In doing so, I explore how script theory and script analysis, emblematic of the retrospective compositional analysis (of technology), metamorphose in its encounter with design practices. Script theory is particularly germane here, not least since at its inception it concerned how 'technical objects and people are brought into being in a process of reciprocal definition' (Akrich 1992: 222). More significantly, script theory has passed through various encounters with design, from its inception in relation to the analysis of the design and use of electricity generating and metering technologies in West Africa, to its reformulation – by way of an encounter between Dutch and Norwegian feminist debates and 'constructivism' – in the identification of genderscripts (Oudshoorn 1996) that operate in, for example, digital design (Rommes et al. 1999) and product design practices (van Oost 2003). More recently, script analysis has been modified for design pedagogy and design research (Wilkie et al. 2018) and software design (Allhutter 2012) where the identification and analysis of compositions form the starting point.

Procomposing smart energy monitors

The 'Re-scripting Energy' workshop took place as part of the early design research phases of the 'Sustainability Invention and Energy-demand Reduction: Co-Designing Communities and Practice (ECDC for short) project, one of seven projects funded under the Research Councils United Kingdom (RCUK) Energy Programme. The scope of the ECDC project was to explore and understand individual's and local communities' own understanding and effective use of energy-demand reduction technologies and regulation. In so doing, the project would inform UK government efforts to meet energy and environmental policy targets for reducing carbon emissions and tacking climate change. The attention placed on local energy communities – a principal requirement of the RCUK funding – complemented and diversified existing government policy objectives for addressing energy-demand reduction by 'nudging' individual consumption (e.g. smart monitors) and macroeconomic (e.g. cap and trade emissions trading) intervention. As a small part of these emerging practices and technologies for managing energy and ameliorating climate change, the three-year ECDC project brought together design researchers and STS scholars in order to engage with a number of local energy communities, based in England, and, in doing so, to investigate, raise issues and speculatively experiment with the nature of social and technical community practices.

Over the three-year course of the project, various kinds of engagement with the issue of community energy-demand reduction were conducted including participant observation with community members, probe workshops (an energy community variant of cultural probes; Gaver et al. 1999), social media 'bots' (Wilkie et al. 2015) and exhibitions. Much of this effort directly informed the centrepiece of the project, namely the conception, development and three-month deployment of 36 Energy Babble research devices (Gaver et al. 2015). The 'Re-scripting' energy workshop took place during the early stages of the project and its purpose was to examine the interplay between the 'frameworks of action' and competencies inscribed into existing 'smart metering' technology and the actual practicalities of their installation, as well as the potential for some kind of designed intervention into the monitor's script. All of this was conducted by way of group-based analysis and visualisation (by team members and invited design and STS practitioners). The choice of smart meters here was certainly not arbitrary given the UK government's huge financial investment and political commitment to the promises of a well-defined techno-behavioural fix with feedback solutions that make visible – and supposedly calculable – energy consumption to the individual end user.[3] There was also growing evidence that, in practice, smart monitors were failing to live up to expectations, whether as a result of the complex domestic settings they were installed in (Hargreaves et al. 2010) or the efficacy of the energy consumer figured as an economic–behavioural actor, troubled, in part, by the 'boomerang effect' (Schultz et al. 2007).

Practically speaking, the workshop involved participants having to instal (or at least try to) and use a smart energy monitor for a period of time (typically one to two weeks) prior to the workshop. Three different monitors were used here, which were seen as indicative of two distinct consumer-based approaches to monitoring and feedback available in the UK at the time: The first as an exemplar of indirect and impressionistic and the latter as utilitarian and quantitative. Participants then brought their energy monitor – and experience of using it – to the workshop. During the first part of the workshop, participants were asked, in groups, to identify and visually diagram the programme of action inscribed into the monitor. The second part of the workshop then involved using the diagram to identify entry points for intervening in and proposing redesigns of the monitors, hence the title of the workshop – rescripting.

For my purposes in this chapter, then, the workshop serves as a – somewhat crude – model of how social scientific practices of retrospective analysis can be combined with the prospective practices of design and its obligation to proposition formulation. In the following, I discuss two practices, which I tentatively call retroscription and procomposition, that, on the one hand involves an active participation in lived reality of using a smart energy monitor and transferring this experience, by way of visualisation, into a diagram of use in which the efficacy of the appliance is laid bare. Crucial here is that the diagram differs from the aims of the traditional STS scholar to produce a (somewhat closed) written analysis (Akrich and Latour 1995: 259). Instead, what is produced is a visual analysis primed for procomposition: Which is, at base, a lure for the rearrangement of the composition and acts of invention where elements are modified, realigned, added or removed, thus effecting a relational change in the patterning of collective.

Retroscription

During the workshop, Liliana, a designer on the project, narrated an account of her failed attempts to install and use an Alertme Starter Kit in the rented accommodation she was living in at the time in North London. The story of Liliana's encounter with the electricity meter is quite ordinary, and yet it is precisely the unremarkable nature of having to find a key to a locked door in a communal area, circumvent warning signs, identify the meter belonging to her flat amongst others using manuals left by her landlord and then finally, blindly attach the transmitter to a cable hidden from view by a duct. Back in her flat, Liliana found that the signal from the transmitter was too weak and thus invisible to the display device.

Whilst Liliana narrated her experience, other participants in the workshop worked together to produce a visual diagram of the Alertme's script elicited by Liliana's account as well as including aspects of their own experiences to the mix – often aligned with Liliana's experience. In addition, the monitor, its packaging and manuals were examined for indications of its putative energy-monitoring and saving script. The visualisation rendered three versions of the monitor. First, wherein the script 'worked' under somewhat restricted conditions, producing a discrete energy-user as a calculative individual concerned with minor variations in cost saving and dependent on online connectivity. Second, where the monitoring set-up (transmitter and display) minimally worked in displaying data sourced from the meter which remained somewhat unintelligible due to the lack of accurate live kWh unit supplier costs or any sense of connection to the UK national grid. Third, where the kit failed to work due to irresolvable conflicts between the script and the situated reality of its point, or site, of application.

In retroscripting the active participation with Alertme smart energy monitor, the workshop participants are doing more than merely tracing an issue or network. Here, retroscription connotes an active involvement with, or dramatisation of, a device, its script and therefore enactments of – in this case – the politics of energy consumption. Second, the processes of transfer and transcription that retroscription involves aim not to remove or downplay the techniques and technologies of capture, such as visualisation, that enable the rendering of experience but rather seek to reflexively include these in the process. Finally, the prefix 'trans' denotes the potential for thoroughly changing the script in combination with its elements – the possibility of transversing, transcending and misaligning the script as part of the process of analysis.

Procomposition

If retroscription involved the rendering of past situated and collective experiences of using an energy monitor into a visualisation, then procomposition included the mobilisation of the experiences of monitoring-in-use and of demand-related appliances as part of the design and deployment of the Energy Babble (see Figure 37.1 and Gaver et al. 2015) speculative research device. The Energy Babble, a radio-like interactive research device that 'spoke' audible energy and climate-related content drawn from the Internet, social media and community member voice messages, deliberately aped generic aspects of smart energy monitor scripts in form and function.

As Figures 37.1–37.3 show, the Energy Babble is relatively small appliance that is designed to be placed into a domestic or workspace and connect to the Internet either by Wi–Fi or Ethernet. Once connected, the Energy Babble emits an algorithmically constructed stream of energy-related information sourced from the Internet including scrapes from Twitter feeds, results from searching Twitter, samples from URLs mentioned in tweets, reports of (any) things being switched off (including 'lights,' 'heat' and 'the internet') and current energy-demand updates from the UK National Grid. In addition, the spoken content of the Energy Babble also included replays of messages left by energy community members invited by automated prompts to respond to questions about their energy-related views and activities. Finally, the Babble included a Markov algorithm that produced an assortment of locally sensible and nonsensical spoken messages by rewriting grammatically correct (but typically semantically

Figure 37.1   The Energy Babble (© Alex Wilkie)

Alex Wilkie

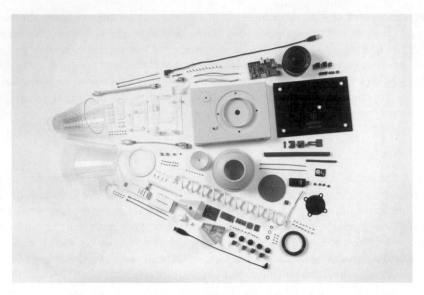

Figure 37.2 An exploded view of the Energy Babble (© Interaction Research Studio)

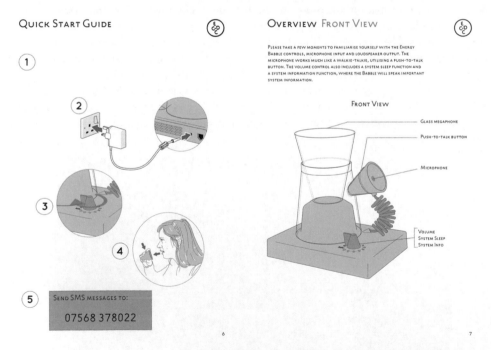

Figure 37.3 Pages from the Energy Babble manual showing the quick start guide for the device (© Interaction Research Studio)

incorrect) strings from the corpus of content it collects and utters. As such, the Energy Babble played on and put into play new ways of monitoring and eliciting energy-demand reduction and environmental feedback. In short, a new appliance-based feedback procomposition in which the solution to energy-demand reduction is yet to be determined.

396

The process of designing the Energy Babble also suggests that procomposition work – a point where design and STS converge – is complex, circuitous and involves combinations of skilled practices. As the aforementioned figures indicate, this involves, at base, the intricate interplay between and coordination of interaction design, industrial design, graphic communication, photography and sound design. The interplay in this case also suggests that, although practical design action may be embodied, it also routinely cuts across or transverses the collective involved in designing where skills and craft are shared and outcomes co-produced.

Locating procomposition only in such activities not only betrays a relation between orthodox design practices (three-dimensional, graphic, etc.) and procomposition but also on how a multiplicity of novel design techniques and approaches are enacted as well as emerge during this engagement. To realise the final Energy Babble device involved any number of encounters, experiments, proposals and trials involving both humans and non-humans. Here, Probe Workshops (Michael et al. 2018) were held with members of energy communities to investigate the meaning and configurations of community, elicit members' expectations of climate change and environmental futures and explore affective or aesthetic aspects of energy use; automated algorithmic software robots (Bots) were designed and deployed on Twitter to detect and identify energy-related practices and actors (Wilkie et al. 2015).

Conclusion

To return to the conversation between Deleuze and Parnet (and Stengers) and the inter-related notions of double and reciprocal capture, the question of equipping designers is transfigured into the possibility of design mutually becoming-with ANT. As sketched out earlier, this entails the rendering, or retroscription, of political collectives, in this case UK energy communities, AND the addition of new speculative propositions – procompositing these collectives in the form of an interactive material-semiotic research device to excite and prehend the possibilities of new energy-demand reduction practices and make such things matter. The upshot of the reciprocal capture of design and ANT does not rest with matters of fact nor description: It animates compossibility in the making of speculative propositions as lures for procomposing new political collectives.

Acknowledgements

This chapter draws on research conducted under the project grant 'Sustainability invention and energy-demand reduction: Co-designing communities and practice' funded by the RCUK and led by the EPSRC (project code ES/I007318/1). With thanks to Daniel López Gómez, Mike Michael and Martin Savransky for invaluable comments and special thanks to Celia Roberts for her comments and editorial care.

Notes

1 See (Fraser 2010: 75) for clarification of 'becoming together.'
2 On the nature of possibles, it was Gottfried Wilhelm Leibniz (cited in Messina and Rutherford 2009: 962) who asserted that not all possibles are compossible and that compossibility entails the possibility of actual existence in a common world. In other words, there are situated and contingent restrictions on speculative relations. See Deleuze 2004: 100–101 on the relation between the compossible and the incompossible.
3 See Darby 2006, Dece 2009 for the rationale, reasoning and UK government policy on the rollout of smart-metering technology.

References

Akrich, M. 1992. The De-scription of Technical Objects. In: Bijker, W. and Law, J. eds. *Shaping Technology/Building Society: Studies in Sociotechnical Change*. Cambridge, MA: MIT Press, 205–224.

Akrich, M. and Latour, B. 1995. A Summary of a Convenient Vocabulary for the Semiotics of Human and Nonhuman Assemblies. In: Bijker, W., E. and Law, J. eds. *Shaping Technology/Building Society: Studies in Sociotechnical Change*. Cambridge, MA; London: MIT Press, 341p 323cmpbk.

Allhutter, D. 2012. Mind Scripting A Method for Deconstructive Design. *Science, Technology & Human Values*, 37(6), 684–707.

Armstrong, L., Bailey, J., Julier, G. and Kimbell, L. 2014. Social Design Futures: HEI Research and the AHRC.

Barry, A., Born, G. and Weszkalnys, G. 2008. Logics of Interdisciplinarity. *Economy and Society*, 37(1), 20–49.

Berg, M. 1998. The Politics of Technology: On Bringing Social Theory into Technological Design. *Science, Technology, & Human Values*, 23(4), 456–490.

Binder, T., Ehn, P., De Michelis, G., Jacucci, G. and Linde, G. 2011. *Design Things*. Cambridge, MA; London: MIT Press.

Button, G. 2000. The Ethnographic Tradition and Design. *Design Studies*, 21(4), 319–332.

Calvillo González, N. 2014. Sensing Aeropolis. Urban air monitoring devices in Madrid, 2006–2010. Arquitectura.

Cartwright, N., Cat, J., Fleck, L. and Uebel, T. E. 2008. *Otto Neurath: Philosophy between Science and Politics*. Cambridge: Cambridge University Press.

Clarke, A. 1998. *Disciplining Reproduction: Modernity, American Life Sciences, And "The Problems of Sex"*. Berkeley; London: University of California Press.

Cooper, G. and Bowers, J. 1995. Representing the User: Notes on the Disciplinary Rhetoric of Human-Computer Interaction In: Thomas, P. J. ed. *The Social and Interactional Dimensions of Human-Computer Interfaces*. Cambridge: Cambridge University Press, 48–66.

Darby, S. 2006. The Effectiveness of Feedback on Energy Consumption: A review for DEFRA of the Literature on Metering, Billing and Direct Displays. Environmental Change Institute: University of Oxford.

Decc 2009. Smarter Grids: The Opportunity. Department of Energy and Climate Change.

Deleuze, G. 2004. *Desert Islands: And Other Texts, 1953–1974*. Los Angeles, CA; New York, NY: Semiotext(e).

Deleuze, G. and Parnet, C. 2002. *Dialogues II*. London: Continuum.

Dunne, A. and Raby, F. 2013. *Speculative Everything: Design, Fiction, and Social Dreaming*. Cambridge, MA: MIT Press.

Ehn, P. 2017. Learing Participatory Design as I Found It (1970–2015). In: Disalvo, B., et al. eds. *Participatory Design for Learning: Perspectives from Practice and Research*. Abingdon, Oxon; New York, NY: Routledge, 7–21.

Ehn, P. and Kyng, M. 1991. Cardboard Computers: Mocking it Up or Hands on the Future. In: Greenbaum, J. M. and Kyng, M. eds. *Design at Work: Cooperative Design of Computer Systems*. Hillsdale, NJ: L. Erlbaum Associates, 169–195.

Fraser, M. 2010. Facts, Ethics and Event. In: Jense, C. B. and Rödje, K. eds. *Deleuzian Intersections: Science, Technology and Anthropology*. New York, NY; Oxford: Berghahn Books, 57–82.

Gaver, W., Dunne, T. and Pacenti, E. 1999. Cultural Probes. Interactions, 6(1), 21–29.

Gaver, W., Michael, M., Kerridge, T., Wilkie, A., Boucher, A., Ovalle, L. and Plummer-Fernandez, M. 2015. *Energy Babble: Mixing Environmentally-Oriented Internet Content to Engage Community Groups*. In *Proceedings of the 33rd Annual ACM Conference on Human Factors in Computing Systems*, 1115–1124.

Hargreaves, T., Nye, M. and Burgess, J. 2010. Making Energy Visible: A Qualitative Field Study of How Householders Interact with Feedback from Smart Energy Monitors. *Energy Policy*, 38(10), 6111–6119.

Jaques, A. 2018. Knowing-Mies as Rendered Society. In: Marres, N., Guggenheim, M. and Wilkie, A. eds. *Inventing the Social*. Manchester: Mattering Press, 148–170

Jensen, C. B. 2001. Cscw Design Reconceptualised through Science Studies. AI & Society.

Kerridge, T. 2015. Designing Debate: The Entanglement of Speculative Design and Upstream Engagement. (PhD). Goldsmiths, University of London.

Kimbell, L. 2009. The Turn to Service Design. In Julier, G. and Moor, L. eds. *Design and Creativity: Policy, Management and Practice*. Oxford; New York: Berg, 157–173.

Kimbell, L. 2015. *Applying Design Approaches to Policy Making: Discovering Policy Lab*. Discussion Paper. Brighton: University of Brighton.

Latour, B. 2008. A Cautious Prometheus? A Few Steps Toward a Philosophy of Design (with Special Attention to Peter Sloterdijk. University College Falmouth, UK, 3–6 September, 2–10.

Messina, J. and Rutherford, D. 2009. Leibniz on Compossibility. *Philosophy Compass*, 4(6), 962–977.

Michael, M., Wilkie, A. and Ovalle, L., 2018. Aesthetics and Affect: Engaging Energy Communities. *Science as Culture*, 27(4), 439–463.

Neurath, M. and Cohen, R. S. 2012. *Empiricism and Sociology*. Springer Science & Business Media.

Oudshoorn, N. 1996. Genderscripts in technology. Fate or challenge. Inaugural Speech. Enschede: University of Twente.

Rogers, R. 2000. *Preferred Placement: Knowledge Politics on the Web*. Maastricht: Jan Van Eyck Akademie.

Rommes, E., van Oost, E. and Oudshoorn, N. 1999. Gender and the Design of a Digital City. *Information Technology, Communication and Society*, 2(4), 476–495.

Star, S. L. 1995. *The Cultures of Computing*. Oxford: Blackwell.

Schultz, P. W., Nolan, J. M., Cialdini, R. B., Goldstein, N. J. and Griskevicius, V. 2007. The Constructive, Destructive, and Reconstructive Power of Social Norms. *Psychological Science*, 18(5), 429–434.

Stengers, I. 2008. A Constructivist Reading of Process and Reality. *Theory, Culture & Society*, 25(4), 91–110.

Stengers, I. 2010. *Cosmopolitics I*. Minneapolis: University of Minnesota Press.

Storni, C. 2015. Notes on ANT for Designers: Ontological, Methodological and Epistemological Turn in Collaborative Design. *CoDesign*, 11(3–4), 166–178.

Strathern, M. 1992. *Reproducing the Future: Essays on Anthropology, Kinship and the New Reproductive Technologies*. Manchester: Manchester University Press.

Suchman, L., A. 1999. Working Relations of Technology Production and Use. In: MacKenzie, D. A. and Wajcman, J. eds. *The Social Shaping of Technology*. 2nd ed. Milton Keynes, [Eng.]; Philadelphia, PA.: Open University Press.

Suchman, L. A. 2006. *Human-Machine Reconfigurations: Plans and Situated Actions*. 2nd ed. Cambridge: Cambridge University Press.

Valsecchi, F., Ciuccarelli, P., Ricci, D. and Caviglia, G. The DensityDesign Lab. ShangHai, 180.

van Oost, E. 2003. Materialized Gender: How Shavers Configure Users' Femininity and Masculinity. In: Oudshoorn, N. and Pinch, T. J. eds. How Users Matter: The Co-Construction of Users and Technologies. Cambridge, MA: MIT Press, 67–80.

Venturini, T., Jacomy, M., Meunier, A. and Latour, B. 2017. An Unexpected Journey: A Few Lessons from Sciences Po médialab's Experience. *Big Data & Society*, 4(2), 1–11.

Wilkie, A. 2010. User Assemblages in Design: An Ethnographic Study. (PhD). Goldsmiths, University of London.

Wilkie, A., Farías, I. and Sanchez-Criado, T. 2018. For a Speculative Aesthetics of Description: Interview with Alex Wilkie. Diseña, 12, 70–87.

Wilkie, A., Michael, M. and Plummer-Fernandez, M. 2015. Speculative method and Twitter: Bots, Energy and Three Conceptual Characters. *The Sociological Review*, 63(1), 79–101.

Wilkie, A., Rosengarten, M. and Savransky, M. 2017. Speculative Techniques. In: Wilkie, A., Savransky, M. and Rosengarten, M. eds. *Speculative Research: The Lure of Possible Futures*. Abingdon, Oxon; New York, NY: Routledge, 111–113.

38

How to run a hospital with ANT?

Yuri Carvajal Bañados

~Humus

In October 2014, the 40-year-old Puerto Montt Hospital, designed for a Chilean town of 150,000 inhabitants, moved into new premises, emerging as the institution with the highest concentration of professionals and specialists in the region, with equipment comparable to the best available in the country and a building of a unique style that you would only expect to find in the most elite areas of the country's private sector. I started to work there seven months later. In the midst of a national controversy about building hospitals under a system of concessions, where private companies obtain a licence to provide services (e.g. laundry, catering and building maintenance) and build clinics as a means of resolving the crisis of an inefficient and backward public sector, the opportunity to work in this place, completely public, well equipped and with an indisputable advantage over the local private sector, was something that, as a public health worker seeking new experiences, I could not ignore. I had, by that stage, completed a doctoral thesis on health statistics in which I deployed ANT ideas and developed postdoctoral work in two books (Carvajal, 2017; Carvajal and Gaete, 2016).

Although I entered as just another doctor trying to discover something new and to be part of an interesting experiment, I had some idea of what I was rummaging for. I had previously worked for six months in a good office, as an average bureaucratic official. Then, the position of Director of the Hospital, as mandated by the schedule of the National Civil Service, was called to a contest as a position of Senior Public Management. Several hospital officials whispered that I should run for the position and I did so. I was selected in the final list and on March 1, 2017, started working as the director. I did not conceal my interests or my visions of what a hospital is supposed to be. I acted according to the rules and I was the candidate with the highest score.

Can this be an ANT story?

In what way can an intellectual approach that does not consider organisations to be stable structures nor accept that some sites inherently have cognitive or hierarchical privileges, be useful to govern an institution? It is not easy, as 99.99% of those who work in the hospital would argue, against ANT, that organisations have an 'up' and a 'down'; that there are vertices and bases; that organisational charts are a faithful representation of the organisation and that a person whose door says 'Director' effectively directs it. And some of them – as I will tell you later – are willing to wage a religious war to defend the flow diagram of the institution as a faithful reflection of the organisation.

In my own field, Public Health, I am a solitary navigator who maintains that ANT is a valuable tool for this practice. I use the definition of the discipline 'training in social sciences for health workers' as a wide flag. Despite its immaculate simplicity, it has not been waved by any of my colleagues. Perhaps, if a small conglomerate of physicians shared this summary version of the discipline, questions about my approach would become more difficult. I might be asked to explain what social sciences I speak of and why they are relevant for health workers. So far, in fact, my closest colleagues and friends have been irritated by my homologation of the doctors with Callon's heteronomous engineers. Few believe me when I say medicine is a technique, an adequate understanding of which can only arise from the sociologies of that technique.

Beyond the position I hold and the functions that I may or may not assume, the question of the hospital itself arises. How does an approach in which hospitals are places of collective, hybrid and technical life can help make it technically rigorous, collectively democratic, economically viable and anthropocenically responsible? How do my epistemological concerns and the intrigue of knowing what statistics are made of and how they work (one of my favourite ANT areas) help to produce knowledge that is more grounded, statistically more comprehensible, interdisciplinarily more real and ethically more debated? How, in other words, can you run a hospital with ANT?

In this chapter, I will talk first about the 'hospital metaphors' that I coined during a year of work, and then describe some events and problems that have disturbed me, in order to explore how a heteronomous doctor (ANT) has been directed by a hospital.

Hospital metaphors

Labyrinths and the multicoloured city

Often, during my time as the director, I have felt that I am in a labyrinth of epistemological paths that lead to dead ends, technical detours that offer shortcuts to new deviations, ethical dilemmas that cross alternate roads that appear to hold similar results, psychologies that lead to sociologies and vice versa and engineering failures that summon common sense as an aid and vice versa. How can one direct a labyrinth? How is it possible to walk tirelessly, without the magic of Daedalus and Icarus to look at it from above? How can I be something other than the subject of a behavioural experiment, who busily travels the immediate corridor in front of him, knowing that the experimenter's purpose is only to measure how long it takes me to get to an exit? (Figure 38.1)

The second metaphor behind the labyrinth is that of a city: A world in which there are possible colours and sociologies, in which disease is an encounter of science and technology, psychologies and negotiations, uncertainties, biological folds, ethical problems, legal norms, economies, administrative rules and professions. A city where there is civility, where some have city rights (i.e. a life more of their own) than others.

This metaphor can be sweet. If it were true, we would be unable to distinguish the city from the hospital-city, and could live in it as if it were a small-scale version of everything that the big city holds. It would therefore not be necessary to direct it. Certain urban planning

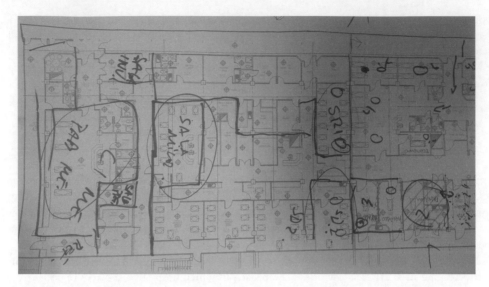

Figure 38.1   The map is not the territory: planning the simulation centre (taken by the author)

rules, traffic lights at intersections, a tax system and a police regulation would suffice. It is almost as if it regulates itself. That sweetness becomes elusive when the hospital becomes a fiefdom, a walled zone, an impregnable territory, a space that is pleased to maintain unbreakable frontiers, to govern itself with relentless sovereign rules, to coexist under the Kafkaesque weight of an authority without limits, in the state of exception of a permanent war, which makes bacteria, neoplastic cells, inflammation, autoimmunity and death the absolute enemies against which there is no truce.

But the city also has something that Elman Service called chiefdoms, groups of a few thousand humans with a centralised command and a couple of hierarchical levels (cited in Diamond, 2017). The regular need to intervene in the daily running of the hospital puts into question all the subtlety of a self-regulated city, in this peculiar place where there are beds for patients and officials who work during the night, a city that never stops. Although it was useful for Goffman to talk about total institutions (Goffman, 2012), the truth is that there are many gaps and conditions for patients or officials to carry out secondary adjustments. That is what I will talk about below: My experience of the immanence of the hospital as a technical place and the production of truths about disease.

Hospital stories

Story 1: The war of professions

I owe a lot of my fortune and my sorrows to ANT. Armed with an understanding of the clinic as the fundamental pillar of the hospital, I seek to understand technical practices as regimes of true diagnosis production: This pathology, this prognosis, this functional alteration, this anatomical or histological problem. I increasingly transfer the weight of decisions to the technicians of the professions, placing physicians as an obligatory step. I am not vehement or bold: I hope that conditions appear gradually, although from the first day I declare my purpose. I used three verbs in my inauguration speech: 'to develop,' referring to clinical medicine and technical truth; 'to articulate,' talking about the seamless cloth, the work of ants on the sociologically continuous surface of hospital life; and 'to value,' delineating the two ethical issues that I relay (emphasis on what is important and the calculation of prices as a secondary effect, and the environmental sustainability of the hospital).

Four months after I arrived, I called a contest to appoint new heads of medical services, and unleashed a war of nurses against me. I found myself in the middle of the war of professions that the hospital incubates in its womb. The flag – with all due respect to Mol – is care management, a specific category that is raised as a humanist flag (but with a 'managerial' preamble) against the technical inhumanity of doctors. Ah, Foucault, I curse the day your spirit entered my soul and carved in me that

> if some event of which we can at the moment do no more than sense the possibility – without knowing either what its form will be or what it promises – were to cause them to crumble, as the ground of Classical thought did, at the end of the 18th century, then one can certainly wager that man would be erased, like a face drawn in sand at the edge of the sea.
>
> *(Foucault, 1993, p. 375)*

A profession rises, denouncing inhumanity and protecting its autonomy as the guarantee against the excesses of the technique. In front of me, nothing but the flags I loathe: Denunciation, autonomy, humanism and anti-technique. I step back from this great epistemological stumble. What is it that I do not know? How can I get to move towards other truths? I dive into the war of professions. I feel that this 'omnium bellum contra omnes' is nourished by the difficulties of cognition in the wild, which is always the unique and original body in which we try to capture the disease, as a confluence of 'immutable mobiles.' The hospital tries to be a standardised space and, at the same time, to standardise patients as much as possible, to make them sleep, to sedate them, to sit them up.

But in this war, there is an increasing gap between the myth of authorship, which comes from antediluvian times of medicine, and the technical modernity of a technical network that produces diagnoses, treatment and prognoses, therapeutic successes or failures. How to make professions understand that today they are part of a cloth that is increasingly folded, bigger, but made with fewer stitches? Another war of a wrong knowledge, of the poverty of the life of social sciences in this sociologically overheated world, of the radical lobotomy of collective knowledge performed on the students of the clinical world, amplified by a practice that is silenced by ethnographic knowledge.

Wars are simple solutions to tangled problems. Loosening the knots requires some serenity. War is quick, but its results, as we already know, are evanescent. In the coming week, as I write this, the new heads of service will assume their roles. The local Court of Appeal has rejected an appeal filed by 268 nurses against the hospital, represented by me, and we await the decision of the Supreme Court in about a month. The impregnable juridical reality against the autonomous knowledge, whether it validates them or not, the organisation of a hospital belongs, partially, to the judges. A gap between the tradition of diagnosis and care authorship, of professions as protective cells, guardians of a knowledge that today is produced collectively. After the war, come the spoils. What objects are stolen by the victors?

Story 2: Conjugating the verb 'LTE'

In the law on 'Duties and rights of patients' (Law 20.584, April 24, 2012), there is the expression 'limitation of therapeutic effort.' It is not a central phrase nor is it defined in its preamble.

It is only considered in article 17, second paragraph, that 'if the insistence on the indication of the treatments or the limitation of the therapeutic effort is rejected by the person or by their legal representatives, the opinion of said committee may be requested.' The aforementioned body is the Healthcare Ethics Committee, an institution from each establishment that must discuss the requirements of professionals and patients regarding the ethical aspects of a clinical decision.

'Patients who have decided to not be resuscitated,' 'proportional measures' and 'limitation of therapeutic effort' are expressions that circulate in any hospital. The question is how to provide a formal and accepted deliberative mechanism for these situations. I take a step back and agree to exclude the word euthanasia from this reflection. Therefore, I speak of a state of proximity of death, of extreme indignity, of insufferable pain. My bioethics professor will get sick of my ambiguous compositions, of the imprecision of my criteria or their inability to become objective parameters.

However, I think that this problem arises from that same insufficiency. When what predominates is uncertainty and the questions are fragmented around life and death, like two insurmountable abysses, the search for stable parameters faces a desert. I see the public dialogue of facts and decisions as the only viable path: In the increasingly public debate about what happens when a patient is connected to a ventilator and has a multiple organ failure; what to do in such an unfortunate situation. This is a debate that should involve laymen and experts, patients, caregivers and relatives, in real time and with scrupulous sovereignty (Figure 38.2).

I believe we should let the acronym LTE ('limitation of therapeutic effort') obtain the full right of a city in the ethical life of hospitals. Let it stop being a jargon used only in semi-private areas, and allow its transformation into a verb and its conjugation be passed into public speech, a world in which any of us could fortunately be 'lte'd.'

Figure 38.2 Statistical map (public image, Puerto Montt Hospital)

Story 3: Multiple parliaments

If the hospital is an intensely technical place, a pragmatic arena of turns and results, of effects that demolish any result anticipated by the theory, an intellectually prospective place, a place in which 'a step of real movement is worth more than a hundred programs,' where does arbitrariness give way to logic? I do not know of such places other than discussion sites, whether magazines, books, congresses, parliaments of things or hybrid forums. This hospital has more than 60 committees where its actions are discussed, excluding the innumerable meetings of services, units, departments, teams, offices and centres. It is an intense political life, because perhaps this is where the new protocols obtain their city rights, the ever-changing institutionality that governs the sick, the diseases and the caregivers.

Tracking the life of one of these committees involves entering the seamless cloth, visiting the Oncology Committee, for example, in charge of discussing cancer patients and defining their treatment on a case-by-case basis. We might observe a documented and recorded discussion involving surgeries, radiotherapies and cytotoxic drugs, and understand the committee as a driving force of the hospital regarding the use of surgery rooms, pathological anatomy, medication acquisition, development of specialties and sub-specialties, and imaging requirements. Who is directing, then? Who is the sole author? There is no privileged place to narrate other stories: Breastfeeding committee, integrative medicines, welfare, acquisition, pharmacy, adverse events, healthcare ethics, emergency operation and so on. Coordinating these parliaments requires more than one official journal. We have established two statistical bulletins and a hospital clinical journal (http://www.hospitalpuertomontt.cl/index.php/hpm/90- letras-cl %C3 %ADnicas; Figure 38.3).

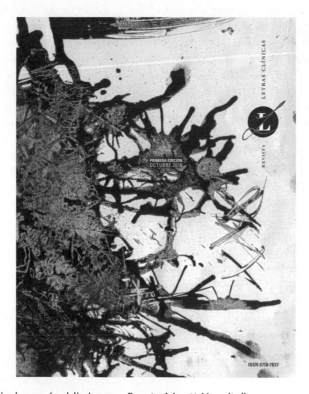

Figure 38.3   Clinical map (public image, Puerto Montt Hospital)

Story 4: A hectic morning

Although the hospital lives in a state of tension and its movements always feel urgent, it also has valleys, secular and geological rhythms. Sometimes, those rhythms come together, synchronise and express their unity in a brusque way. It is difficult then to disentangle what is urgent, what is secular and what is geological. Here are my notes:

Tuesday 13. It starts with a coffee of sensitivities and specificities. In the prenatal control of a pregnant woman, there is a positive HIV test result. The protocol states that this delivery must be by caesarean section, as to avoid vertical transmission of the virus. Breastfeeding is also to be suspended. We strictly comply with the protocol, but the mother shows us a negative HIV test, taken previously. While waiting for the confirmation of the diagnosis (unlikely with the current data), the patient questions our decisions.

I then listen to a hospital worker, son of a hospital worker who had hip surgery. A catheter remained inside her after the surgery. We punctured her artery instead of her vein, which is why that device was left longer than usual. The son is calm because there were no later complications and the surgery was successful. It was a nice conversation with a worker whom I already knew. Injured by a TBI, he uses a cochlear implant. He disconnects it to sleep and his audible alarm clock is useless. He solved his problem by inventing, with his father, a mechanical alarm that is connected to his bed. A technological innovation of daily life of which I only learned due to this feudal role of a worker-rating supreme judge.

Later, I see the daughter of an elder who was brought by her brothers into our emergency room. The woman's sons ask for a voluntary discharge, although the diagnosis is serious. The difference between a liberal autonomist vision and a sociologically inclusive vision is minimal. How to decide what is better in the midst of the signs of an over-demanded service in these days of respiratory syncytial virus, overwhelmed by older people brought by family members who are also overwhelmed in their home, and who seek out more experienced caregivers – us? How can I think about a consultation that cannot take more than a few minutes, whose decision should be quick? How to respond to these voices that now resonate calmly in my office, in the rural world, the family dispute over a cow or a fence that should be placed 100 metres farther to the west?

Our voluntary discharge protocol is still pending. A decision that is not easy, but that just as LTE, should involve more conversations, notifications and recording systems.

I then receive a family whom we did not notify of the death of their father and husband. They confirm that they noticed we removed his wedding ring, and we do not know where it is. I only have apologies and the investigation protocol, but I try to understand who this man was. His son tells me that he is one of the three most notable citizens of the city who died that year, an exiled Spanish man from Vigo. A day later, one of my best friends shares a similar story, and I tell him about this. It is the same family. The hospital that I direct has brought grief to the family of my close friend.

I write here of adverse events, events in which reality suddenly reveals an unfamiliar face, an unexpected, unpredictable or probabilistically negligent course. Where professions lose a little grip and go into technically wild areas. The dialogue between professions, sometimes interrupted, jumps into an area of dialogue and camaraderie. Engineers rub shoulders with technologists, doctors with electricians and computer technicians with accountants.

Adverse events have an ambiguous definition, because they just try to capture a fact that only retrospectively turns out to be such. The timeline breaks it. It was neither an event nor an adverse one, but it existed. From a specific moment, it is recognised as such. That epistemology collapses what we understand as facts in medical positivism: Things are always what they are; the name is always the thing.

Adverse events speak of a counterpart, of the production of facts as a systematic objective that organises the purpose of the hospital world, such as the diagnosis, treatment and documentation of the healthy – often a battery of exams without any remarkable findings. This reveals that even the technical world is a sociologically organised world, a piece of plasma in which norms are constituted and where material supports are instituted to hold those rules together. Protocols, records, auditing systems, checklists, and golden rules are all part of those instituting forms. Promoted as the soul of quality, they displace the force of cooperation, of the continuum, of the symbiosis, to recordable rules that can be verified in the future.

The technique to come in medicine, if we are at a threshold … 'a new experience of the disease is about to be born' (Foucault, 1991: 9), will have to turn on itself and dismantle its machinery. I foresee that the biome, the holobionts as our problems, will lead this new technique. Only symbiosis and ties can make us free. Understanding this present can be part of inventing the hospital of the 21st century.

★ ★ ★

Coda

Sunday morning. Five days ago, I stopped being the Director of the Hospital. The editorial processes are mysterious too. My sedentary, compact, unitary, self-contained version, reviewed over and over with Ignacio and Celia, now seems dislocated. When I told them the news, well aware of my delays as an author, they proposed a new deadline to include this coda of sorts.

The arrival of a government that seeks to expand microeconomic spaces in its administration practices is particularly critical for the workings of a Public Hospital. Chile has been part of a massive economic experiment, but the hospitals of the National Health Service have remained scattered islands in that ocean, at the mercy of tides and waves. Surrounded and harassed by increasing sea levels, they have survived in a dignified manner. These last two months have revealed symptoms of things that are moving in a dangerously different direction. The recent law of the voluntary interruption of pregnancy has been modified in a regulatory and illegal manner, a crude expression of the new wave (cms.colegiomedico.cl/2018/05/1, in Spanish).

Someone on the right side of the spectrum tells me that it is 'disgraceful' that I am a Director under this new administration, and I deeply breathe in the discomfort of their culture. I resign, in a gesture that seeks to avoid a political collision that involves a public hospital and our beloved ANT. I escape.

I now write a new summary of these two years trying to run a hospital with ANT. This is the version of the person moving to a different place, not the person sitting on the chair of the Director's office. My first lesson is:

At least in third world countries, an ANT practice will be considered leftist.

(Perhaps, like Rimbaud, we could say 'I always belonged to an inferior race,' although few or no one knows of our radical dissidence with Marx, Trotsky or Adorno.)

The second ANT lesson is a global summary of everything:

ANT is lived as a kind of public happiness for those it touches.

I have already said that the description of a hospital, as a sociological fact from ANT, in this experience unleashed forces that took me by surprise. Many of them are narrated earlier. Thus, I will mention others here.

Many mornings, very early, the head of neurosurgery approached the Director's office to talk about Alfonso Asenjo, creator of the specialty in Chile in the 1940s, founder of the Public Institute of Neurosurgery of the country (which bears his name), whose scientific production reaches 531 articles. He 'converts' me into his enterprise to restore the documentary *Parkinsonismo y cirugía* (1962) directed by Sergio Bravo, narrated by Francisco Coloane (National Prize of Literature in 1964), with music by Gustavo Becerra (National Prize of Musical Arts of Chile in 1971), camera by Pedro Chaskel, Enrique Rodríguez and Fernando Bellet (http://www.cinetecavirtual.cl/fichapelicula.php?cod=33). This process is currently underway through an agreement with the Neurosurgery Service, represented by the hospital, to finance the restoration of the documentary and schedule exhibitions in the main building and the hospital.

My neurosurgeon colleague surprises me with the books he brings as a gift. The last one is about Spinoza and the neurosciences. This hospital is, perhaps, the last place where I would suspect that someone shares my philosophical interest in the Dutch glassmaker. The fact that his interest in Spinoza comes from neuroscience and mine from Deleuze is but a detail. What I experience is the surprise that a specialist delves so deep into the roots of his profession and practice. My colleague is also a martial artist and 8th dan. His sense of worth of the mastery and respect that every disciple has for his Sensei is also present in this challenge. Even as the approach to the process of restoring a film as an expression of a world of collaboration and creation, a happy encounter between scientists and artists, under the wing of the University, in a Chile whose values matter to us.

The other happy manifestation that I bring up is that of a paediatric dentist whose hobby is to calm her patients down by telling stories and performing little clown tricks. Gathered in a ritual lunch every Tuesday along with other dislocated people and a small tree that we carried with us to the table, today there is a hospital narrative medicine committee. Her practice of telling stories to children, crowned with giraffe horns and flanked by a case with small gifts, has grown rapidly to involve other colleagues, in the recognition of the narrative truth of hospital medicine. Another fruitful intellectual dialogue has been carried out in the area of biomedical engineering (Bavestrello and Carvajal, 2018), so I can say:

The most specialised technicians can be proper intellectuals and closer to the philosophical depths of ANT than we may think.

The fourth ANT lesson comes from breaking with sedentarism. The hospital exercises its edge at every moment. However, less than 200 metres from the magnetic resonator, there are houses without sewerage. Just third world things, technology and its coexistence.

I cross the street and participate in a neighbourhood meeting. Disturbed by the presence of this monster, whose proportions far exceed more than a thousand times the size of their houses and families, and the effects of an unusual and disturbing vehicular proliferation, street trading and garbage, the residents listen to me with respect. I call them to take a joint action against those undesirable results. The next day, I attend a meeting of the teachers' council of the Cayenel School, an indigenous name by which this marginal neighbourhood – that used to be beyond the walls of the city with a prison of the 1960s, now empty – was known. Today, everyone calls this place 'the hospital district,' in a symbolic and unfortunate degradation of a toponym that indicates the presence of the absent, the non-extinction of the anthropologically extinct.

I propose to collaborate with the public school, inviting the children to our laboratories, to our auditorium, to hold exhibitions in our playgrounds. They reply to me proposing a chess tournament. The following week, a fluid movement begins between the public hospital and the public school, whose recent expressions are the images of this text.

Therefore, my fourth lesson would be:

Crossing a street means eroding an ANT border.

Once my resignation was known, an official from the REAS Unit (Spanish for Waste from Health Care Establishments) approached me to organise the official start of glass recycling at the hospital in the last week of my duties. The city in which we live does not have public waste containers; using the hospital to that end will be a remarkable contribution. If clinical medicine is the source of introduction of anthropocentric forces under the epithet of the disposable or the anti-: plastics, ecological and/or endocrine disruptors, antibiotics, antineo-plastic, antiparasitic (Bernhardt et al., 2017, Kolpin et al., 2002, Richmond et al., 2017), it is also possible to summon from ANT the forces that lead to a non-anthropogenic medicine (Figure 38.4).

Figure 38.4   Nurses walk through the Cayenel neighbourhood (image taken by Maria Elenea Flores)

ANT is an important un-ANThropogenic force.

Finally, a pragmatic question: numbers. We are under growing rules of government through numbers and figures. Process indicators, result indicators, integral control balances. I will only say that the Puerto Montt Hospital is in 24th place out of 57 similar hospitals throughout the country, achieving over 75% of its goals, a threshold that allows it to be recognised as a Self-managed Network Establishment. The evaluation of the Director's performance was 99.93%, in terms of compliance with the agreement with Senior Public Management. Finally, although its production has been growing steadily, its levels of debt last year were 65% lower than in 2016. This is not a miracle. If the hospital performed more health actions, it is obvious that we spent more. However, we were able to show the value of what we were doing and obtain financing for our services. Although ANT does not govern by and for numbers, our results are backed up by them. Our results are not backed up by a government that goes by universal rules, removed from local conditions and a situated and dialoguing reflexivity (http://www.hospitalpuertomontt.cl/images/cuentaspublicas/CUENTA%20 PUBLICA-2017.pdf). Finally, then,

ANT worked for the Puerto Montt Hospital as a valuable enzyme.

Acknowledgements

This text has also been written by many hands: Verónica Thoms, Yohanna Guzmán, Katy Bosnic, Claudia Pérez, Brigitte Sievers, Paola Olave, Italo Bavestrello, Marco Balkenhol, Gonzalo de Toro, Javier Scheele, José Luis Calisto, Benjamín Rocha, Jorge Andrade, Jorge Tagle, César González and Eduardo Bravo. All of them are completely innocent – but somehow, not so much.

References

Bavestrello, I. and Carvajal, Y. (2018). Ingeniería para los hospitales públicos: el caso de la ingeniería biomédica. *Cuadernos Médico-Sociales*, 58(1): 85–86.

Bernhardt, E., Rosi, E., and Gessner, M. (2017). Synthetic Chemicals as Agents of Global Change. *Front Ecol Environ*, 84–90. DOI:10.1002/fee.1450.Carvajal, Y. (2017). *Pensando la práctica médica. Volver al Hospital II*. Editorial Universitaria, Santiago de Chile.

Carvajal, Y. and Gaete, J. (2016) i. Editorial Universitaria, Santiago de Chile.

Diamond, J. (2017). *Armas, gérmenes y acero. Breve historia de la humanidad en los últimos trece mil años*. Penguin Random House, Barcelona.

Foucault, M. (1991). *El nacimiento de la clínica. Una arqueología de la mirada médica*. Siglo veintuno editores, Buenos Aires.

Foucault, M. (1993). *Las palabras y las cosas. Una arqueología de las ciencias humanas*. Siglo veintuno editores, Buenos Aires.

Goffman, E. (2012). *Internados. Ensayos sobre la situación social de los enfermos mentales*. Amorrortu, Buenos Aires.

Kolpin, D., Furlong, E., Meyer, M., Thurman, M., and Zaugg, S. (2002). Pharmaceuticals, Hormones, and Other Organic Wastewater Contaminants in U.S. Streams, 1999–2000: A National Reconnaissance. *Environ. Sci. Technol.*, 36:1202–1211.

Richmond, E., Grace, M., Kelly, J., Reisinger, A., Rosi, E., and Walters, D. (2017). Pharmaceuticals and Personal Care Products (PPCPs) are Ecological Disrupting Compounds (EcoDC). *Elem Sci Anth* 5: 66. DOI: https://doi.org/10.1525/elementa.252.

Index

Note: *Italic* page numbers refer to figures and page numbers followed by "n" denote endnotes.

Index